Energy-Efficient Electrical Systems for Buildings

Energy-Efficient Electrical Systems for Buildings, Second Edition offers a systematic and practical approaches to design and analyze electrical distribution and utilization systems in buildings. It considers safety and energy efficiency, while also focusing on sustainability and resiliency, to design electrical distribution systems for buildings. In addition, the second edition provides guidelines on how to design electrified and energy-resilient buildings.

Utilizing energy efficiency, sustainability, and resiliency as important criteria, this book discusses how to meet the minimal safety requirements, set by the National Electrical Code (NEC), to select electrical power systems for buildings. It also considers the impact of building electrification on the design of electrical power systems. The second edition features a new chapter on the optimal design energy-efficient and resilient power systems. In addition, this book includes new end-of-chapter problems, examples, and case studies to enhance and reinforce student understanding.

This book is intended for senior undergraduate mechanical, civil, and electrical engineering students taking courses in Electrical Systems for Buildings and Design of Building Electrical Systems.

Instructors will be able to utilize an updated solutions manual and figure slides for their course.

Mechanical and Aerospace Engineering

A Series of Textbooks and Reference Books

Founding Editor

Frank Kreith

Power Plant Synthesis
Dimitris Al. Katsaprakakis

Air Distribution in Buildings
Essam E. Khalil

Nanotechnology: Understanding Small Systems, Third Edition
Ben Rogers, Jesse Adams, and Sumita Pennathur

Introduction to Biofuels
David M. Mousdale

Energy Conversion
D. Yogi Goswami and Frank Kreith

CRC Handbook of Thermal Engineering
Raj P. Chhabra

Energy Efficiency and Renewable Energy Handbook, Second Edition
D. Yogi Goswami and Frank Kreith

Energy, the Environment, and Sustainability
Efstathios E. Michaelides

Principles of Sustainable Energy Systems, Third Edition
Charles F. Kutscher, Jana B. Milford, and Frank Kreith

Fluid Power Circuits and Controls
Fundamentals and Applications, Second Edition
John S. Cundiff and Michael F. Kocher

For more information about this series, please visit: https://www.crcpress.com/Mechanical-and-Aerospace-Engineering-Series/book-series/CRCMECAERENG

Energy-Efficient Electrical Systems for Buildings

2nd Edition

Moncef Krarti

CRC Press
Taylor & Francis Group
Boca Raton London New York

CRC Press is an imprint of the
Taylor & Francis Group, an **informa** business

Cover image Credit: Hajer Tnani Krarti

Second edition published 2024
by CRC Press
6000 Broken Sound Parkway NW, Suite 300, Boca Raton, FL 33487-2742

and by CRC Press
4 Park Square, Milton Park, Abingdon, Oxon, OX14 4RN

CRC Press is an imprint of Taylor & Francis Group, LLC

First edition published by CRC Press 2017

ISBN: 978-1-032-23383-3 (hbk)
ISBN: 978-1-032-23384-0 (pbk)
ISBN: 978-1-003-27699-9 (ebk)
ISBN: 978-1-032-23403-8 (eBook+)

DOI: 10.1201/9781003276999

Typeset in Times
by codeMantra

Access the Instructor Resources: https://routledgetextbooks.com/textbooks/instructor_downloads/

Contents

Preface

Worldwide, buildings are responsible for over 40% of the total primary energy use and related greenhouse emissions. Through standards and energy efficiency programs, several countries are attempting to improve the energy performance of new and existing buildings. Designing and retrofitting electrical power systems to be reliable, safe, energy-efficient, and resilient are the main goals of professionals in the built environment. Most energy end-use systems for both residential and commercial buildings including lighting, air conditioning equipment, and appliances require electrical power to operate. In particular, electricity has to be readily available throughout the building in order to ensure people can live comfortably and work productively. However, if not designed safely, electrical distribution systems can cause serious injury and even death. Therefore, the main objective, when designing and retrofitting electrical distribution systems within buildings, is safety for both humans and equipment.

The second edition of this book outlines the fundamental principles and methods to design safe, flexible, reliable, accessible, energy-efficient, and resilient electrical power systems for both residential and commercial buildings. In particular, this edition presents simplified but effective calculation and analysis methods to design and evaluate safe and energy-efficient distribution electrical systems suitable for residential and commercial buildings. These simplified methods are based on well-established engineering principles. In addition, several innovative yet proven energy efficiency technologies and strategies are presented to improve the energy performance of existing electrical systems and make them buildings more integrated and interactive with the grid. The second edition of this book is designed to be a self-contained textbook aimed at seniors and/or first-year graduate students interested in designing energy-efficient and resilient distribution of electrical systems for buildings. The contents of this book can be covered in a one-semester course for building electrical systems. However, this book can also be used as a valuable reference for practitioners. The users of this book are assumed to have a basic understanding of basic electrical circuits including single-phase and three-phase power systems. Basic knowledge of general concepts of engineering economics and building mechanical systems is also recommended.

The second edition of this book is organized in 14 self-contained chapters with several worked-out examples and design case studies. Moreover, several problems are provided at the end of most chapters to serve as review or homework assignment problems for the users of the book. As the instructor of a course on building electrical systems, you may find that the best approach for the students to understand and apply various design and analysis methods and tools discussed in this book is either through individual or through group projects. These projects include (1) the design of electrical systems of new residential and commercial buildings and (2) audit and redesign of distribution power systems for existing buildings.

Chapter 1 provides a basic overview for the basic components of electrical distribution systems specific to both residential and commercial buildings including an overview of electrification systems and a description of the current trend in deploying direct current (DC) distribution systems. Moreover, the general approach and the main objectives of designing building power distribution systems are outlined. In Chapters 2–4, a basic review is presented for electric circuits, transformers, and motors. Chapter 5 provides the basic operation of protection devices while Chapter 6 summarizes the design criteria for wiring systems including branch circuits and feeders. Chapters 7 and 8 present detailed design requirements as well as specific case studies for residential and commercial buildings, respectively. In particular, the impacts of electrification of buildings on sizing the distribution systems are outlined in Chapter 7. Then, Chapters 9–11 sequentially outline the principles of economic analysis, typical energy efficiency measures for electrical systems, and general power quality issues and the means to avoid or eliminate them. Chapters 12 and 13 present the components as well as typical design procedures for PV systems and electrical generation systems. Finally, Chapter 14 introduces optimization-based design methods to integrate renewable electricity generation technologies in designing electrical systems for energy-efficient, net-zero energy, and resilient buildings. In addition, Chapter 14 introduces the concepts of grid-interactive efficient buildings and innovative technologies that can be used to enhance the abilities of the built environment to response to the grid needs and variability of power generation through renewable energy resources.

When using this book as a textbook, the instructor should start from Chapter 1 and proceed through Chapter 14 in order. However, some of the chapters can be skipped or covered lightly depending on the time constraints and the background of the students. A special effort has been made to use metric (SI) units throughout this book. However, in several chapters, English (IP) units are also used since they are still the standard set of units used in the United States. Conversion tables between the two unit systems (from English to metric and metric to English units) are provided as part of the appendix to this book.

I acknowledge the assistance of several people in the conception and preparation of the second edition of this book. Special thanks to the input of several of my students at the University of Colorado at Boulder, Finally, I am greatly indebted to my wife, Hajer, and my three children for their continued patience and support throughout the preparation of this book.

Author Biographies

Moncef Krarti is a Professor and Coordinator in Building Systems Program, Civil, Environmental, and Architectural Engineering Department at the University of Colorado at Boulder. He is also the director of the Building Energy Smart Technologies (BEST) center that fosters collaboration between the industry and universities to enhance energy efficiency, sustainability, and resilience of buildings. Prof. Krarti has vast experience in designing, testing, and assessing innovative energy efficiency and renewable energy technologies applied to buildings. He also directed several projects in designing energy-efficient buildings including innovative mechanical and electrical energy systems. Prof. Krarti has published over 300 technical journals and handbook chapters in various fields related to energy efficiency and energy management of the built environment. Moreover, he has published several books on building energy-efficient systems. He taught courses related to building electrical systems for over 25 years in the United States and abroad. As part of his activities at the University of Colorado, he has been managing the research activities of an energy management center with an emphasis on testing and evaluating the performance of mechanical and electrical systems for residential and commercial buildings. He has also helped the development on similar energy efficiency centers in other countries including Brazil, Mexico, and Tunisia. Dr. Krarti has extensive experience in promoting building energy efficiency technologies and policies overseas, including the development of building energy codes and energy efficiency training programs in several countries, including Tunisia, Sri Lanka, and Egypt, and collaborative research with over 10 countries in Europe, Africa, Asia, and South America.

1 Introduction

This chapter provides an overview of a general design approach for electrical distribution systems in residential and commercial buildings. As noted throughout this chapter, safety is the main objective when designing building distribution systems. The relevant codes and standards that assist in designing safe electrical systems are outlined. In addition to safety, other objectives should be considered when specifying building electrical systems, as discussed in this chapter. In particular, building power distribution systems should be designed to be reliable, flexible, accessible, and energy-efficient. Moreover, typical design phases for building projects and the main roles of the architectural engineering design team members are briefly presented.

1.1 INTRODUCTION

In residential and commercial buildings, most energy end-use systems such as lighting, air conditioning equipment, and appliances require electrical power to operate. In particular, electricity has to be readily available throughout the building to ensure people can live comfortably and work productively. However, electricity can be dangerous since it can result in serious injury or death for people and cause significant damages for equipment and property. Therefore, the main objective when designing electrical distribution systems within buildings is safety for both humans and equipment. In the United States, the National Electrical Code (NEC) provides the minimum requirements for electrical distribution systems to ensure safety within building premises (NEC, 2014). NEC is often used as the basis for local codes enforced by counties and municipalities throughout the United States. NEC has developed the National Fire Protection Association (NFPA), an international nonprofit organization dedicated to promote fire protection methods. NEC, first published in 1897, is currently updated every three years and is approved by the American National Standards Institute (ANSI). It should be noted, however, that NEC is not a design guide and cannot be used as a manual to specify electrical power distribution systems. The latest version of NEC is organized into nine chapters, as briefly outlined here:

- Chapter 1: Definitions of electrical terms and general requirements for installations
- Chapter 2: Wiring and protection of electrical systems
- Chapter 3: Wiring methods and materials
- Chapter 4: General electrical equipment (including lighting, motors, transformers, HVAC systems, and generators)
- Chapter 5: Requirements for special occupancies (recreational vehicles, floating buildings, and petrochemical facilities)
- Chapter 6: Special electrical equipment (such as elevators and signs)

DOI: 10.1201/9781003276999-1

- Chapter 7: Special systems (emergency lighting, control circuits, and fiber optic cables)
- Chapter 8: Communications systems
- Chapter 9: Tables and examples

It should be noted that electrical code enforcement and inspection of electrical systems are typically performed by an authority having jurisdiction (AHJ), commonly an electrical inspector or a fire marshal. Specially for challenging cases, the interpretation of code requirements is the sole responsibility of the AHJ. In particular, the AHJ may waive specific NEC requirements and permit alternate design solutions as stated in NEC 90–4 (NEC, 2014).

In addition to NEC, other codes and standards should be consulted to design electrical power distribution systems for a wide range of applications including

- National Electrical Safety Code (NESC) for any electrical installations outside building premises (i.e., utility wiring)
- Occupational Safety and Health Administration (OSHA) for any requirements related to workplace
- Underwriters Laboratory (UL) for testing product safety
- National Electrical Manufacturers Association (NEMA) for developing standards related to electrical equipment
- Other codes and standards by American Society of Heating, Refrigerating, and Air Conditioning (ASHRAE); Illuminating Engineering Society (IES); Institute of Electrical and Electronics Engineers (IEEE); Factory Mutual; and ANSI.

1.2 OVERVIEW OF BUILDING ELECTRICAL DISTRIBUTION SYSTEMS

To safely distribute electricity within buildings, several components have to be specified, including wiring and protection equipment. A basic description of the typical power distribution systems for residential and commercial buildings is provided in this section.

1.2.1 Residential Buildings

Power distribution systems for small residential and commercial buildings are rather simple to design and install with a limited number of components, as illustrated in Figure 1.1. In particular, a typical electrical distribution system consists of a meter connected to one panel that serves a set of branch circuits to provide electricity to various loads (lighting and receptacles) located within the building. In the US, small buildings are served using 240/120 V system (dwellings and detached homes) or 208Y/120-V system (small commercial buildings or apartment buildings). These voltages are obtained directly from a utility transformer that is served by a 13.8 kV distribution voltage.

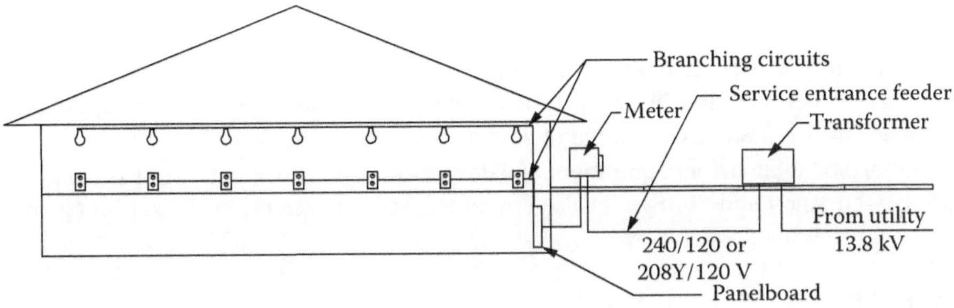

FIGURE 1.1 Typical electrical distribution system for a residential building.

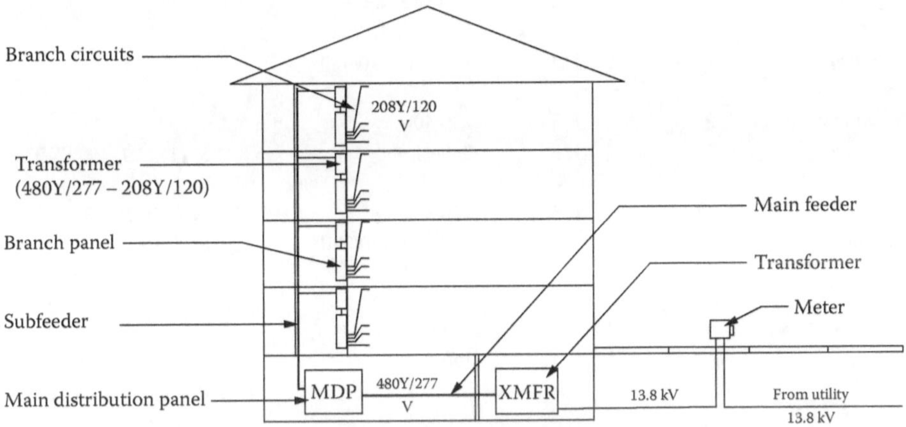

FIGURE 1.2 Typical electrical distribution system for a large commercial building.

1.2.1.1 Commercial Buildings

The electrical power distribution systems for large commercial buildings are more complex than those of residential buildings and utilize several components, including a network of step-down transformers, lighting and power panels, protection devices, grounding systems, and wiring methods. Figure 1.2 shows an example of an electrical distribution system for a large building. The electricity is supplied at a high voltage of 13.8 kV and is distributed to the building at lower levels of 480Y/277-V and/or 208Y/120-V. A main distribution panel (MDP), served by the main feeder from the main step-down transformer (13.8-kV–480Y/277-V), provides electricity safely to various loads through subfeeders, low-voltage step-down transformers (480Y/277-V–208Y/120-V), and panels. While the MDP is protected by power circuit breakers as well as a grounding system, the panels include several molded-case breakers that protect branch circuits serving building end-use loads such as plug-loads (receptacles) and lighting fixtures, or motors.

1.2.2 Distribution Voltages

The specific voltages distributed and served to buildings vary significantly worldwide and depend on the building types. Table 1.1 summarizes typical distribution voltages and frequencies available in several countries (DOD, 1999).

Importantly, electrical equipment and devices are generally sensitive to changes in supply frequency and voltage levels. Therefore, specific attention should be given to

TABLE 1.1
Voltages and Frequency Used in Select Countries

Country	Frequency (Hz)	Number of Phases	Low Voltages (V)	Medium Voltages (kV)
Afghanistan	50	1, 3	220/380	3.2, 6, 10, 15,20
Algeria	50	1, 3	127/220, 220/380	5.5, 6.6, 10, 30
American Somoa	60	1, 3	120/240, 240/480	
Argentina	50	1, 3	230/400	6.6, 13.2, 33
Australia	50	1, 3	240/415	6.6, 7.6, 11, 12.7, 19, 22, 33, 66
Austria	50	1, 3	220/380	3, 5, 6, 10, 20, 25, 28, 40
Belgium[a]	50	1, 3	127/220, 130/220, 220/380	6.6, 10, 15, 36, 70
Brazil[a]	60	1, 3	110/220, 125/216, 127/220, 220/380	6, 11.4, 13.8, 22, 25, 34.5
Canada	60	1, 3	120/240	2.4, 4.16, 7.2, 8, 12.47, 13.8, 14.4, 20, 25, 34.5, 44, 49
Chile	50	1, 3	220/380	12, 13.2, 13.8, 15, 23
Denmark	50	1, 3	220/380	6, 10, 20, 30
Egypt	50	1, 3	220/380	3, 6.6, 11, 20, 33, 66
France[a]	50	1, 3	115/220, 127/220, 220/380	3.3, 5.5, 10, 15, 20, 30
Germany	50	1, 3	220/380	3, 6, 10, 20, 30, 45, 60
Greece	50	1, 3	220/380	6.6, 15, 20, 22
Guam	60	1, 3	110/220, 120/208, 200/346	4, 13.8
Hong Kong	50	1, 3	200/346	11, 33
Iceland	50	1, 3	220/380	6, 11, 22, 33
India[a]	50	1, 3	230/380, 230/400, 230/415, 250/440	2.2, 3.3, 6.6, 11, 15
Indonesia	50	1, 3	127/200, 220/380	3, 20
Iran	50	1, 3	220/380	11, 20, 33, 63, 66
Iraq	50	1, 3	220/380	6.6, 11
Ireland	50	1, 3	220/380	5, 10, 20, 38
Italy	50	1, 3	127/220, 220/380	3.6, 10, 15, 20, 30, 45, 66
Jamaica	50	1, 3	110/220	6.9, 13.8, 24
Japan	50	1, 3	100/200	3, 6, 6.6, 11, 20, 22, 60

(Continued)

TABLE 1.1 (*Continued*)
Voltages and Frequency Used in Select Countries

Country	Frequency (Hz)	Number of Phases	Low Voltages (V)	Medium Voltages (kV)
Jordan	50	1, 3	220/380	6.6, 11, 33
Korea	60	1, 3	110/220, 120/240, 120/208, 220/380	22.9
Malaysia	50	1, 3	240/415	6.6, 11, 22, 33
Mexico	60	1, 3	127/220	6.6, 13.2, 13.8, 23, 34.5, 44, 69
Morocco[a]	50	1, 3	127/220, 220/380	5.5, 20, 22
Nepal	50	1, 3	220/440	11, 33
New Zealand	50	1, 3	230/400	11
Nigeria	50	1, 3	230/415	11, 33
Pakistan[a]	50	1, 3	220/380, 230/400	11, 33
Portugal	50	1, 3	220/380	6, 10, 15, 30, 40, 60
Puerto Rico	60	1, 3	120/240	4.16, 13.2
Russia	50	1, 3	220/380	NA
Saudi Arabia[a]	50, 60	1, 3	127/220, 220/380, 230/400	13.8, 33, 34.5, 69
Senegal	50	1, 3	127/220	5.5, 16.6, 30
Singapore	50	1, 3	230/400	6.6, 22
South Africa[a]	50	1, 3	220/380, 230/400, 240/415, 250/433	6.6, 11, 22, 33
Spain[a]	50	1, 3	127/220, 220/380	3, 6.6, 10, 11.6, 15, 20, 33
Sweden	50	1, 3	220/380	3, 6, 7, 10, 20, 30
Thailand	50	1, 3	220/380	3.5, 11, 12, 22, 24, 33
Tunisia[a]	50	1, 3	127/220, 220/380	10, 15, 30
Turkey[a]	50	1, 3	110/220, 220/380	6.3, 10.5, 15, 34.5
UAE[a]	50	1,3	220/380, 230/400, 240/415	6.6, 11, 33
United Kingdom[a]	50	1, 3	220/380, 230/400, 240/415, 240/480	3.5, 6.6, 11, 22, 33, 66
United States[a]	60	1, 3	120/240, 120/208, 277/480	2.4, 4.16, 4.8, 6.9, 8.32, 12, 12.47, 13.2, 13.8, 14.4, 19.9, 20.8, 22.86, 23, 24.94, 46, 69
Uruguay	50	1, 3	220	6, 15, 30, 60
Venezuela	60	1, 3	120/240	2.4, 4.16, 4.8, 12.47, 13.8
Vietnam[a]	50	1, 3	120/208, 127/220, 220/380	6.6, 10, 15, 22, 35
Zimbabwe[a]	50	1, 3	220/380, 230/400	11, 22, 33, 66

[a] The listed voltages may change within a country and may not be available in all cities.

specify the appropriate levels of voltage and frequency for various electrical systems used in buildings. In most cases, electrical systems are designed to operate within certain ranges of tolerances of specific values of frequencies and voltages. Any increase in the supply voltage level can cause higher currents to flow through electrical devices. Indeed, the current through an electrical device follows Ohm's law and is equal to the voltage across the device divided by its impedance, as reviewed in Chapter 2. A larger current flow can result in higher heat to be dissipated in the device. Specifically, the dissipated heat is proportional to the square of the current flow. For instance, doubling the voltage will typically double the current, resulting in the device dissipating four times the heat. Most devices cannot tolerate large increases in heat generation and may be significantly damaged. Most electrical devices cannot operate reliably with supply voltage levels that are higher than 10% of their rated voltage.

Moreover, some devices depend on magnetic fields to transfer and convert electrical energy to operate (such as motors and transformers) and are thus affected by any changes in frequency levels (DOD, 1999). Chapter 3 discusses the operation of transformer, while Chapter 4 overviews how a motor operates and how the magnetic field is utilized to convert electrical energy into mechanical energy. When a 60 Hz rated motor is operated using 50 Hz power supply, the electrical motor shaft speed is reduced by the ratio 5/6. As discussed in Chapter 4, a motor speed is directly proportional to the electricity frequency. Thus, a pump driven by a 60 Hz electrical motor transfers less fluid when operated with 50 Hz source voltage. Therefore, the output of direct-driven systems (such as HVAC equipment including pumps and fans) should be derated, typically by a factor of 5/6. It should be noted, however, that the 60 Hz motor can be operated to deliver the same mechanical power even when operated at a 50 Hz source. In this case, the torque has to be increased when operated at 50 Hz compared to when it is supplied by a 60 Hz source since the mechanical power is the product of the torque and the shaft speed (refer to Chapter 4). Under these operating conditions, the motor may require more current and may be operating at levels that can cause overloading and overheating. Similarly, operating a 60 Hz transformer using a 50 Hz source may cause saturation of its core resulting in overheating conditions. Other electrical systems can be sensitive to changes in frequencies from 60 to 50 Hz. For instance, circuit breakers have different tripping curves depending on the frequency level. It is important to ensure that adequate trip curves with the proper frequency value are utilized when coordinating protection devices. Moreover, reading meters may lose their accuracy when operating at different frequency systems.

1.3 ELECTRIFICATION OF BUILDINGS

To specify the components for electrical distribution systems for buildings, it is important to determine all the end-use loads that need to be served by electricity as well as their rated voltages and frequencies. Typically, the design procedure of building electrical systems follows a bottom-up approach. First, the loads are estimated, then the branch circuits are selected, and finally the panels as well as the feeders and subfeeders are specified along with any transformers needed to supply the required voltages throughout the building. In this section, the design objectives are first discussed when sizing electrical distribution systems.

Electrification refers to the use of all-electric systems within buildings including the replacement of any fossil-fuel-based equipment for space heating, water heating, and cooking with electric devices. The primary benefit of electrification is to lower the carbon emissions of buildings since electricity can be generated from renewable energy resources (i.e., solar and wind). It has been estimated that replacing a natural gas furnace with an electric heat pump could reduce carbon emissions by 46% for most US households (McKenna et al., 2020). Moreover, electrification can lower the construction costs of new buildings due to the avoided costs of gas connections and services (McKenna et al., 2020). For retrofit applications, electrical systems can be more cost-effective than fuel-based counterparts when these systems reach their useful life and must be replaced. Often, heat pumps are more energy efficient than fossil-fuel and electric-resistance heating systems with the ability to provide both heating and cooling capabilities to maintain indoor thermal comfort within buildings (Nadel, 2016).

The building sector is responsible for 40% of global carbon emissions, more than any other sector including industry, transportation, and agriculture. Greenhouse gas emissions are typically divided into three main categories including:

- Operation emissions consist of mainly energy consumption either directly through burning fuels on-site (i.e., for space heating through furnaces and boilers, for instance) or indirectly through using electricity generated from fossil fuels.
- Embodied emissions representing the life cycle emissions for extracting and manufacturing materials (i.e., concrete and steel) and systems (i.e., boilers and air conditioning systems) used in the construction of buildings.
- Refrigerant emissions correspond to emissions related to the use of refrigerants in heating, ventilating, and air conditioning (HVAC) equipment as well as refrigeration systems.

The indirect emissions represent 60% while the direct emissions and the embodied emissions contribute to 25% and 15% of the total emissions for the US building sector. Of the total direct emissions specific to the US residential and commercial buildings, space heating is responsible for 68%, water heating for 19%, and cooking for 13%.

An important pillar to achieve the goal of keeping the raise in global temperature below 1.5°C set by the Intergovernmental Panel for Climate Change is to lower the carbon emissions of the building sector. For the US building sector, it is estimated that 70% and 100% of the 2005 operating emissions must be reduced by 2030, and 2050, respectively to achieve the goal of maintaining the global temperature raise at only 1.5°C. Some US states, like New York, have already approved and implemented specific policies to reach between 85% and 93% of carbon emissions by 2050.

Five fundamental principles for decarbonizing buildings including (RMI, 2021):

- Efficiency for both new and existing buildings to reduce energy demands and carbon emissions during the construction and operation phases.

- Electrification to ensure that all needs are served without the reliance on fossil fuels including meeting any space heating and domestic hot water needs.
- Grid-interactive so building energy systems can respond to the utility signals and accommodate the variability of renewable energy generation.
- Use of low carbon fuels to meet end-uses that are difficult and cost-prohibitive to electrify especially in some specific regions and building types.
- Deployment of low-embodied carbon materials and systems especially for new constructions and deep retrofit applications.

Chapter 10 includes an overview of the electrical heating systems that can be used to replace fuel-based equipment. In Chapter 14, the benefits of electrification of buildings to lower carbon emissions are estimated and compared to the traditional energy efficiency improvement of fossil-fuel-based systems.

1.4 DC DISTRIBUTION SYSTEMS

The electrical systems for buildings are based almost exclusively so far on alternating current (AC) rather than on direct current (DC). However, there is an increasing interest in deploying DC distribution systems for the built environment especially those that are standalone or those that have on-site renewable generation systems. Indeed, DC electrical distribution systems permit readily to integrate on-site energy storage and renewable generation without the need for power electronics converters required by the AC systems. DC systems can meet energy efficiently for most end-user loads including LED lighting, DC fans, electronic appliances, and variable frequency drive (VFD) operated heating and cooling systems. Moreover, DC systems have lower wire losses due to the absence of the skin effects characterizing the flow of AC power and have been adopted for high-voltage power transmission lines (Lin and Xiao, 2019). Thus, DC distribution systems can be more energy-efficient than traditional AC systems, especially with on-site power generation and energy storage (Vossos et al., 2014).

In the last decade, DC low-voltage distributions have been adopted for specific applications including data centers using 380-V and telecommunication power systems using 48-V due to their higher energy efficiency and reliability features (Usui et al., 2016). For instance, DC-powered data centers have been found to save 28% of energy consumption compared to AC-supplied facilities (Prabhala et al., 2018). A simulated-based analysis has indicated that a DC distribution with 48-V is 9% more energy efficient that the conventional AC system for a two-story house (Siraj and Khan, 2020). However, to date, there are no universally accepted standards and guidelines to design DC distribution systems for buildings including dwellings including the voltage levels to be used. Studies have indicated that the energy efficiency of the DC systems depends on the distribution voltage with some recommending the use of 48-V as the optimal value for residential buildings (Prabhala et al., 2018). Moreover, there are limited commercially available DC-based equipment and devices as well as consensus on design issues related to safety, grounding, and fault protection.

NEC considers circuits as having low voltage when the distribution occurs at voltage lower than 60 V and referred to these as Class-2 systems with less stringent safety requirements. On the other hand, high-voltage circuits have voltage greater than 60 V but less than 1,000 V and need to meet stricter NEC safety requirements. Low DC voltages include USB (5-V), 12-V, 24, and 48-V and are used to serve devices with less 100-W power such as computers, monitors, and hand-held devices. High DC voltages consist mainly of 380-V commonly used in North America to power data centers (Emerge Alliance, 2022), 350-V used in Europe, and 375-V in China (Prabhala et al., 2018).

Chapter 12 outlines one of the applications of DC distribution systems for wiring PV arrays and batteries for both standalone and grid-connected buildings.

1.5 GENERAL DESIGN APPROACH

To specify the components for electrical distribution systems for buildings, it is important to determine all the end-use loads that need to be served by electricity as well as their rated voltages and frequencies. Typically, the design procedure of building electrical systems follows a bottom-up approach. First, the loads are estimated, then the branch circuits are selected, and finally the panels as well as the feeders and subfeeders are specified along with any transformer needed to supply the required voltages throughout the building. In this section, the design objectives are first discussed when sizing electrical distribution systems. The role of the design team, including the electrical engineer, is then briefly described for typical building projects. Finally, the general design approach, including typical tasks and expected deliverables for building electrical distribution systems, is presented.

1.5.1 Design Objectives

1.5.1.1 Safety

As noted earlier, safety should be the most important objective for specifying various components of the power distribution systems for buildings. Through an example, Figure 1.3 illustrates the importance of design specifications in ensuring the safety using two location options for a unit substation within an electrical room. As discussed in Chapter 8, a unit substation typically includes a high-voltage transformer section. In the case of Design I in Figure 1.3, the unit substation is located in the middle of the electrical room. In Design II, the unit substation is placed against one of the walls of the electrical room. While there are some exceptions as noted in NEC, the safest design is Design I. Indeed, when any problem occurs, such as a fire due to short-circuiting, arcing, or melting within the unit substation, any person located near corner B of the electrical room can have a safe pathway to the exit door in Design I. In the case of Design II, the same person located near corner B of the electrical room would be completely trapped and would have no chance to escape from a potential fire hazard.

In addition to safety, other objectives should also be considered when designing electrical distribution systems for buildings as outlined in Sections 1.3.1.2–1.3.1.5.

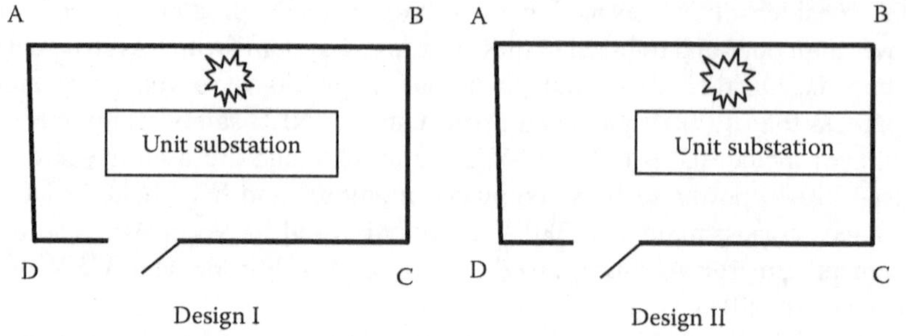

FIGURE 1.3 Two location options for a unit substation within an electrical room.

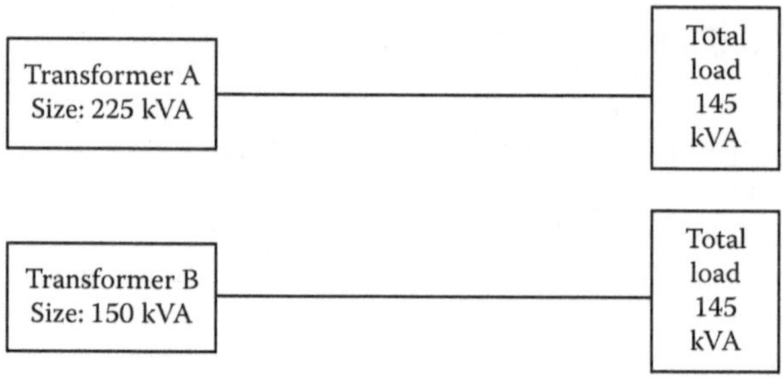

FIGURE 1.4 Two design options for a transformer serving the load of an office building.

1.5.1.2 Flexibility

The design specifications of any electrical distribution system should allow for some flexibility. In particular, the system should be able to handle additional electrical loads due to future expansion and/or change of end-use equipment or loads (i.e., lighting, appliances, or motor loads). Figure 1.4 shows an example of two design specifications for a transformer specific to an office building having a total load of 145 kVA. In the first case (Transformer A), the size of the transformer is specified to be 225 kVA. In the second design option (Transformer B), the size of the transformer is limited to 150 kVA, just 5 kVA above the actual load of the building. While both transformers meet the current load requirements, transformer A provides more flexibility for any future expansion of the building electrical system.

1.5.1.3 Accessibility

The components of the power distribution systems should be designed to be easily accessible in order to facilitate their maintenance, repair, and replacement. Figure 1.5 indicates two potential locations (Panel A and Panel B) of a panelboard, which, as outlined in Chapters 7 and 8, includes most of the protection devices specific to the branch circuits serving various loads within a building. Along one of the walls, Panel A is located at a height of 1.5 m (5 ft) from the ground level. Panel B is placed at a height of 3.0 m (10 ft). It is clear that Panel A is more easily accessible than Panel B for recommended regular maintenance work.

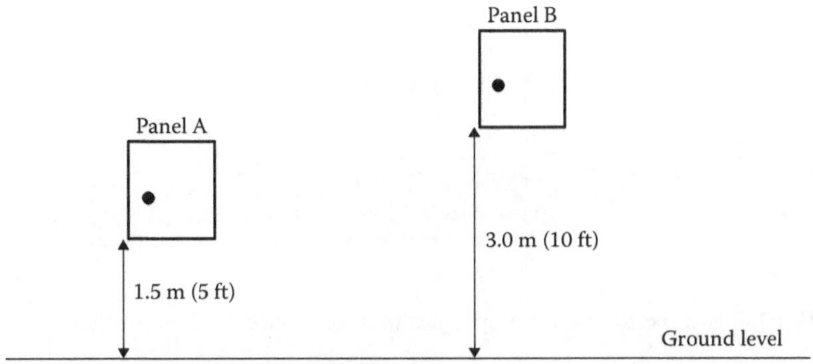

FIGURE 1.5 Two height options for a panelboard.

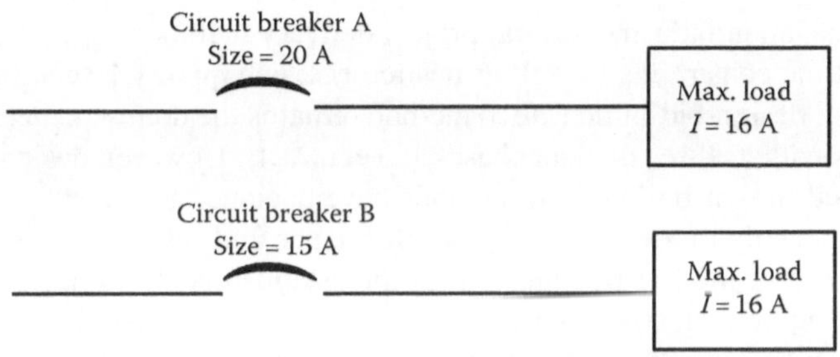

FIGURE 1.6 Two rating options for a circuit breaker serving a branch circuit.

1.5.1.4 Reliability

Electrical distribution systems should be designed to ensure that they operate reliably without interruption under normal loading conditions. Figure 1.6 presents two potential design options (A and B) for a circuit breaker to protect a branch circuit serving a variable load with a maximum current of 16 A. In the case of circuit breaker B, with a rating of 15 A, there is a high probability that the breaker trips and the load would not be served as soon as the current exceeds 15 A. Thus, selecting a rating of 20 A for the circuit breaker (i.e., circuit breaker A) provides more reliability (and safety) than using a 15 A circuit breaker (i.e., circuit breaker B).

1.5.1.5 Energy Efficiency

To ensure the efficient use of electricity, specifying high energy-efficient components for the power distribution system should be considered. Indeed, it is typically cost-effective to invest in energy efficiency measures that would improve the overall energy performance of the building. In particular, selecting high-efficiency transformers, as noted in Figure 1.7, is becoming a common and cost-effective practice in designing electrical systems for buildings. A wide range of energy efficiency measures specific to building electrical systems is presented and discussed in Chapter 10.

Transformer A
Size: 200 kVA
Efficiency: 99.3%

Transformer B
Size: 200 kVA
Efficiency: 98.5%

FIGURE 1.7 Two energy efficiency specifications for an electrical transformer.

1.5.2 Design Team

For the construction industry, several project delivery methodologies depend on the budget and time constraints as well as desired risk and quality levels. In particular, the traditional design-bid-build (DBB) method remains the dominant delivery option for buildings with its three distinct phases and contracts. However, design-build (DB) and integrated-design-build (IDB) methods have become more common in the last few years especially for delivering high performance and sustainable buildings. The main advantage of the last two methods is the establishment of one single contract for the building project, so responsibility and reliability are shared between various partners of the design and construction teams. There is some evidence that both DB and IDB have the potential to enhance communication and collaboration between design and construction team partners resulting in the completion of the project in a timely manner. For all project delivery methods, the electrical engineer is typically part of the architectural and engineering design team, which is structured as noted in Figure 1.8. While the architect typically plays a role of the main coordinator of the design team, the engineers have crucial roles in various phases of building design and construction.

Specifically, the architect typically develops the overall aesthetic form and structure of the building, including its shell and material requirements, while accounting for factors, concepts, and practices that ensure functional, safe, and economical design as per client specifications. For most current building projects, the architect delivers the design through plans and drawings completed using computer-aided drafting (CAD) and building information modeling (BIM) tools. In particular, the architect coordinates the project and communicates the design specifications to various engineers through these drawings. After the design phase, the architect continues overseeing the project during the construction phase through site visits and sometimes makes revisions to the design based on any changes in client needs, budget constraints, and other unexpected factors. Through consultations with the architect, engineers provide detailed design specifications for various systems within the building from the conception to the construction phases.

In particular, the mechanical engineer develops design solutions and specifications for energy systems including heating, ventilation, and air conditioning (HVAC),

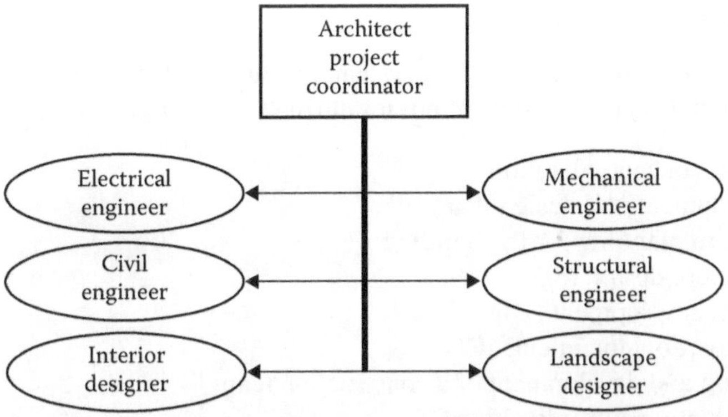

FIGURE 1.8 Typical architectural and engineering design team for building projects.

plumbing, and fire protection. Moreover, the mechanical engineer assists the architect in specifying thermal properties of materials to be used in the building structure as well as any energy systems in order to improve the energy efficiency performance of the building to meet any relevant codes and standards and/or labeling systems.

The electrical engineer provides design options for power service and distribution as well as lighting, communication, fire detection and alarm, and general electrical equipment and space requirements. Chapters 5–8 provide specific design approaches and case studies of power distribution systems for residential and commercial buildings. Moreover, the electrical engineer collaborates closely with the architect and the mechanical engineer to meet the requirements of any applicable codes and standards and any energy performance targets by integrating within buildings cost-effective energy efficiency technologies as well as cogeneration and distributed generation systems such as roof-mounted photovoltaic (PV) panels, as discussed in Chapters 9–14.

The structural engineer assists the architect in the planning, design, and construction follow-up of the building structure and its components such as slabs, beams, columns, and foundations. Moreover, the structural engineer specifies the various materials to be considered for the building shell in pursuance of all relevant codes and standards.

The civil engineer evaluates and provides recommendations for the geotechnical specifications of the building site such as ground soil properties and design solutions for earthworks (i.e., grading, drainage, and pavement). In addition, the civil engineer provides assistance to the structural engineer to design building foundations to ensure that these elements meet the applicable codes and standards.

The interior designer and the landscape designer work mostly with the client and the architect. The main focus of the interior designer is recommending layouts and types of furnishings and decorations within the interior building spaces, while the main concern of the landscape designer is proposing layouts and types of greenery to be located just outside the building structure. Both designers may also have to communicate their plans and service needs to the mechanical and electrical engineers.

1.5.3 General Design Procedure

When specifying buildings and their systems, including electrical distribution components, there are typically several phases during the design process:

- Project conceptualization
- Development of a design team
- Program planning for the building
- Schematic design (SD)
- Design development (DD)
- Construction documents (CD)
- Bidding and development of a construction team
- Construction administration

The main achievements as well as rough estimations of the effort required by the electrical engineer to complete each of the design phases are briefly outlined here using the traditional DBB delivery method:

Project conceptualization

- Develop an idea (concept) for a facility by an owner/developer.
- Procure funding through bank loans or donations.

Development of a design team

- Select a short list of design teams (architects and engineers) through a request of information (RFI) via business journals.
- Complete the interview process with the selected design teams.
- Review the proposal and complete fee negotiation (power distribution engineering design fee typically represents 5%–10% of electrical construction cost).
- Prepare and sign final contracts.

Program planning for the building

- Gather information from various stakeholders of the building, including potential users and occupants.
- Define main objectives of the building, including types of activities and functions as well as space requirements.
- Specify the time and budget constraints as well as design targets, such as energy efficiency ratings and applicable codes and standards.
- Prepare and document a detailed program plan for the building.

SD

- Review program plan.
- Study relevant codes and standards (NFPA, UBC, etc.) based on building type (emergency, life safety egress lighting, and exit sign requirements).

- Coordinate with utilities (power, TV, internet, telephone).
- Coordinate with the architect for required spaces (transformers, MCC, electrical room).
- Complete SD deliverables (typically include a one-line diagram, load analysis, and a report).
- Review SD deliverables with the owner and all design team members.
- After this phase, the design project is 30% completed with 15% hours spent.

DD

- Meet with utility representatives to determine specific utility routing options.
- Meet with the mechanical engineer to determine specific HVAC loads.
- Perform lighting calculations (for typical areas).
- Finalize the DD deliverables (including site-plan, revised one-line diagram, floor plans for power and lighting, schedule sheets, cost estimates, and book specifications).
- Review of DD deliverables with the owner and other design team members.
- Project 50% completed (with 35% budgeted hours spent).

CD

- Coordinate with all stakeholders and involved parties through meetings and exchanges of drawings.
- Revise and finalize CD deliverables (revised site-plan, revised one-line diagram, revised floor plans, complete schedule sheets, revised cost estimate, and book specifications).
- Review of CD deliverables with owner and design team.
- After this phase, the design project is 100% completed (with 90% budgeted hours spent).

Bidding and development of a construction team

- Approve the list of potential bidders (preferred contractors!).
- Answer any questions from contractors.
- Assist in the evaluation of bids.
- Rebid, if necessary, after value engineering and some modifications of the design specifications.

Construction administration

- Plan and complete the groundbreaking ceremony.
- Finalize intermediate and final observations (i.e., punch-lists).
- Prepare and finalize as-built documents.

TABLE 1.2
Typical Symbols Used for Some Components of an Electrical Power Distribution System

Symbol	Meaning	Symbol	Meaning
	Duplex receptacle		Recessed light
	Ground fault current interrupter duplex receptacle		Recessed light served by an emergency branch circuit
	Weatherproof duplex receptacle	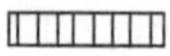	Recessed linear light
	Duplex receptacle served by an emergency branch circuit		Recessed linear light served by an emergency branch circuit
	One of the duplex receptacles is controlled by a switch		Track light
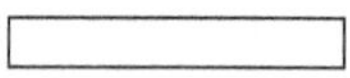	Simple switch		Recessed can light
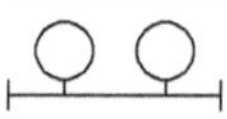	Three-way switch		Wall-mounted light
	Switch with built-in dimmer		Recessed wall wash light
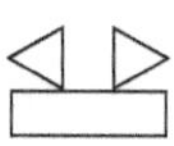	Power panel		Battery-powered emergency light
	Lighting panel	S	Ceiling-mounted exit sign—arrow for direction
S_3	Junction box	S_{DIM}	Wall-mounted exit sign—arrow for direction

Table 1.2 shows typical symbols used in various drawings and deliverables of design specifications for electrical distribution systems. It should be noted that a legend similar to Table 1.2 has to be defined and be part of the design deliverables for each project.

Figure 1.9 illustrates an example of a floor plan for an office building that shows the branch circuiting for lighting fixtures. Figure 1.10 shows a section of a one-line diagram for a commercial building's electrical distribution system. An example of an office building power panel schedule is presented in Figure 1.11.

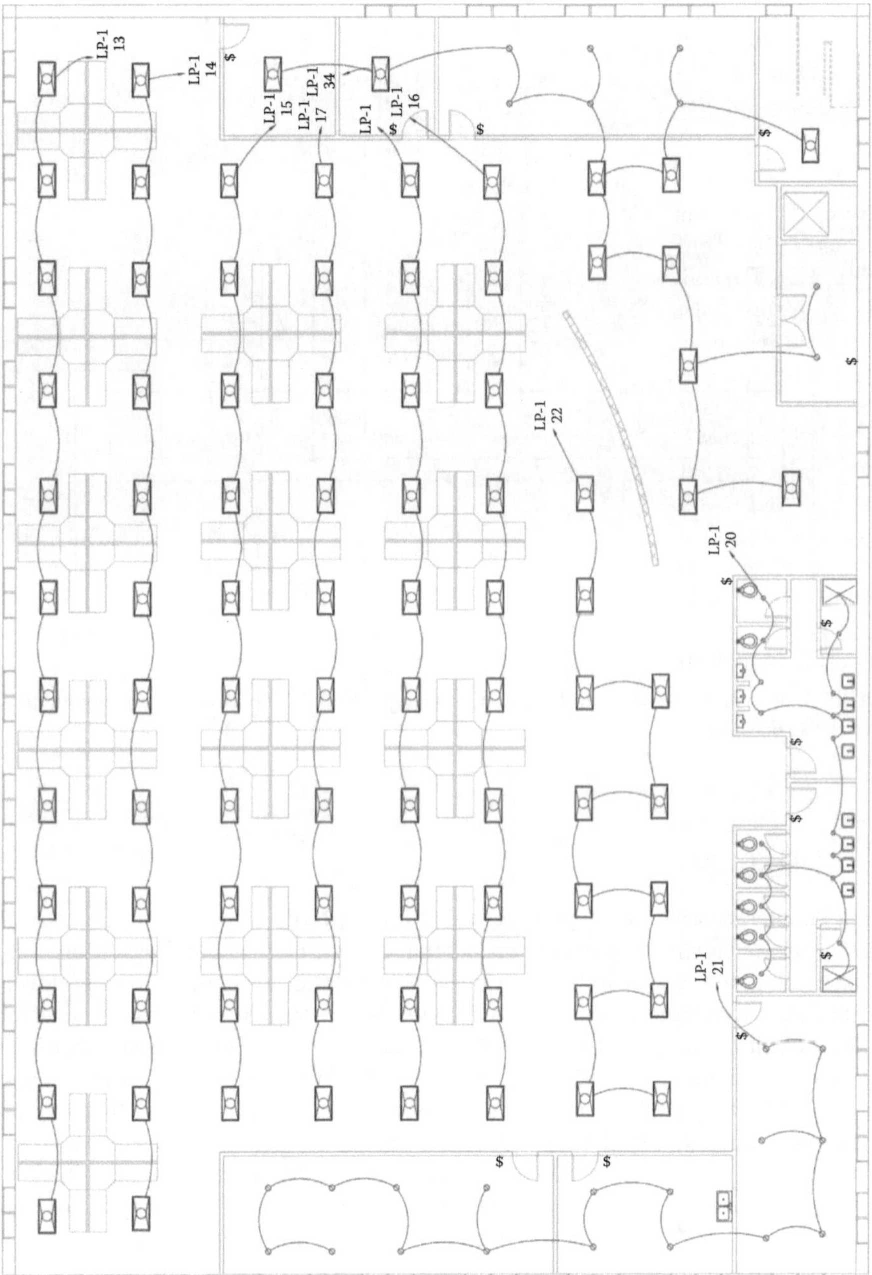

FIGURE 1.9 Circuiting diagram for a lighting system within an office building floor.

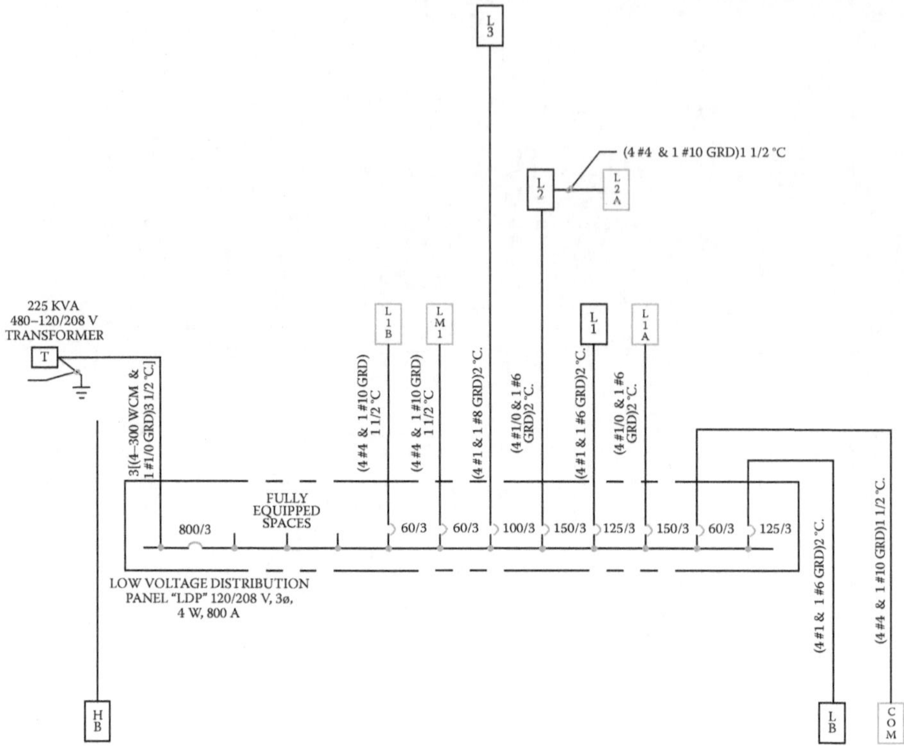

FIGURE 1.10 One-line diagram for a part of an electrical distribution system specific to a commercial building.

1.6 SUMMARY

In this chapter, a basic overview is provided for the basic components of electrical distribution systems specific to both residential and commercial buildings. Moreover, the general approaches as well as the main objectives of designing building power distribution systems are outlined. In Chapters 2–4, a basic review is presented for electric circuits, transformers, and motors. Chapter 5 provides the basic operation of protection devices, while Chapter 6 summarizes the design criteria for wiring systems including branch circuits and feeders. Chapters 7 and 8 present the detailed design requirements as well as specific case studies for residential and commercial buildings, respectively. Then, Chapters 9–11 outline sequentially the principles of economic analysis, typical energy efficiency measures for electrical systems, and general power quality issues and the means to avoid or eliminate them. Chapters 12 and 13 present the components as well as the typical design procedures for PV systems and electrical generation systems. Finally, Chapter 14 introduces optimization-based design methods to integrate renewable electricity generation technologies in designing electrified, net zero energy, and resilient buildings.

PANEL•LOOC	
225	A, bus
	Full neutral bus and full ground bus
208	Y/ 120 volt, 3 phase, 4 wire
10,000	ampere panel short-circuit rating
Panel type:	Square D NQOD
Panel mounting:	Recess
Note:	* Shunt trip

Description	Total (VA)	Breaker	A/P	CCT	PH	CCT	Breaker	A/P	Total (VA)	Description
Plug-in 00-5 *	3000	100	/	1	A	2	20	/1	720	RCPT 00-50
—	3000		/	3	B	4	20	/1	720	RCPT 00-50
—	3000		/3	5	C	6	20	/1	720	RCPT 00-50
Lathe	1000	20	/	7	A	8	20	/1	720	RCPT 00-50 *
—	3000		/	9	B	10	20	/1	720	RCPT 00-50 *
—	3000		/3	11	C	12	20	/1	720	RCPT 00-50 *
Space	0		/	13	A	14	20	/1	0	F,A, damper
—	1000		/	15	B	16	20	/1	0	Spare
—	1000		/3	17	C	18	20	/1	0	Spare
RCPT 00-50	300	100	/	19	A	20		/	0	Space
RCPT 00-50	300		/	21	B	22		/	0	Space
RCPT 00-50	300		/3	23	C	24		/	0	Space
TVSS	0	15	/	25	A	26		/	0	Space
—	0		/	27	B	28		/	0	Space
	0		/3	29	C	30	20	/1	100	EPD
TVSS	0	15	/	37	A					
—	0		/	39	B					
—	0		/3	41	C					

	Panel	Loading		Summary	
Load Type	Connected Load	Power Factor	KVA	Load Factor	NEC Calculated Load
LIGHTING					
Incandescent	0.0 kW @	100%=	0.0 @	125%=	0.0 kVA
Fluorescent	0.0 kW @	95%=	0.0 @	125%=	0.0 kVA
Receptacles					
First 10 KW	4.3 kW @	100%=	4.3 @	100%=	4.3 kVA
Remainder	0.0 kW @	100%=	0.0 @	50%=	0.0 kVA
Motors					
Largest	0.0 kW @	80%=	0.0 @	125%=	0.0 kVA
Remainder	0.0 kW @	80%=	0.0 @	100%=	0.0 kVA
Other	21.1 kW @	100%=	21.1 @	100%=	21.1 kVA
Total	25.4 kW		25.4 kVA		25.4 kVA
			Minimum panel ampacity =		71 A

Phase "A"	8.4 kW	Panel		A-B	100%	Phase
Phase "B"	8.4 kW	Power		B-C	99%	balance
Phase "C"	8.5 kW	Factor	100%	C-A	99%	

FIGURE 1.11 Schedule for a power panel for an office building.

2 Overview of Electrical Circuits

This chapter provides a review of basic concepts in power systems and electrical circuits. These basic concepts provide the fundamental knowledge needed to carry out any design calculations for building electrical systems. First, a review of the basic characteristics of an electric system operating under direct current is provided to introduce the basic properties of electricity. Then, a summary of alternating current properties is presented for both single- and three-phase systems commonly used in buildings. Throughout this chapter, several calculation examples are presented to further explain the basic principles of electrical circuits in buildings and show the applications of these principles to determine the characteristics and ratings of various equipment and loads.

2.1 INTRODUCTION

Electricity has become the main energy source that is used in most buildings. Therefore, electricity is required for lighting, air conditioning, transportation, and operation of numerous appliances in all residential and commercial buildings. The use of electrical power in buildings has increased significantly over the last decades. For instance, the electrical power usage in office buildings has grown from 10–33 W/m^2 (1–3 W/ft^2) in the 1940s to 54–108 W/m^2 (5–10 W/ft^2) in the 2000s.

The main feature of electricity is that it can be easily and quickly transmitted and distributed over long distances using wires and cables without significant losses. The size of the required wires to transmit and distribute electricity depends mainly on the voltage level. The higher the voltage, the smaller the required wire size to deliver a given amount of electrical power. Unfortunately, high voltage levels are not safe for human life and property. For safe utilization, low voltages are typically used in buildings. In the United States, small appliances are rated at 120 V and large equipment are designed to operate at voltage levels of up to 480 V. In other countries, other low voltages between 100 and 500 V are typically used, as outlined in Chapter 1.

In this chapter, basic electricity concepts and principles are reviewed and applied. Since almost all buildings use alternating current (AC), an overview of the fundamental principles of single- and three-phase systems is provided. The concepts presented in this chapter are essential to perform the required design calculations properly to select the proper electrical systems for buildings, as outlined in the subsequent chapters of this textbook.

DOI: 10.1201/9781003276999-2

2.2 REVIEW OF DC AND AC CIRCUITS

2.2.1 Direct Current

When electric current remains constant and does not change significantly over time, it is typically called direct current (DC). To generate DC, batteries or DC generators are used. In the following sections, basic definitions and principles related to DC electrical circuits are presented. As will be illustrated later, most of these principles apply to AC systems as well. A good analogy to an electrical circuit and its properties is a hydraulic circuit as illustrated in Figure 2.1.

Basic definitions related to DC electrical systems are summarized here.

Voltage, E, is the potential difference or the electromotive force (emf) that forces the electrons to flow in an electrical circuit in a similar manner that the gravitational forces (i.e., potential energy) push water to flow downward in the pipes, as illustrated in Figure 2.1.

Current, I, is the flow rate of electricity and is defined as the number of electrons (measured in Coulomb) flowing per second in an electrical circuit. Coulomb is the SI unit of electric charge, Q, and represents 6.0×10^{18} electrons. Thus, the current, I, can be expressed as a function of the quantity of electricity, Q, flowing in an electrical circuit over time, t:

$$I = \frac{Q}{t} \tag{2.1}$$

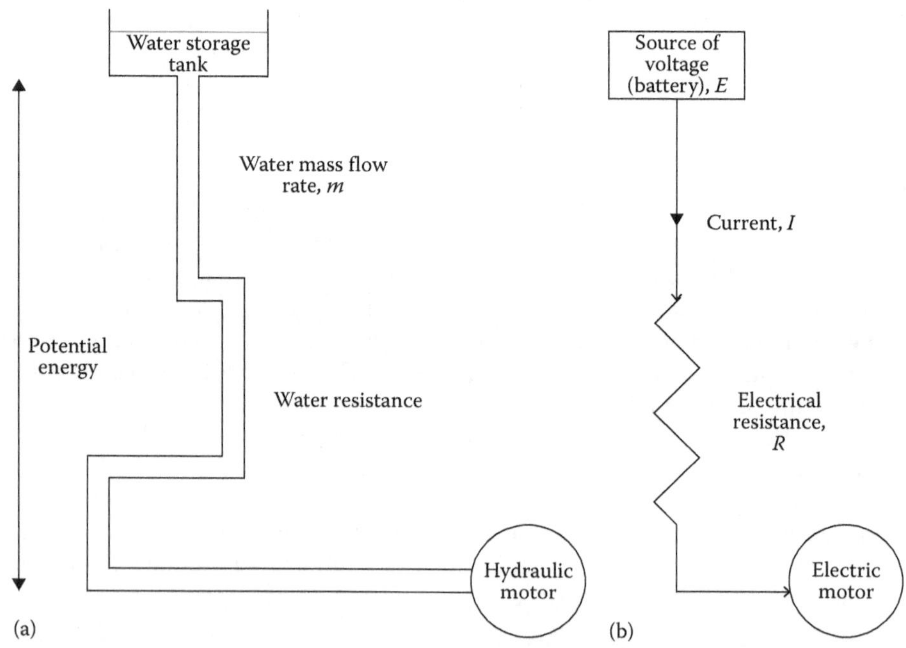

FIGURE 2.1 Analogy between (a) a hydraulic circuit and (b) an electrical circuit.

The unit of current, I, is *ampere* (A). The direction of the flow is conventionally considered positive when the electrons are flowing from negative to positive voltage.

Resistance, R, is the property of a circuit to resist the flow of electrons in a similar manner that the pipe resists the flow of water due to friction. The unit of resistance is *Ohm* (Ω) named after the scientist who first discovered Ohm's law.

The resistance of any wire of a cross-section A, and a length L, can be expressed as follows:

$$R = \rho \frac{L}{A} \tag{2.2}$$

where ρ is the resistivity of the material that makes up the wire. Materials with low resistivity to electrical current are typically called "conductors," while materials with high resistivity are called "insulators." Pure metals are generally good conductors, while synthetic materials are good insulators. Materials with limited conductivity are generally referred to as semiconductors. Table 2.1 provides values of the electrical resistivity of selected materials.

As indicated in Table 2.1, silver, copper, and aluminum are examples of materials considered to be good conductors. Only aluminum and copper are sufficiently inexpensive to be considered for general wiring applications in buildings. Aluminum is too soft to be used in small wires even though it is cheap. Therefore, copper is the preferred material for most wiring applications in buildings. Table 2.2 presents common conductor sizes used in the United States for building electrical systems. Specifically, conductor sizes are expressed using either the American wire gauge (AWG) scale for small wires (typically used for branch circuits or feeders of small buildings) or the 1,000 circular mil (MCM) scale for large wires (typically considered for feeders and subfeeders of commercial buildings). A circular mil, A_{cmil}, is defined as the cross-sectional area of a conductor having a diameter $d = 0.001$ in. (or 1 mil):

$$A_{cmil} = \pi \times \frac{d^2}{4} = \frac{\pi}{4} \times 10^{-6} = 0.785 \times 10^{-6} \text{in.}^2 \tag{2.3}$$

TABLE 2.1
Resistivity for Selected Materials

Material	Resistivity (Ω-cmil/ft)	Resistivity (Ω-cmil/m)
Silver	9.8	32.1
Copper	10.4	34.1
Aluminum	17	55.8
Tungsten	33	108
Nickel	50	164
Iron	60	197
Manganin	290	951
Nichrome	660	2,165

TABLE 2.2
Cross-Sectional Area and Resistance for Electrical Conductors

S (AWG/MCM)	Cross-Sectional Area (cmil)	Number of Wires	Resistance of Copper Wire (Ω/1,000 ft)	Resistance of Aluminum Wire (Ω/1,000 ft)
		AWG		
18	1,620	1	6.51	10.7
16	2,580	1	4.10	6.72
14	4,110	1	2.57	4.22
12	6,530	1	1.62	2.66
10	10,380	1	1.018	1.67
8	16,510	1	0.6404	1.05
6	26,240	7	0.410	0.674
4	41,740	7	0.259	0.424
3	52,620	7	0.205	0.336
2	66,360	7	0.162	0.266
1	83,690	19	0.129	0.211
0	105,600	19	0.102	0.168
00	133,100	19	0.0811	0.133
000	167,800	19	0.0642	0.105
0000	211,600	19	0.0509	0.0836
		MCM		
250	250,000	37	0.0431	0.0708
300	300,000	37	0.0360	0.0590
350	350,000	37	0.0308	0.0505
400	400,000	37	0.0270	0.0442
500	500,000	37	0.0216	0.0354
600	600,000	61	0.0180	0.0295
700	700,000	61	0.0154	0.0253
750	750,000	61	0.0144	0.0236
800	800,000	61	0.0135	0.0221
900	900,000	61	0.0120	0.0197
1000	1,000,000	61	0.0108	0.0177

The area expressed in circular mil, A, of a conductor with any arbitrary diameter, D, (in inches) is obtained by dividing its area in in.^{2} by A_{cmil}:

$$A = \frac{\pi \times \frac{D^2}{4}}{A_{\text{cmil}}} = \frac{\pi \times \frac{D^2}{4}}{\pi \times \frac{d^2}{4}} = \frac{D^2}{d^2} = D^2 \times 10^6 = \left[D \times 10^3\right]^2 \tag{2.4}$$

TABLE 2.3
Diameters and Cross-Sectional Areas for American Wire Gauge and 1,000 Circular Mil Conductors in IP and SI Units

Wire Size	Diameter		Cross-Sectional Area	
	(in.)	(mm)	(in.²)	(mm²)
		AWG		
18	0.0402	1.0223	0.0013	0.8209
16	0.0508	1.2902	0.0020	1.3073
14	0.0641	1.6284	0.0032	2.0826
12	0.0808	2.0525	0.0051	3.3088
10	0.1019	2.5878	0.0082	5.2596
8	0.1285	3.2637	0.0130	8.3657
6	0.1620	4.1145	0.0206	13.2960
4	0.2043	5.1893	0.0328	21.1500
3	0.2294	5.8265	0.0413	26.6629
2	0.2576	6.5432	0.0521	33.6251
1	0.2893	7.3480	0.0657	42.4063
0	0.3250	8.2540	0.0829	53.5083
0	0.3648	9.2666	0.1045	67.4428
0	0.4096	10.4047	0.1318	85.0255
0	0.4600	11.6840	0.1662	107.2193
		MCM		
250	0.5000	12.7000	0.1963	126.6769
300	0.5477	13.9122	0.2356	152.0122
350	0.5916	15.0268	0.2749	177.3476
400	0.6325	16.0644	0.3142	202.6830
500	0.7071	17.9605	0.3927	253.3537
600	0.7746	19.6748	0.4712	304.0245
700	0.8367	21.2512	0.5498	354.6952
750	0.8660	21.9970	0.5890	380.0306
800	0.8944	22.7185	0.6283	405.3660
900	0.9487	24.0966	0.7069	456.0367
1000	1.0000	25.4000	0.7854	506.7075

Therefore, the area of the conductor in circular mils is simply its diameter in mils squared. Table 2.3 lists the conductor sizes, both in IP units (i.e., in.^{2}) and SI units (i.e., mm^2). The actual dimensions of the wires depending on the type of insulation and protection layers used for the conductors will be discussed in Chapter 6. Example 2.1 illustrates the concept of circular mil for copper wire.

Example 2.1

Problem

Determine the overall resistance of 1,500 ft of copper wire with 0.021 in. diameter.

Solution

First, the cross-sectional area of the wire is determined in circular mils using Equation 2.4, since its diameter $D = 0.021$ in. $= 21$ mils:

$$A = (21)^2 = 441 \text{ cmil}$$

From Table 2.1, the resistivity of copper is $\rho = 10.4$ Ω-cmil/ft. Thus, the resistance of 1,500 ft wire based on Equation 2.2 is

$$R = \frac{(10.4\,\Omega - \text{cmil/ft}) \times (1500\text{ ft})}{(441\text{ cmil})} = 35.3\Omega$$

2.2.1.1 Ohm's Law

A simple yet fundamental relationship between voltage, current, and resistance was established by Georg Simon Ohm in 1827. This relationship is known as Ohm's law and can be expressed as

$$E = R \times I \tag{2.5}$$

There is a good analogy between Ohm's law in an electrical circuit and Darcy's law in a hydraulic circuit as illustrated in Figure 2.2. In a hydraulic circuit, the water flow rate is directly proportional to the pressure differential produced by a pump and inversely proportional to the resistance of the pipes. Similarly, in an electric current, the current (i.e., the flow rate of electric charges) is directly proportional to the voltage supplied by a battery or a generator and inversely proportional to the electrical resistance of the wires.

Based on Ohm's law, electrical power and energy consumed by an electrical circuit can be estimated using one of the three relationships depending on the known characteristics of the electrical circuit (i.e., voltage, current, and/or resistance) as summarized in Table 2.4.

2.2.1.2 Kirchhoff's Laws

Two important laws, known as Kirchhoff's laws, are used in almost any analysis of electrical circuits. These laws are illustrated in Figure 2.3 and can be easily visualized and understood using the analogy between electrical circuits and hydraulic circuits as illustrated in Figures 2.1 and 2.2.

The first Kirchhoff's law states that the algebraic sum of all currents in one node is equal to zero:

$$\sum_{j=1}^{n_c} I_j = 0 \tag{2.6}$$

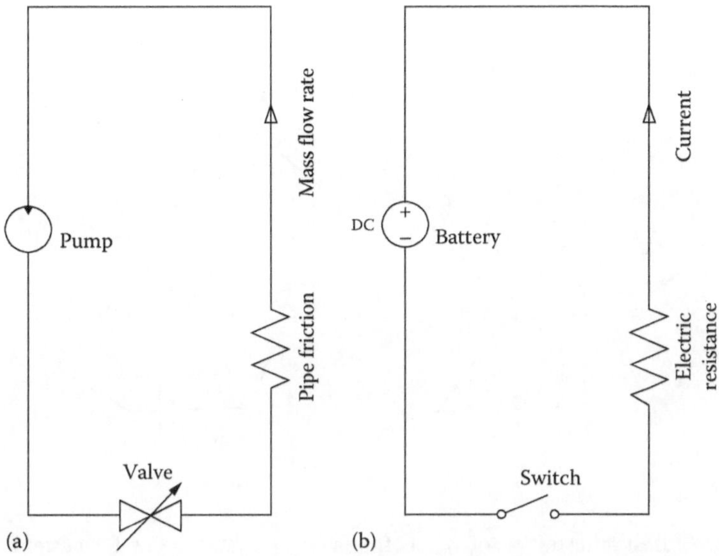

FIGURE 2.2 Comparison between (a) Darcy's law for hydraulic circuits and (b) Ohm's law for electrical circuits.

TABLE 2.4
Summary of Basic Expressions of Electrical Power and Energy

Known Circuit Characteristics	Power (W)	Energy[a] (J or Wh)
Voltage and current (E, I)	EI	EIt
Voltage and resistance (E, R)	E^2/R	$E^2 t/R$
Current and resistance (I, R)	RI^2	RI^2t

[a] Energy consumed during a time period t.

where n_c is the number of electrical circuits that meet in one node, as illustrated in Figure 2.3a.

The second Kirchoff's law states that the algebraic sum of all the voltages around a closed loop in an electrical circuit is zero:

$$\sum_{j=1}^{n_V} E_j = 0 \tag{2.7}$$

where n_V is the number of voltages measured in the closed loop as presented in Figure 2.3b.

A common application of Kirchoff's laws and Ohm's law consists of the determination of the equivalent resistance for electrical circuits formed by several resistances connected either in parallel or in series as illustrated in Figure 2.4.

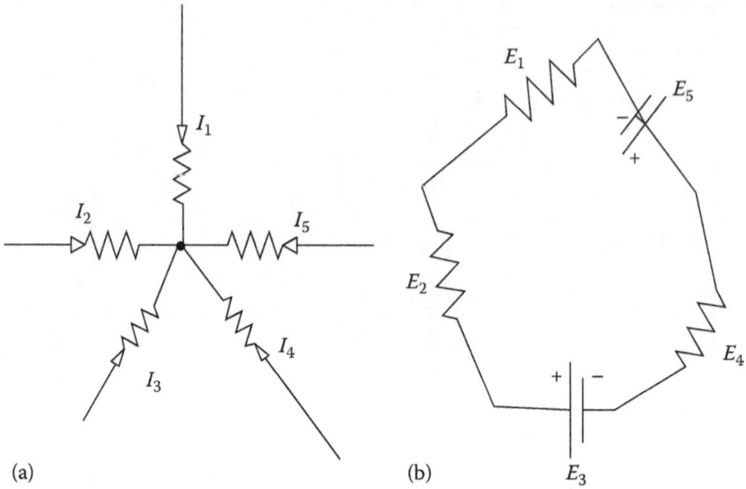

FIGURE 2.3 Basic illustration of Kirchoff's laws: (a) first, the law for currents, and (b) second, the law for voltages.

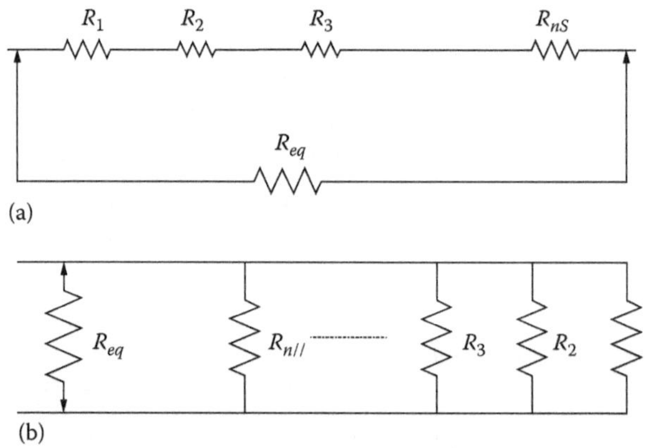

FIGURE 2.4 Equivalent resistance for a combination of resistances connected in (a) series and (b) parallel.

2.2.1.2.1 Resistances Connected in Series

The equivalent resistance for a circuit, made up of several electrical resistances connected in series, is given by

$$R_{eq} = \sum_{j=1}^{n_S} R_j \tag{2.8}$$

where n_S is the number of resistances connected in series.

2.2.1.2.2 Resistances Connected in Parallel

When the electrical resistances are connected in parallel, they can be approximated by one equivalent resistance, R_{eq}, which can be obtained using the following equation:

$$\frac{1}{R_{eq}} = \sum_{j=1}^{n_{\parallel}} \frac{1}{R_j} \tag{2.9}$$

where $n_{\parallel}$ is the number of resistances connected in parallel.

It should be noted that the total power, P_{tot}, consumed by the electrical resistances connected either in series or in parallel is simply equal to the sum of all the powers, P_j, consumed individually by the resistances. Therefore, the power in electrical DC circuits is additive, independently of the connection configuration of the loads:

$$P_{tot} = \sum_{j=1}^{n} P_j \tag{2.10}$$

Example 2.2 illustrates a typical application of Ohm's law and Kirchoff's laws to analyze an electrical branch circuit for a building.

Example 2.2

Problem

Consider a 120 V branch circuit that serves three loads connected in parallel: (1) one 1,000 W electric heater, (2) one 100 W lamp, and (3) a 75 W lamp. Determine (1) the current flowing to each load and the total current flowing in the branch circuit, (2) the resistance of each load and the equivalent resistance of the branch circuit, and (3) the total annual energy use as well as the annual cost when the branch circuit is used to power all three loads during 1,000 h/year (assume the electricity cost is $0.10/kWh).

Solution

a. Using the expression of electrical power as a function of voltage and current (as summarized in Table 2.4), the current of each load can be determined since the voltage is known to be $E = 120$ V for all three loads (since they are connected in parallel): For the 1,000 W electric heater, the current, I_1, is

$$I_1 = \frac{P_1}{E} = \frac{1{,}000 \text{ W}}{120 \text{ V}} = 8.33 \text{ A}$$

For the 100 W lamp, the current, I_2, is

$$I_2 = \frac{P_2}{E} = \frac{100 \text{ W}}{120 \text{ V}} = 0.83 \text{ A}$$

For the 75 W lamp, the current, I_3, is

$$I_3 = \frac{P_3}{E} = \frac{75\ \text{W}}{120\ \text{V}} = 0.63\ \text{A}$$

Finally, the total current, I_{tot}, flowing in the branch circuit when all three loads are connected can be determined using the first Kirchoff's law:

$$I_{tot} = I_1 + I_2 + I_3 = 8.33\ \text{A} + 0.83\ \text{A} + 0.63\ \text{A} = 9.79\ \text{A}$$

It should be noted that the total current, I_{tot}, can be also determined by using, first, Equation 2.10 to calculate the total connected power, P_{tot}:

$$P_{tot} = P_1 + P_2 + P_3 = 1175\ \text{W}$$

Then, the same relation between power, voltage, and current used for each load can be applied to calculate the total current flowing through the branch circuit:

$$I_{tot} = \frac{P_{tot}}{E} = \frac{1{,}175\ \text{W}}{120\ \text{V}} = 9.79\ \text{A}$$

b. The resistance for each load can be determined using Ohm's law since both the voltage and the current are known:
The resistance, R_1, for the 1,000 W electric heater is

$$R_1 = \frac{E}{I_1} = \frac{120\ \text{V}}{8.33\ \text{A}} = 14.41\ \Omega$$

The resistance, R_2, for the 100 W lamp is

$$R_2 = \frac{E}{I_2} = \frac{120\ \text{V}}{0.83\ \text{A}} = 144.10\ \Omega$$

The resistance, R_3, for the 75 W lamp is

$$R_3 = \frac{E}{I_3} = \frac{120\ \text{V}}{0.63\ \text{A}} = 192.0\ \Omega$$

Finally, the equivalent resistance, R_{eq}, for the entire branch circuit is

$$R_{eq} = \frac{E}{I_{tot}} = \frac{120\text{V}}{9.79\text{A}} = 12.26\Omega$$

It can be shown that the same equivalent resistance can be calculated using Equation 2.9.

c. The annual total energy use, kWh_{year}, by the branch circuit is determined from the total power, P_{tot}, and the number of hours, N_h, when the circuit is fully used ($N_h = 1{,}000\,\text{h}$):

$$\text{kWh}_{\text{year}} = P_{tot} \times N_h = 1{,}175 \quad \text{W} \times 1{,}000\ \text{h} = 1{,}175{,}000\ \text{Wh} = 1{,}175\ \text{kWh}$$

Thus, the annual energy cost, $\text{Cost}_{\text{year}}$, attributed to the use of the branch circuit is

$$\text{Cost}_{\text{year}} = 1175\ \text{kWh} \times \$0.10/\text{kWh} = \$117.5$$

2.2.2 Alternating Current

Unlike DC systems, the electricity in AC systems changes both direction and magnitude. Almost all power provided by U.S. electrical utilities is produced by AC generators. The operating principles of a single-phase AC generator are illustrated in Figure 2.5. In summary, the generation process involves the principle of electromagnetism. When a rotating coil is moved across a magnetic field (created by a permanent magnet or an electromagnet), an induced voltage or often called electromotive force (emf) is generated in the coil. As the coil rotates, it meets the magnetic field at various angles (four angles are shown in Figure 2.5b) and the induced voltage changes direction and magnitude over time, as illustrated in Figure 2.6.

2.2.2.1 Instantaneous Voltage and Current

When a linear electrical system is subjected to AC, the time variation of the voltage and current can be represented as a sine function:

$$e(t) = E_m \cos \omega t \tag{2.11}$$

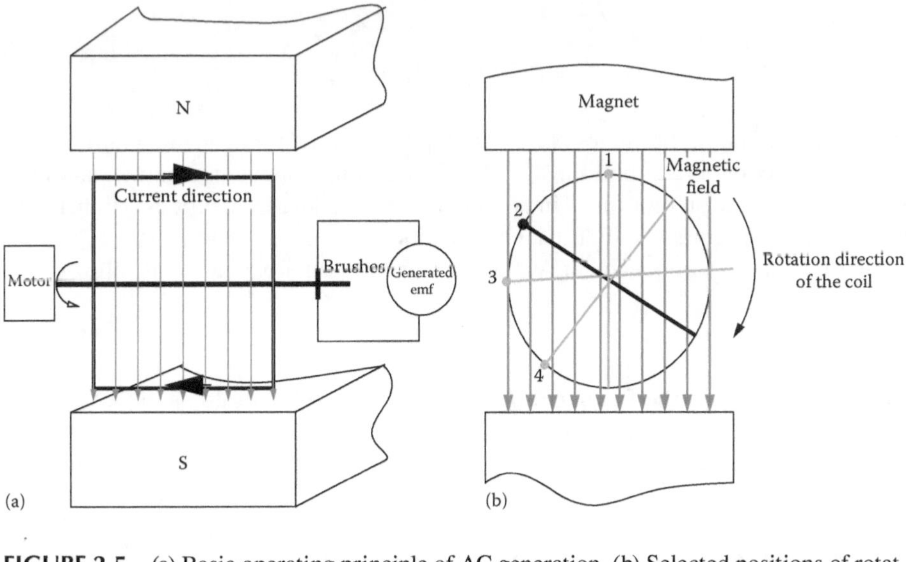

FIGURE 2.5 (a) Basic operating principle of AC generation. (b) Selected positions of rotating coil relative to the magnetic field during one cycle.

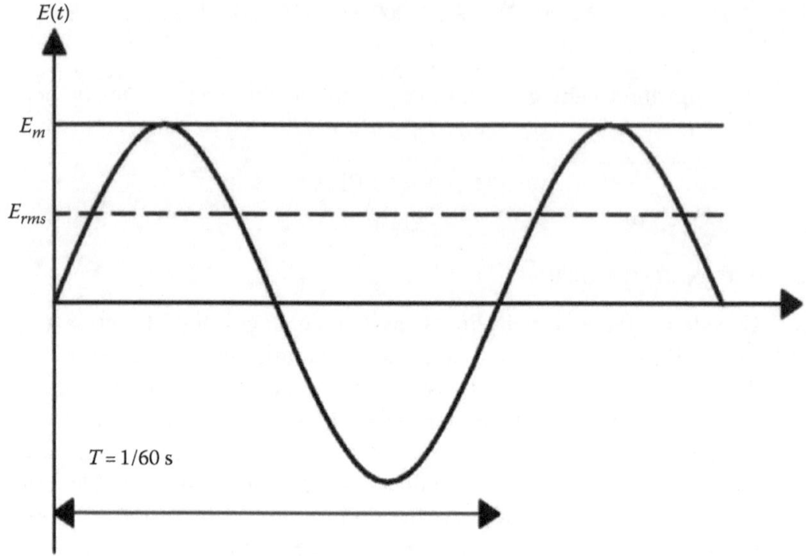

FIGURE 2.6 Illustration of the voltage waveform and the concept of E_{rms}.

$$i(t) = I_m \cos(\omega t - \phi) \tag{2.12}$$

where E_m and I_m are the maximum instantaneous values of voltage and current, respectively. These maximum values are related to the effective or root mean square (rms) values as follows:

$$E_m = \sqrt{2} \times E_{\text{rms}} = 1.41 \times E_{\text{rms}}$$

$$I_m = \sqrt{2} \times I_{\text{rms}} = 1.41 \times I_{\text{rms}}$$

In the United States, the values of E_{rms} are typically 120 V for residential buildings or plug-load in commercial buildings, 277 V for lighting systems in commercial buildings, and 480 V for motor loads in commercial and industrial facilities. Higher voltages can be used for some industrial applications.

ω is the angular frequency of AC and is related to the frequency f as follows:

$$\omega = 2\pi f$$

In the United States, the frequency, f, is 60 Hz, that is, 60 pulsations in 1 second. In other countries, the frequency of AC is $f = 50$ Hz.

ϕ is the phase lag between the current and the voltage. In this case, the electrical system is a resistance (an electric heater), the phase lag is zero, and the current is in phase with the voltage. If the electrical system consists of a capacitance load (such as a capacitor or a synchronous motor), the phase lag is negative and the current is in advance relative to the voltage. Finally, when the electrical system is dominated by an inductive load (such as a fluorescent fixture or an induction motor), the phase lag is positive and the current lags the voltage.

Figure 2.6 illustrates the time variation of the voltage for a typical electric system. The concept of root mean square (also called effective value) for the voltage, E_{rms}, is also indicated in Figure 2.6. It should be noted that the cycle for the voltage waveform repeats itself every 1/60 second (since the frequency is 60 Hz).

2.2.2.2 Impedance of AC Systems

For the analysis of AC systems, it is sometimes convenient to introduce the vector presentation of voltage and current using the theory of complex numbers. The vector presentation is often called the phase diagram. For instance, the time variation of the voltage and the current presented, respectively, by Equations 2.11 and 2.12 can be expressed as follows:

$$e(t) = E_m \cos \omega t = \text{Re}\left[E_m \times e^{j\omega t}\right] \tag{2.13}$$

and

$$i(t) = I_m \cos(\omega t - \phi) = \text{Re}\left[I_m \times e^{(j\omega t - \phi)}\right] \tag{2.14}$$

where

Re is the real part of a complex number
j is the complex number such that $j^2 = -1$

By eliminating the time variable in Equations 2.13 and 2.14, the voltage and the current can be presented more conveniently by two complex numbers, E and I, as indicated in the following (using the polar form representation of a complex number):

$$E = E_m \times e^{j0} = E_m < 0 \tag{2.15}$$

and

$$I = I_m e^{-j\phi} = I_m < -\phi \tag{2.16}$$

Using this representation of voltages and currents, it can be shown that for electrical linear systems, the voltage and the current are related using Ohm's law through an impedance, Z, which is the response coefficient (a complex number) that characterizes each electric system:

$$E = Z \times I \tag{2.17}$$

In most buildings, linear electrical systems are made up of a combination of three basic systems: resistance, inductance, and capacitance. For instance, electrical heaters and incandescent lamps behave like pure resistances. Induction motors and ballasts for fluorescent lamps can be considered as combinations of resistances and inductances. On the other hand, synchronous motors act like capacitances, as will be discussed in Chapter 4.

For any linear system, it can be shown that the impedance can be expressed in the following form:

$$Z = R + j \times X \tag{2.18}$$

where

R is the resistance (in Ω)
X is the reactance (also in Ω) of the system

Figure 2.7 summarizes the three basic electrical systems (i.e., resistance, inductance, and capacitance), the governing equation (in the time domain), the expression for the impedance (complex number), and the phasor diagram (vectorial representation).

As indicated in Figure 2.7, the current is in-phase with voltage for pure resistive loads. However, the current leads voltage for capacitive loads and lags voltage for inductive loads.

The use of impedances facilitates significantly the analysis of the AC systems. In particular, Kirchoff's laws for AC systems can be formulated using the same expressions in Equations 2.6 and 2.7 with the exception that the currents and the voltages are represented by complex numbers (or vectors) instead of simple algebraic values as stated for the DC systems.

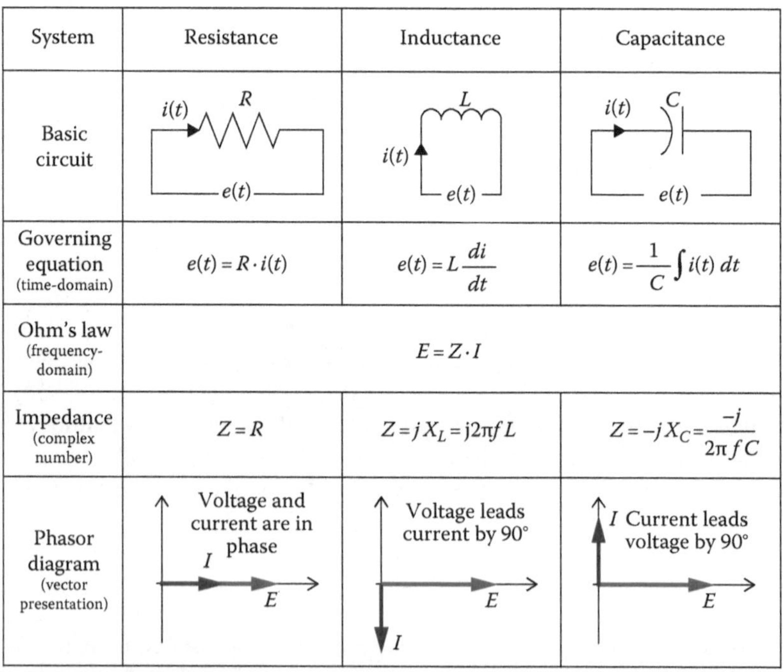

FIGURE 2.7 Governing equations, impedances, and phase diagrams for three basic AC systems: resistance, inductance, and capacitance.

2.2.2.3 Power Triangle and Power Factor

The instantaneous power, $p(t)$, consumed by the electrical system operated on one-phase AC power supply can be calculated using Ohm's law:

$$p(t) = e(t) \times i(t) = E_m I_m \cos \omega t \times \cos(\omega t - \phi) \tag{2.19}$$

This equation can be rearranged using some basic trigonometry and the definition of the rms values for voltage and current (i.e., $E_{rms} = E_m / \sqrt{2}$ and $I_{rms} = I_m / \sqrt{2}$):

$$p(t) = E_{rms} \times I_{rms} \left(\cos\phi (1 + \cos 2\omega t) + \sin\phi \times \sin 2\omega t \right) \tag{2.20}$$

Two types of power can be introduced as a function of the phase lag angle ϕ: the real power P_R and the reactive power P_X as defined in the following:

$$P_R = E_{rms} \times I_{rms} \cos\phi \tag{2.21}$$

$$P_X = E_{rms} \times I_{rms} \sin\phi \tag{2.22}$$

For convenience, a complex power is introduced to represent the real power and the reactive power as follows:

$$P_T = P_R + j \times P_X \tag{2.23}$$

P_T is called the total or apparent power and represents the vectorial sum of the real power and the reactive power. To help understand the meaning of real and reactive power, it is useful to note that the average of the instantaneous power consumed by the electrical system over one period is equal to P_R:

$$\bar{p} = \frac{1}{T} \int_0^T p(t)\,dt = P_R \tag{2.24}$$

Therefore, P_R is the actual power consumed by the electrical system over its operation period (which consists typically of a large number of periods, T). As noted earlier, P_R is typically called real power and is measured in kW. P_X is the power required to produce a magnetic field to operate the electrical system (such as an induction motor) and is stored and then released; this power, typically called reactive power, is measured in kVAR. A schematic is provided in Figure 2.8 to help illustrate the meaning of each type of power.

While the user of the electrical system consumes only the real power, the utility or the electricity provider has to make available to the user, both the real power, P_R, and the reactive power, P_X. The vectorial sum of P_R and P_X constitutes the total power, P_T, and is measured in kVA. Therefore, the utility has to know, in addition to the real power needed by the customer, the magnitude of the reactive power, and thus the total power.

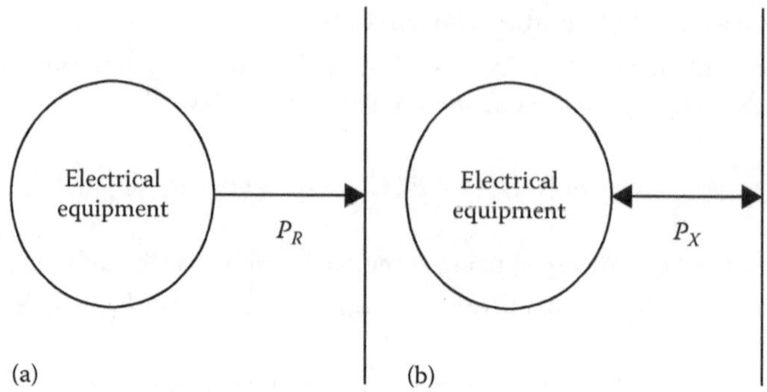

FIGURE 2.8 Illustration of the direction of electricity flow for (a) real power and (b) reactive power.

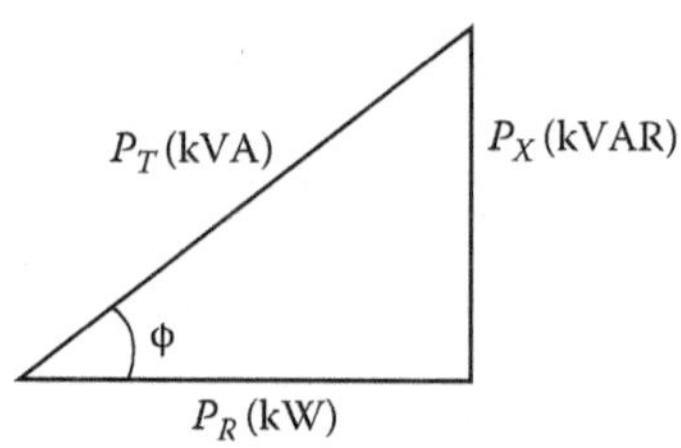

FIGURE 2.9 Power triangle for an electrical system.

As mentioned earlier, for a resistive electrical system, the phase lag is zero and thus the reactive power is also zero (refer to Figure 2.7). Unfortunately, for commercial buildings and industrial facilities, the electrical loads are not strictly resistive, and the associated reactive power can be significant. The higher the phase lag angle ϕ, the more important the reactive power P_X. To illustrate the importance of the reactive power relative to the real power P_R and the total power P_T consumed by the electrical system, a power triangle is used to represent the power flow as shown in Figure 2.9.

In Figure 2.9, it is clear that the ratio of real power to total power represents the cosine of the phase lag. This ratio is widely known as the power factor, *pf*, of the electrical system:

$$pf = \frac{P_R}{P_T} = \cos\phi \tag{2.25}$$

Ideally, the power factor has to be as close to unity as possible (i.e., $pf = 1.0$). Typically, power factors above 90% are considered to be acceptable. If the power factor is low, that is, if the electrical system has a high inductive load, capacitors can be added in parallel to reduce the reactive power as illustrated in Figure 2.10.

Examples 2.3 and 2.4 illustrate how to calculate the impedance of an AC load and the effect of improving power factor on the magnitude of the current for single-phase loads.

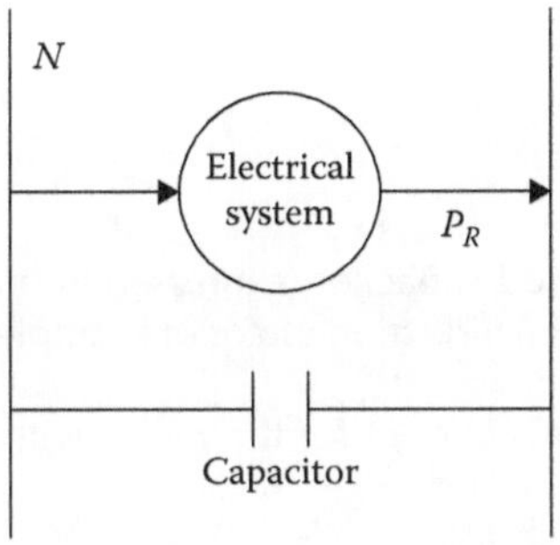

FIGURE 2.10 The addition of capacitor can improve the power factor of an electrical system.

Example 2.3

Problem

Determine the impedance of an induction motor rated at 120 V and 960 W (electrical power required) with a power factor of 0.80.

Solution

First, the phase angle of the impedance can be determined from the power factor using Equation 2.25:

$$\cos\phi = 0.80$$

Thus,

$$\phi = \cos^{-1}0.80 = 36.9°$$

Then, the magnitude of the current, rms, is determined from Equation 2.21 since the real power is known:

$$I = \frac{P_R}{E_{rms} \times \cos\phi} = \frac{960 \text{ W}}{(120\text{ V})(0.80)} = 10\text{ A}$$

Since the induction motor is an inductive load, the current lags the voltage (refer to Figure 2.7). Therefore, the current can be written using the complex numbers as follows:

$$I = 10 \text{ A} < -36.9°$$

Using Ohm's law for AC circuits provided by Equation 2.17, the impedance of the induction motor can be calculated:

$$Z = \frac{E}{I} = \frac{120 \text{ V} < 0°}{10 \text{ A} < -25.8°} = 12.0 \ \Omega < 36.9°$$

Example 2.4

Problem

I. Determine the required capacitor (expressed in both VAR and μF) to be added in parallel to the induction motor of Example 2.3 so that the power factor is unity.
II. Calculate the new rms current for the motor with the added capacitor. Comment.

Solution

a. Using the power triangle concept of Figure 2.9, the reactive power for the motor can be estimated as a function of the phase angle ϕ and the real power P_R:

$$P_X = P_R \times \tan\phi = 960 \text{ W} \times \tan 36.9° = 576 \text{ VAR}$$

Therefore, we should add a capacitor that has a power, P_C, equal to the reactive power of the motor so that the new power factor of the load (motor with the capacitor) is unity:

$$P_C = 576 \text{ VAR}$$

The reactance X_C of the capacitor can then be estimated:

$$X_C = \frac{E_{rms}^2}{P_C} = \frac{(120 \text{ V})^2}{(576 \text{ VAR})} = 25\Omega$$

Using the expression of the reactance as a function of the capacitance C (in farads) provided in Figure 2.7:

$$C = \frac{1}{2\pi f X_C} = \frac{1}{2\pi(60 \text{ Hz})(25 \ \Omega)} = 106 \times 10^{-6} \text{ F} = 106 \ \mu\text{F}$$

b. For the motor with the capacitor, the rms current can be calculated using Equation 2.21 with the real power remaining the same, P_R=960 W, but with a power factor of one ($\cos\phi = 1$):

$$I = \frac{P_R}{E_{rms} \times \cos\phi} = \frac{960 \text{ W}}{(120 \text{ V})(1.0)} = 8 \text{ A}$$

Thus, the improved factor for the motor allows the selection of a smaller conductor size for the branch circuit serving the motor. Moreover, the lower load current reduces the heat losses dissipated by the conductor (i.e., RI^2t).

2.2.3 Advantages of AC Systems

Currently, almost all building electrical loads are supplied by AC power even though only DC power was available in the late 1900s when electricity was first discovered. Their inherent advantages are the main reasons for the dominance of AC over DC systems. Among these advantages are the following:

- *Lower generation costs*: As illustrated in Figure 2.5, it is relatively easy to construct AC generators. In particular, AC generators do not require frequent maintenance as DC generators.
- *More efficient voltage transformations*: AC power voltages can be easily reduced or increased by the use of transformers without significant energy losses. In particular, transformers permit the transmission of large amounts of AC power over long distances using small transmission cables.

It should be noted, however, there is an increasing interest in the use of DC power for electrical distribution systems due to the recent developments in high-voltage DC (HDV) distribution lines and a greater reliance on renewable energy systems such as photovoltaic and wind technologies to generate electricity for buildings (refer to Chapters 13 and 14). But there are some technical and regulatory challenges that need to be resolved before wider adoption of DC power for building applications is possible.

In large commercial and industrial facilities, three-phase AC systems are preferred over single-phase AC systems since they allow larger electrical power to be distributed with less wires. In buildings with small power requirements, single-phase AC systems are generally used, especially in residential buildings. However, even in these buildings, the AC power is not truly single phase, as will be discussed in Section 2.3.

2.3 MULTIPHASE AC SYSTEMS

In almost all buildings, two-phase and/or three-phase power systems are used. It is therefore important to review the basic characteristics of multiphase systems.

2.3.1 Two-Phase AC Systems

Figure 2.11 illustrates the basic principles of the generation of AC two-phase electrical power. The generation process is similar to that outlined for the one-phase power except that two coils (instead of one coil) are rotated within the magnetic field. Among all the potential options for the placement of two coils, two possibilities are of interest: (1) the two coils form an angle of 90° and (2) the two coils are placed in the opposite direction to each other to form an angle of 180°.

The time variation of the phase voltages generated by the two-phase power generators as illustrated in Figure 2.11 can be expressed as follows:

For the configuration of Figure 2.11a,

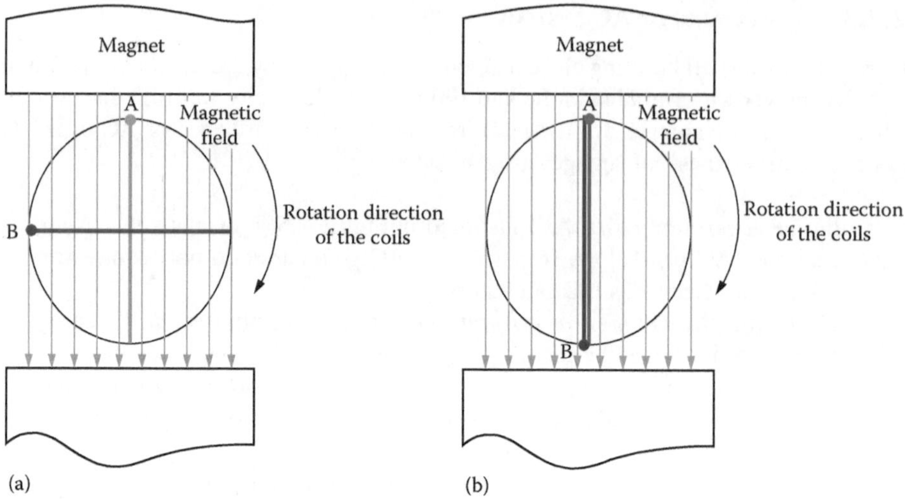

FIGURE 2.11 Principles of generation of two-phase AC power using two coils placed (a) 90° and (b) 180° relative to each other.

$$e_A(t) = E_m \cos \omega t \text{ and } e_B(t) = E_m \cos(\omega t + 90°) \tag{2.26}$$

For the configuration of Figure 2.11b,

$$e_A(t) = E_m \cos \omega t \text{ and } e_B(t) = E_m \cos(\omega t + 180°) \tag{2.27}$$

When each phase is connected to the same load with an impedance Z, the currents flowing in the circuits for both configurations vary with time according to the following expressions:

For the configuration of Figure 2.11a

$$i_A(t) = I_m \cos(\omega t - \theta) \text{ and } i_B(t) = I_m \cos(\omega t - \theta + 90°) \tag{2.28}$$

For the configuration of Figure 2.11b

$$i_A(t) = I_m \cos(\omega t - \theta) \text{ and } i_B(t) = I_m \cos(\omega t - \theta + 180°) \tag{2.29}$$

where θ is the phase angle of the impedance (i.e., $Z = Z < \theta$) and $I_m = E_m / Z$.

By combining the two circuits of both phases as illustrated in Figure 2.12, a common wire is created. This common wire carries the algebraic sum of phase currents:

$$i_C = i_A + i_B \tag{2.30}$$

When the load is balanced between the two phases (i.e., the same impedance is applied to both phases), the current that flows in the common wire can be calculated for both configurations of two-phase power generation (i.e., cases of Figure 2.11a and b).

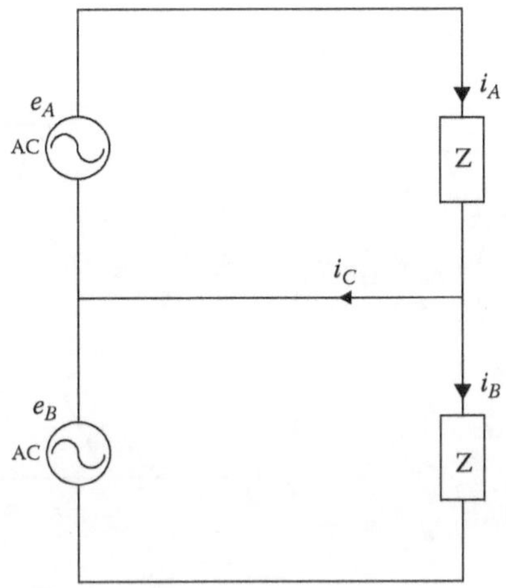

FIGURE 2.12 Two-phase system with a common wire and balanced phase loads.

For the two-phase AC power generation configuration of Figure 2.11a,

$$i_C(t) = I_m\sqrt{2} \times \cos(\omega t - \theta + 45°) \tag{2.31}$$

For the two-phase AC power generation configuration of Figure 2.11b,

$$i_C(t) = 0 \tag{2.32}$$

Therefore, there is no current flowing in the common wire when the two-phase AC power is generated according to the configuration of Figure 2.11b. When the loads are balanced between the two phases, the phase currents have the same magnitude but opposite direction, as can be clearly seen from Equation 2.28. The size of the common wire can be selected to be smaller than the other two wires since it will always carry less current. In the ideal case of balanced loads, the common wire carries no current and is for this specific reason called the "neutral" wire.

The two-phase AC system generated with two coils placed at 180° angle is often called one-phase system, three-wire system, or the Edison system. This type of AC power system is commonly used in residential and small commercial buildings. The phase voltage used in the United States is 120 V. Thus, the one-phase and three-wire systems can supply single-phase loads rated at either 120 or 240 V (240 V = 2 × 120 V).

2.3.2 Three-Phase AC Systems

For commercial buildings, three-phase AC power systems are typically used. Two options are commonly used to generate three-phase AC power as indicated in Figure 2.13. In the first option, the coils are placed in a "star" or "wye" connection with an angle of 120° between any coil, as depicted in Figure 2.13a. In the second option, the coils are placed on a "delta" connection with the coils placed to form a triangle of equal sides, as illustrated in Figure 2.13b.

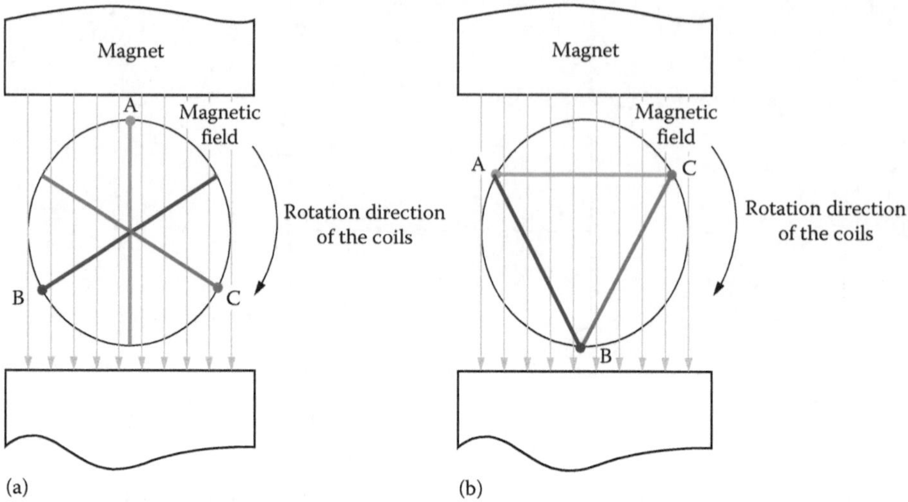

FIGURE 2.13 Principles of generation of three-phase AC power using three coils placed on a (a) wye connection and (b) delta connection.

In both wye and delta connections, the phase voltages generated from AC power systems connected in wye configuration can be expressed as a function of time as follows:

$$\begin{aligned} e_A(t) &= E_m \times \cos \omega t \\ e_B(t) &= E_m \times \cos(\omega t - 120°) \\ e_C(t) &= E_m \times \cos(\omega t + 120°) \end{aligned} \tag{2.33}$$

Using the notation with complex numbers, the phase voltages can also be expressed as shown in the following equation:

$$\begin{aligned} E_A &= E_m < 0 \\ E_B &= E_m < -120° \\ E_C &= E_m < 120° \end{aligned} \tag{2.34}$$

Therefore, the phase-to-phase voltages, commonly known as line voltages, can be determined as the difference between two phase voltages:

$$\begin{aligned} E_{AB} &= E_A - E_B = (E_m < 0) - (E_m < -120°) = \sqrt{3}E_m < 30° \\ E_{BC} &= E_B - E_C = (E_m < 120°) - (E_m < 120°) = \sqrt{3}E_m < -90° \\ E_{CA} &= E_C - E_A = (E_m < 120°) - (E_m < 0) = \sqrt{3}E_m < -210° \end{aligned} \tag{2.35}$$

Thus, the magnitude of the line voltages is higher by a factor of $\sqrt{3}$ relative to the magnitude of the phase voltages:

$$E_L = \sqrt{3}E_m = \sqrt{3}E_\phi \tag{2.36}$$

The characteristics of both wye and delta three-phase systems are summarized in the following sections to help illustrate the specific advantages and the common applications of each system in buildings.

2.3.2.1 Wye Systems

Figure 2.14 shows basic three-phase wye-connected load. In addition to the currents flowing through each phase, there is a current that may flow through a wire that connects the center of wye connection for the power sources to the center of wye connection for the loads. This current is called the neutral current and is equal to the algebraic sum of all the phase currents:

$$i_N = i_A + i_B + i_C \tag{2.37}$$

The phase currents can be calculated using Ohm's law for loads with known impedances Z_A, Z_B, and Z_C. In the particular case of balanced loads (i.e., $Z_A = Z_B = Z_C = Z = Z < \theta$), the phase currents can be readily determined using complex numbers:

$$\begin{aligned} I_A &= \frac{E_A}{Z} = \frac{E_m < 0}{Z < \theta} = \frac{E_m}{Z} < -\theta \\ I_B &= \frac{E_B}{Z} = \frac{E_m < -120^\circ}{Z < \theta} = \frac{E_m}{Z} < -120^\circ - \theta \\ I_C &= \frac{E_C}{Z} = \frac{E_m < 120^\circ}{Z < \theta} = \frac{E_m}{Z} < 120^\circ - \theta \end{aligned} \tag{2.38}$$

Therefore, the current in the neutral wire is

$$I_N = I_A + I_B + I_C = 0 \tag{2.39}$$

When the loads are balanced, the neutral current in three-phase wye-connected system is zero. In distributing the loads among the three phases, it is therefore important

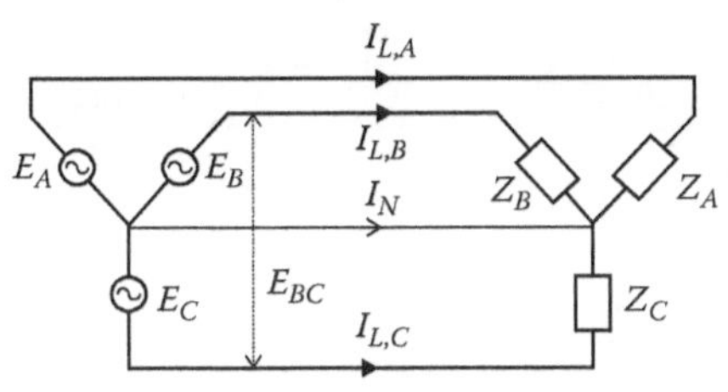

FIGURE 2.14 Basic wye connections for both the voltage sources and the loads.

to maintain as much as possible similar load characteristics between the three phases, so that the neutral carries a small amount of current.

It should be noted that the phase current, I_ϕ, is the same as the line current, I_L, in wye-connected system:

$$I_L = \frac{E_m}{Z} = I_\phi \tag{2.40}$$

The total power consumed by each phase can be calculated using the complex power representation introduced in Equation 2.23 for balanced three-phase wye-connected system:

$$\begin{aligned} P_{T,A} &= E_A \times I_A = (E_m < 0)\left(\frac{E_m}{Z} < -\theta\right) = \frac{E_m}{Z} < -\theta \\ P_{T,B} &= E_B \times I_B = (E_m < -120°)\left(\frac{E_m}{Z} < -120° - \theta\right) = \frac{E_m}{Z} < 120° - \theta \\ P_{T,C} &= E_C \times I_C = (E_m < 120°)\left(\frac{E_m}{Z} < 120° - \theta\right) = \frac{E_m}{Z} < -120° - \theta \end{aligned} \tag{2.41}$$

Since the magnitude of the line voltage $E_L = \sqrt{3} \times E_m$ (see Equation 2.36) and the magnitude of the line current $I_L = E_m/Z$ (refer to Equation 2.40), the total power consumed by each phase in wye-connected systems can be expressed as

$$\begin{aligned} P_{T,A} &= \frac{1}{\sqrt{3}} E_L I_L < -\theta \\ P_{T,B} &= \frac{1}{\sqrt{3}} E_L I_L < 120° - \theta \\ P_{T,C} &= \frac{1}{\sqrt{3}} E_L I_L < -120° - \theta \end{aligned} \tag{2.42}$$

2.3.2.2 Delta Systems

Figure 2.15 illustrates basic connections for three-phase delta systems. Typically, the delta system has only three wires (i.e., three phases) and no neutral. The loads are directly connected between the phase voltages.

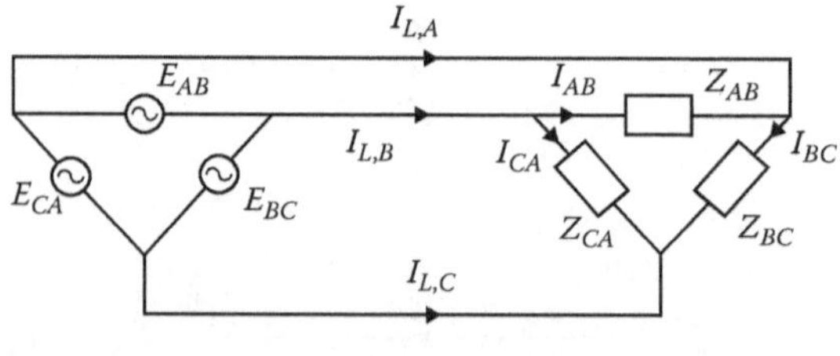

FIGURE 2.15 Basic delta connections for both the voltage sources and the loads.

The current in each phase can be calculated using Ohm's law. In the particular case of balanced loads (i.e., $Z_A = Z_B = Z_C = Z < \theta$), the phase currents are determined to be

$$
\begin{aligned}
I_{AB} &= \frac{E_{AB}}{Z} = \frac{\sqrt{3}E_m < 30^\circ}{Z < \theta} = \frac{\sqrt{3}E_m}{Z} < 30^\circ - \theta \\
I_{BC} &= \frac{E_{BC}}{Z} = \frac{\sqrt{3}E_m < 90^\circ}{Z < \theta} = \frac{\sqrt{3}E_m}{Z} < -90^\circ - \theta \\
I_{CA} &= \frac{E_{CA}}{Z} = \frac{\sqrt{3}E_m < -210^\circ}{Z < \theta} = \frac{\sqrt{3}E_m}{Z} < -120^\circ - \theta
\end{aligned}
\tag{2.43}
$$

Using Kirchoff's law at each node, the line currents I_A, I_B, and I_C can be determined:

$$
\begin{aligned}
I_A &= I_{AB} - I_{CA} = \frac{3E_m}{Z} 3 < -\theta \\
I_B &= I_{BC} - I_{AB} = \frac{3E_m}{Z} < -120^\circ - \theta \\
I_C &= I_{AC} - I_{BC} = \frac{3E_m}{Z} < 120^\circ - \theta
\end{aligned}
\tag{2.44}
$$

Unlike the case for wye-connected systems, the phase currents are not the same as the line currents for delta-connected systems. If the loads are balanced, the magnitude of the line current is higher by a factor of $\sqrt{3}$ relative to the magnitude of the phase current:

$$
I_L = \sqrt{3} I_\phi = \frac{3E_m}{Z} \tag{2.45}
$$

The complex power consumed by each phase can be calculated for balanced three-phase delta-connected systems:

$$
\begin{aligned}
P_{T,AB} &= E_{AB} \times I_{AB} = (\sqrt{3}E_m < 30^\circ)\left(\frac{\sqrt{3}E_m}{Z} < 30^\circ - \theta\right) = \frac{3E_m^2}{Z} < 60^\circ - \theta \\
P_{T,BC} &= E_{BC} \times I_{BC} = (\sqrt{3}E_m < -90^\circ)\left(\frac{\sqrt{3}E_m}{Z} < -90^\circ - \theta\right) = \frac{3E_m^2}{Z} < 180^\circ - \theta \\
P_{T,CA} &= E_{CA} \times I_{CA} = (\sqrt{3}E_m < -210^\circ)\left(\frac{\sqrt{3}E_m}{Z} < -210^\circ - \theta\right) = \frac{3E_m^2}{Z} < -60^\circ - \theta
\end{aligned}
\tag{2.46}
$$

Using the fact that the magnitude of the line voltage $E_L = \sqrt{3} \times E_m$ (refer to Equation 2.36) and the magnitude of the line current $I_L = 3E_m/Z$ (refer to Equation 2.45), the total power consumed by each phase in a delta system can be expressed as follows:

$$P_{T,AB} = \frac{1}{\sqrt{3}} E_L I_L < 60° - \theta$$

$$P_{T,BC} = \frac{1}{\sqrt{3}} E_L I_L < 180° - \theta \quad (2.47)$$

$$P_{T,CA} = \frac{1}{\sqrt{3}} E_L I_L < -60° - \theta$$

Thus, the power used in each phase by delta- or wye-connected systems has the same magnitude but different phase lag.

2.3.2.3 Power in Three-Phase Systems

The instantaneous power used by the entire three-phase system can be estimated as the sum of the power used by each phase. For wye-connected loads, the instantaneous three-phase power can be expressed as

$$p_{3-\phi}(t) = p_A(t) + p_B(t) + p_C(t) \quad (2.48)$$

Using similar analysis to derive Equation 2.20, the instantaneous power consumed in each phase can be calculated as a function of both E_L and I_L. Specifically, consider the following:

First, for phase A, the phase angle Φ is replaced by θ:

$$p_A(t) = \frac{1}{\sqrt{3}} E_L I_L (\cos\theta + \cos\theta \times \cos 2\omega t + \sin\theta \times \sin 2\omega t \quad (2.49)$$

For phase B, the phase angle Φ is replaced by $\theta - 120°$:

$$p_B(t) = \frac{1}{\sqrt{3}} E_L I_L \left(\cos\theta + \cos(\theta - 120°)\cos 2\omega t + \sin(\theta - 120°)\sin 2\omega t\right) \quad (2.50)$$

For phase C, the phase angle Φ is replaced by $\theta + 120°$:

$$p_B(t) = \frac{1}{\sqrt{3}} E_L I_L \left(\cos\theta + \cos(\theta + 120°)\cos 2\omega t + \sin(\theta + 120°)\sin 2\omega t\right) \quad (2.51)$$

since

$$\sin\theta + \sin(\theta - 120°) + \sin(\theta + 120°) = 0 \quad (2.52)$$

and

$$\cos\theta + \cos(\theta - 120°) + \cos(\theta + 120°) = 0 \quad (2.53)$$

The instantaneous power consumed in the three-phase system is simply

$$p(t) = 3 \times \frac{1}{\sqrt{3}} E_L I_L \cos\theta = \sqrt{3} E_L I_L \cos\theta \tag{2.54}$$

Therefore, the instantaneous power consumed by the three-phase system remains constant over time as long as the phase loads are balanced. The same result can be demonstrated for delta-connected systems with balanced loads.

Similar to the one-phase systems, three types of power can be defined for three-phase systems: total power, P_T (expressed in VA or kVA), real power, P_R (expressed in W or kW), and reactive power, P_X (expressed in VAR or kVAR). The magnitudes of the total power, the real power, and the reactive power used by any three-phase system can be easily determined from the line voltage, E_L, and line current, I_L, values using the following equations:

$$\begin{aligned} P_T &= \sqrt{3} E_L I_L \\ P_R &= \sqrt{3} E_L I_L \cos\theta \\ P_X &= \sqrt{3} E_L I_L \sin\theta \end{aligned} \tag{2.55}$$

These equations are particularly useful to determine the rated load current, I_L, from the voltage and power ratings. Examples 2.5 and 2.6 illustrate some applications of the power relations summarized by Equation 2.55 for three-phase systems.

Example 2.5

Problem

a. Determine the line current drawn by 7.50 kW load with a power factor of 75% from a 480 Y / 27 V power source.
b. Estimate the real power and the total power used by each phase if all the phases are balanced.

Solution

a. From Equation 2.55, the line current can be determined as a function of the real power, the line voltage, and the power factor:

$$I_L = \frac{P_R}{\sqrt{3} \times E_L \times \cos\theta} = \frac{7.50\text{ kW}}{\sqrt{3} \times (480\text{ V})(0.75)} = 12.0\text{ A}$$

For a balanced wye-connected load, the phase current is the same as the line current.

b. If the phase loads are balanced, the real power used by each phase is simply— $P_{R,\varphi} = P_{R,3\varphi}/3 = 2.50$ kW. Thus, the total power consumed by each phase is

$$P_{T,\phi} = \frac{P_{R,\phi}}{\cos\theta} = \frac{2.5\text{ kW}}{0.75} = 3.33\text{ kVA}$$

Example 2.6

Problem

A three-phase load has an impedance of $Z = 1.5 + j0.5\ \Omega$ for each phase. Determine the magnitude of phase and line currents as well as the real power drawn by the three-phase load from (1) 480 Y / 277 V source (wye connection) and (2) 480 V power source (delta connection).

Solution

First, the impedance can be put in a polar form to directly obtain its magnitude and its phase angle:

$$Z = 1.5 + j0.5\ \Omega = 1.581\ \Omega\ \angle 18.4°$$

Then, the magnitude of the phase current drawn by the load can be determined using Ohm's law (see Equation 2.17) using single-phase systems.

a. For wye connection, the magnitude of the phase voltage is $E_\varphi = 277$ V:

$$I_\phi = \frac{E_\phi}{Z} = \frac{277\ \text{V}}{1.581\ \Omega} = 175.2\ \text{A}$$

b. For delta connection, the magnitude of the phase voltage is $E_\phi = 480$ V:

$$I_\phi = \frac{E_\phi}{Z} = \frac{480\ \text{V}}{1.581\ \Omega} = 303.5\ \text{A}$$

Finally, the magnitudes for line current, real power, and total power can be calculated as follows:

a. For wye connection, the line current is the same as the phase current based on Equation 2.40:

$$I_L = I_\phi = 175.2\ \text{A}$$

b. The real power consumed by the three-phase system is based on Equation 2.55:

$$P_R = \sqrt{3} E_l I_L \cos\theta = \sqrt{3}(480\ \text{V})(175.2\ \text{A})\cos 18.4° = 138.2\ \text{kW}$$

For delta connection, the line current is obtained by multiplying the phase current by $\sqrt{3}$ as indicated in Equation 2.45:

$$I_L = \sqrt{3} I_\phi = 525.6\ \text{A}$$

The real power consumed by the three-phase system is based on Equation 2.55:

$$P_R = \sqrt{3}E_l I_L \cos\theta = \sqrt{3}(480\text{V})(525.6\text{ A})\cos 18.4° = 414.6\text{kW}$$

Thus, the power consumed by the delta-connected load is three times higher than the power used by the wye-connected load.

2.4 POWER FACTOR CORRECTION

As mentioned in the previous section, the reactive power has to be supplied by the utility even though it is not typically monitored by the power meter that measures generally the real power used by the building. The magnitude of this reactive power increases as the power factor decreases. To account for the added total power required due to the reactive power, most utilities have established rate structures that penalize any user that has low power factor. Therefore, significant savings in the utility costs can be achieved by improving the power factor. As illustrated in Figure 2.10, this power factor improvement can be obtained by adding a set of capacitors to the entire electrical system. The size of these capacitors, P_C, is typically measured in kVAR (the same unit as the reactive power) and can be determined as indicated in Figure 2.16 using the power triangle analysis:

$$P_C = P_{Xe} - P_{Xr} = P_R(\tan\phi_e - \tan\phi_r) \tag{2.56}$$

where

- P_{Xe} and P_{Xr} are reactive powers, respectively, before retrofit (existing conditions) and after retrofit (retrofitted conditions).
- P_C is the reactive power of the capacitor to be added.
- ϕ_e and ϕ_r are phase lag angles, respectively, before retrofit (existing conditions) and after retrofit (retrofitted conditions).

Using the values of power factor before and after the retrofit, the size of the capacitors can be determined:

$$P_C = P_R\left[\tan\left(\cos^{-1} pf_e\right) - \tan\left(\cos^{-1} pf_r\right)\right] \tag{2.57}$$

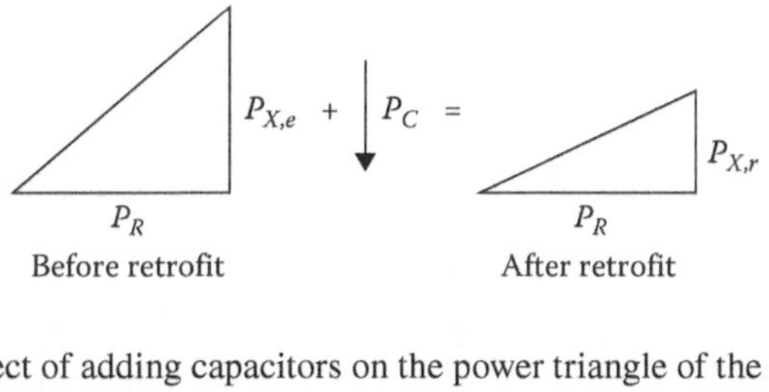

FIGURE 2.16 Effect of adding capacitors on the power triangle of the electrical system.

The calculations of the cost savings due to power factor improvement depend on the utility rate structure. In most rate structures offered by utilities to their customers, one of three options is used to assess the penalty for low power factor. These three options are summarized as follows. Basic calculation procedures are typically needed to estimate the annual cost savings in the utility bills:

1. *Modified billing demand*: In this case, the demand charges are increased in proportion with a fraction by which the power factor is less than a threshold value. The size for the capacitors should be selected so the system power factor reaches at least the defined threshold value.
2. *Reactive power charges*: In this rate, charges for reactive power demand are included as part of the utility bills. In this option, the size of the capacitors should be ideally determined to eliminate this reactive power (i.e., so that the power factor is unity).
3. *Total power charges*: This rate is similar to the option described earlier, but the charges are set for the building/facility total power. Again, the capacitors should be sized so that the power factor is equal to unity.

The calculations of the cost savings due to power factor improvement are illustrated in Example 2.7.

Example 2.7

Problem

Consider a building with a total real power demand of 500 kW with a power factor of $pf_e = 0.70$. Determine the required size of a set of capacitors to be installed in parallel with the building service entrance so that the power factor becomes at least $pf_r = 0.90$.

Solution

The size in kVAR of the capacitor is determined using Equation 2.57:

$$P_C = 500\left[\tan\left(\cos^{-1}0.70\right) - \tan\left(\cos^{-1}0.90\right)\right] = 268\text{kVAR}$$

Thus a capacitor rated at 275 kVAR can be selected to maintain a power factor for the building electrical system above 0.90.

2.5 SUMMARY

In this chapter, an overview is provided for the basic characteristics of electrical systems. In particular, the generation principles of AC power and the properties of single- and three-phase systems are emphasized. Moreover, illustrative examples are presented to provide some insights into the applications of the basic concepts for both DC and AC power systems. Moreover, the chapter has provided suggestions to improve the power quality and increase the power factor in buildings. These suggestions are presented to illustrate the wide range of issues that an engineer should address when designing, analyzing, or retrofitting electrical systems for buildings.

2.6 PROBLEMS

1. A coil of # 12 AWG copper wire wound in a single layer on a cylindrical form so that the mean diameter of each turn is 6.0 in., and it has a resistance of 4.0 Ω. How many turns of wire are there in the coil?
2. Consider the following circuit setup shown below. Determine the voltage drop across the 7 Ω resistor.
3. An automobile generator with an internal resistance of 0.5 Ω develops an open-circuit emf of 9.0 V. The storage battery has an internal resistance of 0.03 Ω and open-circuit emf of 7.0 V. Both sources are connected in parallel to 0.8 Ω load. Determine the generator current, battery current, and load current.
4. A 5 A current flows when a neon lighting fixture is supplied by a 120 V power source. If the current lags the voltage by 75°, determine the power factor and the real power consumed by the fixture.
5. Determine the rating of a generator in kVA for a home in order to ensure that the following systems and appliances can be operated in case of a long-term power outage:
   a. Air conditioner rated at 9.5 A, 120 V, and *pf* = 0.90 (lagging)
   b. Freezer rated at 4.5 A, 120 V, and *pf* = 0.85 (lagging)
   c. Refrigerator at 3.5 A, 120 V, and *pf* = 0.80 (lagging)
6. A 240 V, single-phase load has a resistance of 3 Ω and an inductive reactance of 4 Ω. Calculate
   a. The impedance of the load
   b. The power factor of the load
   c. The power consumed by the load
7. For a single-phase circuit served by an AC power source:
   a. Determine the total impedance at 60 Hz of a circuit consisting of a 0.25 Hz inductor with a 100 Ω resistance and a 50 μF capacitor, all connected in parallel.
   b. If the resistance, inductor, and capacitor are connected to 120 V AC source, determine the magnitude as well as the phase angle (relative to the voltage) of the source current drawn.
8. A 1 horsepower motor is connected to 120 V, 60 Hz AC source and has a *pf* = 0.80 (lagging) and an efficiency of 0.90. A capacitor is connected in parallel with the motor in order to correct its *pf* to 0.95. Determine
   a. The current drawn from the source with and without the capacitor
   b. The rating of the capacitor in farads needed to make the *pf* correction
9. A motor for a blender can be modeled as a resistance of 10 Ω and an inductance of 25 mH connected in a series and supplied by a 120 V, 60 Hz source. Determine
   a. The real power consumed by the motor

b. The power factor of the motor
c. The capacitor needed to be connected in parallel to the motor in order to change its *pf* to 0.90

10. A single-phase induction motor draws 250 W at a power factor of $pf = 0.80$ (lagging) when connected to 120 V, 60 Hz source. Find the capacitance needed to correct the pf of the motor to unity (i.e., $pf = 1.0$).
11. Find the real power and the power factor for a load whose impedance is $60\ \Omega < 60°$ when connected to an AC source of 120 V and 60 Hz.
12. Determine the capacitance that can increase the overall power factor to 95% lagging when connected in parallel with a load that draws 1.0 kW at a 70% lagging power factor from a 240 V, 60 Hz source.
13. Find the capacitance that minimizes the source current when connected in parallel with a single-phase induction motor drawing 10 A at a power factor of $pf = 0.80$ (lagging) when connected to 120 V, 60 Hz source.
14. A synchronous motor capable of operating with a leading power factor draws 10 kW from a transformer while driving a fan. The remainder load on the transformer consists of 100 kW with 0.80 lagging power factor. Determine
    a. The reactive power needed from the motor for the overall power factor to be 0.95 lagging
    b. The power factor of the synchronous motor for the overall power to be 0.90 lagging
15. Consider a motor rated 840 W with 70% power factor operating from a 120 V, 60 Hz, one-phase source. Determine
    a. The source current if a 308 μF capacitor is added to the motor in series
    b. The value of the capacitance that must be placed in parallel with the motor to increase the overall power factor to unity
    c. The source current in the case of (b).

    Conclude on the benefits of improving the power factor.
16. A 200 HP induction motor with a power factor of 0.65 and an efficiency of 90% is supplied by a 3-φΔ-connected 480 V AC source.
    a. Show the power triangle for the motor.
    b. Determine the impedance of the motor (assume three equal impedances, one per phase).
    c. Calculate the capacitance in kVAR required to improve the power factor of the motor to 0.85. Solve this question using power triangles.
    d. Compare the line currents of the motor before and after improving the power factor from 0.65 to 0.85.
17. A building has an electrical load of 500 kVA at 0.70 power factor, lagging. Among the load is a 50 HP induction motor with an efficiency of 0.93 and a power factor 0.90 (lagging). It has been decided to replace the induction motor with a synchronous motor of the same size, efficiency, and power factor (leading). Determine the impact of this replacement on:
    a. The change of the power factor of the building.
    b. The variation in the total line current that the building draws if it served by 208 Y/120-V source.

18. A manufacturing plant has an electrical load of 5,000 kW at 0.80 power factor, lagging. It has been decided to replace a 500 HP induction motor that drives a pump. This motor operates at an efficiency of 96% and a power factor of 0.90, lagging. A synchronous motor is purchased as the replacement and is capable of operating at 0.80 power factor, leading, with the same efficiency as the induction motor. Determine
    a. The new plant power factor.
    b. The percentage decrease in line current is a result of the improved power factor.
19. A one-phase motor rated at 5-hp and 240-V with an efficiency of 0.90, and a power factor of 0.85, lagging. Determine the equivalent impedance for the motor.
20. A three-phase motor rated at 100-hp and 460-V with an efficiency of 0.92, and a power factor of 0.90, lagging. Determine the equivalent impedance for the motor.

3 Electrical Transformers

This chapter describes some general concepts of electrical transformers. First, a review of the basic characteristics of a transformer operating under alternating current is provided. Then, commonly used types and connections of electrical transformers are briefly discussed. Finally, testing procedures to characterize some of the properties of transformers are outlined. Throughout this chapter, calculation examples are provided to determine the design specifications and benefits of selected measures and strategies for improving the energy efficiency of electrical transformers.

3.1 INTRODUCTION

The transformer is a device that allows to change the voltage level of an alternating current. In particular, it is common to use transformers at generating stations to increase transmission voltages to high levels (13,800 V) and at near or inside buildings to reduce distribution voltages to low levels for utilization (480 or 208 V).

A typical transformer consists of two windings: primary and secondary. The primary winding is connected to the power source while the secondary winding is connected to the load. Between the primary and the secondary windings, there is no electrical connection. Instead, electric energy is transferred through inductance within the core, which is generally made up of laminated steel. Therefore, transformers operate only on alternating current.

There are basically two types of transformers: (1) liquid-filled transformers and (2) dry-type transformers. In liquid-filled transformers, the liquid acts as a coolant and an insulation dielectric. Dry-type transformers are constructed so that the core and coils are open to allow cooling by the free movement of air. In some cases, fans may be installed to increase the cooling effect. The dry-type transformers are widely used because of their lighter weight and simpler installation method compared to the liquid-filled transformers.

In this chapter, the basic operation of transformers is first overviewed. Then, the types and common connections of electrical transformers used in buildings are discussed. Finally, testing protocols suitable to evaluate the performance of transformers are presented.

3.2 FUNDAMENTALS OF TRANSFORMERS

A transformer generally consists of a core made up of laminated steel and two windings. The windings that are connected to the power source are generally called primary windings. The other windings are connected to the load and are generally labeled as the secondary windings. A schematic diagram for a single-phase transformer is illustrated in Figure 3.1. A photo of an uncovered single-phase transformer is shown in Figure 3.2. A three-phase transformer can be constructed from a set of

DOI: 10.1201/9781003276999-3

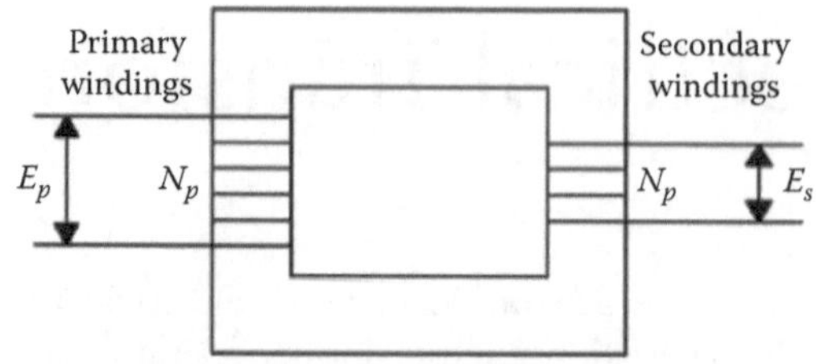

FIGURE 3.1 Simplified model for a single-phase transformer.

FIGURE 3.2 An uncovered single-phase transformer.

three single-phase transformers electrically connected so that the primary and the secondary windings can be wye or delta configurations.

The primary windings create a magnetic flux that induces a voltage at the secondary windings. The laminated steel core serves as a path for the magnetic flux to travel from the primary to the secondary windings. It can be shown that the primary and the secondary voltages, E_p and E_s, are directly proportional to the respective number of turns, N_p and N_s, in the windings:

$$\frac{E_p}{E_s} = \frac{N_p}{N_s} = a \tag{3.1}$$

where a is the turns ratio of the transformer. As indicated by Equation 3.1, the turns ratio can be determined directly from the voltages without the knowledge of the actual number of turns on the transformer windings.

Transformers are rated by their volt-ampere capacity of the secondary windings. For large transformers, the power output in kilo volt-ampere or kVA rating is generally used. For single-phase transformers, the power rating is simply the product of the voltage, E_s, and the current, I_s, at the secondary windings:

$$kVA = \frac{E_s \times I_s}{1{,}000} \tag{3.2}$$

The power rating for three-phase transformers can be determined by the following expression:

$$kVA = \frac{\sqrt{3}E_s \times I_s}{1{,}000} \tag{3.3}$$

where E_s and I_s are the rated line-to-line voltage and the rated line current of the secondary windings, respectively.

Transformers are typically very efficient with energy losses representing only 1%–3% of the transformer capacity. For most design calculations, ideal transformers are considered. For ideal transformers, it is assumed that the windings have no resistance and that there is no flux leakage along the core. Thus, ideal transformers are 100% efficient. In particular, the power input of an ideal transformer is equal to the power output. Therefore, the primary and secondary currents can be determined if the transformer rating is known along with the primary and secondary voltages using Equation 3.4 for single-phase transformers

$$\begin{aligned} I_p &= \frac{kVA}{E_p} \\ I_s &= \frac{kVA}{E_s} \end{aligned} \tag{3.4}$$

and Equation 3.5 for three-phase transformers

$$\begin{aligned} I_p &= \frac{kVA}{\sqrt{3}E_p} \\ I_s &= \frac{kVA}{\sqrt{3}E_s} \end{aligned} \tag{3.5}$$

Example 3.1 illustrates the calculation procedure for primary and secondary currents as well as the turns ratio of both single-phase and three-phase transformers using Equations 3.1, 3.4, and 3.5.

Example 3.1

Problem

Determine the turns ratio, the primary current, and the secondary current for the following transformers:

a. 10 kVA single-phase transformer with a primary voltage of 2,400 V and a secondary voltage of 120 V
b. 10 kVA single-phase transformer with a primary voltage of 2,400 V and a secondary voltage of 120 V
c. 1,500 kVA three-phase transformer with a primary voltage of 13.8 kV and a secondary voltage of 480 Y/277 V

Solution

a. For the single-phase transformer, the turns ratio is determined from Equation 3.1:

$$a = \frac{E_p}{E_s} = \frac{2400 \text{ V}}{120 \text{ V}} = 20$$

The primary and secondary current can be estimated from Equation 3.4:

$$I_p = \frac{\text{kVA}}{E_p} = \frac{10{,}000 \text{ VA}}{2{,}400 \text{ V}} = 4.17 \text{ A}$$

$$I_s = \frac{\text{kVA}}{E_s} = \frac{10{,}000 \text{ VA}}{120 \text{ V}} = 83.3 \text{ A}$$

Note that the ratio I_s/I_p is equal to the turns ratio.

b. For the three-phase transformer, the same Equation 3.1 can be used to determine the turns ratio if phase voltages are used:

$$a = \frac{E_p}{E_s} = \frac{13{,}800 \text{ V}}{277 \text{ V}} \approx 50$$

The primary and secondary current for three-phase transformers can be estimated from Equation 3.5.

$$I_p = \frac{\text{kVA}}{\sqrt{3}E_p} = \frac{1{,}500{,}000 \text{ VA}}{(1.732)\times(13{,}800 \text{ V})} = 62.76 \text{ A}$$

$$I_s = \frac{\text{kVA}}{\sqrt{3}E_s} = \frac{1{,}500{,}000 \text{VA}}{(1.732)\times(480 \text{ V})} = 1{,}804.27 \text{ A}$$

Real transformers have inherent power losses in the windings (often called copper losses) and leakage of magnetic flux in the steel core (also called core losses).

The efficiency of real transformers is defined as the ratio of the power output, P_{out} (at the secondary windings) over the power input, P_{in} (at the primary windings):

$$\eta = \frac{P_{out}}{P_{in}} \tag{3.6}$$

It can be shown that maximum efficiency is obtained when the losses from the windings are equal to those from the magnetic flux leakage. Manufacturers attempt to design their transformers so the optimum efficiency is achieved at a specific average load which may not necessarily be the nominal rating. It is generally cost-effective to invest in high-efficiency transformers, especially if they are used continuously at their rated capacity, as illustrated in Example 3.2.

Example 3.2

Problem

Determine the cost-effectiveness of selecting a unit with an efficiency of 99.95% rather than 99.90% for a 500 kVA rated transformer. Assume that

- The cost of electricity is $0.10/kWh.
- The installed costs of 99.0% and 99.5% efficient transformers are respectively $7,000 and $9,500.
- The average power factor of the load is 0.90.
- The no-load losses are the same for both transformers.

For the analysis, consider two cases for the length of time during which the transformer is used at its rated capacity:

a. 10 h/day and 250 days/year
b. 16 h/day and 300 days/year

Solution

To determine the cost-effectiveness of installing an energy-efficient transformer, a simplified economic analysis based on the simple payback period method was used (refer to Chapter 9). The savings in energy losses in kWh for the high-efficiency transformer can be calculated as follows:

$$\text{kWh}_{\text{saved}} = N_h \times \text{kVA} \times pf\left(\frac{1}{\eta_{std}} - \frac{1}{\eta_{eff}}\right)$$

where

N_h is the total number of hours (per year) during which the transformer is operating at full load (for case (a) $N_h = 10 \times 250 = 2{,}500$ h/year; for case (b) $N_h = 16 \times 300 = 4{,}800$ h/year) *kVA* is the rated transformer power output (500 kVA)

pf is the annual average power factor of the load ($pf=0.90$)

η_{std} and η_{eff} are the efficiency of the standard transformer and the efficient transformer, respectively (0.990 and 0.995)

The energy savings and the payback period for each case are presented earlier.
For case (a) $N_h=2{,}500$ h/year
The energy savings in kWh is calculated as follows:

$$kWh_{\text{saved}} = 2500 \times 500 \times 0.90\left(\frac{1}{0.990} - \frac{1}{0.995}\right) = 5710 \text{ kWh/year}$$

Therefore, the simple payback period, *SPP*, for investing on the efficient transformer is

$$SPP = \frac{\$9{,}000 - \$7{,}000}{5{,}710 \text{ kWh} \times \$0.07 \text{ kWh}} = 5.0 \text{years}$$

For case (a) $N_h=4{,}800$ h/year

The energy savings in *kWh* is calculated as follows:

$$kWh_{\text{saved}} = 4{,}800 \times 500 \times 0.90\left(\frac{1}{0.990} - \frac{1}{0.995}\right) = 10{,}964 \text{ kWh/year}$$

Thus, the simple payback period, *SPP*, for investing on the efficient transformer is

$$SPP = \frac{\$9{,}000 - \$7{,}000}{10{,}964 \text{ kWh} \times \$0.07 \text{ kWh}} = 2.6 \text{years}$$

It is clear that it can be cost-effective to consider investing in a more energy-efficient transformer especially when the load is supplied during longer periods of time. It should be noted that additional energy savings can be expected during no-load conditions for the energy-efficient transformer.

3.3 TYPES OF TRANSFORMERS

There are several types of transformers depending on their applications. Among the transformers commonly used in building installations, three categories can be identified.

3.3.1 Power Transformers

For electricity transmission and distribution, power transformers are used. For residential buildings, a utility-owned power transformer is generally installed outdoors so low voltages (i.e., 240 V or lower) are available to all neighboring dwellings. For large commercial buildings, power transformers are part of unit substations to reduce the distribution voltage (typically 13.8 kV) to voltages below 600 V (specifically 600 V in almost all commercial buildings).

Within buildings, there are several applications of power transformers. In particular, power transformers are cost-effective devices to make available voltages needed for lighting applications. For instance, when the only readily available voltage in an industrial facility is 480 V (which is typically required to operate three-phase motors), it is better to use power transformers rather than request the utility to provide voltages of 120 or 277 V for lighting connections.

3.3.2 Instrument Transformers

To monitor and meter electricity use, instrument transformers are typically used. There are two basic types of instrument transformers: voltage or potential transformers (PT) and current transformers (CT). These instrument transformers allow the measurement of high voltage and current with low-scale voltmeters and ammeters. Figure 3.3 illustrates the basic construction and operation of PT (Figure 3.3a) and CT (Figure 3.3b). Example 3.3 presents the basic calculation procedure to determine the relationship between actual voltage and current (E_a and I_a) and measured voltage and current (E_m and I_m).

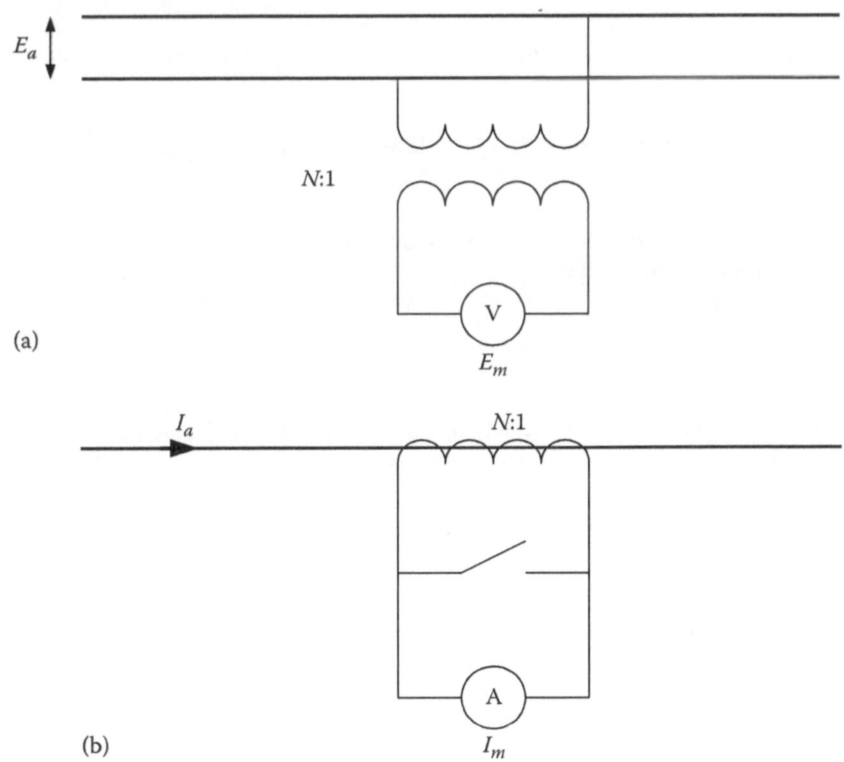

FIGURE 3.3 Typical connection and operation of (a) potential transformer (PT) and (b) current transformer (CT).

Example 3.3

Problem

To determine the electrical energy consumption for a building, a utility may elect to install a meter in the high-voltage side of the service transformer. In particular, PTs and CTs are installed to form wattmeters. One method that is commonly used to measure energy use in buildings (kWh) consists of using three wattmeters (one per phase). A wattmeter is made of a PT and a CT with a common turns ratio of 100. Estimate the actual current supply to a building supplied by 208 Y/120 voltage system if a phase wattmeter measures a power of 1.35 W and a power factor of 0.90. Then, calculate the actual real power used by the building in kW.

Solution

a. Since the turns ratio for the PT is 100 and the actual phase voltage for the building is 120 V, then the measured voltage is determined from Equation 3.1:

$$E_s = \frac{E_P}{a} = \frac{120\ \text{V}}{100} = 1.2\ \text{V}$$

The secondary current measured by the CT can then be estimated from Equation 3.4:

$$I_s = \frac{P_R}{E_s pf} = \frac{1.35\ \text{W}}{1.2\ \text{V}(0.9)} = 1.25\ \text{A}$$

Thus, the actual current supplied to the buildings is determined using the turns ratio of the CT from Equation 3.1:

$$I_p = a \times I_s = 100 \times 1.25\ \text{A} = 125\ \text{A}$$

The actual real power used by the building in kW can be estimated from Equation 3.5 since the line current and voltage are known:

$$P_R = \sqrt{3} E_L \times I_L pf = \frac{1}{1{,}000} 1.732 \times 125\ \text{A} \times 208\ \text{V} \times 0.90 = 40.5\ \text{kW}$$

3.3.3 AUTOTRANSFORMERS

To increase the safety of using certain electrical systems such as appliances and lighting fixtures, it is desirable to operate at low voltages (i.e., lower than 120 V) using small and relatively inexpensive transformers, which typically consist of autotransformers. Typical connections of autotransformers are illustrated in Figure 3.4. Unlike normal transformers, autotransformers do not have separate primary and secondary windings, but they are characterized by common windings. Figure 3.5 shows

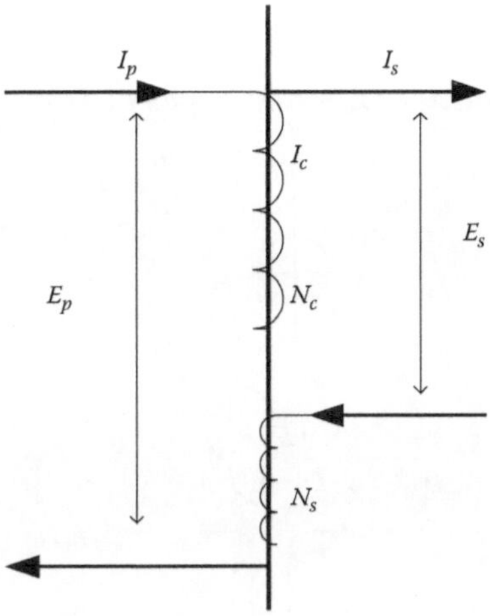

FIGURE 3.4 Typical connection for an autotransformer.

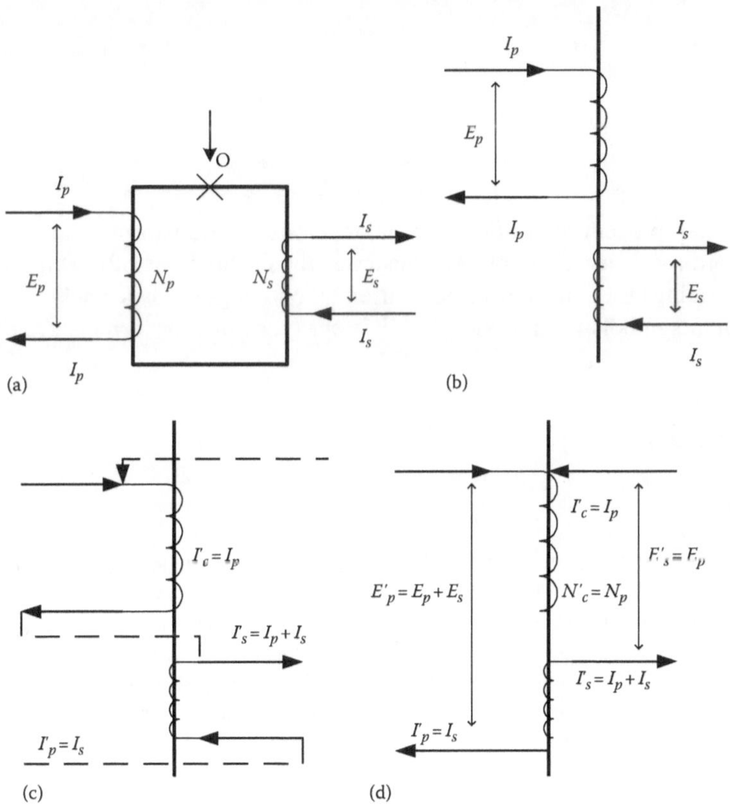

FIGURE 3.5 Transition from (a) a single transformer to (b) a broken transformer, (c) a reconnected transformer, and (d) an autotransformer.

TABLE 3.1
Commonly Used Symbols for Different Types of Transformers

Symbol	Alternative Symbol	Description
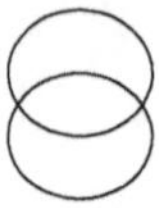	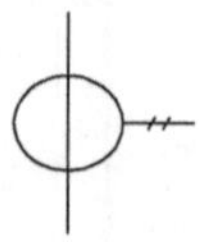	Power transformer
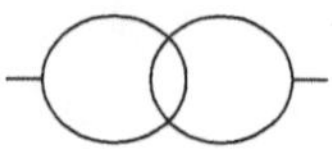	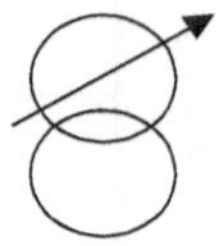	Current transformer
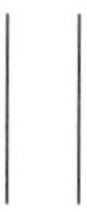	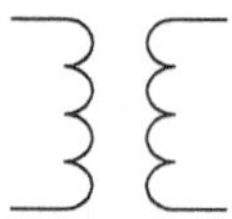	Potential transformer
		Adjustable transformer
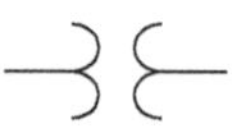		Magnetic core

a step-by-step procedure of how a normal single-phase transformer can be changed to an autotransformer. Example 3.4 illustrates the calculations of the currents, voltages, and ratings for autotransformers based on one-phase transformers.

Common symbols used to identify different types of transformers are presented in Table 3.1.

Example 3.4

Problem

A one-phase 100 kVA 240/120 V transformer is reconnected as an autotransformer:

a. Calculate the currents in each part of the autotransformer and determine its power rating if it is reconnected as a step-down 360/240 V autotransformer. Show the connection details of the autotransformer.
b. Repeat (a) if the transformer is reconnected as a step-up 240/360 V autotransformer.

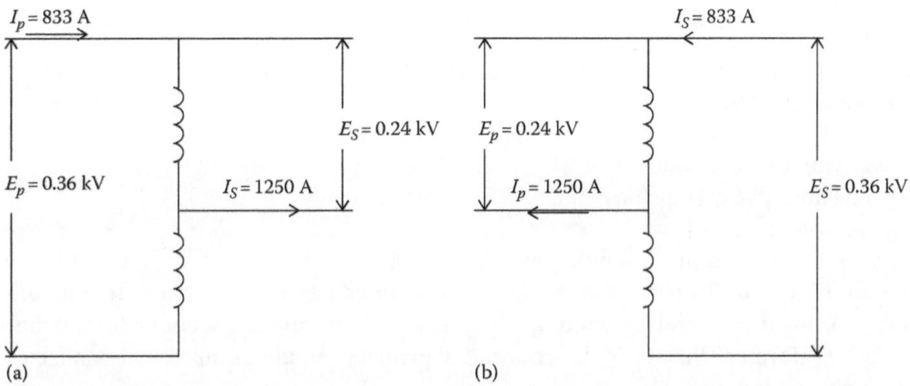

FIGURE 3.6 Connections, currents, and voltages for the autotransformers of Example 3.4 for cases (i) and (ii).

Solution

For the one-phase transformer

$$I_{240} = \frac{P_T}{E_P} = \frac{100 \text{ kVA}}{0.24 \text{ kV}} = 417A$$

$$I_{120} = \frac{P_T}{E_S} = \frac{100 \text{ kVA}}{0.12 \text{ kV}} = 833A$$

For the 360/240 V autotransformer

$$E_P = 0.24 + 0.12 = 0.36 \text{ kV}$$

$$I_P = 833 \text{ A}$$

$$I_S = 417 + 833 = 1250 \text{ A}$$

The rating for the 360/240 V autotransformer, P_{AT}, can then be estimated:

$$P_{AT} = E_P \times I_P = 0.36 \text{ kV} \times 833 \text{ A} = 300 \text{ kVA}$$

A similar analysis and results can be shown for the 240/360 V autotransformer. Figure 3.6 illustrates the connections and the values of current and voltages for both 360/240 and 240/360 V autotransformers.

3.4 TRANSFORMER CONNECTIONS

There are basically two types of connections for electrical transformers used for building systems:

1. Single-phase transformers
2. Three-phase transformers

A typical connection for single-phase transformers is illustrated in Figure 3.1. For three-phase transformers, wye or delta connections can be used either for the primary or for the secondary windings. Figure 3.7 shows the connections for a delta–wye transformer, that is, delta-connected primary windings and wye-connected secondary windings.

Standard sizes for both single-phase and three-phase transformers that are typically used in buildings are listed in Tables 3.2 and 3.3, respectively. In particular, Tables 3.2 and 3.3 provide the average values of the percentage impedance and the ratio X/R for each transformer rating. These values will be required for the

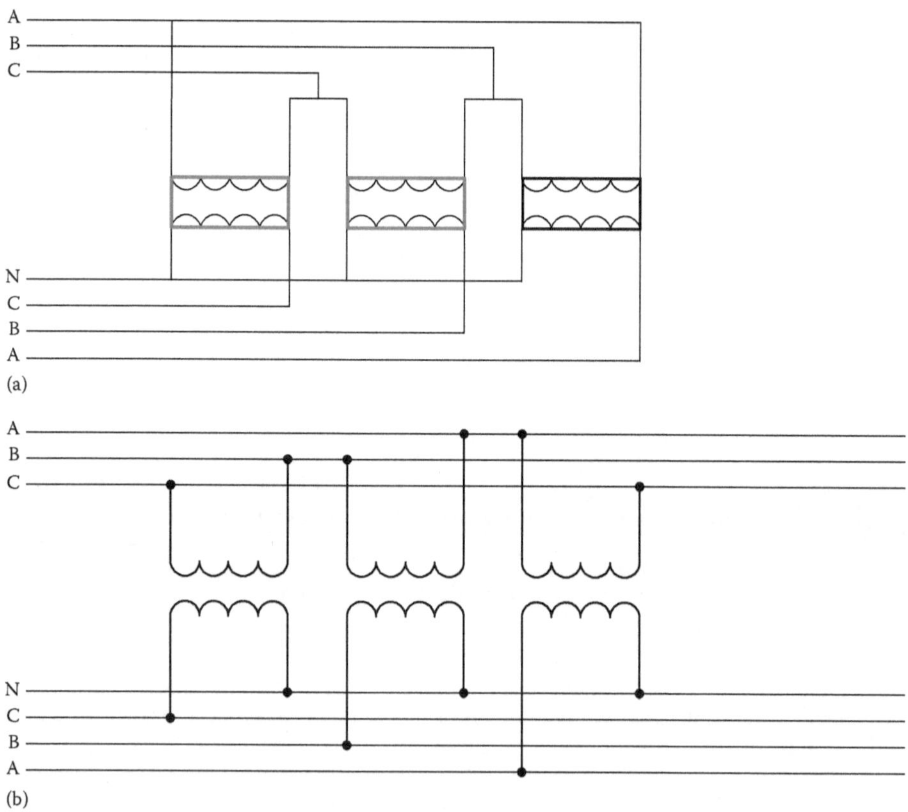

FIGURE 3.7 Connections for delta–wye three-phase transformer, (a) wiring details and (b) representation of the connections.

TABLE 3.2
Typical Ratings of Single-Phase General-Purpose Transformers

Power Rating (kVA)	Height (in.)	Width (in.)	Depth (in.)	Weight (lb)	Average Impedance (%*Z*)	Average *X*/*R* Ratio
3.0	14.25	7.50	7.75	48	—	3.53
5.0	15.75	9.38	9.00	75	—	2.64
7.5	16.00	12.00	10.63	102	—	2.54
10.0	19.00	12.00	10.63	128	—	2.18
15.0	23.00	16.00	15.00	190	5.2	0.60
25.0	23.00	16.00	15.00	230	6.4	1.06
50.0	30.00	20.00	20.00	430	5.2	1.21
75.0	37.00	20.00	20.00	450	6.1	1.64
100.0	42.00	24.00	24.00	665	6.5	1.65

TABLE 3.3
Typical Ratings of Three-Phase General-Purpose Transformers

Power Rating (kVA)	Height (in.)	Width (in.)	Depth (in.)	Weight (lb)	Average Impedance (%*Z*)	Average *X*/*R* Ratio
15.0	23.00	22.25	15.00	230	3.6	1.94
30.0	23.00	22.25	15.00	285	6.4	0.92
45.0	26.00	24.00	15.00	369	6.6	1.13
75.0	30.00	30.00	20.00	590	5.7	1.38
112.5	37.00	30.00	20.00	690	6.1	1.51
150.0	42.00	36.00	24.00	1,050	5.5	1.53
225.0	42.00	36.00	24.00	1,350	6.6	2.00
300.0	48.00	48.00	29.50	2,000	3.6	1.81
500.0	58.00	48.00	29.50	2,700	5.0	2.89
750.0	90.00	72.00	54.00	5,200	5.0	1.98
1,000.0	90.00	72.00	54.00	6,000	5.8	2.38

short-circuit analysis, as described in Chapter 8. The geometric dimensions are useful to determine the space minimum requirements for the enclosure that houses the transformer. Space and protection requirements for transformers are outlined in section 450 of the NEC (2014).

3.5 TESTING METHODS FOR TRANSFORMERS

The performance of transformers can be evaluated through testing procedures. Some of these testing procedures are briefly described in this section. While testing is not part of their responsibilities, designers of building electrical systems need to be familiar with the testing protocols required for transformers. In particular, tests to measure transformer efficiency, percentage impedance, and transformer regulation are often required before and during the installation of transformers.

3.5.1 Transformer Efficiency

The power output from any transformer is always less than the input power. The transformer power losses are typically due to copper losses and core losses. The copper losses, also known as winding losses or I^2R losses, are attributed to the electrical resistance of the transformer windings. While it is possible to reduce the resistance of the transformer windings, some power losses are unavoidable. Almost all the copper losses are dissipated in the form of heat. The core losses are the result of the leak of magnetic flux within the core of the transformer. Thus, core losses are always present independently of the transformer loading. To determine the transformer efficiency, both core and copper losses need to be measured, as illustrated in Figure 3.8.

Figure 3.8a indicates the basic test commonly used to measure core losses. The test requires an open circuit in the secondary windings of the transformer:

1. The rated voltage is applied at the source (i.e., primary windings).
2. The power loss is obtained by multiplying the voltmeter and ammeter readings, as shown in Figure 3.8a (or using wattmeter connected near the voltage source). The measured power is the core losses.

The copper losses are estimated using the short-circuit test as illustrated in Figure 3.8b:

1. The voltage source is adjusted until the ammeter connected in the short-circuited windings (i.e., secondary windings) indicates the rated current for the secondary windings.

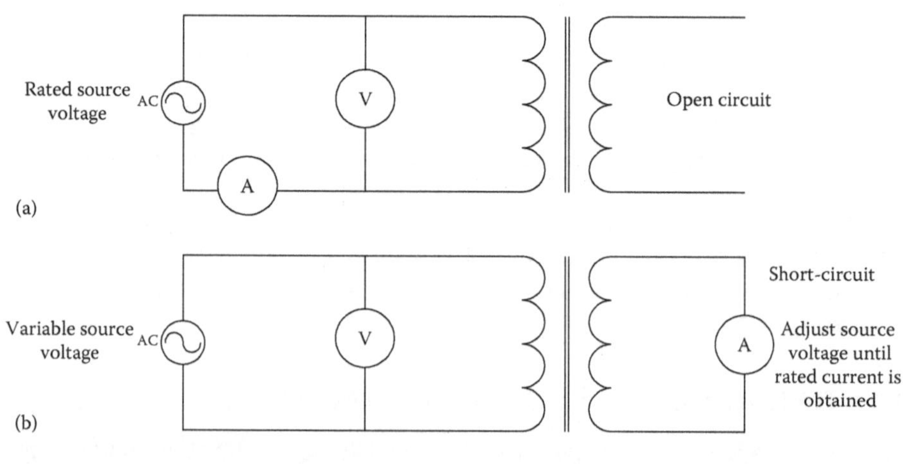

FIGURE 3.8 Measurement of the transformer efficiency: (a) test for core losses and (b) test for copper losses.

2. The power used by the primary windings is measured by multiplying the voltmeter and ammeter readings, as shown in Figure 3.8b (or using wattmeter connected near the variable voltage source). This power measurement represents the copper losses.

The efficiency of the transformer can be estimated from its power output rating P_{out} and both core losses, P_{core}, and copper losses, P_{copper}, as follows:

$$\eta = \frac{P_{out}}{P_{out}} = \frac{P_{out}}{P_{out} + P_{core} + P_{copper}} \tag{3.7}$$

Example 3.5 illustrates the procedure for estimating transformer efficiency based on the results of the test measurements of copper and core losses.

Example 3.5

Problem

Determine the efficiency of 10 kVA, single-phase transformer based on the following test results:

- The open test indicates a core loss of 170 W.
- The short-circuit test provides a copper loss of 162 W.

Assume a power factor of unity for the transformer load.

Solution

By definition, the transformer efficiency is the ratio of the power out over the power in as indicated by Equation 3.7. The power out is 10 kVA or 10,000 W for a load with a power factor of unity:

$$\eta = \frac{P_{out}}{P_{out} + P_{core} + P_{copper}} = \frac{10{,}000\ \text{W}}{10{,}000\ \text{W} + 170\ \text{W} + 162\ \text{W}} = 96.78\%$$

3.5.1.1 Transformer Regulation

Typically, the output voltage of transformers vary slightly with the load. The ability of a transformer to maintain constant voltage is measured by its regulation. The transformer regulation is expressed as a percentage and is defined as a function of the secondary voltages at no-load and load conditions, as shown in Equation 3.8:

$$\text{Regulation} = \frac{E_s(\text{no}-\text{load}) - E_s(\text{load})}{E_s(\text{no}-\text{load})} \tag{3.8}$$

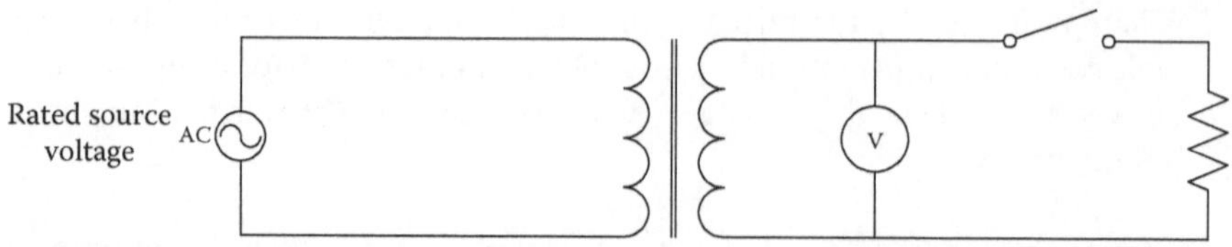

FIGURE 3.9 Regulation test for transformers.

Figure 3.9 illustrates a test that measures the transformer regulation. The procedure of the regulation test is summarized here:

1. The rated source voltage is applied at the primary windings.
2. The voltage obtained at the secondary windings is measured using a voltmeter, as shown in Figure 3.8, under both no-load conditions (when the switch is open) and loading conditions (when the switch is closed).
3. The percent regulation is then calculated using Equation 3.8.

In general, the regulation is tested under full-load conditions but other loads can also be considered. Example 3.6 illustrates the estimation procedure of the percent regulation of a transformer.

Example 3.6

Problem

Determine the percent regulation of a transformer based on the following test results:

- The secondary voltage for no-load conditions is 120 V.
- The secondary voltage for full-load conditions is 114 V.

Solution

Using Equation 3.8, the percent regulation of the transformer can be calculated:

$$\text{Regulation} = \frac{E_s(\text{no} - \text{load}) - E_s(\text{load})}{E_{st}(\text{no} - \text{load})} \times 100\%$$

$$= \frac{120 \text{ V} - 114 \text{ V}}{120 \text{ V}} \times 100\% = 5.0\%$$

3.5.1.2 Transformer Percent Impedance

To estimate the short-circuit current under fault conditions in both the primary and the secondary windings of any transformer, its percent impedance should be

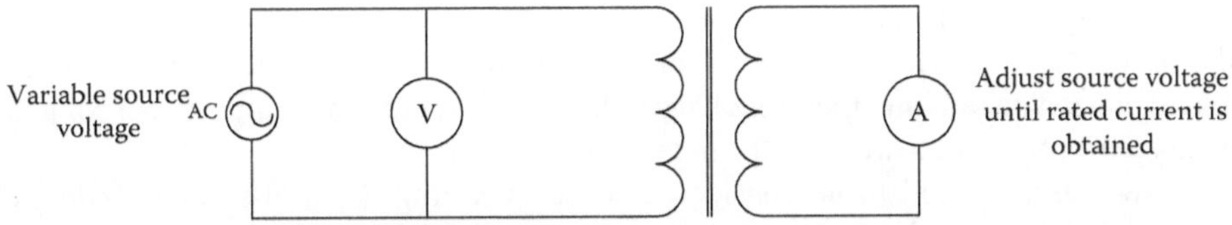

FIGURE 3.10 Percent impedance test for transformers.

obtained. Figure 3.10 illustrates a simple test commonly used to measure the percent impedance (often indicated by $Z\%$ on transformer nameplate):

1. The variable source voltage in the primary windings is increased until the rated current at the secondary windings (short-circuit) is obtained and is indicated in the ammeter.
2. The percent impedance is then estimated as the ratio of the voltage indicated in the ammeter in step 1 over the rated primary source voltage:

$$Z\% = \frac{E_p(\text{short} - \text{circuit})}{E_p(\text{rated})} \times 100\% \tag{3.9}$$

The calculation of the percent impedance is shown in Example 3.7.

Example 3.7

Problem

Determine the percent impedance of a transformer based on the following test results:

- The rated primary voltage is 240 V.
- The short-circuit test provides a primary voltage of 10 V.

Solution

Using Equation 3.9, the percent impedance of the tested transformer can be easily calculated:

$$Z\% = \frac{E_p(\text{short} - \text{circuit})}{E_{Pt}(\text{rated})} \times 100\% = \frac{10\text{ V}}{240\text{ V}} \times 100\% = 4.17\%$$

3.6 DESIGN SPECIFICATIONS FOR TRANSFORMERS

For building applications, two types of transformers are typically used:

1. Dry-type transformers for indoor locations
2. Oil-filled transformers for outdoor locations

The dry-type transformers are more expensive, typically by more than 200%, than the oil-filled transformers. However, oil-filled transformers can create a severe fire and explosion hazard and therefore have to be mounted outdoors unless fireproof and explosion-proof vaults are used for indoor installations.

Several issues need to be considered when installing the transformers indoors. Among these issues are

1. *Noise problems*: To reduce noise problems, it is recommended to specify transformers with minimum sound level (50 dB or lower). Moreover, it is desirable to place the transformer in the least valuable space of the building away from any quiet areas. In some cases, the electrical room where the transformer is located may need to be soundproofed using adequate acoustical materials.
2. *Vibration problems*: To reduce the vibration associated with the operation of transformers, isolators should be placed beneath the transformer mounting pads. Moreover, flexible conduits should be installed to house the secondary and primary feeders of the transformer. Without these measures, vibration may be transferred to other building parts and cause structural damages.
3. *Heat problems*: Heat is dissipated by transformers as part of the copper losses. This heat can represent 0.5%–2% of the total transformer power rating depending on the transformer efficiency. To avoid overheating of the transformer and the room housing the transformer, ventilation should be provided or adjustable louver should be installed in the wall or the door adjacent to the transformer.

NEC (2014) provides several general principles for the design and the installation of transformers. In particular, NEC indicates that no separate room is required for transformers with a power rating that is less than 112.5 kVA. However, even for transformers with a rating less than 112.5 kVA, the designer has to provide at least 12 in. separation from any combustible wall unless a fire-resistant heat-insulated barrier is used. For high-voltage transformers (above 35 kV), NEC requires a vault to protect the general public.

3.7 SUMMARY

In this chapter, the basic characteristics of electrical transformers are outlined. In particular, commonly available types of transformers and their connections are presented. Throughout this chapter, examples are provided to illustrate procedures for estimating performance indicators of electrical transformers. Moreover, illustrative examples are presented to evaluate the cost-effectiveness of selected energy efficiency measures. For instance, it was shown that the use of energy-efficient transformers can provide significant energy savings and thus can be cost-effective.

3.8 PROBLEMS

1. For one-phase 10 kVA transformer, determine the turns ratio if the primary and secondary voltages are 120 and 240 V, respectively. Estimate the primary and secondary currents.
2. A one-phase transformer has 220 V in the primary and 110 V and 125 A in the secondary. Assuming it as an ideal transformer (i.e., 100% efficient), determine the rating of the transformer in kVA.
3. If the secondary phase current of a 480 Y/277–208 Y/120 V transformer, a three-phase transformer, is 10 A, estimate the primary line current and the total primary power used by the transformer in kW if the load has a power factor of 90%. Assume the efficiency of the transformer to be 99.0%.
4. A three-phase transformer 13.6 kV–480 Y/277 V serves a building. Determine the primary and secondary line currents if the rating of the transformer is 500 kV. Estimate the turns ratio of the transformer.
5. A three-phase transformer in a large building steps down the primary voltage of 13.8 kV (line-to-line) to a secondary voltage of 208 V (line-to-line). The transformer rating is 5,000 kVA. Specify the line and phase voltages and currents of the transformer's primary and secondary windings for the following connections:
   a. Low-voltage windings Δ, high-voltage windings Y
   b. Low-voltage windings Y, high-voltage windings Δ
   c. Low-voltage windings Y, high-voltage windings Y
   d. Low-voltage windings Δ, high-voltage windings Δ
6. A one-phase 500 kVA 240/120 V transformer is reconnected as an autotransformer:
   a. Calculate the currents in each part of the autotransformer and determine its power rating if it is reconnected as a step-down 360/240 V autotransformer. Show the connection details of the autotransformer.
   b. Repeat (a) if the transformer is reconnected as a step-up 240/360 V autotransformer.
7. In a building, it is found that 60 fluorescent luminaires are needed as well as 90 receptacle outlets. Each receptacle has a load of 180 VA and operates at 120 V with a power factor of 0.90 while each fluorescent luminaire has four lamps and is designed to operate at 120 V and requires 2.5 A. Each lamp uses 40 W, and the ballast supplying all four lamps requires an additional 25 W. The entire load is served by 480 Y/277–208 Y/120 V transformer. Determine the following:
   a. Estimate the power factor for each fluorescent luminaire.
   b. Calculate the total load in kVA connected to the transformer.
   c. Show the transformer connections and the number of receptacles and luminaires connected to each phase to ensure that the system is balanced.
   d. Determine the phase and the line currents in both the secondary and the primary windings of the transformer.

8. Estimate the cost savings for replacing a 750 kVA transformer of an efficiency of 99.2% by an energy-efficient transformer of 99.6% if the cost of electricity is $0.12/kWh. The total annual full load hours for operating the transformer is 7,500 h/year. If the cost for the energy-efficient transformer is $15,000, assess if it is cost-effective to replace the existing transformer (consider a simple payback period of less than 5 years for the replacement to be cost-effective).
9. Determine the cost-effectiveness using the net present value method for replacing a 1,000-kVA transformer of an efficiency of 99.2% with an energy-efficient transformer of 99.7% if the average power factor of the transformer load is 65% and the cost of electricity is $0.15/kWh. Consider both replacement options: (1) the existing transformer is in a good condition, (2) the existing transformer is at the end of its useful life. The cost for the standard and energy-efficient transformers are respectively, $20,000 and $26,000.
10. Determine the efficiency of 500 kVA, three-phase transformer if the open test indicates a loss of 0.900 kW, and the short-circuit test provides a loss of 2.550 KW.
11. Determine the percent regulation of a three-phase 480 Y/277-V transformer if the secondary voltage for full-load conditions is measured to be 269 V.
12. Determine the percent impedance of a 480 Y/277–208 Y/120 V transformer if the short-circuit test provides a primary voltage of 25 V.

4 Electrical Motors

This chapter reviews some general concepts related to electric motors. First, a review of the basic characteristics of an electric system operating under alternating current is provided. Then, the impact of various design specifications on the performance of three-phase and single-phase induction motors, commonly used in buildings, is discussed with illustrative examples. Throughout this chapter, several measures and guidelines are provided to ensure safe operation and improved energy performance of electrical motors.

4.1 INTRODUCTION

According to an International Energy Agency (IEA) study, electric motors and the systems they drive constitute the single largest electrical end use worldwide (IEA, 2011). Specifically, electric motor-driven systems account for about 45% of the global electricity consumption, twice as much as lighting, which is the second largest end use. Electric motors are utilized for a wide range of applications and by all economic sectors, including industry, buildings, transport, and agriculture. Table 4.1 summarizes the annual global energy consumption attributed to electric motor-driven systems for various sectors based on an IEA study (IEA, 2011).

As noted in Table 4.1, electrical motors are widely used in the industry followed by buildings, then transport and agriculture. While large motors of more than 375 kW dominate the energy consumption in the industrial sector, the vast majority of electrical motors used throughout the world have small capacity drawing less than 0.75 kW of electrical power and are used to drive equipment in commercial and residential buildings (IEA, 2011).

In general, electric motors convert electrical energy to mechanical energy and are typically used to drive machines. For the building sector, electric motor-driven machines serve several purposes, including moving air (supply and exhaust fans), moving liquids (pumps), moving objects or people (conveyors, elevators), and compressing gases (air compressors, refrigerators). To select the type of motors to be used for a particular application, several factors have to be considered, including

1. *The form of the electrical energy that can be delivered to the motor*: Direct current (DC) or alternating current (AC), single or three phase
2. The requirements of the driven machine such as motor speed and load cycles
3. *The environment in which the motor is to operate*: Normal (where a motor with an open-type ventilated enclosure can be used), hostile (where a totally enclosed motor must be used to prevent outdoor air from infiltrating inside the motor), or hazardous (where a motor with an explosion-proof enclosure must be used to prevent fires and explosions)

DOI: 10.1201/9781003276999-4

TABLE 4.1
Annual Global Electricity Consumption of Electrical Motors by Sector

Sector	Electricity Consumption (THW/Year)	Percent of Sector Electricity Use (%)
Industry	4,488	69
Commercial buildings	1,412	38
Residential buildings	948	22
Transport and agriculture	260	39

Motors can operate using either DC or AC depending on the applications. However, equipment and electrical systems in buildings almost exclusively use AC motors. DC motors are typically categorized by the type of connections between the field windings and the armature windings and include shunt, series, and compound motors. Shunt motors are typically used for machine tools while series motors are utilized as traction motors for trains and electrical cars. DC motors are also suitable for variable speed drives for fans and pumps, which are integral parts of heating and cooling systems in buildings. However, the vast majority of electrical motors used for buildings include single-phase or three-phase AC motors.

This chapter discusses in detail the basic operation, typical performance, and general characteristics of AC motors. In addition, simple measures and techniques are described throughout this chapter to improve the energy efficiency of motors. Both three-phase AC motors, commonly used for heating and cooling of commercial buildings, and single-phase AC motors, used in residential buildings, are overviewed in this chapter.

4.2 OPERATION OF THREE-PHASE MOTORS

There are two types of electric AC motors used in buildings as well as industrial facilities: (1) induction motors and (2) synchronous motors. Induction motors are the more common type, accounting for about 90% of the existing motor mechanical power output. Both types of motors use a motionless stator and a spinning rotor to convert electrical energy into mechanical power. The operation of both types of motors is relatively simple and is briefly described in this section. Figure 4.1 illustrates the two main components of an electrical motor: a stator and a rotor.

4.2.1 Rotating Magnetic Field

For the three-phase AC motors, alternating current is applied to the stator through the armature windings generating a rotating magnetic field. This magnetic field causes the rotor to spin in trying to align with the rotating stator magnetic field. Specifically, Figure 4.2a illustrates for two-pole three-phase motor, the connections of the stator windings to generate a rotating magnetic field. In particular, each winding carries one-phase current and is arranged to be 120° from the other two windings

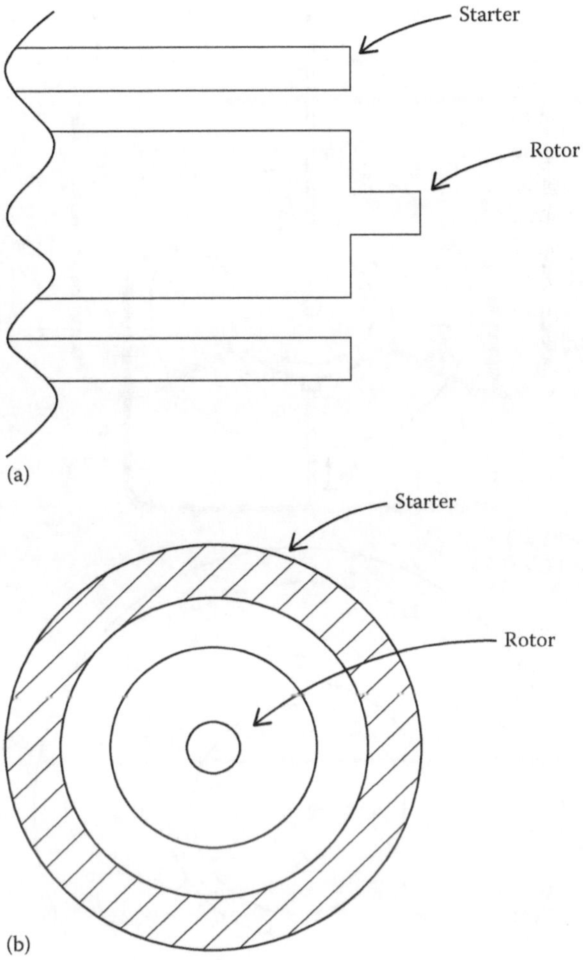

FIGURE 4.1 Basic components of an electrical motor: (a) a stator and (b) a rotor.

of the stator. One end of each winding enters slot E and leaves slot S for a total of six slots around the stator. The time variations for all three phase currents are shown in Figure 4.2b. At the start, the current in one phase is positive and at its maximum value, I_m, while the currents at the other two phases are negative with an absolute value of $I_m/2$. As a result, two sections are formed, as shown in Figure 4.2a, with the top section having only positive currents and the bottom section being made up of negative currents resulting in a north–south pole magnetic field vector. When the currents vary with time, as shown in Figure 4.2b, the magnetic field vector remains with the same magnitude but rotates, as shown in Figure 4.3, during four time periods: $t=T/6$, $t=T/3$, $t=T/2$, and $t=2T/3$. Specifically, the magnetic field rotation is 60° at $t=T/6$, 120° at $t=T/3$, 180° at $t=T/2$, and 240° at $t=3T/2$. For a four-pole three-phase motor, Figure 4.4 shows the connections for the stator windings while

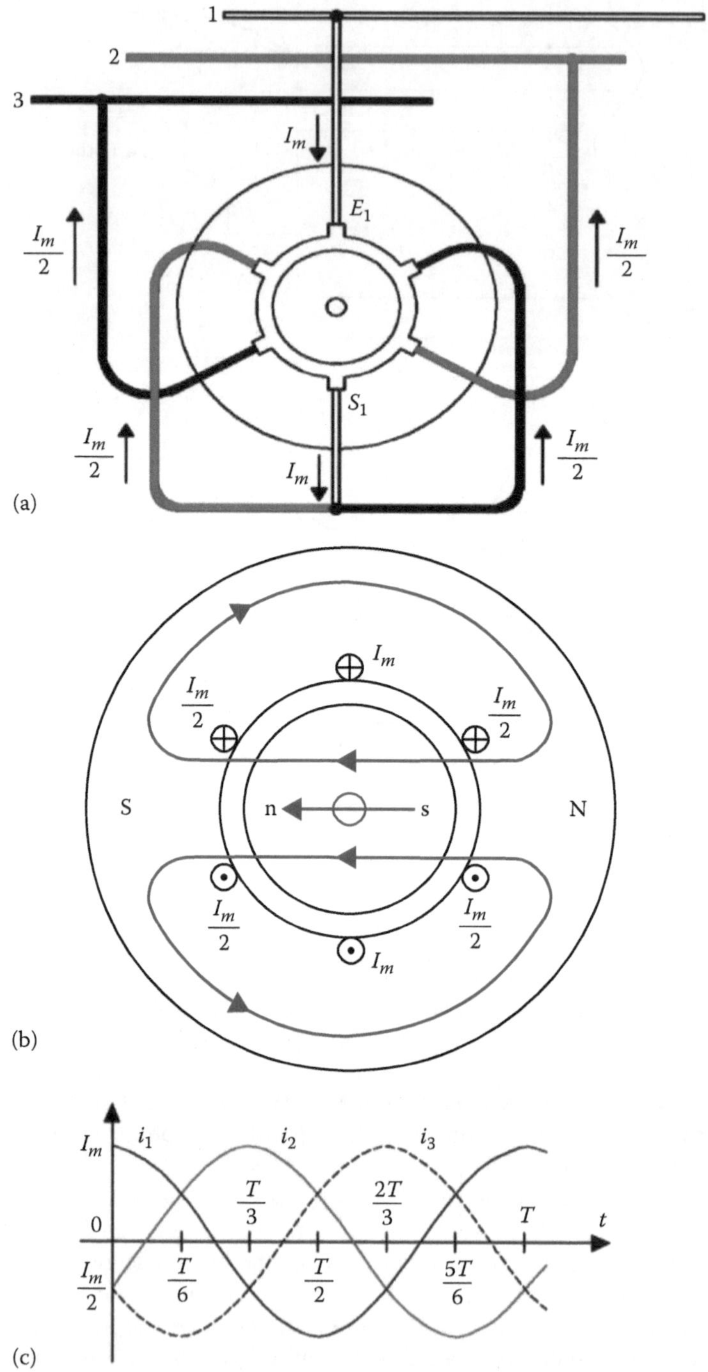

FIGURE 4.2 (a) Stator connections for two-pole, three-phase AC motor to (b) generate a rotating magnetic field at (c) the initial time period (i.e., $t=0$).

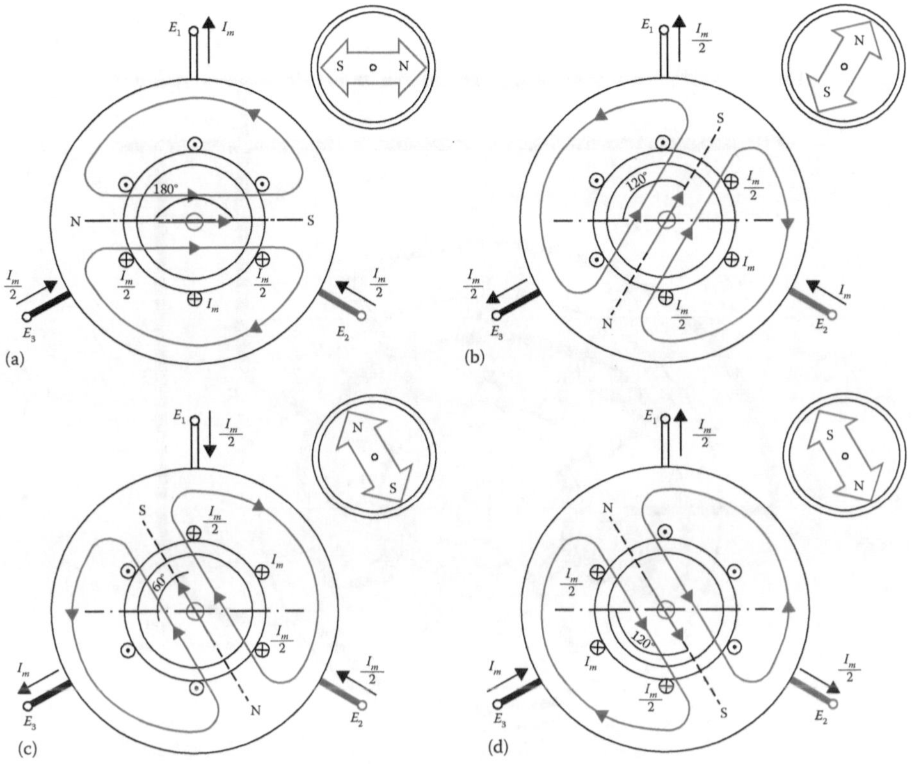

FIGURE 4.3 Rotating magnetic field for two-pole, three-phase AC motor at different time periods. (a) $t = T/6$, (b) $t = T/3$. (c) $t = T/2$, and (d) $t = 2T/3$.

Figure 4.5 indicates the magnetic field vector rotation at $t = 0$ and $t = T/6$. It is clear that the four-pole motor magnetic field rotation is slower than that of the two-pole motor by a factor of 2.

For any three-phase AC motor, it can be shown that the rotation of the magnetic field of the stator has an angular speed that is a function of both the number of poles, N_P, and the frequency, f, of the AC current as expressed by Equation 4.1 using SI units (i.e., rad/s):

$$\omega_{mag} = \frac{4\pi f}{N_P} \tag{4.1}$$

Equation 4.2 indicates the magnetic field rotation in IP units (i.e., rpm or rotations per minute):

$$\omega_{mag} = \frac{120f}{N_P} \tag{4.2}$$

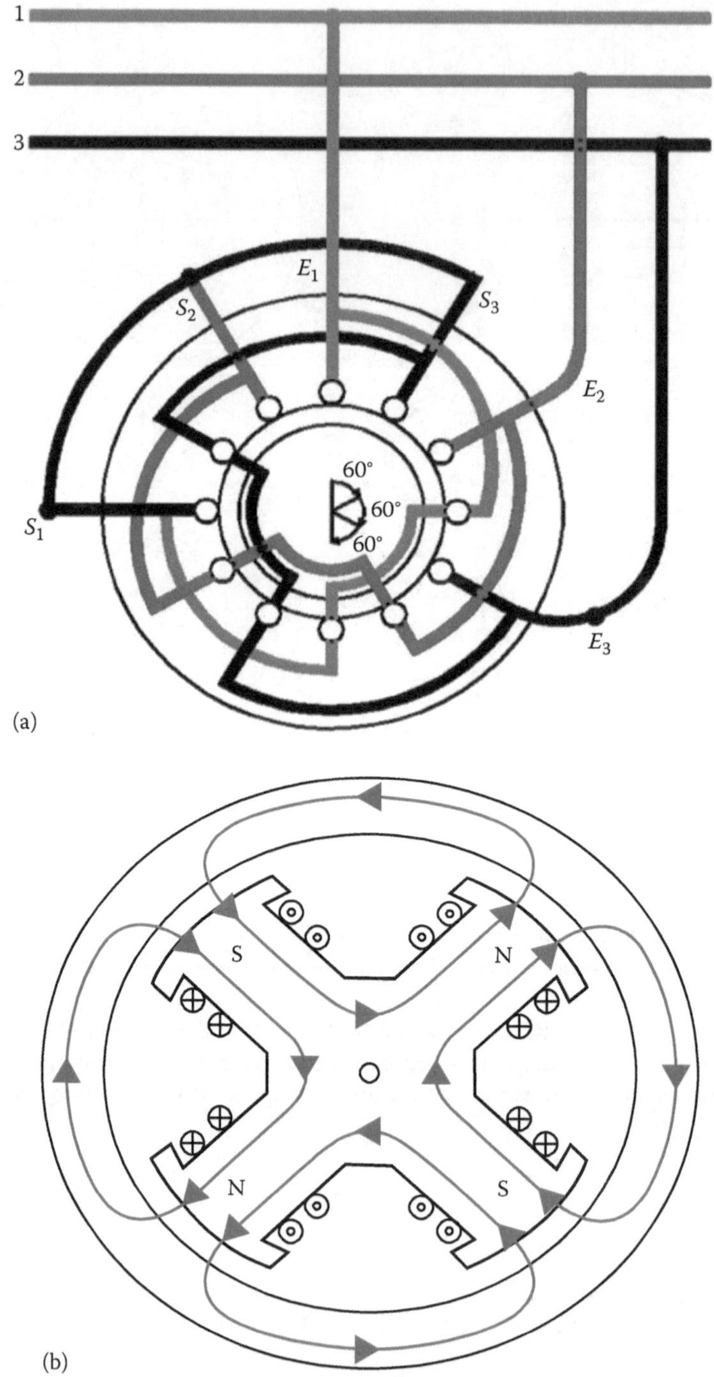

FIGURE 4.4 (a) Stator connections for four-pole, three-phase AC motor to (b) generate a rotating magnetic field.

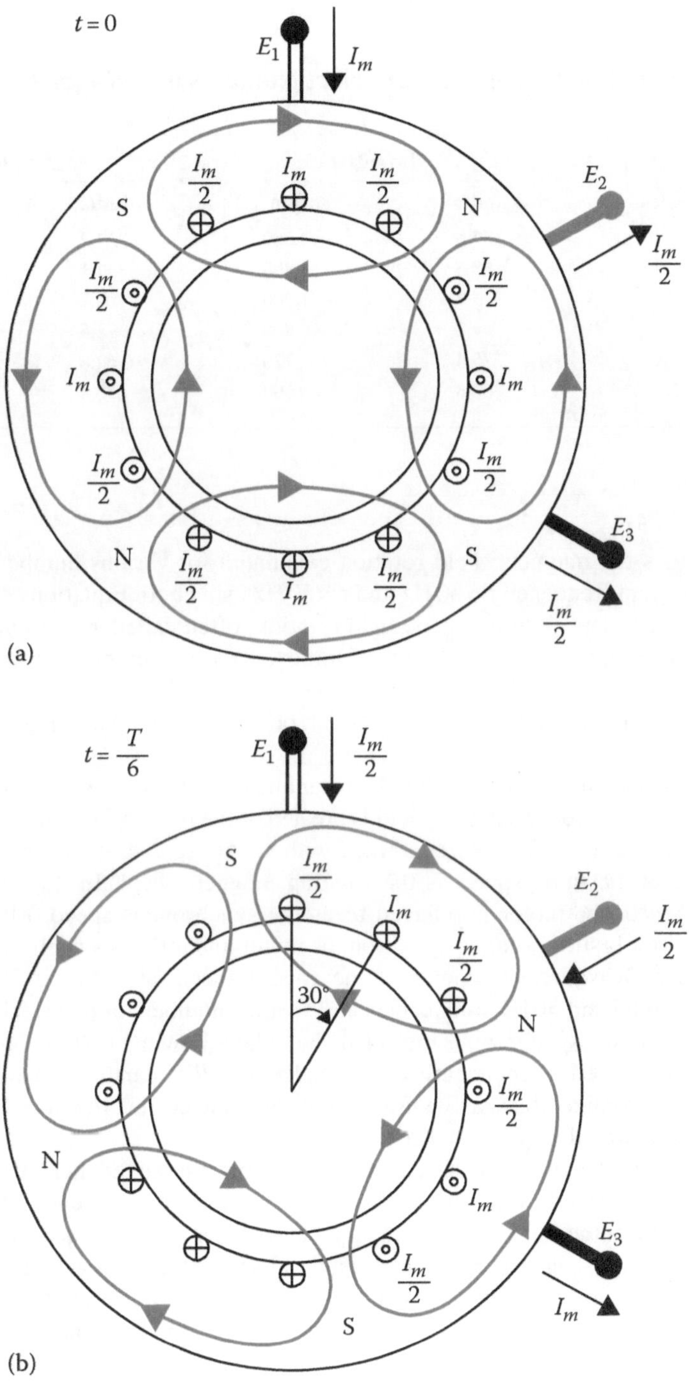

FIGURE 4.5 Rotating magnetic field for four-pole, three-phase AC motor at (a) $t=0$ and (b) $t=T/6$.

TABLE 4.2
Magnetic Field Rotation Speeds for Select Numbers of Poles for Three-Phase AC Motors

	$f = 60$ Hz		$f = 50$ Hz	
Number of Poles	rad/s	rpm	rad/s	rpm
2	377.0	3,600	314.2	3,000
4	188.5	1,800	157.1	1,500
6	125.7	1,200	104.7	1,000
8	94.2	900	78.5	750
10	75.4	720	62.8	600
12	62.8	600	52.4	500

Table 4.2 lists the magnetic field rotation calculated for various numbers of poles when the current frequency $f=60$ Hz and $f=50$ Hz using both Equations 4.1 and 4.2. For a given frequency, the magnetic field rotation, often called synchronous speed, decreases with higher number of poles. Moreover, for a given number of poles, the rotation is higher at $f=60$ Hz than at $f=50$ Hz.

The rotor of the three-phase AC motor can be rotated using two options:

1. Field windings are mounted on the rotor and energized using DC source. In this case, a second magnetic field is created at the rotor. When the armature windings at the stator are energized with an AC source, the rotor starts to rotate at the same speed as the rotating magnetic field. In this case, the rotation of the motor is referred to as the synchronous speed. Moreover, the motor is simply called synchronous motor and utilizes a combination of AC and DC sources to operate. Synchronous speeds are listed in Table 4.2 for both 60 and 50 Hz frequencies and various numbers of poles. The main advantages of synchronous motors include the following: (1) they operate at the same speed independently of the load and (2) they can operate at lagging power factor and thus can be used as capacitors to correct the overall power factor of an inductive electrical system.
2. Instead of imposing a DC current on the rotor-mounted field windings, induced currents can be used to start the rotation of the motor. The most common method of producing induced current for a three-phase motor is to use a squirrel-cage rotor with parallel thin bars connected together by rings. When the motor is started with energized stator windings, the rotating magnetic field induces currents that flow through the bars and rings of the squirrel-cage rotor and thus ensures that the motor starts and continues to rotate. However, the speed of the rotor never reaches the speed of the magnetic field. Indeed and if the rotor is assumed to attain the magnetic field speed, the induced current stops to flow since there is no relative movement

between the rotor and the magnetic field. The absolute difference between the magnetic field speed and the induction motor speed is called the slip factor:

$$\alpha(\text{RPM}) = \omega_{mag} - \omega_R \tag{4.3}$$

The relative speed difference is denoted as the percent slip factor of the induction motor:

$$\alpha(\%) = \frac{\omega_{mag} - \omega_R}{\omega_{mag}} \times 100 \tag{4.4}$$

In Equations 4.3 and 4.4, ω_R denotes the angular speed of the motor (i.e., rotor).

Motors that utilize induced currents to operate are called induction motors. They are the most common types of motors used to drive electrical systems in buildings, including fans and pumps for air-conditioning systems. The main advantage of induction motors is that they are generally cheaper to construct than synchronous motors. Indeed, induction motors do not require commutator bars, slip rings, and brushes. Moreover and because of their construction characteristics, induction motors are inductive loads and thus have a lagging power factor.

Example 4.1 illustrates the calculation of the rotation speed and the slip factor of a three-phase AC induction motor with known number of poles.

Example 4.1

Problem

Determine the motor speed in rpm and rad/s for a three-phase AC induction motor with a 3% slip and four-pole design when the current frequency is (1) $f=60\,\text{Hz}$ and (2) $f=50\,\text{Hz}$.

Solution

Using Table 4.2, the magnetic field speed can be determined. Then, the motor speed can be estimated from Equation 4.4.

For the case of (1) $f=60\,\text{Hz}$, Table 4.2 indicates that for four-pole motor, $\omega_{mag}=1{,}800$ rpm $=188.5$ rad/s. Thus, the motor speed in RMP can be calculated using Equation 4.4:

$$\omega_R = (1-\alpha)\omega_{mag} = (1-0.03)\times 1{,}800 = 1{,}746\text{rpm}$$

Similarly using SI units

$$\omega_R = (1-\alpha)\omega_{mag} = (1-0.03)\times 188.5 = 182.9\text{rad/s}$$

For the case of (2) $f = 50$ Hz, Table 4.2 indicates that for four-pole motor, $\omega_{mag} = 1,500$ rpm = 157.1 rad/s. Thus, the motor speed can be calculated using Equation 4.4:

$$\omega_R = (1 - \alpha)\omega_{mag} = (1 - 0.03) \times 1,500 = 1,455 \text{rpm}$$

The motor speed can also be expressed in SI units:

$$\omega_R = (1 - \alpha)\omega_{mag} = (1 - 0.03) \times 157.1 = 152.4 \text{ rad/s}$$

4.2.2 Torque–Speed Relationship

It should be noted from Equation 4.4 that when the slip factor approaches zero (i.e., $\alpha = 0$), the motor speed approaches the synchronous speed. However, when the slip factor is close to unity (i.e., $\alpha = 1$), the motor speed is close to zero (i.e., $\omega_R = 0$), indicating the rotor is stationary, often referred to as the locked-rotor condition. For low slip factors, it can be shown that the torque generated by the motor is proportional to the slip value. As the motor becomes loaded, the rotor slows down and the slip factor increases inducing higher current and torque. The rotor speed continues to decrease until the induced torque matches the load torque (i.e., full-load condition). Figure 4.6 shows the typical speed–torque relationship for an induction AC motor. In particular, it can be seen at high speeds (close to the synchronous speed), that is, at low slip factors, the torque increases significantly without any loss of speed (i.e., slip). As the speed decreases (i.e., the slip factor increases), the torque reaches a maximum value, called the breakdown torque, as noted in Figure 4.6. It can be shown that the maximum torque occurs when the slip factor is such that the rotor inductive reactance becomes equal to the rotor resistance.

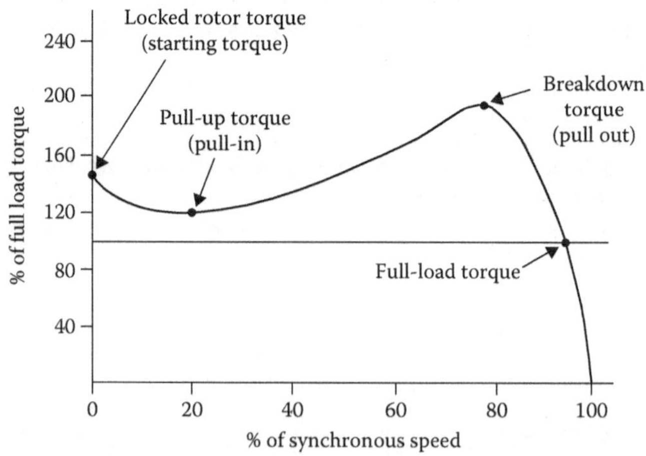

FIGURE 4.6 Typical torque–speed relationship for an induction motor.

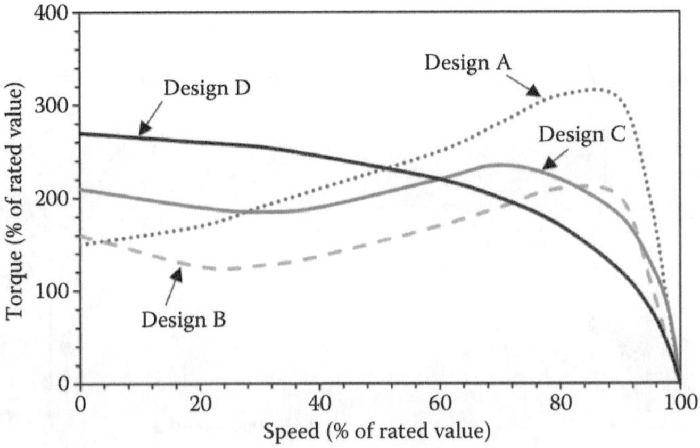

FIGURE 4.7 Torque–speed relationships for common design classes of induction motors.

Four torque–speed relationships are commonly considered in the design of induction motors. In particular, the National Electrical Manufacturers Association (NEMA) has defined classes A–D for squirrel-cage induction motors depending on their starting torque, starting current, breakdown torque, and full-load slip factor. Figure 4.7 shows the torque–speed performance curves for the four design classes of induction motors. As noted in Figure 4.7, the basic characteristics of the four NEMA designs for induction motors can be summarized as follows:

1. *Class A motors*: These motors, typically referred to as low-torque and low-slip motors, have low starting current and torque. Indeed, their starting torque is 125% of the rated value while their maximum inrush current is 800% of the rated value. Moreover, the breakdown torque is only 160% of the rated value while their slip factor does not exceed 5%.
2. *Class B motors*: These motors are referred to as general-purpose motors and are most commonly used in buildings. The starting torque depends on the size of the motors and varies from 80% to 275% of the rated value. The breakdown torque ranges from 175% to 300% of the rated value. The starting or inrush current for these motors can reach up to 600% of the rated value. The maximum slip factor of the class B motors is 5%.
3. *Class C motors*: These motors, typically referred to as high torque motors, are equipped with two sets of rotor bars with different resistance values. The rotor bar with the highest resistance is used to develop the starting torque while the second bar is utilized for normal operation. The starting torques of class C motors range from 200% to 250% of their rated values. The breakdown torque remains about 200% of the rated value for almost all class C motors. Like class B motors, the slip factor and the inrush current for class C motors can reach up to respectively, 5% and 600% of their rated value.

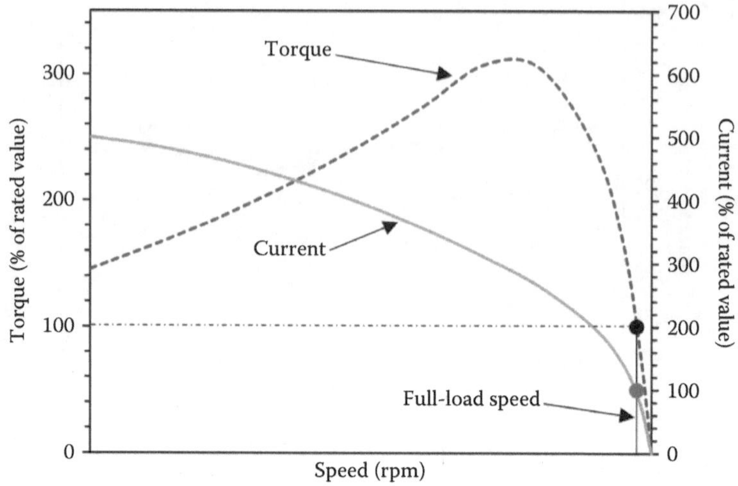

FIGURE 4.8 Typical torque–speed relationship for an induction motor.

4. *Class D motors*: These motors, typically referred to as high-torque and high-slip motors, have high-resistance rotor bars and achieve their highest torque when they start. Indeed, the starting torque is 275% of the rated value. The slip factor of these motors can reach up to 15%. Their maximum inrush current is 600% of the rated value.

4.2.3 Current–Speed Relationship

Another important feature of induction motors is the relationship between current and speed. As the slip factor increases, the current drawn by the motor increases, as illustrated in Figure 4.8. Simultaneously, the power factor of the motor decreases as the slip factor and the current increase. In particular, the starting current (i.e., locked-rotor condition with zero rotor speed) can be as high as five to six times the full-load current. The high starting currents, also called inrush currents, combined with low power factors can cause several problems to the electrical distribution systems when operating induction motors in buildings. To reduce the high inrush currents of induction motors, some soft starting methods are recommended. In Chapter 6, some of the common starting options as well as the components of branch circuits including protection devices for induction motors will be discussed in detail. Moreover, the installation of capacitors can improve the power factor of motors, as outlined later in this chapter and discussed in more detail in Chapter 2.

4.3 OPERATION OF ONE-PHASE MOTORS

In residential buildings as well as in some small commercial buildings, single-phase AC motors are needed for a wide range of applications, including air-conditioning systems, exhaust fans, and appliances. Like the case of three-phase motors, induction

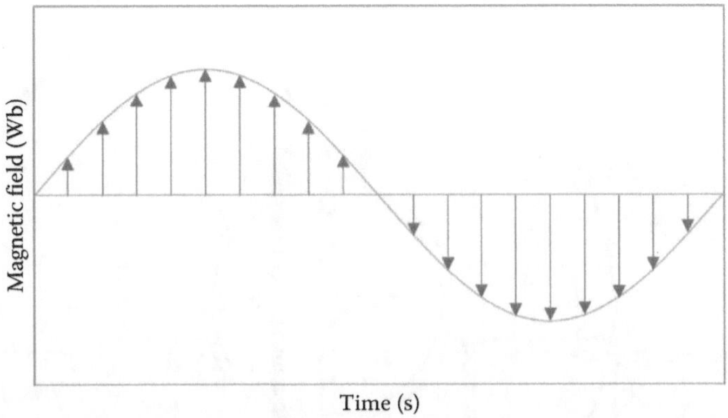

FIGURE 4.9 Magnetic field variation with time for a single-phase induction motor.

motors are the most common types of single-phase AC motors used in buildings. However and unlike three-phase motors, single-phase induction motors are not self-starting and are not smooth when operating. Indeed, the magnetic field generated by one-phase current in the stator remains in the same line but changes its magnitude and direction, as illustrated in Figure 4.9. Instead of rotating magnetic field developed for three-phase AC motors, the magnetic field for single-phase AC motors is pulsating with periodic changes in both magnitude and direction. As a result, the rotor pulsates back and forth depending on the magnetic field variation resulting in a noisy operation for single-phase AC motors.

It should be noted that once started, a single-phase motor continues to rotate even when the magnetic field and subsequently the torque becomes zero. Indeed, the momentum of the rotor would provide sufficient inertia to ensure that the motor continues to rotate.

The basic characteristics of single-phase induction motors are similar to those of three-phase induction motors. In particular, the synchronous speeds are the same as those provided in Table 4.2. The only main difference and challenge for single-phase AC motors is the mechanism to start them. Several approaches are available to start single-phase induction AC motors. Some of the most common starting mechanisms are briefly described:

1. *Resistance split-phase starters*: Two separate windings are considered in a stator: one for starting the motor and the second for normal operation. Figure 4.10 illustrates the resistance split-phase starter mechanism used for a single-phase AC motor. Specifically, the starting windings have higher electrical resistance than the normal windings, also known as running windings. The difference in resistance between the two windings develops a phase lag between the currents flowing in both windings resulting in a starting torque that initiates the rotation of the rotor. Once the speed of the rotor reaches typically 75% of the rated value, a centrifugal switch is opened to disconnect the starting windings from the stator circuit. Typically, the

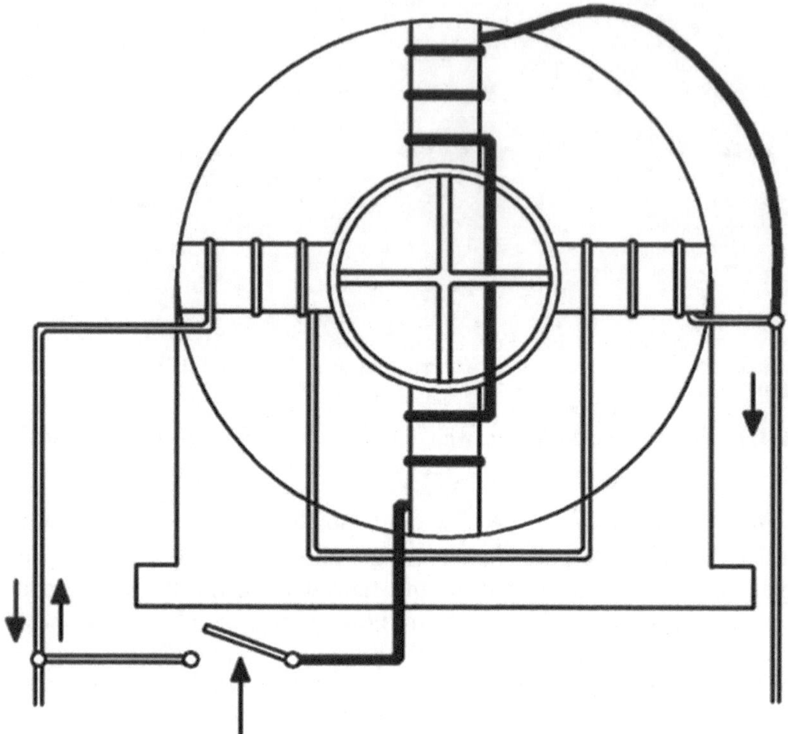

FIGURE 4.10 Resistance split-phase starter for a single-phase AC motor.

resistance split-phase starters develop high starting currents as well as low starting torques.

2. *Capacitance split-phase starters*: Like the resistance split-phase starters, these starters also utilize two separate windings, but the capacitance is connected to the starting windings, as shown in Figure 4.11. The phase difference between the currents flowing into the two windings is more significant than that obtained by applying the resistance split-phase starters resulting in a higher starting torque (up to 4.5 times the full-load running torque) but a lower starting current. The capacitance split-phase starters are typically more expensive than the resistance split-phase starters.
3. *Two-value capacitance split-phase starters*: A variation of capacitance split-phase starter is to include two parallel capacitors as shown in Figure 4.12. One capacitor connected to the starting windings is first used to initiate the rotation of the motor. Then, this capacitor is disconnected from the circuit through a centrifugal switch leaving the second capacitor connected to the running windings. These types of split-phase starters are typically used for large single-phase induction motors.

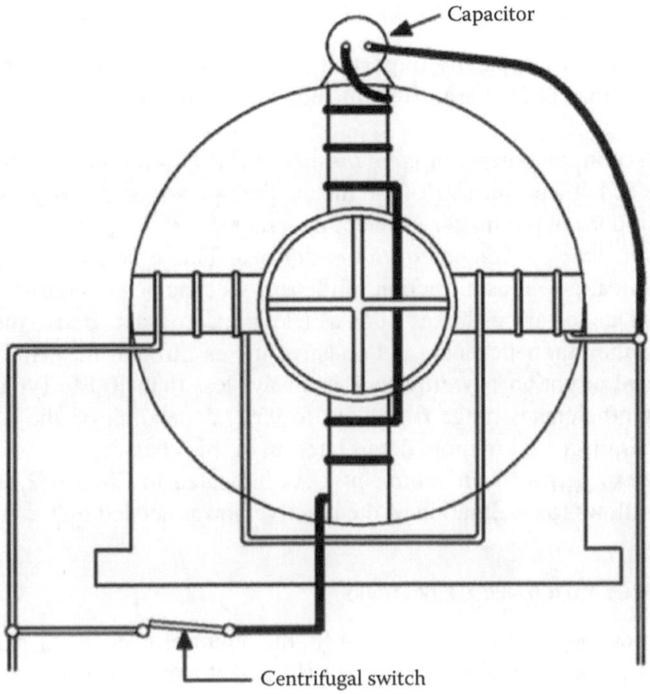

FIGURE 4.11 Capacitance split-phase starter for a single-phase AC motor.

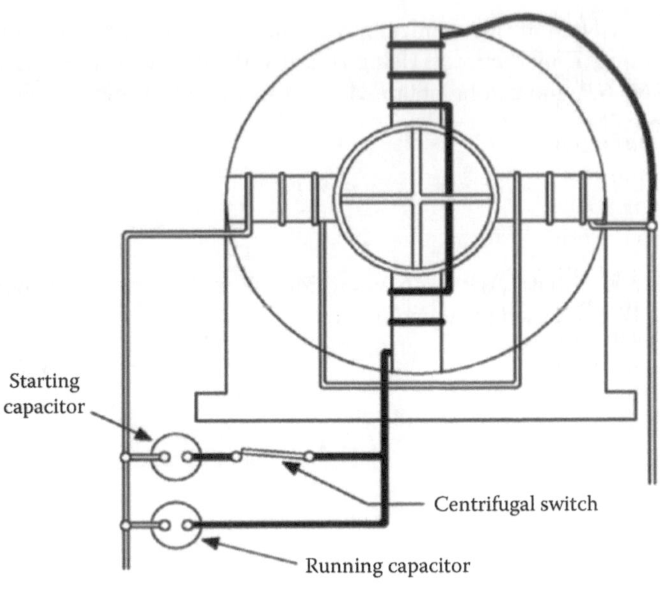

FIGURE 4.12 Two parallel capacitors split-phase starter for a single-phase AC motor.

4.4 PERFORMANCE CHARACTERISTICS OF MOTORS

Three parameters are typically important to characterize an electric motor during full-load operation. These parameters include the following:

1. *The mechanical power output of the motor, P_M*: This power can be expressed in kW or horsepower (HP). The mechanical power output is generally the most important parameter in selecting a motor.
2. *The conversion efficiency of the motor, η_M*: This efficiency expresses the mechanical power as a fraction of the real electric power consumed by the motor. Due to various losses (such as friction, core losses due to the alternation of the magnetic field, and resistive losses through the windings), the electrical motor energy efficiency is always less than 100%. Typical motor energy efficiencies range from 75% to 95%, depending on the size of the motor as discussed in more detail later on in this chapter.
3. *The power factor of the motor, pf_M*: As indicated in Chapter 2, the power factor allows the estimation of the reactive power needed by the AC motor.

4.4.1 Energy Efficiency of Motors

Using the schematic diagram of Figure 4.13, the motor efficiency, η_M, can be defined as a ratio of the mechanical power output, P_M, and the real electrical power input of the motor, P_R:

$$\eta_M = \frac{P_M}{P_R} \tag{4.5}$$

It should be noted that the mechanical power output, P_M, is generally a function of the developed torque, T, by the rotor. Using IP units, the mechanical power is expressed in horsepower, HP, and can be obtained as a function of torque, T, and motor speed expressed in RPM:

$$HP = \frac{T \times \text{RPM}}{5{,}252} \tag{4.6}$$

It should be noted that 1 HP = 746 W = 0.746 kW and 5,252 is a conversion factor based on the IP units used in Equation 4.6.

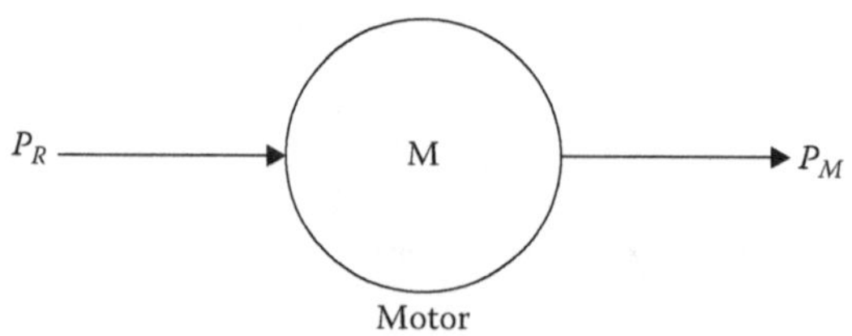

FIGURE 4.13 Definition of the efficiency of a motor.

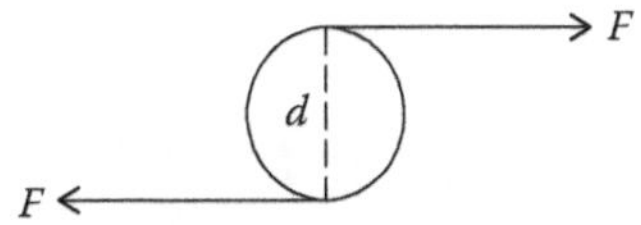

FIGURE 4.14 Torque developed by a motor rotor.

In SI units, the motor mechanical power output, P_M, is expressed in W or kW and can be obtained as a function of the rotor angular speed given in rad/s, $\omega \times_R$, simply as follows:

$$P_M = T \times \omega_R \tag{4.7}$$

Figure 4.14 illustrates that a torque developed by a motor is a function of the generated force and the rotor diameter using both SI and IP units:

$$T = Fd(\text{Nm})\text{or}(\text{lbf} \times \text{ft}) \tag{4.8}$$

As alluded to earlier, the conversion of the electrical power input to mechanical power output is not perfect for any electrical motor. Indeed, several types of losses are inherent to this conversion as summarized in Figure 4.15 for a 15 HP, three-phase, four-pole induction motor (Baldor, 2010). The main losses are associated to the stator and rotor resistance losses (i.e., RI^2) as well as the core losses, also known as hysteresis and eddy current losses, related to the magnetic field. Representing about 13% of the total losses, the friction losses are relatively small and are associated to the motor bearings as well as the windage losses related to the drag and cooling effect of air moving around the rotor. Finally, other losses related to the loading of the motor are referred as stray losses and represent about 18% of the total motor losses.

Based on the types of losses outlined in Figure 4.15, the following design modifications can be considered to improve the energy efficiency of AC motors:

- Increase the thickness of the windings for the stators so the resistance losses can be reduced (indeed, thicker wires have less resistance and thus less RI^2 losses for the same current flows). More details about wires and conductors are provided in Chapter 6.
- Moreover, the use of copper instead of aluminum for the stator windings can also reduce the resistance losses.
- Similarly increasing the thickness of the windings and bars for the rotor can reduce the resistance losses.
- Reduce hysteresis and eddy current losses through better steel material.

Table 4.3 summarizes the average efficiencies for both standard and energy-efficient, commonly known as premium, motors commercially available in the United States (Krarti, 2010). As noted earlier, the improved efficiency for high or premium motors is mainly due to better designs with the use of alternative materials

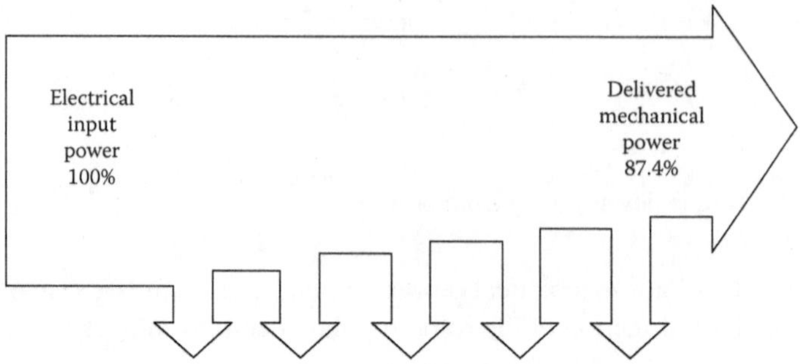

	Stator Resistance Loss	Rotor Resistance Loss	Core Loss	Friction and Windage Loss	Stray Load Loss	Total
Percent of losses (%)	30	20	19	13	18	100
Percent of input (%)	3.8	2.5	2.4	1.6	2.3	12.6

FIGURE 4.15 Typical losses from converting electrical power to mechanical power for a 15 HP three-phase AC motor.

TABLE 4.3
Typical Standard and Premium Motor Efficiencies

Motor Mechanical Power OutputkW (HP)	Average Nominal Efficiency for Standard-Efficiency Motors	Average Nominal Efficiency for Premium-Efficiency Motors
0.75 (1.0)	0.730	0.830
1.12 (1.5)	0.750	0.830
1.50 (2.0)	0.770	0.830
2.25 (3.0)	0.800	0.865
3.73 (5.0)	0.820	0.876
5.60 (7.5)	0.840	0.885
7.46 (10)	0.850	0.896
11.20 (15)	0.860	0.910
14.92 (20)	0.875	0.916
18.65 (25)	0.880	0.926
22.38 (30)	0.885	0.928
29.84 (40)	0.895	0.930
37.30 (50)	0.900	0.932
44.76 (60)	0.905	0.933
55.95 (75)	0.910	0.935
74.60 (100)	0.915	0.940
93.25 (125)	0.920	0.942
111.9 (150)	0.925	0.946
149.2 (200)	0.930	0.953

TABLE 4.4
Comparison of Minimum Energy Efficiency Requirements for EU and U.S. Four-Pole Motors

Motor Mechanical Power OutputkW (HP)	EU IE2 Energy Efficiency Level (50 Hz)	EU IE3 Energy Efficiency Level (50 Hz)	U.S. NEMA Premium Efficiency (60 Hz)
0.75 (1.0)	0.796	0.825	0.855
1.12 (1.5)	0.814	0.841	0.865
1.50 (2.0)	0.828	0.853	0.865
2.25 (3.0)	0.843	0.867	0.895
3.73 (5.0)	0.866	0.886	0.895
5.60 (7.5)	0.877	0.896	0.910
7.46 (10)	0.887	0.904	0.917
11.20 (15)	0.898	0.914	0.930
14.92 (20)	0.906	0.921	0.930
18.65 (25)	0.912	0.926	0.936
22.38 (30)	0.916	0.930	0.941
29.84 (40)	0.923	0.936	0.941
37.30 (50)	0.927	0.939	0.945
44.76 (60)	0.931	0.942	0.950
55.95 (75)	0.935	0.946	0.950
74.60 (100)	0.940	0.950	0.954
93.25 (125)	0.942	0.952	0.954
111.9 (150)	0.945	0.954	0.958
149.2 (200)	0.951	0.960	0.958

for the rotor and stator windings to reduce resistance and hysteresis losses. However, this energy efficiency improvement comes typically with higher prices for premium motors which are about 10%–30% more expensive than standard-efficiency motors of similar sizes.

Table 4.4 compares the minimum energy efficiency requirements for both European Union (EU) and United States for four-pole three-phase electrical motors (IEA, 2011). In particular, EU Ecodesign Directives state that all electrical motors of rated output between 0.75 and 375 kW and sold after January 1, 2015, have to meet IE3 minimum energy efficiency levels. However, motors with IE2 minimum energy efficiency levels can be used when equipped with variable speed drives, also known as adjustable speed drives (ASDs). The operation and the advantages of ASDs are discussed later on in this chapter. In the United States, the listed premium energy efficiency requirements are specific to general-purpose electrical motors.

In Example 4.2, basic energy efficiency calculations are described for an induction motor when the resistance and core losses are considered.

Example 4.2

Problem

A four-pole squirrel-cage motor has the following characteristics:

- Power input at a rated load is 4,200 W with a power factor of 0.60.
- Rated speed is 1,150 rpm.
- Resistance of each stator windings (there are three) is $Rs = 0.16\ \Omega$.

The motor is supplied by a three-phase source with a line-to-line voltage of 240 V and a frequency of 50 Hz. The windings of the motor have a Y-connection. Determine

a. The slip factor of the motor in percent
b. The line current drawn by the motor
c. The resistance losses in the stator
d. The efficiency losses of the motor if the core losses at the stator are 350 W (neglect the friction losses and other losses, including the resistance and core losses at the rotor)
e. The torque delivered by the motor in N × m

Solution

a. The slip factor of the motor in percent

$$\omega_{mag} = \frac{120f}{N_{poles}} = \frac{120 \times 50\ \text{Hz}}{4} = 1,200\ \text{rpm}$$

$$\alpha = \frac{\omega_{mag} - \omega_{motor}}{\omega_{mag}} = \frac{1,200 - 1,150}{1,200} = 4.2\%$$

b. The line current

$$P_R = \sqrt{3} I_L \times E_L \times pf$$

$$I_L = \frac{P_R}{\sqrt{3} E_L \times pf} = \frac{4,200\ \text{W}}{\sqrt{3}(240\ \text{V})(0.60)} = 16.8\ \text{A}$$

c. The resistance losses in the stator (i.e., the I^2R losses):
For a Y connection—$I_L = I_\Phi$

$$P_{\text{losses},\Phi} = (I_\Phi)^2 R_S = (16.8\ \text{A})^2 (0.16\ \Omega) = 45.4\ \text{W}$$

$$P_{\text{losses,resistance}} = 3P_{\text{losses},\Phi} = 3(45.4\ \text{W}) = 135\ \text{W}$$

d. Energy efficiency of the motor:
First the total losses are estimated:

$$P_{\text{losses,total}} = P_{\text{losses,core}} + P_{\text{losses,resistance}} = 350 + 135 = 485 \text{ W}$$

$$\eta_{\text{motor}} = \frac{P_{\text{input}} - P_{\text{losses,total}}}{P_{\text{input}}} = \frac{4,200 - 485}{4,200} = 88.5\%$$

$$\eta_{\text{losses}} = 1 - \eta_{\text{motor}} = 11.5\%$$

e. The torque delivered by the motor:

$$P_{\text{motor}} = T \times \omega$$

$$P_{\text{motor}} = P_{\text{input}} \times \eta = 4,200 \text{ W} \times 0.885 = 3715 \text{ W}$$

$$\omega_{mag} = 4\pi \frac{f}{N_p} \text{ rad/s} = 4\pi \frac{50Hz}{4} = 157 \text{ rad/s}$$

$$\omega_{\text{actual}} = \omega_{mag}(1 - \alpha) = 157(1 - 0.042) = 150.5 \text{ rad/s}$$

$$T = \frac{P_{\text{motor}}}{\omega_{\text{actual}}} = \frac{3715 \text{ W}}{150.5 \text{ rad / s}} = 24.7 \text{ Nm}$$

It should be noted that since the resistance losses are related to the rotor and stator windings that do not vary significantly with the motor size, the energy efficiency of large motors is generally higher than that of small motors. Moreover, the energy efficiency of any electrical motor depends on its loading level, as illustrated in Figure 4.16, for both standard efficiency and premium 5 HP motors. At low loads,

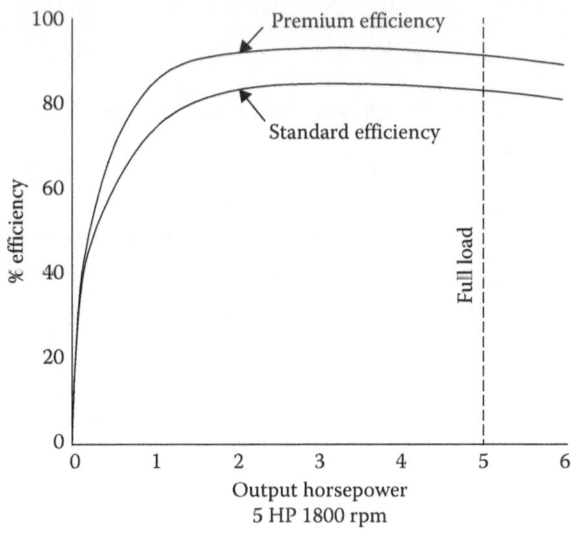

FIGURE 4.16 Variation of energy efficiency as a function of the load level of both standard and premium 5 HP motors.

the energy efficiency is close to zero because the motor has to generate the magnetic field and overcome friction. At higher loads, the energy efficiency increases until leveling off at around 50% of the rated mechanical output of the motor. The peak energy efficiency of electrical motors occurs typically between 50% and 100% of their rated power output depending on their design specifications.

4.4.2 Power Factor for Motors

The real electrical power, P_R, used with an electrical motor can be calculated as a function of the mechanical power output, P_M, and the efficiency, η_M, of the motor:

$$P_R = \frac{1}{\eta_M} P_M \tag{4.9}$$

Therefore, the total power, P_T, and the reactive power, P_X, needed to operate the motor are respectively

$$P_T = \frac{P_R}{pf_M} = \frac{1}{pf_M \times \eta_M} P_M \tag{4.10}$$

$$P_X = P_R \tan\phi = \frac{1}{\eta_M} P_M \times \tan\left(\cos^{-1} pf\right) \tag{4.11}$$

To improve the power factor of an induction motor, a capacitor can be placed in parallel to the motor as noted in Figure 4.17. The reactive power of the capacitor can be estimated as discussed in Chapter 2. In particular, the capacitor reactive power, P_C, needed to improve the existing power factor, pf_e, to a retrofitted value, pf_r, can be determined as a function of the mechanical power, P_M, and the energy efficiency, η_M, of the motor:

$$P_C = \frac{P_M}{\eta_M}\left[\tan\left(\cos^{-1} pf_e\right) - \tan\left(\cos^{-1} pf_r\right)\right] \tag{4.12}$$

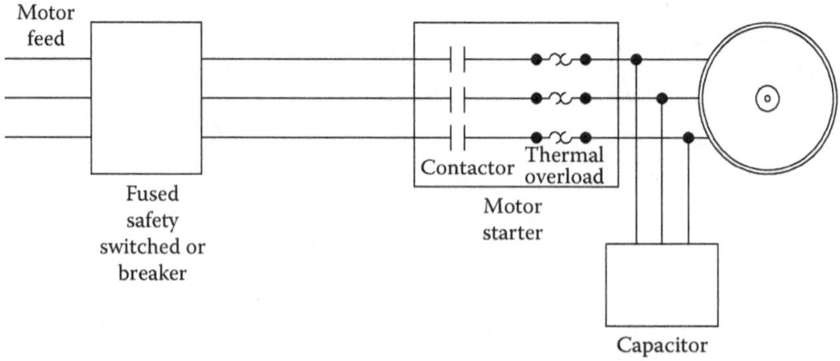

FIGURE 4.17 Placement of a capacitor to improve the power factor of a three-phase inductor motor.

Example 4.3 illustrates the estimation of the size of a capacitor needed to improve the power factor of an induction motor.

Example 4.3

Problem

Consider a 100 HP induction motor with a nameplate-rated energy efficiency of 91.5% and a power factor of $pf_e = 0.87$. Determine the required size in kVAR of a capacitor to be installed in parallel with the induction motor so that the power factor becomes at least $pf_r = 0.95$.

Solution

The size in kVAR of the capacitor is determined using Equation 4.12:

$$P_C = \frac{100 \times 0.746}{0.915}\left[\tan\left(\cos^{-1} 0.87\right) - \tan\left(\cos^{-1} 0.95\right)\right] = 19.4 \text{ kVAR}$$

Thus, a capacitor of 20 kVAR can be selected to ensure that the power factor of the induction motor to be at least 0.95.

4.4.3 Effect of Voltage Variation

Motors are sensitive to changes in source voltage. Typically, a 10% tolerance band for the source voltage is recommended to ensure that during extremely low and high values of the voltages, motors perform within an acceptable range. In particular, high voltages result typically in higher currents causing overheating which can shorten motor life. Figure 4.18 illustrates the typical effects of voltage variations on the performance indicators of an induction motor, including starting current, full-load (F.L.) current, power factor (P.F.), energy efficiency (EEF.), and the starting and maximum torques.

Some general guidelines on the impact of voltage variations on the performance of motors can be summarized from data outlined in Figure 4.18 as well as from observations:

- Small motors are typically more sensitive to overvoltage than large motors.
- Single-phase motors are more sensitive to overvoltage than three-phase motors.
- Premium-efficiency motors are less sensitive to overvoltage than standard-efficiency motors.
- Two-pole and four-pole motors are typically less sensitive to high voltage than six-pole and eight-pole motors.
- Overvoltage can increase amperage and temperature even on lightly loaded motors resulting in shorter life.
- Full-load energy efficiency is reduced by both high and low voltages.

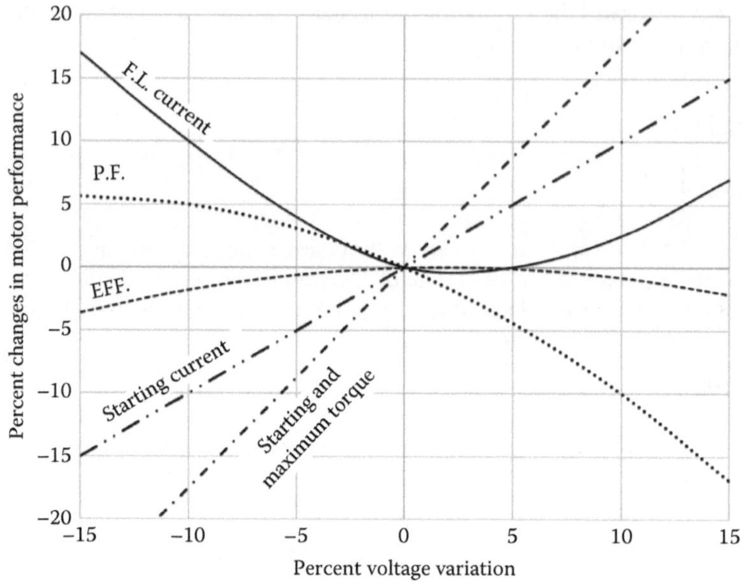

FIGURE 4.18 Effect of voltage variation on a typical induction motor performance.

- Power factor can improve with lower voltage and is reduced sharply with higher voltage.
- Inrush current increases with higher voltage.

Example 4.4 shows the impact of an overvoltage on the energy efficiency and power factor of an induction motor.

Example 4.4

Problem

Consider a 40 HP three-phase induction motor with a nameplate-rated voltage of 460, an energy efficiency of 91.5%, and a power factor of $pf_e = 0.90$. Determine the impact of 10% overvoltage on both the energy efficiency and the power factor of the motor. When experiencing this overvoltage, estimate the real power consumed by the motor to provide the same 40 HP mechanical power output.

Solution

Based on the data shown in Figure 4.18, the 10% overvoltage results in (1) 10% reduction in power factor and (2) 1.5% reduction in energy efficiency.

Thus, the new power factor of the motor when operating at 10% higher voltage of 506 V (instead of 460 V) is

$$pf_{\text{new}} = (1 - 0.10)\, pf_{\text{rated}} = 0.90 \times 0.90 = 0.81$$

The new energy efficiency of the motor when operating at 506 V voltage is

$$\eta_{\text{new}} = (1 - 0.015)\eta_{\text{rated}} = 0.985 \times 91.5\% = 90.13\%$$

When the voltage is 460 V, the rated real power consumed by the motor is

$$P_{R,\text{rated}} = \text{HP} \times \frac{0.746}{\eta_{\text{rated}}} = 40 \times \frac{0.746}{0.915} = 33.6\ \text{kW}$$

When the motor is supplied with 506 V, the new real power consumed by the motor can be estimated based on the new energy efficiency value with the motor providing the same 40 HP rated mechanical power output:

$$P_{R,new} = \text{HP} \times \frac{0.746}{\eta_{new}} = 40 \times \frac{0.746}{0.9013} = 33.1\ \text{kW}$$

4.4.4 Specifications of Motors

When selecting an electrical motor, several factors and specifications should be considered depending on the application. In addition to the mechanical power output, the energy efficiency, the frequency, and the voltage requirements, other features can be used to define the characteristics of electrical motors, including the following:

- *Service factor*: Some electrical motors can be slightly overloaded beyond their rated mechanical power output. The overloading capacity of motors is referred to as the service factor. Thus, a service factor of 1.10 indicates that the motor can handle 10% additional mechanical load than its maximum rated value. Typically, service factors of electrical motors range from 1.10 to 1.15.
- *Enclosure type*: Motors can operate in a wide range of environmental conditions depending on the applications and uses ranging from corrosive and wet locations, such as the case of several industrial facilities, to clean and dry spaces common in buildings. Two basic enclosures are available for electrical motors, including open type and closed type. The open-type motors are ventilated through natural or forced circulation of ambient air. For the closed-type motors, two main categories are available: nonventilated or fan-cooled.
- *Frame size*: The physical dimensions of the motors have been standardized in the United States by NEMA. Thus, motors made by various manufacturers have the same dimensions or frame sizes. Typically, the frame size includes three numbers and one letter at the end. The first two numbers refer to the height of the shaft above the motor mounting plate while the third number provides a code related to the length of the motor. The letter can be (1) U if the motors meet ratings set after 1953 but before 1964 or (2) T if the motors are rated based on a system initiated in 1964. If no letter is used, then the motor frame size is set based on the original rating system of 1928. Moreover, a letter S can be added to indicate that the motor has a short shaft.

TABLE 4.5
Temperature Ratings for Common Motor Insulation Classes

	Temperature Rating	
Insulation Class	**(°C)**	**(°F)**
A	105	221
B	130	266
F	155	311
H	180	356

- *Insulation class*: Most electrical motors are rated to operate when ambient air temperature is below 40°C (104°F). However, heat generated by the rotor and stator windings increases the motor temperature especially during full-load operation. The temperature difference between the starting and the full-load conditions is referred to as the temperature rise of the motor. Depending on the insulation type and rating, specific maximum temperature rise is permissible during its average life span typically set at 20,000 hours. Table 4.5 defines the four commonly available insulation classes and their temperature ratings, which are defined as ambient temperature plus maximum temperature rise allowed. It should be noted that most motors with frame size T have class B insulation.

4.5 MOTOR ENERGY EFFICIENCY IMPROVEMENTS

In addition to better design, the energy efficiency of electrical motors can be improved through the installation of variable or ASDs especially when they drive fans, pumps, and compressors. The operation and the potential benefits of energy efficiency improvements of ASDs are first discussed. Then, general calculation methods are described with some illustrative examples to estimate energy use reduction due to the improving energy efficiency of electrical motors.

4.5.1 Adjustable Speed Drives

With more emphasis on energy efficiency, an increasing number of designers and engineers are recommending the use of variable speed motors for various building heating, ventilation, and air conditioning (HVAC) systems. Indeed, the use of ASDs is now becoming common especially for supply and return fans in variable air volume (VAV) systems and for hot and chilled water pumps in central heating and cooling plants.

Electronic ASDs convert the fixed-frequency AC power supply (50 or 60 Hz) first to a DC supply and then to an AC power supply with a new frequency value, as illustrated in Figure 4.19. Therefore, the ASDs can change the speed of AC motors with no moving parts, presenting high reliability and low maintenance requirements.

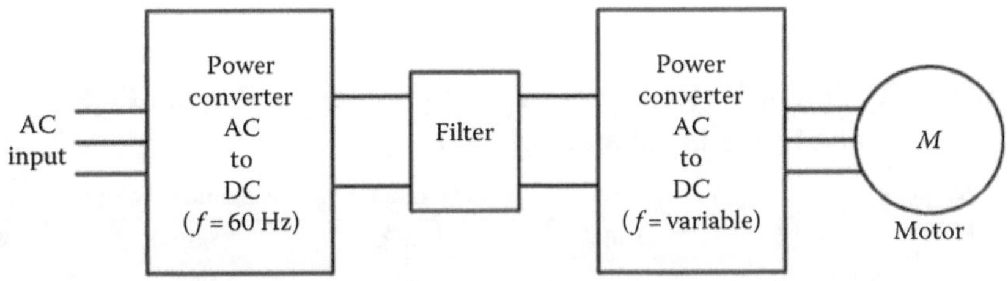

FIGURE 4.19 Basic concept of ASD power inverter.

To achieve optimal energy savings potential for any HVAC application, the engineer needs to know the actual energy efficiency of the motors. For ASD applications, it is important to separate the losses between the drive and the motor to achieve an optimized HVAC system with the lowest operating cost. Specifically, the engineer would need to know the loss distribution including the iron losses, copper losses, friction and windage losses, and the distribution of losses between stator and rotor. To determine these losses and thus the motor efficiency, accurate measurements are needed. Unfortunately, existent power measurement instruments are more suitable for sinusoidal rather than distorted waveforms (which are typical for ASD applications).

The use of low-cost solid-state power devices and integrated circuits to control the speed has made most of the commercially available ASDs draw power with extremely high harmonic contents, as outlined in Chapter 11. Some investigators have studied the behavior of commercially available power measurement instruments subjected to voltage and current waveforms typical of ASD-motor connections. In particular, three drive technologies used in HVAC industry were investigated by Czarkowski and Domijan (1997) as part of ASHRAE project 770-RP (Domijan et al., 1996), namely, PWM induction, switched reluctance, and brushless DC drives. The main findings of their investigation are that existing power instruments fail to accurately measure power losses (and thus motor energy efficiency) due to high harmonic content in the voltage spectra, especially for brushless DC and switched reluctance motors, which represent a substantial portion of the HVAC market.

Typically, the motor losses are measured indirectly by monitoring the power input and the power output and then by taking the difference. This traditional approach requires extremely high, accuracy for the power measurements to achieve "reasonable" estimation of motor losses especially for premium-efficiency motors (with low losses). However, other approaches to measure the ASD-motor losses directly have been proposed (Krarti, 2010).

To assess the potential energy use savings associated with the installation of ASDs to change the speed of motors, the theoretical relation between the motor's real power change from $P_{R,1}$ to a $P_{R,2}$ due to a change in its speed from $\omega_{R,1}$ to $\omega_{R,2}$ can be considered:

$$\frac{P_{R,1}}{P_{R,2}} = \left(\frac{\omega_{R,1}}{\omega_{R,2}}\right)^3 \tag{4.13}$$

TABLE 4.6
Percent Reduction of Motor Power as a Function of the Percent Reduction in Motor Speed due to the Installation of ASD

Adjustable Speed Drive (ASD)	Energy Use (in % of Maximum Value)			
	100%	**80%**	**60%**	**40%**
Actual performance	100%	64%	36%	16%
Theoretical performance	100%	51%	22%	6%

This equation is often applied to fans in HVAC systems and is referred to as the third of three fan laws. However, due to back pressure effects, the actual reduction in fan motor power consumption is less significant as noted in Table 4.6 as compared to the theoretical and actual impact of energy use as a function of the motor speed (Krarti, 2010). Example 4.5 illustrates the energy use savings associated to the installation of ASD drive on a constant-speed fan motor.

Example 4.5

Problem

A supply fan motor for an air handling unit of a variable air volume system is rated at 50 HP with 90% energy efficiency. The fan is operated for 4,000 h/year according to the following load profile:

- 100% load, 10% of the time
- 80% load, 40% of the time
- 60% load, 40% of the time
- 40% load, 10% of the time

Determine the energy use savings associated with the installation of ASD. Consider both the theoretical and the actual performances of the ASD. Estimate the energy cost savings if the electricity price is \$0.10/kWh. Determine the simple payback period if installing the VSD costs \$18,000.

Solution

The electrical energy input of the supply fan motor is first determined for each control option by calculating a weighted average fan mechanical power rating, expressed in horsepower, using the fan performance data given in Table 4.6.

For the actual VSD fan performance, the average horsepower is

$$k\overline{W}_{e,fan}^{VSD-A} = 50\text{ HP}(1.0\times 0.1 + 0.64\times 0.4 + 0.36\times 0.40 + 0.16\times 0.1) = 26\text{ HP}$$

while for the theoretical VSD fan performance, the average horsepower is

$$k\overline{W}_{e,fan}^{VSD-T} = 50\text{ HP}(1.0\times 0.1 + 0.51\times 0.4 + 0.22\times 0.40 + 0.064\times 0.1) = 20\text{ HP}$$

To determine the annual energy use savings due to the installation of ASD, first, the annual electrical energy use for the existing constant speed (i.e., volume) fan is estimated:

$$\text{kWh}_{e,fan}^{CV} = \left(\frac{50 \text{ HP} \times 0.746 \text{ kW/HP}}{0.90}\right) 4{,}000 \text{ h} = 165{,}778 \text{ kWh/year}$$

For the actual performance of the variable speed drives (VSD)

$$\text{kWh}_{e,fan}^{VSD-A} = \left(\frac{26 \text{ HP} \times 0.746 \text{ kW/HP}}{0.90}\right) 4{,}000 \text{ h} = 86{,}204 \text{ kWh/year}$$

For the theoretical performance of the variable speed drives (VSD)

$$\text{kWh}_{e,fan}^{VSD-T} = \left(\frac{20 \text{ HP} \times 0.746 \text{ kW/HP}}{0.90}\right) 4{,}000 \text{ h} = 66{,}311 \text{ kWh/year}$$

The fan motor energy use cost and the payback period for each fan control option are summarized here.

Motor Speed Option	Annual Energy Use (kWh/Year)	Annual Energy Savings (kWh/Year)	Annual Cost Savings ($/Year)	Simple Payback Period (Years)
CV	165,778	0	0	—
VSD (act.)	86,204	79,574	7,957	2.26
VSD (th.)	66,311	99,467	9,947	1.81

It is clear from the results of the economic analysis presented here that the installation of a variable speed drive is a very cost-effective option even when the actual performance of the fan motor is considered.

4.5.1.1 Energy-Efficient Motors

There are three methods to calculate the energy use savings due to energy-efficient motor replacement. These three methods are outlined in this section.

4.5.1.1.1 Method 1: Simplified Method

This method has been and is still being used by most energy engineers to determine the energy and cost savings incurred by motor replacement. Inherent to this method, two assumptions are made: (1) the motor is fully loaded, and (2) the change in motor speed is neglected. Using this method, the electric power savings due to motor replacement is first computed as follows:

$$\Delta P_R = P_M \left(\frac{1}{\eta_e} - \frac{1}{\eta_r}\right) \tag{4.14}$$

where

P_M is the mechanical power output of the motor

η_e is the design (i.e., full-load) efficiency of the existing motor (e.g., before retrofit)

η_r is the design (i.e., full-load) efficiency of the energy-efficient motor (e.g., after retrofit)

The electric energy use savings incurred from the motor replacement can then be estimated:

$$\Delta kWh = \Delta P_R N_h LF_M \tag{4.15}$$

where

N_h is the number of hours per year during which the motor is operating

LF_M is the load factor of the motor's operation during 1 year

Example 4.6 illustrates the estimation of the energy use savings associated with using a premium-efficiency motor instead of a standard-efficiency motor using the simplified calculation method of Equations 4.14 and 4.15.

Example 4.6

Problem

A 30 HP standard efficiency motor is operated 5,000 h/year. A premium-efficiency motor has an efficiency of 92.8%, compared to 90.0% for the standard-efficiency motor. Calculate the annual energy savings in kWh when using the premium efficiency instead of the standard-efficiency motor. Assume the load factor for the motor is 0.75.

Solution

Using Equation 4.14, the savings in real power used by the motor is first calculated:

$$\Delta P_R = 30\text{ HP} \times 0.746\text{ kW/HP} \times \left(\frac{1}{0.90} - \frac{1}{0.928}\right) = 0.74\text{ kW}$$

Using Equation 4.15, the annual energy savings can be estimated:

$$\Delta kWh = 0.75\text{ kW} \times 5{,}000\text{ h/year} \times 0.75 = 2814\text{ kWh}$$

4.5.1.1.2 Method 2: Mechanical Power Rating Method

In this method, the electrical peak demand of the existing motor is assumed to be proportional to its average mechanical power output:

$$P_{R,e} = \frac{P_M}{\eta_{op,e}} LF_{M,e} \times PDF_{M,e} \tag{4.16}$$

where

- $\eta_{op,e}$ is the motor energy efficiency at the operating average part-load conditions. To obtain this value, the specific energy-efficiency curve for the existing motor should be used. If the efficiency curve for the specific motor is not available, a generic curve can be used.
- $LF_{M,e}$ is the load factor of the existing motor and is defined as the ratio between the average operating load of the existing motor and its rated mechanical power. It should be noted that in most applications, motors are oversized and operate at very low-part load conditions.
- $PDF_{M,e}$ is the peak demand factor and represents the fraction of typical motor load that occurs at the time of the building peak demand. In most applications, $PDF_{M,e}$ can be assumed to be unity since the motors often contribute directly to the total peak demand of the building.

Since the mechanical load does not change after the installation of an energy-efficient motor, it is possible to consider a smaller motor with a capacity $P_{M,r}$ if the existing motor is oversized with a rating of $P_{M,e}$. In this case, the smaller energy-efficient motor can operate at a higher load factor than the existing motor. The new load factor, LF_r, of the energy-efficient motor can be calculated as follows:

$$LF_r = LF_e \frac{P_{M,r}}{P_{M,e}} \tag{4.17}$$

Moreover, the energy-efficient motors often operate at higher speeds than the standard motors they replace since they have lower internal losses. This higher speed has a negative impact since it reduces the effective energy efficiency of the motor by a factor called the slip penalty. The slip penalty factor, $SLIP_P$, is defined as indicated in Equation 4.18:

$$SLIP_P = \left(\frac{\omega_{M,r}}{\omega_{M,e}} \right)^3 \tag{4.18}$$

where

$\omega_{M,e}$ is the rotation speed of the existing motor
$\omega_{M,r}$ is the rotation speed of the energy-efficient motor

Using similar equation than Equation 4.16, the peak electrical demand for the retrofitted motor (e.g., energy-efficient motor) can be determined:

$$P_{R,r} = \frac{P_{M,r}}{\eta_{op,r}} LF_{M,r} \times PDF_{M,r} \times SLIP_P \tag{4.19}$$

The electrical power savings due to the motor replacement can thus be estimated:

$$\Delta P_R = P_{R,e} - P_{R,r} \tag{4.20}$$

The electrical energy use savings can be therefore calculated using Equation 4.15.

4.5.1.1.3 Method 3: Field Measurement Method

In this method, the motor electrical power demand is measured directly on-site. Typically, current, I_M, voltage, E_M, and power factor, pf_M, readings are recorded for the existing motor to be retrofitted. For three-phase motors (which are common in industrial facilities and in most HVAC systems for commercial buildings), the electrical power used by the existing motor can be either directly measured or calculated from the current, voltage, and power factor readings as follows:

$$P_{R,E} = \sqrt{3}E_M \times I_M \times pf_M \tag{4.21}$$

The load factor of the existing motor can be estimated by taking the ratio of the measured current over the nameplate full-load current, I_{FL}, as expressed in Equation 4.22:

$$LF_{M,E} = \frac{I_M}{I_{FL}} \tag{4.22}$$

A study by Biesemeyer and Jowett (1996) has indicated that Equation 4.22 provides a much higher accuracy to estimate the motor load factor than an approach based on the ratio of the motor speeds (i.e., measured speed over nominally rated speed) used by BPA (1990) and Lobodovsky (1994). It should be noted that Equation 4.22 is recommended for load ratios that are above 50% since for these load ratios, a typical motor draws electrical current that is proportional to the imposed load.

The methodology for the calculation of the electrical power and energy use savings is the same as described for the mechanical power rating method using Equations 4.17–4.20.

4.6 SUMMARY

In this chapter, an overview is provided for the basic characteristics of three-phase and single-phase AC motors. In particular, the operation principles and the performance indicators of induction AC motors are emphasized since they are widely used in various building applications. Throughout this chapter, guidelines are described to ensure safe operation and to improve the energy performance of electrical motors. Moreover, illustrative examples are presented to evaluate the impact of various design features and operating conditions on the performance of electrical motors. In addition, suggestions and guidelines are presented to illustrate the wide range of issues that an engineer should address when designing, analyzing, or retrofitting electrical motors used in buildings.

4.7 PROBLEMS

1. A 200 HP induction motor with a power factor of 0.65 and an efficiency of 90% is supplied by a three-phase Δ-connected 480 VAC source.
   a. Show the power triangle for the motor.
   b. Determine the impedance of the motor (assume three equal impedances, one per phase).
   c. Calculate the capacitance in kVAR required to improve the power factor of the motor to 0.85. Solve this question using power triangles.
   d. Compare the line currents of the motor before and after improving the power factor from 0.65 to 0.85.
2. A commercial building has an electrical load of 2,000 kW at 0.80 power factor, lagging. It has been decided to replace a 100 HP induction motor that drives a pump. This motor operates at an efficiency of 94% and a power factor of 0.90, lagging. A synchronous motor is purchased as the replacement and is capable of operating at 0.80 power factor, leading, with the same efficiency as the induction motor. Determine
   a. The new building power factor
   b. The percent decrease in line current as a result of the improved power factor
3. Determine the motor speed in rpm and rad/s for an induction motor with a 2% slip and six pole for two cases of voltage source frequency (a) $f = 60\,\text{Hz}$ and (b) $f = 50\,\text{Hz}$.
4. A four-pole squirrel-cage motor has the following characteristics: (a) power input at rated load is 5,200 W with a power factor of 0.70, (b) a rated speed of 11,500 rpm, and (c) a resistance for each stator winding (there are three) of $R_s = 0.195\ \Omega$. The motor is supplied by a three-phase source with line-to-line voltage 460 V and a frequency of 500 Hz. The windings of the motor have a Y-connection. Determine
   a. The slip factor of the motor in percent
   b. The line current drawn by the motor
   c. The resistance losses in the stator
   d. The efficiency losses of the motor if the core losses at the stator are 475 W (neglect the friction losses and other losses, including the resistance and core losses at the rotor)
   e. The torque delivered by the motor in Nm
   f. Consider a 50 HP induction motor with a nameplate-rated energy efficiency of 92% and a power factor of 0.85. Determine the required size in kVAR of a capacitor to be installed in parallel with the induction motor so that the power factor becomes at least 0.95.
5. A commercial building is subject to an electricity rate schedule that requires a demand charge of $2.50 per kVA per month. The electrical load of the building consists of (a) a group of ten (10) identical induction motors, (b) a plug load of 200 kW with a power factor of 0.90, and (c) of lighting load of 50 kW with a power factor of 0.80. Each motor has the following nameplate rating: 480 V, 30 A, 20 HP, 1,780 rpm, 0.70 power factor at full load. Determine

a. The efficiency, the number of poles, and the slip factor of each motor
b. The capacitor (expressed in kVAR and in μF) needed to raise the power factor of the building to 0.90
c. The annual savings in the power bill for the plant due to the increase in the power factor

6. The load in a building panel is 120 kW at 480 V and is three-phase with a low power factor of 0.75 lagging. This power factor is to be corrected to 0.95 lagging. A motor rated at 60 Hz, 70 HP, 1,200 rpm, and an efficiency of 0.88 is used to correct the system power factor. The panel is operating 5,500 h/year. Determine
   a. The current before the power factor correction
   b. The type, the number of poles, and the power factor of the motor
   c. The real power and the current of the system after the power factor correction
   d. An estimate of the annual cost increase or decrease in supplying power to the panel by adding the motor if the demand charge is $20/kVA/month and the energy charge is $0.05/kWh
7. A building has a total load of 150 kVA and a power factor of 0.85 (lagging). Among the electrical load, there is a 40 HP, three-phase, 60 Hz, 480 V induction motor with an efficiency of 0.92 and a power factor of 0.80 (lagging). Calculate the following:
   a. The line current required by the building.
   b. If a synchronous motor of the same size, the same energy efficiency, and the same power factor (but leading) is used instead of the induction motor, estimate the overall new power factor of the building.
   c. The line current of case (b). Compare this current to the line current obtained for (a). Comment on the advantages of case (b).
8. For a 20 HP three-phase induction motor with a nameplate-rated voltage of 208 V energy efficiency of 90.5% and a power factor of 0.87. Determine the impact of 5% overvoltage on the energy efficiency, the full-load current, and the power factor of the motor.
9. A supply fan motor for an air handling unit of a variable air volume system is rated at 20 HP with 88% energy efficiency. The fan is operated for 5,500 h/year according to the following load profiles:
   a. 100% load, 5% of the time
   b. 80% load, 40% of the time
   c. 60% load, 50% of the time
   d. 40% load, 5% of the time
   e. Determine the energy use savings associated with the installation of ASD. Consider both the theoretical and the actual performance of the ASD. Estimate the energy cost savings if the electricity price is $0.12/kWh. Determine the simple payback period of installing the VSD at a total cost of $13,500.
10. A 50 HP motor is to operate 4,500 h/year at 80% load. A premium-efficiency motor has an energy efficiency of 93.4%, compared to 91.2% for the standard-efficiency motor. Calculate the annual energy use savings

in kWh when using the premium efficiency instead of the standard-efficiency motor.

11. An induction motor is rated 60 HP, three-phase, 60 Hz, 460 V, 1,760 rpm, and a power factor of 90%.
    a. Estimate the efficiency of each motor if its line current is 70 A.
    b. Determine the number of poles, the slip (%), and the torque delivered by the motor.
12. A motor control center (MCC) has three induction motors each rated at 40 HP and 480 V and four induction motors each rated at 30 HP and 480 V. Each 40 HP motor has an energy efficiency of 0.92, a rated speed of 1,772 rpm, and a power factor of 0.85. Each 30 HP motor has an energy efficiency of 0.90, and a power factor of 0.90. Determine
    a. The number of poles, the slip (%), and the torque delivered by each 40 HP motor.
    b. The induction motors (by specifying their HP and their number) that need to be replaced by synchronous motors so that the overall power factor of the MCC is as close to unity as possible without changing the total kW load of the MCC nor the HP ratings of the motors. Assume that the power factor for any synchronous motor is 0.90 leading.
13. A building has a total electrical load of 280-kVA with a power factor of 0.80 (lagging). Among the electrical load, there is a 60-hp, three-phase, 60 Hz, 1,760 RPM, 480-volt motor with an efficiency of 0.90 and a power factor of 0.75 (lagging). Calculate:
    a. Calculate the line current required by the building.
    b. Determine the type, the number of poles, and the percentage slip factor for the 60-hp motor.
    c. If a synchronous motor of the same size, the same efficiency, and the same power factor (but leading) is used instead of the induction motor, estimate the overall new power factor of the building.
    d. Determine for case (c) the rating in kVAR of the capacitors that need to be added in parallel to ensure that the building has a power factor of unity (i.e., $pf = 1.0$). In this case, determine the line current of the building. Compare this current to the line current obtained in case (a). Comment on the main advantages of this case compared to case (a).
14. A building has a total electrical load of 140-kVA with a power factor of 0.80 (lagging). Among the electrical load, there is a 30-hp, three-phase, 60 Hz, 480-volt, 1,170 RPM, motor with an efficiency of 0.90 and a power factor of 0.75 (lagging). Calculate:
    a. Calculate the line current required by the building.
    b. Determine the type, number of poles, and the percent slip factor for the 30-hp motor.
    c. If a synchronous motor of the same size, the same efficiency, and the same power factor (but leading) is used instead of the induction motor, estimate the overall new power factor of the building.
    d. Determine for case (c) the rating in kVAR of the capacitors that need to be added in parallel to ensure that the building has a power factor

of unity (i.e., $pf = 1.0$). In this case, determine the line current of the building. Compare this current to the line current obtained in case (a). Comment on the main advantages of this case compared to case (a)?

15. The load in a plant is 120 kW at 480 V and is three-phase with a low power factor of 0.75 lagging. This power factor is to be corrected to 0.95 lagging. A motor rated at 60 Hz, 70 HP, 1,200 rpm, and an efficiency of 0.88 is used to correct the system power factor. The plant is operating 5,500 hours per year. Determine:
    a. The current before the power factor correction.
    b. The type, the number of poles, and the power factor of the motor.
    c. The real power and the current of the system after the power factor correction. Comment?
    d. An estimate of the annual cost increase or decrease in operating the plant by adding the motor if the demand charge is $25/kVA per month and the energy charge is $0.12/kWh.

5 Protection Systems

To ensure the safety of the electrical systems in buildings, it is important that proper protection devices be utilized. The main role of protection devices is to disconnect and isolate faulty portions of the electrical system in case of overloading and/or short-circuiting conditions. Depending on the type of loads, suitable types and response characteristics of the protection devices should be specified. This chapter outlines different types of protection devices, their performance characteristics, and their design specifications.

5.1 INTRODUCTION

One of the main requirements in the design of electrical systems is to minimize power outages and damages in case of fault conditions. Protective devices provide the means to isolate faulted segments of an electrical distribution system as quickly and safely as possible. Specifically, a protective device has two major functions: detection of faulty conditions and disconnection of the faulted sections from the remainder of the electrical distribution system. Some protective systems combine both functions such as fuses while other types separate the two actions such as high-voltage circuit breakers. Article 240 of the NEC covers the design requirements of the overcurrent protective devices.

Several fault types can occur in electrical distribution systems in a building including overloads, short-circuits, overvoltages, undervoltages, transient surges, reverse power flows, and single phasing in three-phase systems. Some of these faults are related to the quality of the power supply, as discussed in Chapter 11. Other faults are specific to the installation and operation of the electrical systems within buildings. Common faults in buildings are those due to overloads and short-circuit currents. These two fault types result in higher-than-normal flows of currents within the electrical distribution systems. In particular, overloads are caused by excessive current demands from utilization equipment. Since overload currents flow in normal distribution conductors, they can be tolerated even though they are difficult to detect without proper protection devices. However, short-circuit currents are caused by some failure in the electrical system such as insulation breakdown. The magnitude of short-circuit currents is significantly higher than that of overload currents. Typically, overload currents are limited to 600% of the full-load capacity of the electrical system section affected by the fault. Any fault currents higher than the 600% threshold are considered to be short-circuit currents. In general, two types of short-circuit faults can occur:

1. Arcing faults, due typically to unintentional connections between one phase and another phase or between one phase and a grounded equipment such as when conductor insulation fails. Arcing fault currents tend to be small but may lead to significant damage since they may continue to flow due to the slow response of typical protection devices. The use of ground-fault protection systems can reduce the often damaging impact of these faults.

DOI: 10.1201/9781003276999-5

2. Bolted faults are caused by direct and often accidental connections between two phases or one phase and a grounded equipment. Because of the low resistance of direct connections, bolted faults can be very large and thus can easily be detected with properly designed and installed protection devices.

Figure 5.1 illustrates examples of both arcing and bolted faults that can cause short-circuit currents and ultimately thermal and magnetic stresses on the electrical systems. Indeed, high fault currents, when allowed to flow without a rapid response, can cause significant damage to the electrical distribution systems. In particular, heat generated by the flow of fault currents in the conductors can be so intense that it can damage the insulation and even the conductor material itself. Figure 5.2a shows an example of thermal stresses caused by higher fault current on a conductor. Typically, the heat generated in a conductor is proportional to the product of the resistance and the square of the current (i.e., RI^2). As indicated in the example of Figure 5.2a, the heat generated in the conductor can increase by a factor of 10,000 if the current is increased by a factor of 100. Similarly, the high fault currents can result in magnetic stresses by generating strong magnetic fields and ultimately large mechanical forces capable of bending and damaging several electrical system components such as conductors, as indicated in Figure 5.2b.

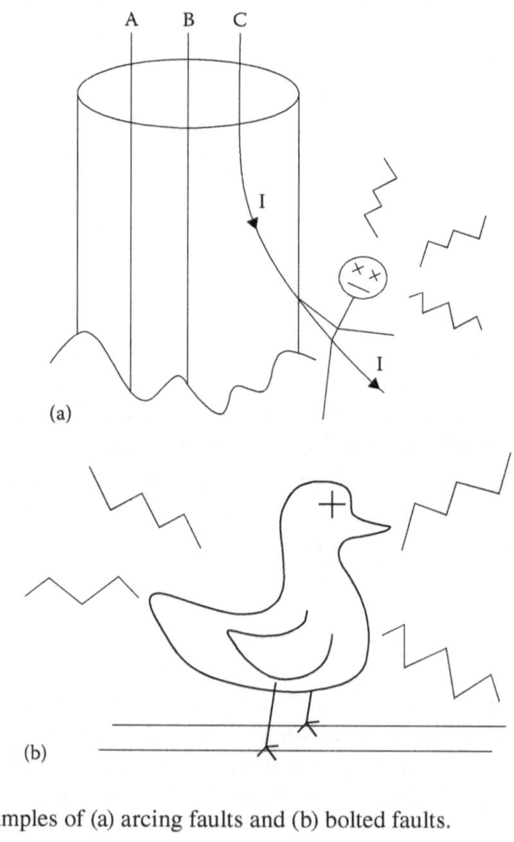

FIGURE 5.1 Examples of (a) arcing faults and (b) bolted faults.

In this chapter, common protection devices are presented to eliminate or reduce the impact of fault currents that may occur within various sections of electrical distribution systems for buildings. Figure 5.3 shows basic protection systems that are specified for power systems in buildings. In particular, molded-case circuit breakers are typically considered to protect lighting and power branch circuits and are housed in panels. On the other hand, power circuit breakers are suitable to protect feeders

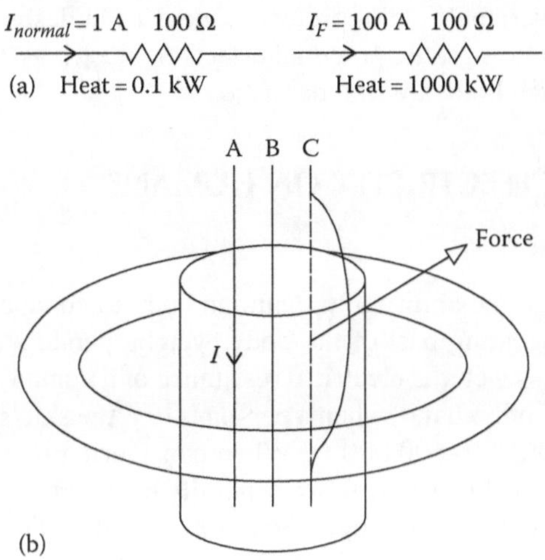

FIGURE 5.2 Impact of fault currents on electrical conductors: (a) thermal stresses and (b) magnetic stresses.

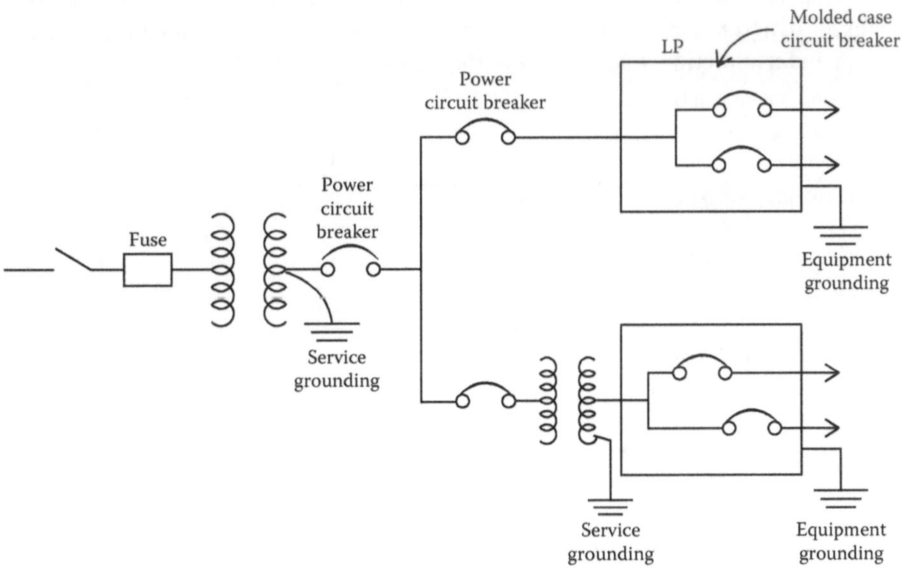

FIGURE 5.3 Typical protection devices in a building electrical distribution system.

and subfeeders since they can handle large currents even under normal conditions. Power circuit breakers are typically housed within main distribution panels or unit substations. Fuses with proper disconnecting means are typically used to protect motor loads as well as primary feeders for transformers. Finally, grounding systems are specified to protect electrical distribution circuits against ground faults flowing either at the equipment surfaces (i.e., panels and motor enclosures) or at the power sources (i.e., transformers or generators).

To emphasize the crucial role of protection devices in designing electrical distribution systems for buildings, the potential effects of electricity flows, including fault currents, on the human body are first described.

5.2 IMPACT OF ELECTRICITY ON HUMANS

5.2.1 Background

It has been shown that the ability of the human body to conduct electricity depends on several factors including part of the body, weight, gender, and level of wetness. Indeed, when a body is wet, the electrical resistance of a human skin can be as low as 500–1,000 Ω. On the other hand, when a person is dry, the skin's electrical resistance can be as high as 100,000–600,000 Ω. When considering only the internal organs of a human, the electrical resistances are generally lower. For instance, the electrical resistance from a hand to a foot ranges from 400 to 600 Ω. The electrical resistance from ear-to-ear is only 100 Ω. Figure 5.4 illustrates the variation of electrical resistance of a human body to conduct currents due to a voltage source.

The impact of current flowing in human bodies has been investigated since the early 1970s by Dalziel using experimental analyses by subject volunteers to low current levels (Dalziel, 1972). In particular, Dalziel found that the impact of electrical shocks is different for men and women and depends mainly on the magnitude of current, I, and its duration, t, according to the following equation:

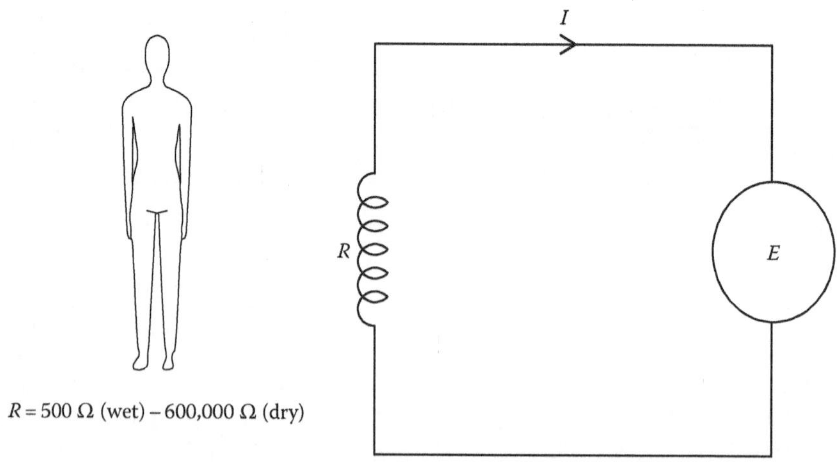

FIGURE 5.4 Electrical resistance model for a human body.

$$I = k\sqrt{t} \tag{5.1}$$

where

I is the threshold level of current that causes fibrillation and is expressed in milliamperes (mA)

k is a constant depending on the shock energy and is determined empirically

t is the duration of the application of the current to the human body and is expressed in seconds (s)

The work of Dalziel expressed by Equation 5.1 is the basis of the safety guidelines and shock hazard information available currently for the electrical power system industry. Table 5.1 illustrates the effects of electrical shocks related to the magnitude and duration of current on both men and women. An electric current as small as 0.3–0.4 mA flowing through a body can provoke an involuntary reaction and may cause accidents as secondary effects. Shocks with greater currents cause increasingly severe muscular reactions. At certain levels, often called let-go currents, victims are unable to control their muscles and cannot release their grasp on any conductor that they are holding. Let-go currents are 16 mA for men and 10.5 mA for women. Heart fibrillation can occur when the shock currents reach a magnitude of 100 mA and a duration of 0.3 seconds. As noted in Table 5.1, ventricular fibrillation depends on the duration and the type of current (i.e., AC vs. DC). For instance, a 50 kg (i.e., 110 lb) man should be able to withstand 67 mA for 3 seconds before going into heart fibrillation and a 70 kg (i.e., 155 lb) man would be able to withstand about 91 mA. Moreover for AC systems, the let-go current depends on the frequency, as noted in Figure 5.5. A higher frequency leads to reduced let-go currents. There are specific current levels at which human heart fibrillation starts depending mainly on the duration of the fault and the type of AC or DC systems. Figure 5.6 shows the time variation of the fibrillation current thresholds for both DC and AC systems. It is clear that humans can

TABLE 5.1
Effects of Electrical Shock for Various Levels and Durations of Currents

Effect of Electrical Shock	Current Threshold Level for Men (mA)	Current Threshold Level for Women (mA)
No sensation in the hands	0.4	0.3
Slight tingling	1.1	0.7
Shock but not painful (no muscle control is lost)	1.8	1.2
Painful shock (no muscle control is lost)	9.0	6.0
Let-go threshold (painful and muscle control is lost)	16.0	10.5
Severe shock (painful, no muscle control, and difficulty in breathing)	23.0	15.0
Possible ventricular fibrillation	1,000	1,000
From short shocks (0.3 seconds)	100	100
Deadly ventricular fibrillation	2,750	2,750
From short shocks (0.3 seconds)	275	275

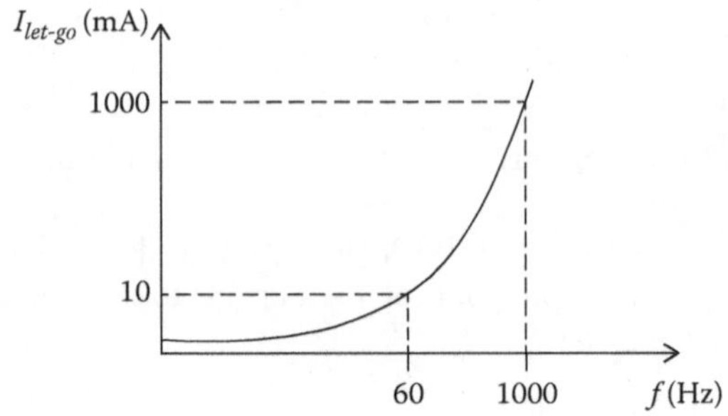

FIGURE 5.5 Let-go current as a function of frequency.

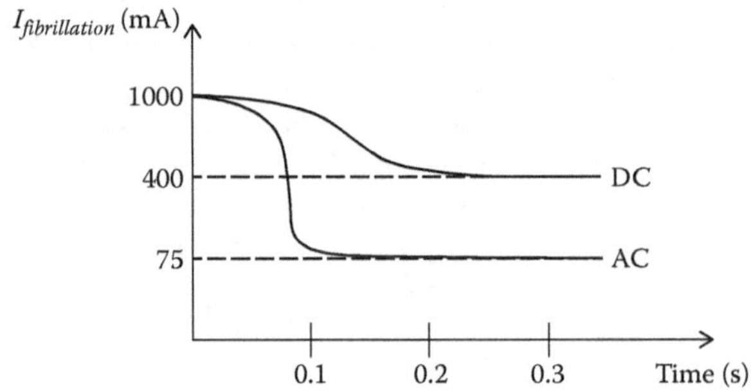

FIGURE 5.6 Time variation of fibrillation current thresholds.

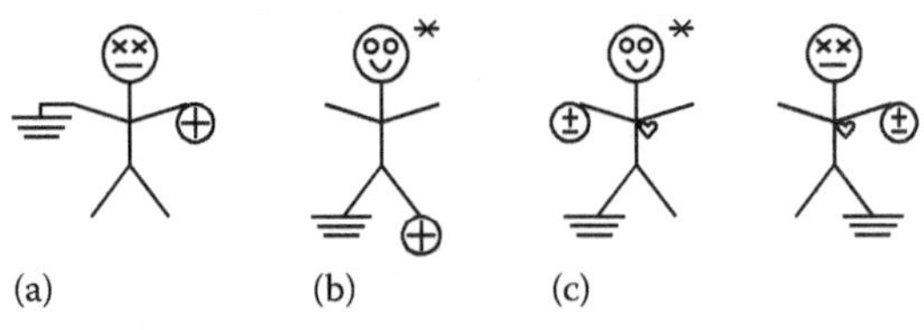

FIGURE 5.7 Configuration for fault paths within a human body: (a) touch potential, (b) step potential, and (c) touch-step potential.

tolerate more DC than AC for a given fault duration. However, this tolerance level is reduced when the fault lasts longer.

5.2.2 Paths of Electrical Shock

In addition to the magnitude and the duration, the impact of fault current on a human body depends significantly on the path of the current relative to critical organs, especially the heart. Indeed, if the current flows through the heart, then potential fibrillation can occur, as noted earlier and illustrated in Figures 5.5 and 5.6, resulting in potential cardiac arrest and ultimately death. The path of fault currents can be divided into three categories, as explained in Figure 5.7:

- *Touch potential*: When the fault current flows from one hand to the other passing through the heart. This touch option may lead to death.
- *Step potential*: When the fault current flows from one foot to another. In this case, injuries to internal organs may occur.
- *Touch-step potential*: When the fault current flows from a foot to a hand or vice versa. Depending on which foot or hand touches the fault source, the current can flow through the heat and may result in death.

5.2.3 Blast Safe Distances

Electrical blasts can generate high temperatures (up to 8,300°C or 15,000°F) and high pressures. Safety distances should be considered when designing electrical rooms that house transformers and unit substations as discussed in Chapter 8. The fatal distance, d_F, has been estimated to be correlated to the total power rating of the transformer, MVA_{rating}, and the duration of the short-circuit fault, t_{SC} (closely related to the response time of the protection device):

$$\begin{aligned} d_F &= \sqrt{39 \times MVA_{\text{rating}} \times t_{sc}}\,(ft) \\ d_F &= \sqrt{3.62 \times MVA_{\text{rating}} \times t_{sc}}\,(m) \end{aligned} \tag{5.2}$$

Similarly, the distance, d_C, at which curable injuries may occur due to the blast is estimated to be

$$\begin{aligned} d_C &= \sqrt{53 \times MVA_{\text{rating}} \times t_{sc}}\,(ft) \\ d_C &= \sqrt{4.92 \times MVA_{\text{rating}} \times t_{sc}}\,(m) \end{aligned} \tag{5.3}$$

Figure 5.8 provides the definition of the fatal and curable distances relative to the location of a blast. Example 5.1 illustrates the application of the correlations defined by Equations 5.2 and 5.3 to estimate the fatal and curable distances.

Example 5.1

Problem

Determine both the fatal and curable distances from a blast source consisting of a transformer rated at 10 MVA and protected by a device with a response time of six cycles.

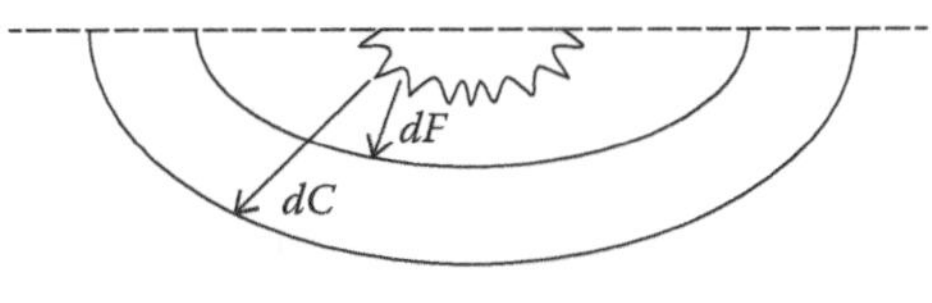

FIGURE 5.8 Fatal and curable distances from a blast source location.

Solution

Using Equations 5.2 and 5.3, the fatal and curable distances can be estimated using the provided values for MVA and t_{sc}:

$$\text{kVA}_{\text{rating}} = 10\ \text{MVA}, \text{and } t_{sc} = 6 \text{ cycles} = 0.1 \text{ s}$$

$$d_F = \sqrt{39 \times 10 \times 0.1} = 6.2 \text{ ft} = 1.90 \text{ m.}$$

$$d_F = \sqrt{53 \times 10 \times 0.1} = 7.3 \text{ ft} = 2.21 \text{ m.}$$

5.3 BASIC OPERATION OF PROTECTION DEVICES

When a fault starts, the current becomes asymmetrical and slowly stabilizes to a symmetrical current (i.e., the current follows a smooth sinusoidal function with time). Indeed, at the start of a fault, it can be shown that the magnitude and the time variation of the asymmetrical current depend on the characteristics of the wiring system (*R* and *L*) and the source voltage. To illustrate the difference between asymmetrical and symmetrical currents, a simple model for a fault is considered, as shown in Figure 5.9, with a source of voltage that varies as a sine function of time as follows:

$$e(t) = E_m \sin(\omega t - \alpha) \tag{5.4}$$

The source voltage serves both the resistance and the inductance and thus using Kirchhoff's law for voltage:

$$e(t) = Ri + L\frac{di}{dt} \tag{5.5}$$

The solution for Equation 5.5 considering the source voltage variation of Equation 5.4 can be shown to be

$$i(t) = \frac{E_m}{\| Z \|}\left\{\sin(\omega t + \alpha - \theta) + \sin(\theta - \alpha)e^{-t/\tau}\right\} \tag{5.6}$$

where $\| Z \| = \sqrt{R^2 + X^2}$, $X = 2\pi f_L$, $\theta = \text{tg}-1(X/R)$, and $\tau = L/R$.

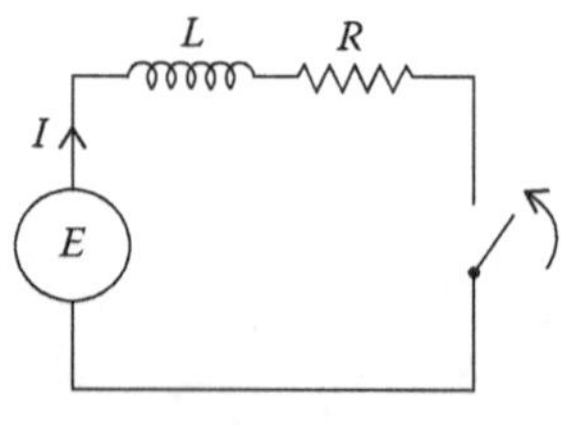

FIGURE 5.9 Simplified model for a fault using *RL* circuit.

Equation 5.6 can be written as a function of the rms value of the current defined as:

$$I_{ac,rms} = \frac{E_m}{\left(\sqrt{2}\,\|Z\|\right)}$$
$$i(t) = \sqrt{2} I_{ac,rms} \left\{ \sin(\omega t + \alpha - \theta) + \sin(\theta - \alpha) e^{\frac{-t}{\tau}} \right\} \tag{5.7}$$

Thus, the current $i(t)$ is made up of two terms: an alternative current, i_{ac}, and a direct current, i_{dc}:

$$i(t) = i_{dc}(t) + i_{ac}(t) \tag{5.8}$$

The AC current that follows the typical sinusoidal variation over time is referred to as the symmetrical fault current:

$$i_{ac}(t) = I_{ac,rms}\sqrt{2} \times \sin(\omega t + \alpha - \theta) \tag{5.9}$$

The DC current follows a decaying exponential function over time:

$$I_{dc}(t) = \sqrt{2} I_{ac,rms} \sin(\alpha - \theta) e^{\frac{-t}{\tau}} \tag{5.10}$$

The sum of the two currents is often called asymmetrical fault current and is a combination of a sinusoidal function (AC part) and a decaying exponential function (DC part), as indicated in Figure 5.10. It should be noted that the asymmetrical current occurs only at the start of the fault, since after a few cycles it converges to the symmetrical current. To estimate the magnitude of the symmetrical current, a quadratic sum of both AC and DC parts is considered:

$$I_{asym} = \sqrt{i_{dc}^2 + i_{ac}^2} = I_{ac,rms}\sqrt{1 + 2\sin^2(\alpha - \theta) e^{\frac{-2t}{\tau}}} \tag{5.11}$$

Thus, the magnitude of the asymmetrical current is proportional to the magnitude of the symmetrical current (represented by $I_{ac,\,rms}$):

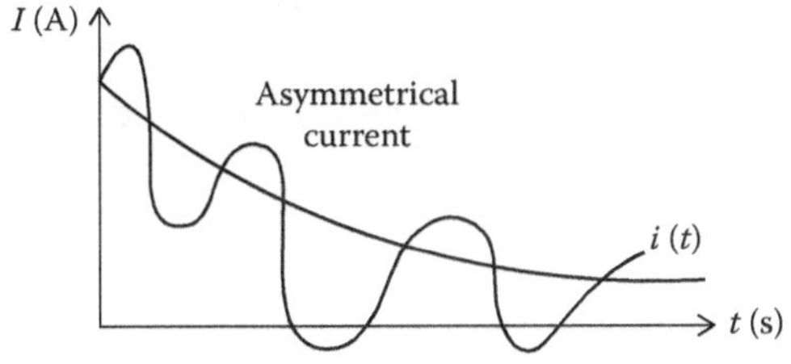

FIGURE 5.10 Asymmetrical fault current variation around its exponentially decaying term.

$$I_{sc} = K(t) \times I_{rms} \tag{5.12}$$

with the constant of proportionality $K(t)$ depending on the time response of the protection device, t:

$$K(t) = \sqrt{1 + 2\sin^2(\alpha - \theta) e^{\frac{-2t}{\tau}}} \tag{5.13}$$

Two limiting cases can be considered:

- Worst case: $t=0$ and $\sin(\alpha-\theta)=1$; in this case, $K = \sqrt{3} = 1.732$.
- Best case: $\sin(\alpha-\theta)=0$ which can occur when the voltage and the current are in phase at the start of the fault; in this case, $K(t)=1$ independently of the response time t.

Thus, in all cases and for all protection devices, the constant K range between 1 and 1.732:

$$1 \le K \le 1.732 \tag{5.14}$$

5.4 TYPES OF PROTECTION DEVICES

The protective devices are generally rated using the following performance characteristics:

- Maximum continuous voltage that can be applied to the electrical system without causing the conductor insulation to fail
- Maximum continuous current that can flow in the electrical system without resulting in overheating
- Rated interrupting current defined as the maximum current up to which the protective device can safely operate to disconnect the electrical system
- Short-time ratings including the momentary current (the maximum current that the protective device can withstand without failure) and the specified time current (the current that the protective device can withstand for a specified time—typically 0.5 seconds—without failure)

In general, the size of the protective device should be less than the ampacity of the conductor being protected. The ampacity is the current that can flow safely in a conductor. Chapter 6 will discuss in more detail the ampacities of conductors.

5.4.1 Fuses

The NEC defines the fuse as an "over-current protective device with a circuit opening fusible part that is heated and severed by the passage of current through it" (NEC, 2014). Currently, there are several types of fuses suitable for various applications.

The basic construction of all fuses remained the same over the years. For current-limiting fuses, the fusible element is packed in quartz filler and hermetically sealed inside a ceramic case. For motor applications with high starting currents (refer to Chapter 4), dual-element time-delay fuses are recommended to prevent tripping the protective device each time the motor is operated.

Figure 5.11 shows the basic construction details of a current-limiting fuse with its fusible element made of a metal and its filler. The filler and the fusible element are encased in a hermetically sealed case. The fusible element has thin sections to facilitate melting when high current flows while the sand filler helps absorb heat resulting from arcing due to the melting of the fusible element. A metal for a fast acting fusible element has to be selected so that it has a high electrical conductivity, an abrupt melting point, and a small mass to reduce the heating time. Typically, zinc, copper, silver, and aluminum or their alloys have been used for fuses. Air, sand, or quartz can be used as a filler to help extinguish electrical arcs. Ceramic, glass, fiberglass, molded mica laminates are used for the casing. Figure 5.12 illustrates four major steps in the operation of a typical fuse when fault occurs. First, the fusible element starts to melt until arcing occurs. At the start of the arcing process, the fault current reaches its peak and a significant heat is generated between the two arcing and melting ends of the fusible element. Due to this heat, the sand around the arcing melts and forms an insulating layer around the two ends of the fusible element completely preventing any flow of current. Figure 5.13 shows the time variation of the current flowing through the protection circuit and the heat generated during the operation of the fuse. The maximum current flowing through the fuse, which occurs during step 3 when arcing phase starts is often called peak let-through current. The heat generated during the operation of the fuse is estimated by the expression I^2t and is represented by the area below the current curve in Figure 5.13. This maximum energy is typically called the let-through energy and is used to rate fuses in terms of the level of protection they can provide to electrical systems. The let-through energy ratings are used to coordinate the selection of load-side and line-side fuses, as will be discussed later

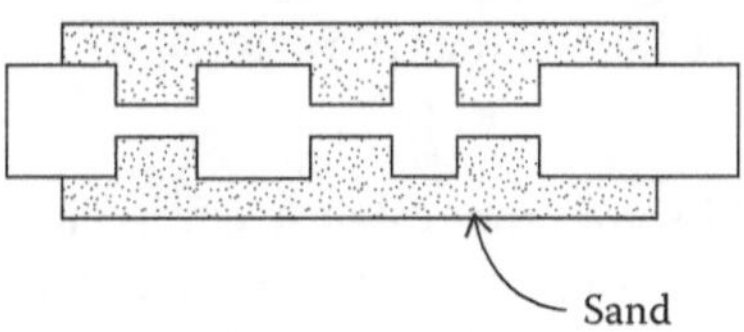

FIGURE 5.11 Construction details of a typical fuse.

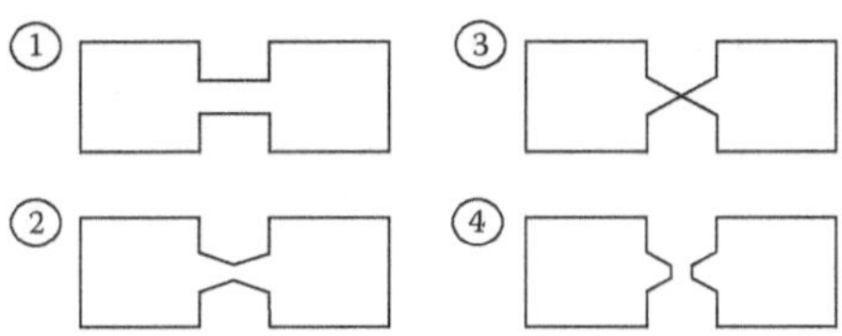

FIGURE 5.12 Main phases for the operation of a fuse when fault current occurs.

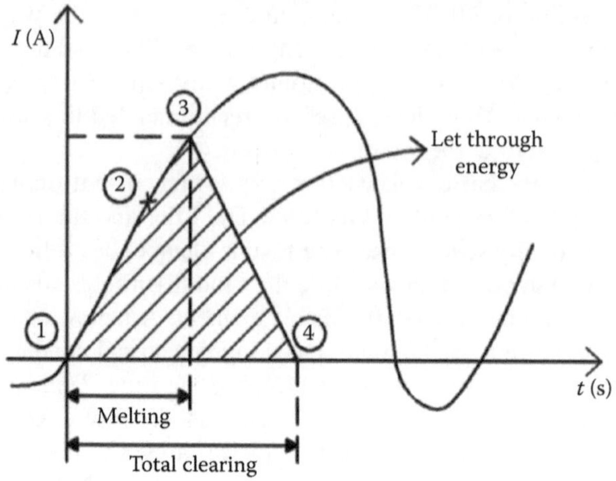

FIGURE 5.13 Current levels at various phases of melting and clearing of a fuse.

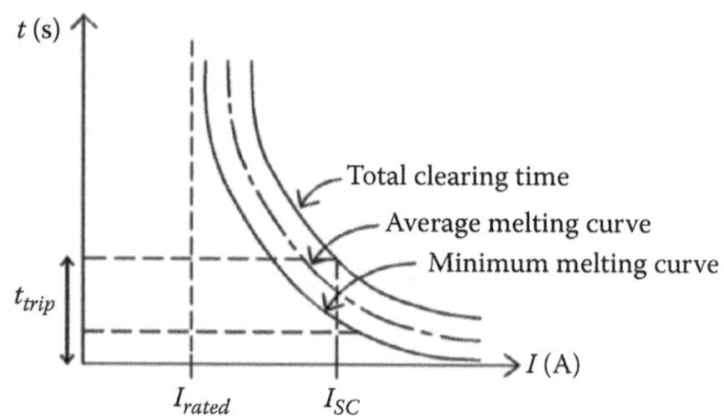

FIGURE 5.14 Characteristic time–current response curves for a typical fuse.

in this section. Figure 5.14 shows the response curve of a typical fuse to determine the time required to melt the fuse and to totally clear the flow of any fault current. Manufacturers typically provide the response time–current curves for their fuses.

As discussed in Chapter 4, induction motors require high in-rush currents for the first few cycles after the start of their operation before their normal rated currents are reached. Thus, slow acting fuses with a time-delay response are needed to protect induction motors, as illustrated in Figure 5.15. However, the use of only one fusible element with slow response to protect against overloads may not be effective to protect against short-circuit currents. Instead, dual-element fuses provide dual protection against both overloads and short-circuit currents. The construction of a typical dual-element fuse is shown in Figure 5.16. In particular, the two elements of the fuse are connected in series within the cartridge. Figure 5.17 shows examples of commercially available low-peak dual-element fuses while Figure 5.18 presents the response curves for the same fuses considering various amperage ratings.

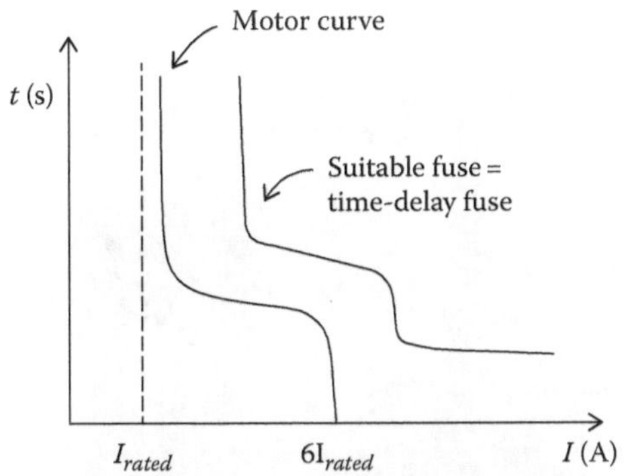

FIGURE 5.15 Response curve for a time-delay fuse suitable for motor protection.

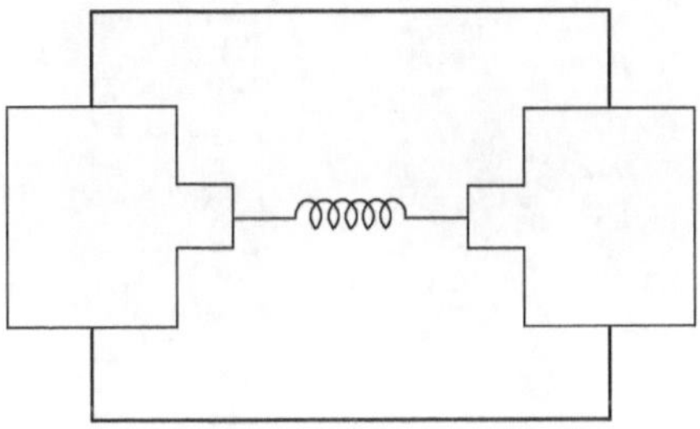

FIGURE 5.16 Construction of a typical dual-element fuse.

Various types, classes, and sizes of fuses depend on their applications and rated voltages. For building applications, typically cartridge fuses are utilized at rated voltages of 250 or 600 V. The main types of fuses include

- Non-time-delay fuses with single type of fusible elements, often called short-circuit elements
- Time-delay fuses constructed with two different types of fusible elements to protect against both overloads and short-circuits
- Current-limiting fuses are fast acting and are able to open the circuit and remove the short-circuit current well before it reaches the peak value
- Medium-voltage fuses including distribution fuse cutouts and power fuses
- Medium-voltage current-limiting power fuses including E-rated, C-rated, and R-rated fuses
- The common classes of low-voltage fuses used in building electrical distribution systems include the following:
- *Class H fuses*: These fuses can be renewable and nonrenewable and have a limited interrupting current rating of only 10,000 A. They include typically low-voltage cartridge fuses and are rated up to 600 A.
- *Class J fuses*: These fuses are nonrenewable and can have interrupting symmetrical current ratings of up to 300,000 A. They are available as non-time-delay fuses and are rated up to 600 A.

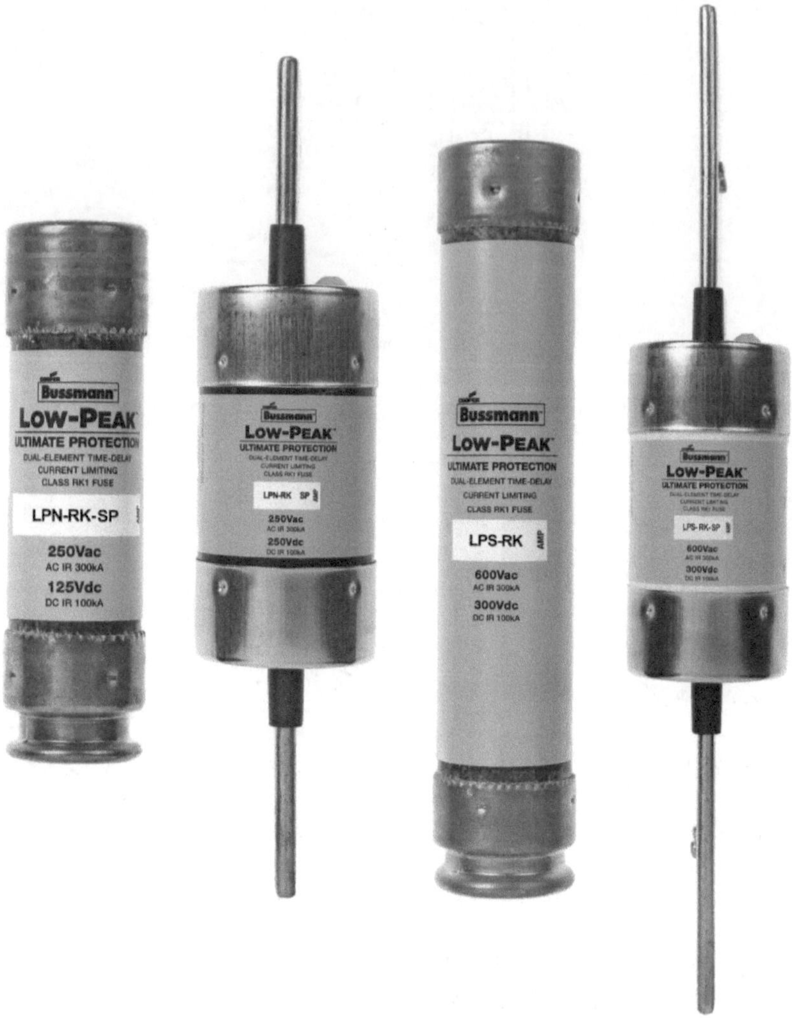

FIGURE 5.17 Low-peak dual-element fuses. (Courtesy of Eaton, Bussmann division, 2016.)

- *Class R fuses*: These fuses are nonrenewable and have interrupting symmetrical current ratings of 200,000 A. They are available in RK-1 and RK-5 categories and are rated up to 600 A. The RK-1 fuses have low let-through peak currents while RK-5 fuses have high peak let-through currents. They can be either non-time-delay or time-delay fuses.
- *Class T fuses*: These fuses are nonrenewable and are current limiting with interrupting symmetrical current ratings of 200,000 A. They are typically non-time-delay fuses and are rated up to 600 A.

Table 5.2 shows the cartridge sizes and standard trip ratings for the fuses and their maximum clearing time ratings.

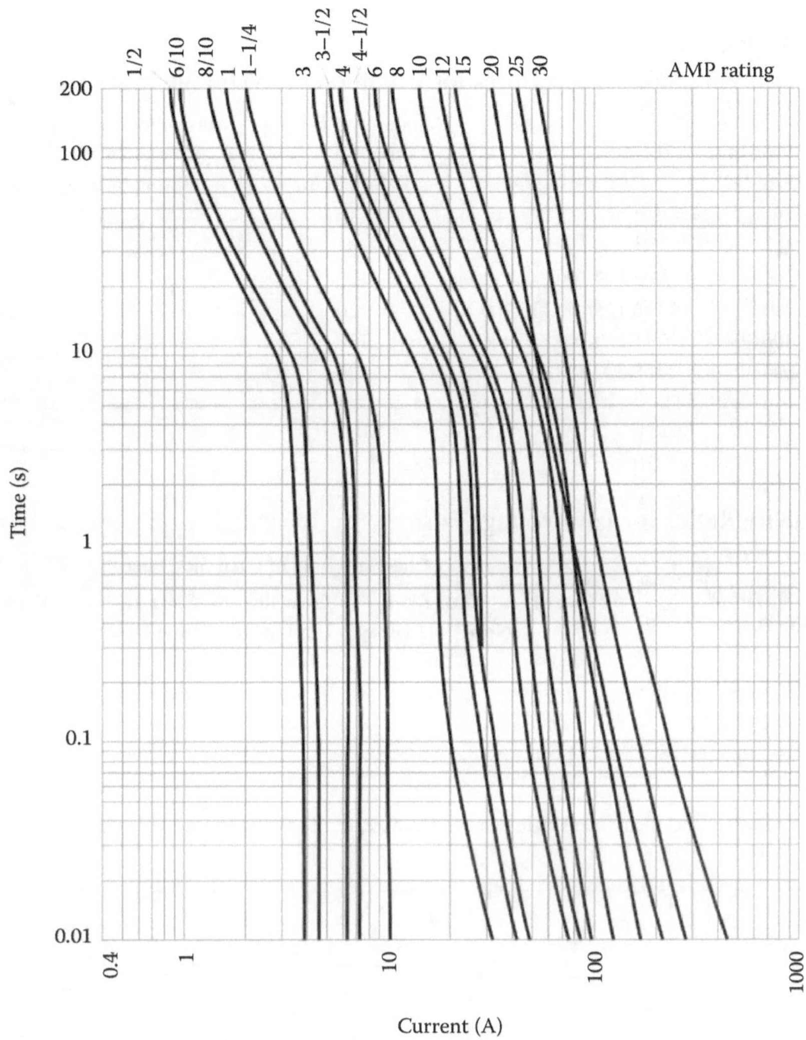

FIGURE 5.18 Average melting time–current characteristic curves for low-peak dual-element fuses. (Courtesy of Eaton, Bussmann division, 2016.)

To properly protect electrical distribution systems, a good coordination between the load-side and the line-side fuses is required. Typically, the load-side fuse has to react first to any abnormal conditions and thus should have lower let-thru energy. A recommended selectivity ratio, defined as the ratio between the size of the line-side fuse and the size of the load-side fuse, is provided in Table 5.3, depending on the types and the classes of the two fuses to be coordinated. Figure 5.19 illustrates the definition of load-side and line-side fuses that need to be coordinated. Example 5.2 shows the selection process of the line-side fuse to coordinate with a load-side fuse.

TABLE 5.2
Ratings and Clearing Times for Common Low-Voltage Fuses

Cartridge Size (A)	Fuse Rating (A)	Maximum Clearing Time at 135% Rating (min)	Maximum Clearing Time at 200% Rating (min)
30	152,030	60	2
60	405,060	60	4
100	708,090,100	120	6
200	110,125,150,175,200	120	8
400	225,250,300,350,400	120	10
600	450,500,600	120	12

TABLE 5.3
Selectivity Ratios for Low-Voltage Fuses[a]

Class and Type of Line-Side Fuse	Class and Type of Load-Side Fuse				
	Class J	Class RK-1 Non-Time Delay	Class RK-1 Time Delay	Class RK-5 Time Delay	Class T
Class J	3:1	3:1	3:1	8:1	3:1
Class RK-1 non-time delay	3:1	3:1	3:1	8:1	3:1
Class RK-1 time delay	3:1	3:1	2:1	8:1	3:1
Class RK-5 time delay	1.5:1	1.5:1	1.5:1	2:1	1.5:1
Class T	3:1	3:1	3:1	8:1	3:1

[a] These ratios are specific to one single manufacturer's fuses. These ratios may be different with other manufacturers.

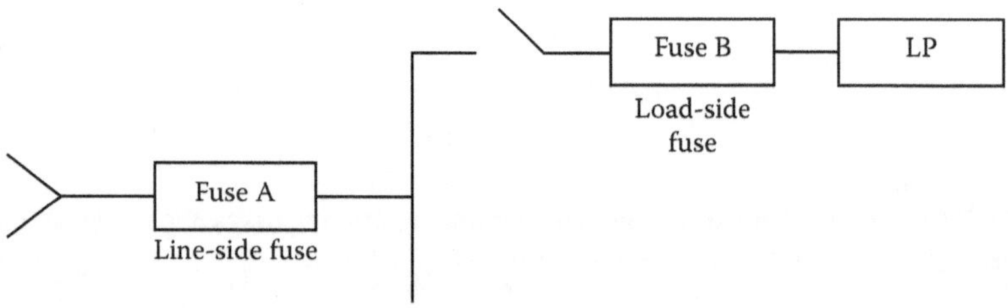

FIGURE 5.19 Coordination of line-side and load-side fuses.

Example 5.2

Problem

Determine the size of the line-side RK-5 t+ime-delay fuse if the load-side fuse type is RK-1 non-time delay and is rated at (1) 100 A, (2) 150 A, and (3) 175 A.

Solution

From Table 5.3, the selectivity ratio for RK-1 non-time-delay load-side fuse (i.e., fuse B in Figure 5.19) and RK-5 time-delay line-side fuse (i.e., fuse A in Figure 5.19) is 1.5:1. Thus, the size of the line-side fuse is determined for the three cases as follows:

For size of fuse B of 100 A, the size of fuse A is $1.5 \times 100 = 150$ A, which is a standard size as noted in Table 5.2.

For size of fuse B of 150 A, the size of fuse A is $1.5 \times 150 = 225$ A, which is again a standard size as noted in Table 5.2.

Finally and for size of fuse B of 175 A, the size of fuse A is $1.5 \times 175 = 262.5$ A, which is not a standard size as indicated in Table 5.2. Instead, a fuse A with a standard size of 300 A should be specified.

The main advantages of fuses compared to other types of protective devices (i.e., circuit breakers) are summarized here:

- Have low initial cost
- Require little maintenance since fuses are simple to construct
- Are generally compact and require little space to be installed
- Have high current interrupting capabilities
- Are inherently fail-safe devices since when fuses fail, they automatically open the circuit

However, fuses also present several disadvantages including

- Can cause single phasing for three-phase systems
- Are not flexible since the time response of the fuses is fixed and is not adjustable
- Must be replaced after each operation.

5.4.2 Circuit Breakers

The NEC defines the circuit breaker as "a device designed to open and close a circuit by non-automatic means and to open the circuit automatically on a predetermined matter without damage to itself when properly applied within its rating" (NEC, 2014). While fuses operate on the mechanism of detect and act, circuit breakers operate by detecting and actuating when faults occur. Figure 5.20 shows the basic operation and components of a typical circuit breaker. Specifically, there are four main components: a moving contact part, a fixed contact part, a sensor, and an actuator with a processing unit. When fault occurs, the sensor detects the current flow and sends a signal to the processing unit. Based on its setting, the processing unit decides to stop the flow of the fault current and send a signal through the actuator to displace the moving contact part from the fixed part to open the circuit. As will be detailed later on in this section, the mechanism of opening the contacts may involve solid-state, thermal, magnetic, or thermal-magnetic trip units.

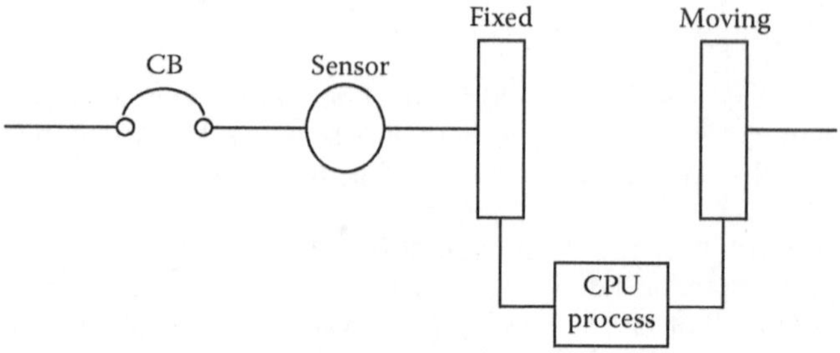

FIGURE 5.20 Operation and components of a typical circuit breaker.

Circuit breakers are available for various voltages and continuous current ratings as well as interrupting current ratings, response characteristics, and methods of operation. For instance, molded-case breakers are compact and are relatively inexpensive, but they have generally low interrupting ratings and thus cannot be applied to large systems. Instead, power circuit breaker should be used for feeders with high current flows. Electronic solid-state trip units are currently commonly used especially for power circuit breakers. Indeed, solid-state trip units provide more flexibility and accuracy than the mechanical dual-magnetic types.

It should be noted that circuit breakers can be installed either in single pole or multipole. Multipole breakers are generally gang operated so that all the poles are closed and opened simultaneously by one common operating mechanism (such as a handle). Therefore, circuit breakers cannot cause single phasing in three-phase systems as can be the case when using fuses as protection devices.

Compared to fuses, circuit breakers have the following inherent advantages:

- Can serve as means of both protecting and switching an electrical circuit
- Does not cause single phasing
- Can be remotely operated
- Can easily incorporate ground-fault protection

However, circuit breakers have some disadvantages compared to fuses. In particular, circuit breakers

- Have higher initial cost.
- Require more space since they are larger.
- Require more maintenance because of their complexity in construction and operation.
- Do not limit the magnitude of fault currents. Thus, an electrical system is subject to higher thermal and magnetic stresses under fault conditions when it is protected by a circuit breaker rather than a fuse.
- Are not fail-safe devices, since their trip mechanism can be damaged and they can be left in a closed rather than open position.

There are several types of circuit breakers with a range of response characteristics. Figure 5.21 compares the time–current response curves for a fuse and a molded-case

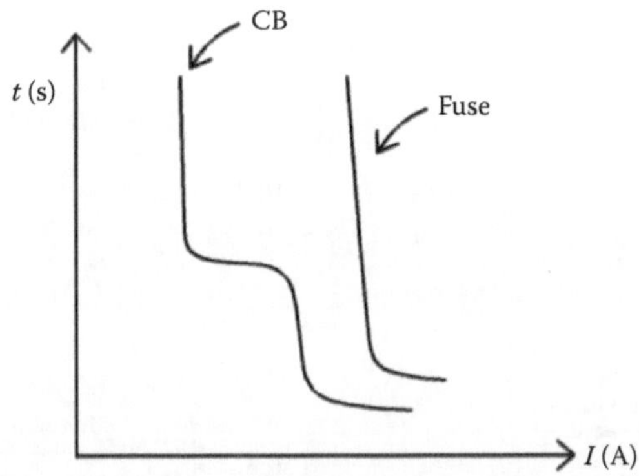

FIGURE 5.21 Comparison of time–current response curves for a fuse and a circuit breaker.

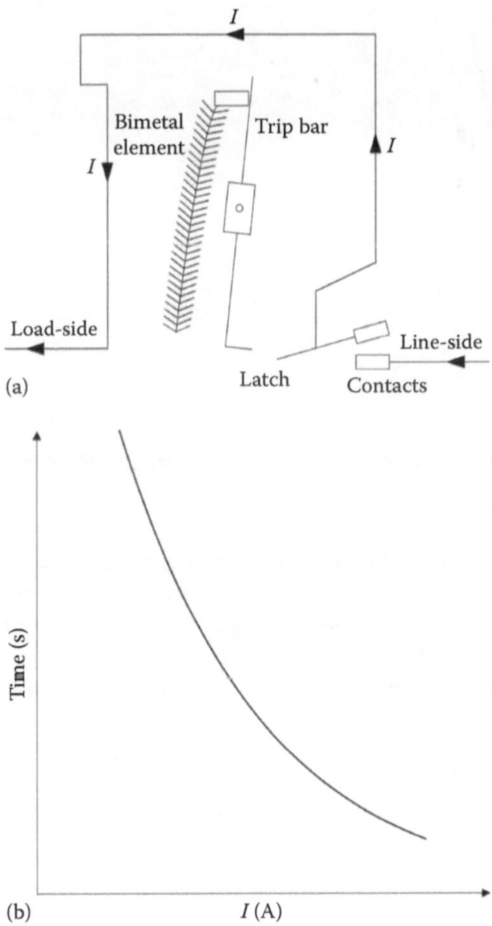

FIGURE 5.22 (a) Operation of circuit breaker thermal trip unit and (b) associated time–current response curve.

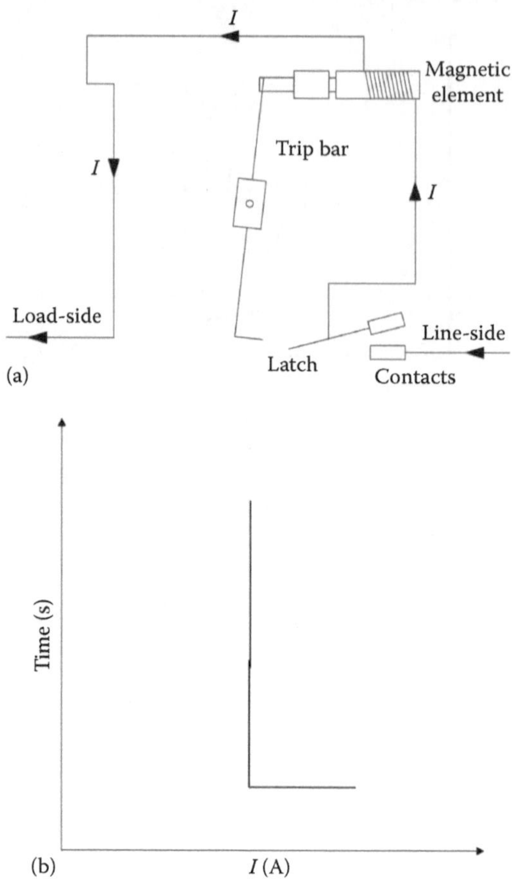

FIGURE 5.23 (a) Operation of circuit breaker magnetic trip unit and (b) associated time–current response curve.

circuit breaker. Typical molded-case circuit breakers operate using thermal-magnetic trip units. The thermal trip action and associated time–current response curve are illustrated in Figure 5.22, while the basic operation and time–current response curve for a magnetic trip unit are shown in Figure 5.23. Finally, the combined thermal-magnetic trip action and the correspondent time–current response curve are illustrated in Figure 5.24.

As shown in Figure 5.22a, for a thermal trip unit, a bimetal element made up of two types of metals with different rates of thermal expansion is placed along the circuit carrying the load-side current. When an overload occurs, heat generated by the high current causes the bimetal element to expand and ultimately to bend and thus to trip a latch that opens the contacts, stopping the current from passing through the circuit. As noted in Figure 5.22b, the response of the thermal trip unit follows a delayed inverse time–current curve.

On the other hand, a magnetic trip unit utilizes a magnetic element and an armature connected to a trip bar, as shown in Figure 5.23a. When short-circuit current flows through the circuit, a sufficient force is generated by the magnetic element to

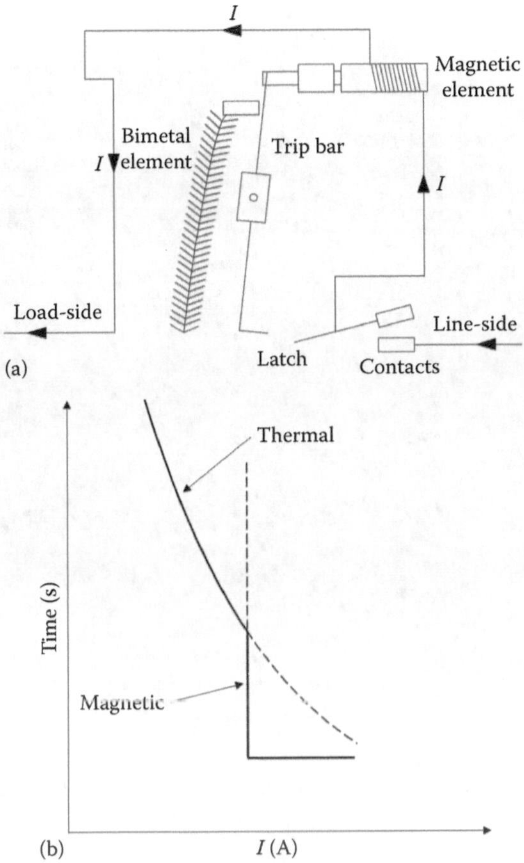

FIGURE 5.24 (a) Operation of circuit breaker–combined thermal-magnetic trip unit and (b) associated time–current response curve.

instantaneously attract the armature and trip the latch that opens the contacts. The magnetic trip unit has instantaneous time–current response, as indicated in Figure 5.23b.

The operation of a combined thermal-magnetic trip unit is illustrated in Figure 5.24a and uses both the thermal trip action for overload currents and magnetic trip action for short-circuit currents. The typical time–current response curve for the thermal-magnetic trip unit is presented in Figure 5.24b.

Currently, several circuit breakers have solid-state trip units and their time–current response curve characteristics can be adjusted depending on the applications. Figure 5.25 shows an example of a commercially available solid-state power circuit breaker.

Circuit breakers have standard trip ratings and operate safely under specific maximum interrupting current ratings, typically related to the short-circuit currents available in the electrical distribution systems. Table 5.4 provides standard frame sizes, trip ratings, and interrupting symmetrical current ratings for low-voltage molded-case circuit breakers. These types of circuit breakers are used to protect nonmotor loads including lighting, appliances, and power systems. For most lighting and power panels used in buildings, molded-case circuit breakers are utilized.

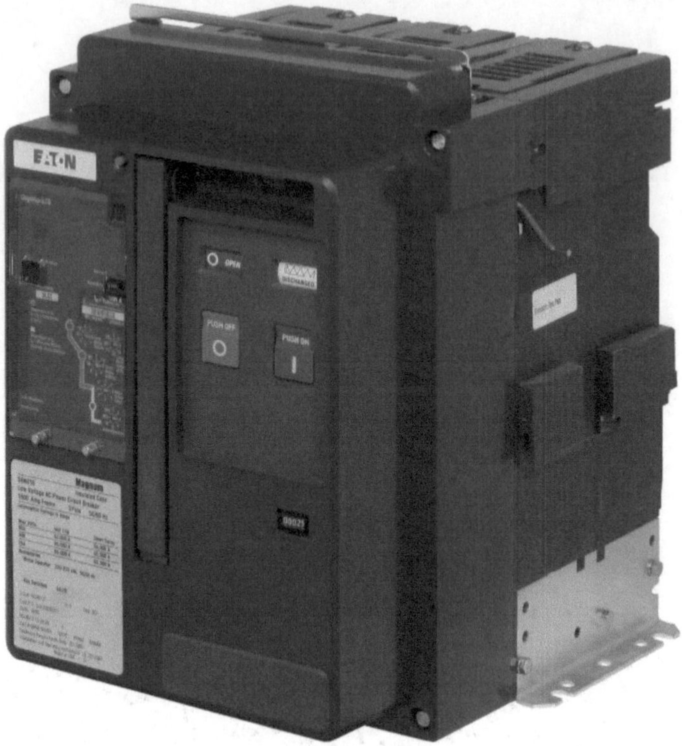

FIGURE 5.25 A solid-state power circuit breaker. (Courtesy of Eaton, Bussmann division, 2016.)

TABLE 5.4
Frame Sizes and Typical Ratings for Low-Voltage Molded-Case Circuit Breakers

Frame Size (A)	Trip Ratings[a] (A)	Rated Voltage (V)	Interrupting Symmetrical Current Ratings (A)		
			240 V	480 V	600 V
100 Std 100 HIC	15–100	240	1,000,065,000	NANA	NANA
100 Std	15–100	480	18,000	14,000	NA
150 Std 150 HIC	15–150	600	1,800,065,000	1,400,025,000	1,400,018,000
225 Std 225 HIC	70–225	600	2,500,065,000	2,200,025,000	1,800,022,000
400 Std 400 HIC	125–400	600	4,200,065,000	3,000,035,000	2,200,025,000
600 Std 600 HIC	250–600	600	4,200,065,000	3,000,035,000	2,200,025,000
800 Std 800 HIC	450–800	600	4,200,065,000	3,000,035,000	2,200,025,000
1,200 Std 1,200 HIC	600–1,200	600	4,200,065,000	3,000,035,000	2,200,025,000

[a] Standard trip ratings include: 15, 20, 25, 30, 35, 40, 50, 60, 70, 80, 90, 100, 125, 150, 175, 200, 225, 250, 300, 350, 400, 450, 500, 600, 800, 1,000, and 1,200 A.

Example 5.3 illustrates the selection approach of the trip rating for a molded-case circuit breaker used to protect a continuous load in a dwelling. Additional examples for sizing circuit breakers for various loads in dwellings and commercial buildings are provided respectively in Chapters 7 and 8.

Example 5.3

Problem

Determine the size and the number of poles of a molded-case circuit breaker to protect a 240 V water heater drawing 5,000 VA power. Assume the water heater is continuous.

Solution

First, the connected load, I_{CL}, expressed in amperes for the water heater is determined:

$$I_{CL} = \frac{5{,}000 \text{ VA}}{240 \text{ V}} = 20.8 \text{ A}$$

Since the load is continuous (i.e., can operate continuously for 3 hours or more), the demand load, I_{DL}, can be obtained by multiplying the actual load by 1.25:

$$I_{DL} = 1.25 \times 20.8 \text{ A} = 26.0 \text{ A}$$

From Table 5.4, a trip rating of 30 A molded-case circuit breaker can be selected. A frame size of 100 A would be needed with a two-pole throw. Thus, a 30 A/2P circuit breaker can be selected.

Table 5.5 provides typical frame size, trip ratings, and interrupting symmetrical current ratings for low-voltage power circuit breakers. These types of circuit breakers are utilized to protect feeders serving panels including lighting and power panels. Example 5.4 illustrates the rating selection of a power circuit breaker for an office building lighting panel.

Example 5.4

Problem

Determine the size and the number of poles of a power circuit breaker to protect a 480Y/277 V lighting panel for an office building. The total panel load is 350 kW with a power factor of 0.90. The instantaneous short-circuit current at the panel is estimated to be 55,000 A.

Solution

First, the connected load, I_{CL}, expressed in amperes for the lighting panel is determined:

TABLE 5.5
Frame Sizes and Trip Ratings for Low-Voltage Power Circuit Breakers

Frame Size (A)	Trip Ratings[a] (A)	Instantaneous Trip Interrupting Symmetrical Current Ratings (A)			Short-Time (0.5 seconds Delay) Trip Interrupting Symmetrical Ratings (A)		
		208/240 V	480 V	600 V	240 V	480 V	600 V
600	50–600	42,000	30,000	30,000	30,000	30,000	30,000
800	50–800	42,000	30,000	30,000	30,000	30,000	30,000
1,600	50–1,600	65,000	50,000	42,000	50,000	50,000	42,000
2,000	50–2,000	65,000	50,000	50,000	50,000	50,000	50,000
3,000	1,200–3,000	85,000	65,000	65,000	65,000	65,000	65,000
4,000	1,600–400	130,000	85,000	85,000	85,000	85,000	85,000

[a] Standard trip ratings include: 50, 100, 150, 200, 225, 300, 400, 600, 800, 1,200, 1,600, 2,000, 2,400, 3,000 and 4,000 A.

$$I_{CL} = \frac{350,000 \text{ W}}{1.732 \times (480 \text{ V})} = 421.0 \text{ A}$$

Since the load is continuous for lighting in commercial buildings, the demand load, I_{DL}, can be obtained by multiplying the actual load by 1.25:

$$I_{DL} = 1.25 \times 421 \text{ A} = 526.2 \text{ A}$$

From Table 5.5, a trip rating of 600 A power circuit breaker can be selected. Since the instantaneous short-circuit current is 55,000 A, a frame size of 2,000 A would be needed with a three-pole throw. Indeed, a frame size of 600 A, for instance, can handle only 30,000 A of instantaneous short-circuit current. Thus, 600 AT/2,000 AF power circuit breaker can be selected.

5.5 GROUNDING AND BONDING

It is important to properly ground electrical systems to minimize fire hazards, electrical shocks and injuries to occupants, and damages to equipment. NEC requires that all derived electrical systems be grounded. There are different types of grounding systems including service grounding and equipment grounding. Moreover, bonding of the metallic parts of an electrical distribution system is required to ensure that there is a continuous path to conduct safely any fault current. Good understanding of grounding and related NEC requirements is important to properly protect electrical distribution systems in buildings. Article 250 of the NEC covers all the requirements associated to both grounding and bonding (NEC, 2014). First, the definition of common terms used in grounding and bonding is provided. Then, a summary is outlined for the NEC requirements specific to grounding and bonding. Then, design guidelines for grounding are presented. Finally, the impact of ground faults on electrical system components is discussed.

5.5.1 Definitions

- *Bonding (Bonded)*: Connection to establish electrical continuity and conductivity.
- *Bonding jumper*: A conductor that connects the metallic parts of an electrical system.
- *Ground*: Earth used as a medium to reject any fault current flowing in the electrical distribution system.
- *Grounded (Grounding)*: Connected to ground or to a conducting body that extends the ground connection.
- *Equipment grounding conductor*: The conductive path installed to connect normally non-current-carrying metal parts of equipment together and to the system grounded conductor or to grounding electrode conductor, or both. It is recognized that the equipment grounding conductor also performs bonding.
- *Grounding electrode*: A conducting object through which a direct connection to earth is established.
- *Grounding electrode conductor*: A conductor used to connect the system grounded conductor or the equipment to a grounding electrode or to a point on the grounding electrode system.
- *Grounded conductor*: A system or circuit conductor that is intentionally grounded.
- *Neutral conductor*: The conductor connected to the neutral point of a system that is intended to carry current under normal conditions.
- *Neutral point*: The common point on a wye connection in a polyphase system or midpoint on a single-phase, three-wire system, or midpoint of a single-phase portion of a three-phase delta system, or a midpoint of a three-wire, direct current system.
- *Ungrounded*: Not connected to ground or to a conductive body that extends the ground connection.
- *Ground fault*: An unintentional, electrically conducting connection between a normally current-carrying conductor of an electrical circuit and the normally non-current-carrying conductors, metallic enclosures, metallic raceways, metallic equipment, or earth.

5.5.2 General Requirements for Grounding and Bonding

The following general requirements identify what grounding and bonding of electrical systems are required to accomplish. The prescriptive methods contained in Article NEC 250 shall be followed to comply with the performance requirements (NEC, 2014).

5.5.2.1 Grounded Systems

1. *Electrical system grounding*: According to NEC, electrical systems that are grounded shall be connected to earth in a manner that will limit the voltage imposed by lightning, line surges, or unintentional contact with higher-voltage lines and that will stabilize the voltage to earth during normal operation.

FIGURE 5.26 Main neutral bonding point and main bonding jumper.

2. *Grounding of electrical equipment*: Normally non-current-carrying conductive materials enclosing electrical conductors or equipment, or forming part of such equipment, shall be connected to the earth so as to limit the voltage to ground on these materials.
3. *Bonding of electrical equipment*: Normally non-current-carrying conductive materials enclosing electrical conductors or equipment, or forming part of such equipment, shall be connected together and to the electrical supply source in a manner that establishes an effective ground-fault current path.
4. *Bonding of electrically conductive materials and other equipment*: Normally non-current-carrying electrically conductive materials that are likely to become energized shall be connected together and to the electrical supply source in a manner that establishes an effective ground-fault current path. Figure 5.26 illustrates the application of a bonding jumper system to connect the panel enclosure to the neutral wires.
5. *Effective ground-fault current path*: Electrical equipment and wiring and other electrically conductive material likely to become energized shall be installed in a manner that creates a permanent, low-impedance circuit facilitating the operation of the protection device or ground detector for high-impedance grounded systems. It shall be capable of safely carrying the maximum ground-fault current likely to be imposed on it from any point in the wiring system where a ground fault may occur to the electrical supply source. The earth shall not be considered as an effective ground-fault current path.

5.5.2.2 System Grounding Connections

Any building electrical system served by a grounded AC service shall have a grounding electrode conductor connected to the grounded service conductor, at each service, in accordance with NEC requirements as noted in section 250 of the NEC (2014).

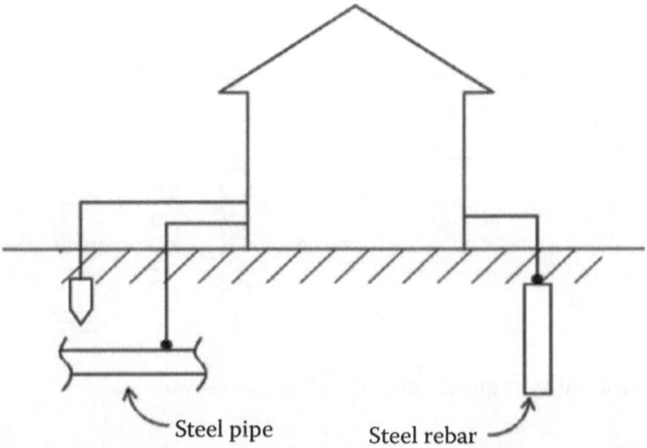

FIGURE 5.27 Common grounding options for building electrical service entrances.

Several options are allowed for grounding building electrical systems. In particular, NEC permits the following grounding electrodes for service grounding:

- Metal underground water pipe
- Metal frame of the building or structure
- Concrete-encased electrode
- Ground ring
- Rod and pipe electrodes
- Other listed electrodes
- Plate electrodes
- Other local metal underground systems or structures

Figure 5.27 illustrates common service grounding options for buildings, including the use of foundation steel rebar, water pipe, or rod electrode.

It should be noted that some connections between the building electrical system and ground are not permitted by NEC as grounding options. Some examples include

- Metal underground gas piping system
- Aluminum electrodes

5.5.2.3 Basic Grounding Guidelines

There are three basic rules for specifying grounding electrodes to protect building electrical systems from fault currents. These rules consider derived electrical systems as those served by generators, transformers, or converters and with no direct connection with other electrical systems. In particular, the number of transformers can affect the number of points where grounding electrodes are specified for a building electrical system.

- *Rule no. 1*: Each derived system has to be grounded, as shown in Figure 5.28. Typically, the secondary windings of each transformer are grounded using service grounding electrodes as indicated by Table 5.5.

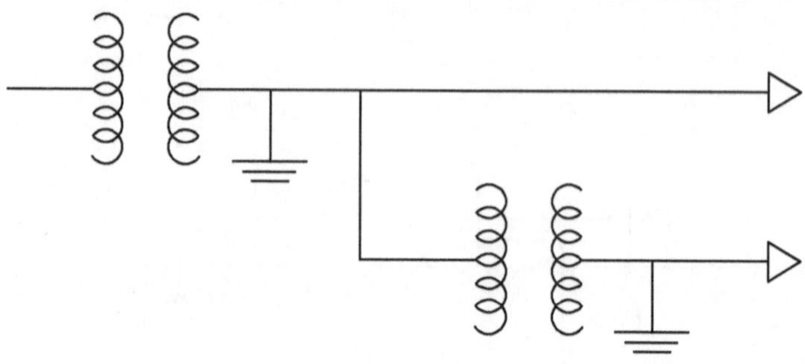

FIGURE 5.28 Grounding requirements for all derived systems.

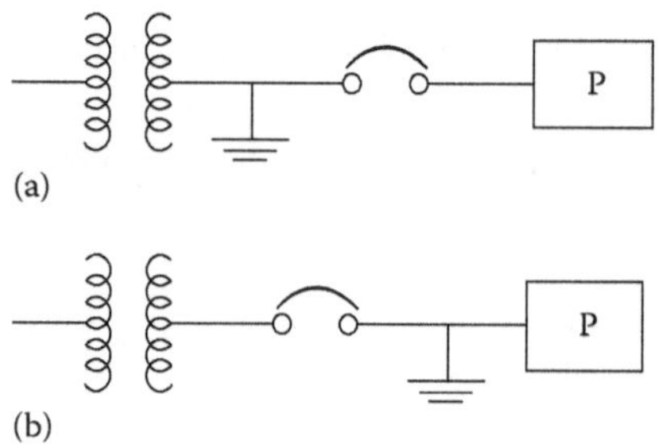

FIGURE 5.29 Grounding before (a) and not after (b) the protection device.

- *Rule no. 2*: Grounding location needs to be as close to the power source of the derived system as possible. In particular, the grounding should be specified before rather than after the protection device. Indeed and when grounding electrode is connected to earth after the protection device, as noted in Figure 5.29, the derived system becomes ungrounded whenever the circuit breaker trips.
- *Rule no. 3*: There should be only one grounding connection per derived system. Indeed, connecting a derived system by two or more locations to the earth may result in return flow of ground-fault current back to the system.

5.5.3 Basic Analysis of Grounding

To facilitate the flow of fault currents into the ground, low resistance path for the grounding conductor is needed. NEC requires that the ground electrode system, connecting the building electrical system to ground, to have an electrical resistance lower than 25 Ω. Several parameters and factors can affect the resistance of the ground electrode system. Figure 5.30 illustrates the part of the earth medium around a grounding rod of a depth, L, when fault or excess currents are rejected to the ground. As indicated in Figure 5.30, the part of earth affected by the rejected electrons from the rod can be approximated by a cylinder of a radius L and a depth of $2L$.

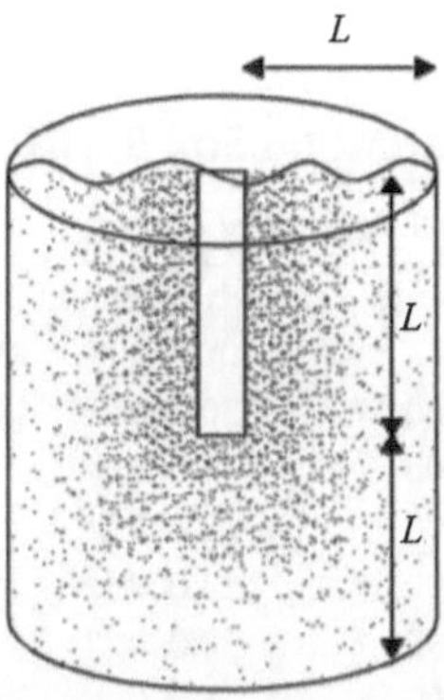

FIGURE 5.30 Part of the earth medium affected by electrons rejected by the ground rod.

TABLE 5.6
Resistivity Range for Selected Materials and Liquids

Soil Material/ Liquid Type	Resistivity Range Ω (m)	Ω (ft)
Clay	200–100,000	65–328,000
Sand	5,000–100,000	16,400–328,000
Limestone	500–400,000	1,640–1,312,000
Shale	500–10,000	1,640–32,800
Sandstone	2,000–200,000	6,560–656,000
Granite	1,000,000	3,280,000
Tap water	1,000–50,000	3,280–164,000
Seawater	20–200	65–656

The total resistance of the cylinder, made up of mostly ground medium, can be estimated as noted in Chapter 2 as

$$R_g = \rho_g \frac{2L}{A} \tag{5.15}$$

where

ρ_g is the ground resistivity expressed in Ω m or Ω ft
A is the cross section of the cylinder of radius, L, so $A = \pi L^2$

Thus, the earth resistance is a function of the ground resistivity as well as the depth of the grounding rod:

$$R_g = \rho_g \frac{2}{\pi L} \tag{5.16}$$

Equation 5.16 indicates that in order to reduce the earth resistance, either the ground resistivity should be lowered or the rod depth should be increased. Table 5.6 lists the

TABLE 5.7
Effect of Moisture Content on Soil Resistivity

Moisture Content (% by Weight)	Ground Resistivity (Ω cm)		Ground Resistivity (Ω ft)	
	Top Soil	Sandy Loam	Top Soil	Sandy Loam
0	10^9	10^9	3.28×10^9	3.28×10^9
2.5	250,000	150,000	820,000	492,000
5	165,000	43,000	541,200	141,040
10	53,000	22,000	173,840	72,160
15	21,000	13,000	68,880	42,640
20	12,000	10,000	39,360	32,800
30	10,000	8,000	32,800	26,240

resistivity of various soil materials and liquids. Among the listed options, seawater has the lowest resistivity. Thus, one approach to reduce ground resistivity is to treat earth around the grounding rod by mixing it with salt. Another approach is to increase the moisture content of the soil.

The impact of moisture content on the resistivity of top soil and sandy loam is illustrated in Table 5.7. It is clear that higher moisture content results in lower ground resistivity.

Example 5.5 indicates the effect of the rod depth on the earth resistance used as a grounding system for a building.

Example 5.5

Problem

Calculate the earth resistance for a rod of 5/8 in. (1.59 cm) diameter when the soil resistivity is 10,000 Ω cm. Consider a rod of a depth (1) 8 ft (244 cm) and (2) 15 ft (457 cm).

Solution

Using Equation 5.16, the earth resistance can be estimated:

1. If the rod depth is $L = 8$ ft $= 244$ cm:

$$R_g = 10{,}000 \frac{2}{\pi(244)} = 26\ \Omega$$

2. If the rod depth is $L = 15$ ft $= 457$ cm:

$$R_g = 10{,}000 \frac{2}{\pi(457)} = 14\ \Omega$$

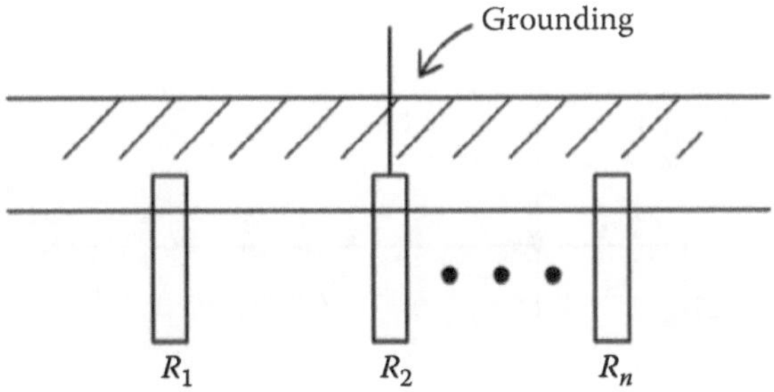

FIGURE 5.31 Grounding rods connected in parallel.

Another approach to reduce the earth resistance in case it is not possible to treat the soil or to dig a deep grounding rod (due to rocky terrain for instance), is to utilize several short grounding rods connected in parallel configured as shown in Figure 5.31.

Indeed, the total resistance, R_g, for n resistances connected in parallel, as shown in Figure 5.31, is estimated as follows:

$$\frac{1}{R_g} = \frac{1}{R_1} + \frac{1}{R_2} + \cdots + \frac{1}{R_n} \tag{5.17}$$

In the special case where all resistances are the same ($R_1 = R_2 = \cdots = R_n = R$), then

$$R_g = \frac{R}{n} \tag{5.18}$$

Thus, the desired value of R_g can simply be obtained by increasing the number of rods connected in parallel.

5.5.4 Selection of Service Grounding Conductors

As noted earlier, every derived electrical system has to be grounded through a service grounding conductor to minimize shocks and fire hazards. The service grounding conductor is typically kept bare and does not carry any current during normal operation. It should be noted that the service grounding conductor is connected to the neutral (often referred to as grounded conductor) to ensure low voltage levels are maintained between the ground and all the energized conductors, as depicted in Figure 5.32, for 240/120 V, one-phase, three-wire systems. When service grounding conductor is connected to phase B, as shown in Figure 5.32a, the voltage between the ground and phase A is 240 V higher than 120 V. However and when service grounding conductor is connected to the neutral, the voltage difference between phases A or B and the ground is only 120 V as indicated in case of Figure 5.32b.

Figure 5.32 Impact of connecting the service grounding conductor to one of the phases (a) and to the neutral (b) for a three-wire one-phase system.

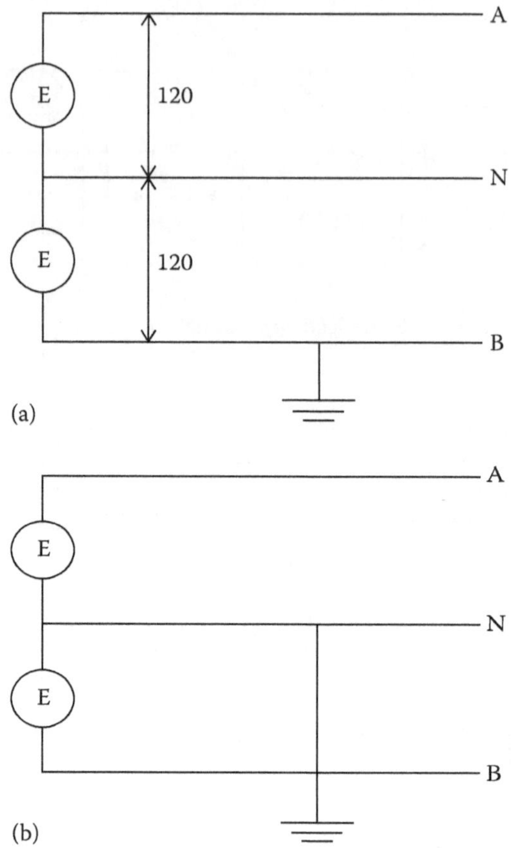

FIGURE 5.32 Impact of connecting the service grounding conductor to one of the phases (a) and to the neutral (b) for a three-wire one-phase system.

Common connections for the service grounding conductors for both wye and delta three-phase systems are shown in Figure 5.33. It should be noted that while service grounding is required for 208Y/120 V and 480Y/277 V systems, it is not typically considered for 480 and 600 V delta systems serving motor loads. One of the reasons for operating ungrounded three-phase systems is the reduction in downtime associated with repeated ground faults. However, ungrounded systems can develop high line-to-line voltages due to resonance conditions. These overvoltages can result in significant damages to the electrical system components. Moreover, identifying the sources of faults can be difficult for ungrounded systems. Therefore, it is highly recommended that all systems be grounded to improve reliability as well as make detection of ground faults easier. With the addition of ground-fault protection devices, the safety of the entire electrical system can also be enhanced significantly.

Table 5.8 provides the size of the service grounding conductor depending on the size of the main feeder conductor. Example 5.6 illustrates the use of Table 5.8 to size service grounding conductors for a derived system.

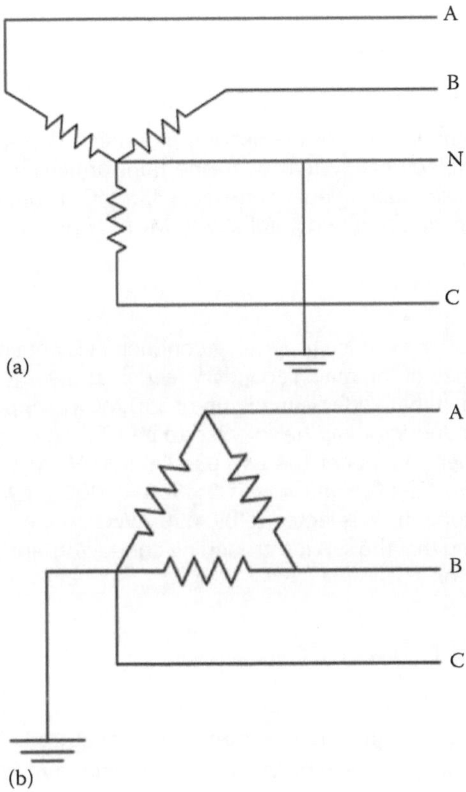

FIGURE 5.33 Service grounding connections for both (a) wye and (b) delta three-phase systems.

TABLE 5.8
Service Grounding Conductor Sizes

Size of Largest Service-Entrance Conductor or Equivalent Area for Parallel Conductors		Size of Service Grounding Conductor	
Copper	**Aluminum or Copper-Clad Aluminum**	**Copper**	**Aluminum or Copper-Clad Aluminum**
2 or Smaller	1/0 or Smaller	8	6
1 or 1/0	2/0 or 3/0	6	4
2/0 or 3/0	4/0 or 250 MCM	4	2
Over 3/0 thru 350 MCM	Over 250 MCM thru 500 MCM	2	1/0
Over 350 MCM thru 600 MCM	Over 500 MCM thru 900 MCM	1/0	3/0
Over 600 MCM thru 1,100 MCM	Over 900 MCM thru 1,750 MCM	2/0	4/0
Over 1,100 MCM	Over 1,750 MCM	3/0	250 MCM

Note: This table is adapted from Table 250.66 of NEC (2014).

Example 5.6

Problem

Determine the minimum size of the service grounding copper conductor for a separately derived electrical system with one transformer for the following two cases: (1) the main secondary feeder consists of 350 MCM copper conductors and (2) main feeder is made up of two parallel 500 MCM copper conductors.

Solution

The minimum size of the service grounding conductors is obtained from Table 5.8 depending on the size of the main secondary feeder. Thus.

For the case when the feeder is made up of 350 MCM copper conductors, the service grounding conductor can be selected to be #2 AWG copper.

For the case when the feeder has two parallel 500 MCM copper conductors, then the overall equivalent feeder size is $2 \times 500 = 1{,}000$ MCM, and the service grounding conductor can be selected to be #2/0 AWG copper.

It should be noted that the service grounding conductors are typically kept bare without any insulation.

5.5.5 Selection of Equipment Grounding

To ensure that all nonelectrical conductive parts of equipment are safe to touch under normal operation, equipment grounding is required. The frames, enclosures, or structural supports of electrical equipment and distribution systems such as transformers, switchboards, panels, motor controllers, motors, generators, cabinets, lighting equipment, and outlet boxes have to be connected to the earth through permanent and continuous paths that may include raceways, cable trays, armors of cables, and equipment grounding busses. For instance, Figure 5.34 illustrates the connection of equipment grounding conductors for a receptacle outlet and a motor. In particular, the receptacle outlet is connected to (1) the enclosure of the panel (through a wire typically identified by a green insulation), (2) the neutral bus (through a wire identified by a white insulation) and (3) the energized conductor (through a wire identified by a red or a black insulation). Figure 5.35 shows the impact of improper equipment grounding connections for a screw shell lamp. In case of Figure 5.35a when the outer shell is properly connected to the neutral wire, the shell has the same voltage as the ground and is safe to touch. However in the case of Figure 5.35b, the outer shell is improperly connected to the energized conductor resulting in a significant and unsafe voltage difference between the outer shell and the ground.

Table 5.9 can be used to determine the size of the equipment grounding conductor depending on the rating of the overcurrent protection for the equipment. Example 5.7 illustrates the use of Table 5.9 to select the proper size of equipment grounding conductor for a panel.

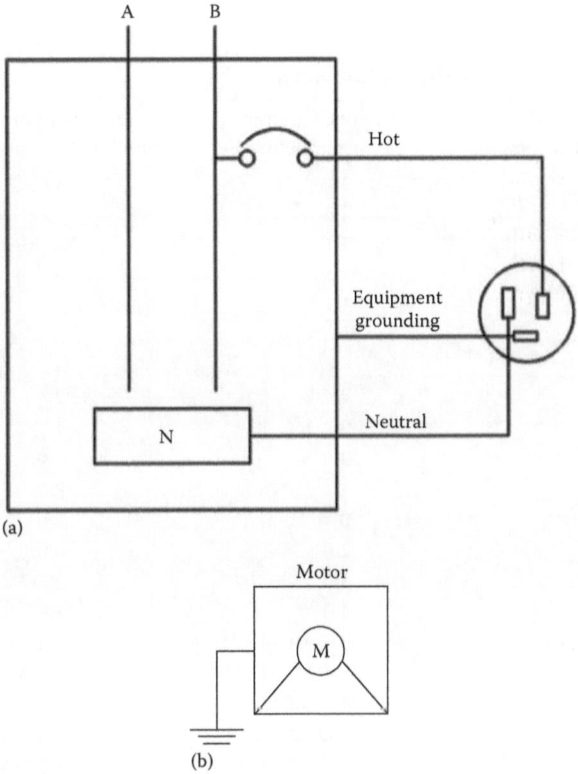

FIGURE 5.34 Equipment grounding connections for (a) a receptacle outlet and (b) a motor.

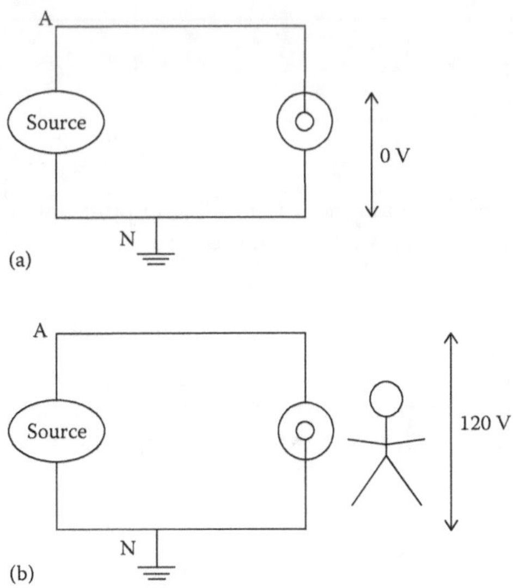

FIGURE 5.35 Impact of equipment grounding for a screw shell lamp with (a) correct and (b) incorrect connections.

TABLE 5.9
Equipment Grounding Conductor Sizes

Rating of Protection Device in Circuit Ahead of Equipment (A)	Copper Grounding Conductor	Aluminum or Copper-Clad Aluminum Grounding Conductor
15	14	12
20	12	10
30	10	8
40	10	8
60	10	8
100	8	6
200	6	4
300	4	2
400	3	1
500	2	1/0
600	1	2/0
800	0	3/0
1,000	2/0	4/0
1,200	3/0	250 MCM
1,600	4/0	350 MCM
2,000	250 MCM	400 MCM
2,500	350 MCM	600 MCM
3,000	400 MCM	600 MCM
4,000	500 MCM	800 MCM
5,000	700 MCM	1,200 MCM
6,000	800 MCM	1,200 MCM

Note: This table is adapted from Table 250.122 of NEC (2014).

Example 5.7

Problem

Determine the size of the equipment grounding copper conductor suitable for a panel protected by a 200 A power circuit breaker.

Solution

The minimum size of the equipment grounding conductor is obtained from Table 5.9 using the rating of the protection device. Thus and since the panel is protected by 200 A circuit breaker, the required equipment grounding conductor is #6 AWG copper conductor.

5.5.6 Estimation of Potential Damages

Typical ground faults are of the arcing type with low magnitudes compared to short-circuit currents and thus are more difficult to detect by standard protection devices such as circuit breakers or fuses. Most of the arcing ground faults occur for

TABLE 5.10
Potential Damages Associated with Arcing Faults

Fault Energy Expressed in kW Cycles	Potential Damages
100	Minimum damages with only some smoke marks. The location of fault can be identified only by observation.
2,000	Little damages with smoke marks and some punctured insulation. The equipment can be restored by cleaning smoke marks and punctured insulation.
10,000	Local damages typically limited to metal enclosures.
20,000	Potential burning of enclosures and damages to some equipment and components of distribution systems.
More than 20,000	Considerable damages depending on the let-through fault energy and may result in severe burning of large sections of the distribution systems.

480Y/277 V systems with a potential peak phase voltage of 390 V (i.e., $\sqrt{2} \times 277\,\text{V}$). For 208Y/120 V systems, the arcing faults extinguish themselves due to a low-peak line-to-ground voltage of 170 V (i.e., $\sqrt{2} \times 120\,\text{V}$). Generally, it is difficult to predict and estimate the damages due to ground-fault currents for a specific electrical system since it depends on the components of the system as well as the location of the fault. However, some attempts have been made to express the damages in terms of the energy dissipated by the arcing fault based on past observations. Specifically, the potential damages due to fault currents can be estimated using the fault energy, E_{Fault}, expressed in W-cycles as indicated in the following equation (Hughes, 1988):

$$E_{\text{Fault}} = \Delta V \times I \times f \times t \tag{5.19}$$

where

ΔV is the line-to-ground voltage (i.e., phase voltage) (V)
I is the ground-fault current (A)
t is the duration of the ground fault (i.e., response time of the protection device) (s)
f is the frequency of the current and voltage (Hz)

Table 5.10 is then used to assess the potential physical damages associated with the fault energy estimated from Equation 5.19. As indicated in Table 5.10, the level of damage can be small and restricted to smoke marks but can be catastrophic with the complete destruction of most of the electrical equipment. Example 5.8 illustrates the level of damages for the example shown in Figure 5.36 as a function of the response time of the protection device.

Example 5.8

Problem

For the electrical system shown in Figure 5.36 with a line-to-ground voltage of 120 V, a ground resistance of 1.5 Ω, and a fault current of 2,400 A, estimate the

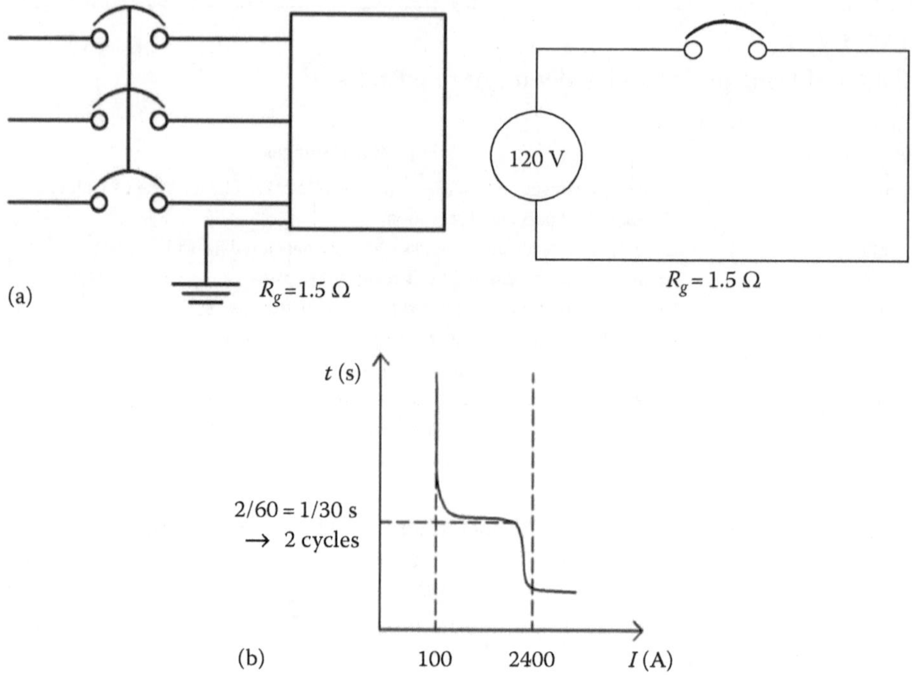

FIGURE 5.36 Estimation of potential damages from ground faults for (a) a known ground resistance and (b) a given response curve of a circuit breaker.

potential damages associated with ground-fault current for two cases: (1) the fault is quickly detected with response time of two cycles and (2) the fault is not detected for over 10 seconds.

Solution

For, a response time of two cycles or $t = \dfrac{1}{30}$ s, the fault energy can be estimated using Equation 5.19:

$$E_{\text{Fault}} = 120\ \text{V}(2400\ \text{A})(2) = 57.6\ \text{kWcycles}$$

For 57.6 kW cycles fault energy, there is little damage with only some smoke marks as noted by Table 5.10.

For, a response time of $t = 10$ seconds, that is 600 cycles, the fault energy can be estimated using Equation 5.19:

$$E_{\text{Fault}} = 120 \times 2400 \times 10 \times 60 = 20{,}000\ \text{kWcycles}$$

For 20,000 kW cycles fault energy, there is a potential for burning enclosures and other electrical equipment as indicated by Table 5.10.

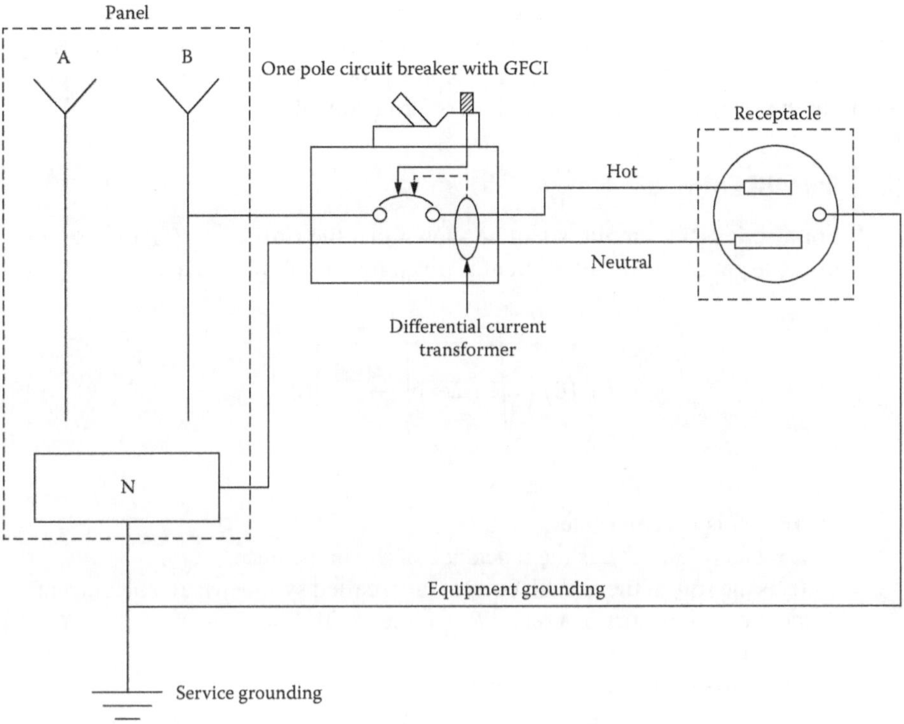

FIGURE 5.37 Operation mechanism of a GFCI circuit breaker.

5.5.7 Ground-Fault Current Interrupters

To reduce and even eliminate the potential damages associated with ground faults, additional protection systems are needed. In particular, the use of ground-fault current interrupters (GFCIs) is required for several locations within buildings. GFCIs allow the detection of small ground faults (as low as 6 mA) and to quickly open the circuit to avoid any harm to humans and damages to electrical equipment. There are several GFCI devices including receptacle outlets and circuit breakers. Figure 5.37 illustrates how one-pole GFCI circuit breaker operates using a differential current transformer to monitor the current flowing through the grounding conductor and consequently trip the circuit breaker. GFCI circuit breakers are commonly installed in panels and provide ground-fault protection to all the loads served by the branch circuits connecting the breakers. Thus, GFCI circuit breakers have a dual purpose: trip when short-circuit current occurs and interrupt when ground fault is detected.

5.6 SUMMARY

In this chapter, various devices used to protect building electrical systems are presented. In particular, the operation and the design guidelines for fuses, circuit breakers, and grounding systems are outlined. The effects of fault currents on both humans and equipment are described and quantified. Moreover, major NEC requirements and specifications to protect the different components of electrical distribution systems are summarized. Specific examples are provided throughout this chapter to illustrate the selection of the proper sizes of protection devices.

5.7 PROBLEMS

1. For a series R–L circuit, it can be shown that the rms (I.E., effective value) of the asymmetrical fault current is given by the following equation

$$I_{rms}(\theta) = \sqrt{1 + 2\exp\left(\frac{-4\pi\theta}{\frac{X}{R}}\right)} I_{ac}$$

   where

   $\theta = f \times t$ is time in cycles

   $X = L/\omega = 2\pi \times f \times L$ is the reactance of the inductance

   I_{ac} is the rms of the ac fault current (also called symmetrical fault current)

   A short-circuit occurs in a series *R–L* circuit with $V = 277$ V, $L = 2$ mH, $R = 0.4$. Determine

   a. The rms symmetrical fault current
   b. The rms asymmetrical fault current at the instant the switch closes, assuming maximum DC offset
   c. The rms asymmetrical fault current five cycles after the switch closes, assuming maximum DC offset
   d. The DC offset as a function of time if the switch closes when the instantaneous source voltage is 300 V
2. A dwelling unit is supplied by a single-phase, 120/240 V service. The electric load of the unit includes. (1) 7,000 VA water heater, 240 V; (2) 800 VA vent fan, 120 V; (3) 1,200 VA compactor, 120 V; (4) 5,000 VA dryer, 240 V; and (5) 6,500 VA range. Each appliance is served by one branch protected by a circuit breaker. Determine the rating and the number of poles of each circuit breaker.
3. Determine the size of molded-case circuit breakers and associated number of poles for the following loads in a dwelling served by 120/240 V system: 1.8 kVA dishwasher, 120 V; 1.2 kVA compactor, 120 V; 1.0 kVA disposal, 120 V; 0.5 kVA vent fan, 120 V; 16 kVA range, 240 V; 4.5 kVA dryer, 240 V; 15 kVA heating unit, 240 V; 9 kVA air conditioning unit, 240 V; and 6.5 kVA water heater, 240 V.
4. Consider a panelboard for a commercial building supplied by a 480Y/277 V, three-phase feeder. The panel has the following loads:
   - A lighting system of a total power 45 kVA with a power factor of 0.93
   - A 15 kVA resistive water heater with a power factor of 1.0
   - A motor rated at 10 kVA and a power factor of 0.82

Determine

a. The size of the molded-case circuit breaker for the water heater. Assume a continuous load.
b. The size of fuses for the motor. Assume a continuous load.
c. The size of the power circuit breaker serving the panel.

5. Consider the distribution system of a commercial building shown in the following diagram. Select the frames and trip ratings for all the power circuit breakers for the various panels and motors as well as the power circuit breaker at the secondary main feeder for the two options:
   a. The secondary transformer is rated at 480Y/277 V.
   b. The secondary transformer is rated at 208Y/120 V.
   c. Assume that the available short-circuit currents are below 25,000 A throughout the entire electrical system.

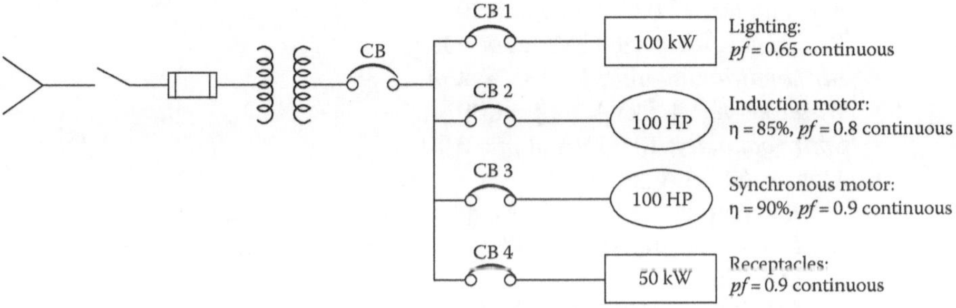

Distribution system and electrical loads for a commercial building

6. Determine the size of the equipment grounding copper conductors for a panel that is served by 2/0 THW (Cu) and 400 A circuit breaker.
7. Consider a 480Y/277 V panelboard that serves the following loads for an office building:
   - Lighting: 15.0 kVA at $pf=0.95$
   - Resistance Heater: 9.0 kVA at $pf=1.00$
     a. Select the proper circuit breaker for the panelboard.
     b. Size the equipment grounding conductor (Cu) for the panelboard.
8. For an office building, a 208Y/120 V power panel has the following loads:
   - Receptacles with a total load of 30.00 kVA at $pf=0.90$
   - Heaters with a total load 12.00 kVA at $pf=1.00$
     a. Determine the proper frame size and trip rating for the power circuit breaker protecting the panel. Assume an available short-circuit current of 35,000 A at the panel.
     b. Size the equipment grounding conductor (Cu) for the panelboard.
9. Determine the copper service grounding conductor required for a service entrance with three parallel 300 MCM copper feeders.
10. Determine the potential damages incurred by an electrical distribution system when 2,000 A ground-fault current occurs for 10 cycles.

11. A 480Y/277 V panelboard for an office building has the following loads:
    - Lighting with a total load of 80.00 kVA and a power factor of 0.95
    - Resistance heater with a total load 25.00 kVA and a power factor of 1.00
      a. Select the proper circuit breaker for the panelboard. Assume only the lighting load is continuous.
      b. Size the equipment grounding conductor (Cu) for the panelboard.
12. A 208Y/120 V panelboard for an office building has the following loads:
    - Receptacles with a total load of 10.00 kVA and a power factor of 0.90
    - An appliance rated at 5-kW with a power factor of 0.92
      a. Size the circuit breaker for the panelboard. The appliance could operate continuously for more than 3 hours.
      b. Size the equipment grounding conductor (Cu) for the panelboard.
13. A 208Y/120-V panelboard serves the following loads for an office building:
    - *Lighting*: 15.0 kVA at $pf = 0.95$
    - *Receptacles*: 22.0 kVA at $pf = 0.90$
    - *Resistance heater*: 9.0 kVA at $pf = 1.00$
    - *Air conditioning unit*: 12.0 kVA at $pf = 0.88$
    - *Water heater*: 4.5 kVA at $pf = 1.00$
    - *Spare capacity*: 10.0 kVA at $pf = 0.90$
    - *Motors*: 6.0 kVA at $pf = 0.88$

      Note that the largest motor load is rated at 16.7-A and 208-V
      a. Determine the size of THW-feeder for the panelboard. Size the conduit.
      b. Select the proper circuit breaker for the panelboard. All the motors are protected by time-delay fuses.
      c. Size the equipment grounding conductor (Cu) for the panelboard.

6 Branch Circuits and Feeders

To ensure the safety of the electrical systems in buildings, it is important to properly select the wires that supply electricity to various loads through the buildings. In this chapter, the types and sizes of conductors and conduits suitable for branch circuits and feeders are presented. The specific requirements to select adequate sizes of conductors to serve building electrical loads are discussed. In particular, ampacity, short-circuit current tolerance, and voltage drop allowance are considered in the design of conductors. Moreover, specific design guidelines for branch circuits and feeders for both nonmotor and motor loads are described with illustrative examples.

6.1 INTRODUCTION

Figure 6.1 shows a typical electrical distribution network for commercial buildings. The network includes a set of wires that supply electricity to various building loads. The wires consist typically of (1) a main feeder that supplies power from the utility line through a distribution transformer, (2) a set of subfeeders serving different panels such as lighting panels and power panels using, as needed, step-down transformers and motor control centers, and (3) branch circuits supplying electricity from the panels to the building loads consisting typically of lighting fixtures, receptacles, and motors.

The electrical wires in building electrical distribution systems consist typically of conductors, cables, and conduits or raceways. The conductors, made up of copper or aluminum, carry the electrical currents and are often insulated. The cables are assemblies of two or more conductors, insulation layers, and any shielding and/or protective coverings, as illustrated in Figure 6.2. The conduits house conductors or cables to protect them from any physical damages or exposure to chemicals. The raceways provide support and space to conductors and cables including feeders and branch circuits to minimize any hazard from electrical faults.

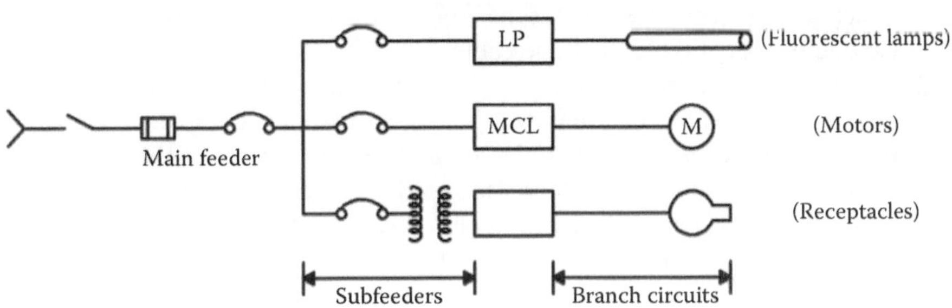

FIGURE 6.1 Typical distribution systems for commercial buildings.

DOI: 10.1201/9781003276999-6

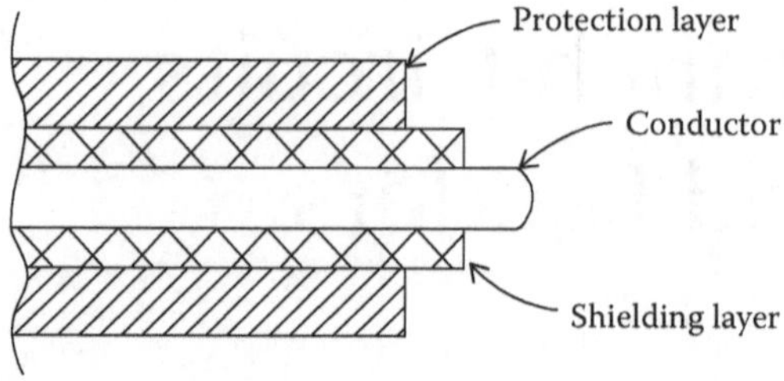

FIGURE 6.2 Typical layers for an electrical conductor.

As discussed in Chapter 2, the size of an electrical conductor represents its cross-sectional area. In the United States, two scales are used to indicate the size of a conductor: the American wire gauge (AWG) for small sizes and 1,000 of circular mils (MCM or kCM) for larger sizes. For the AWG scale, the available sizes are from number 18 AWG to number 4/0 AWG with higher numbers representing smaller conductor sizes. For residential buildings, the smallest size of copper conductor that can be used is number 14 AWG, which is rated for a maximum loading of 15 A. For commercial buildings, number 12 AWG is the smallest size of conductor that is commonly used for supplying lighting fixtures and receptacles. The AWG size designation soon became inadequate after its implementation in the early 1900s due to the ever-increasing electrical loads in buildings. For larger conductors, the cross-sectional area is measured in circular mils. As noted in Chapter 2, a circular mil corresponds to the area of a circle that has a diameter of 1 mil or 1/1,000th of an inch. For instance, a conductor with a diameter of ½ in. (500 mils) has a circular mil area of 250,000, which is designated by 250 MCM or 250 kcmil.

As will be discussed in this chapter, the size of conductors and cables used for feeders and branch circuits in buildings depends on three main criteria including the following:

1. The rating of the continuous current under normal operating conditions. The National Electric Code (NEC, 2014) refers to the continuous current rating as the ampacity of the conductor. The main parameters that affect the ampacity of a conductor include the physical characteristics of the wire such as its cross-sectional area (or size) and its material and the conditions under which the wire operates such as the ambient temperature and the number of conductors installed in the same cable.
2. The rating of short-circuit current under fault conditions. Indeed, high short-circuit currents can impose significant thermal and magnetic stresses not only on the conductor but also on all the components of the electrical system. The conductor has to withstand these relatively high short-circuit currents at least for a limited time period since the protective device generally requires a few cycles before detecting and interrupting the fault currents.
3. The maximum allowable voltage drop across the length of the conductor. Most electrical systems are sensitive to the voltage applied to them. It is therefore important to reduce the voltage drop that occurs across the feeders

and the branch circuits. The NEC recommends a maximum voltage drop of 3% for any feeder or a branch circuit with a maximum voltage drop of 5% from the service entrance to the utilization outlet.

Throughout this chapter, the selection criteria and design approaches for conductors, branch circuits, and feeders are discussed and illustrated.

6.2 SIZE AND RATING OF CONDUCTORS

Two conductor materials are commonly used in building electrical distribution systems: copper and aluminum. Because of its highly desirable electrical and mechanical properties, copper is now the preferred material used for conductors of insulated cables. Aluminum has some undesirable properties and its use is restricted. Indeed, an oxide film, which is not a good conductor, can develop on the surface of an aluminum wire and can cause poor electrical contact, especially at the terminal connections. It should be noted that aluminum can be considered in cases when cost and weight are important criteria for the selection of conductors. However, it is highly recommended even in these cases to use copper conductors for connections and equipment terminals to eliminate poor electrical contact.

To protect the conductors, several types of insulation materials are used. Insulated conductors, as illustrated in Figure 6.2, are identified by their letter designations depending on the type of insulation material, protection coating, and conditions of use. In the United States, the following letter designations are used for conductors:

- *For the insulation material type*: A (asbestos), MI (mineral insulation), R (rubber), SA (silicone asbestos), T (thermoplastic), V (varnished cambric), and X (cross-linked synthetic polymer).
- *For the conditions of use*: H (heat up to 75°C), HH (heat up to 90°C), UF (suitable for underground), and W (moisture-resistant).

Thus, the letter designation THW refers to a conductor that has a thermoplastic insulation rated for maximum operating temperature of 75°C and suitable for use in dry as well as wet locations. Moreover, some types of electrical conductors and cables have outer coverings that provide mechanical/corrosion protection such as lead sheath (L), nylon jacket (N), armored cable (AC), metal-clad cable (MC), and nonmetallic sheath cable (NM). For instance, the designation THHN is for a conductor that is insulated with thermoplastic material, heat resistant up to 90°C, and covered with nylon jacket. As indicated earlier, cables are typically factory assemblies of two or more conductors protected with an outer covering. For a full description of all types of cables and insulated conductors, their letter designations, and their uses, the reader is referred to the NEC, Article 310 (NEC, 2014).

The ampacity is often used as a term to refer to the amount of current that a conductor can carry continuously without exceeding its temperature ratings under normal conditions of utilization. Several parameters affect the continuous rating of conductors including their physical characteristics such as insulation level and cross-sectional area as well as their surrounding environment such as ambient air temperature. Table 6.1 lists the ampacities of various insulated conductors with rated voltage ranging between 0 and 2,000 V (NEC, 2014).

TABLE 6.1
Ampacities for Insulated Conductors Rated 0–2,000 V in Free Air

	Temperature Rating of Conductor						
	60°C (140°F)	75°C (167°F)	90°C (194°F)	60°C (140°F)	75°C (167°F)	90°C (194°F)	
Size AWG or kcmil	Types TW, UF	Types RHW, THHW, THW, THWN, XHHW, USE, ZW	Types TBS, SA, SIS, FEP, FEPB, MI, RHH, RHW-2, THHN, THHW, THW-2, THWN-2, USE-2, XHH, XHHW, XHHW-2, ZW-2	Types TW, UF	Types RHW, THHW, THW, THWN, XHHW, USE	Types TBS, SA, SIS, THHN, THHW, THW-2, THWN-2, RHH, RHW-2, USE-2, XHH, XHHW, XHHW-2, ZW-2	Size AWG or kcmil
Copper				Aluminum or Copperclad Alum			
18	—	—	14	—	—	—	—
16	—	—	18	—	—	—	—
14[a]	20	20	25	—	—	—	—
12[a]	25	25	30	20	20	25	12[a]
10[a]	30	35	40	25	30	35	10[a]
8	40	50	55	30	40	45	8
6	55	65	75	40	50	60	6
4	70	85	95	55	65	75	4
3	85	100	110	65	75	85	3
2	95	115	130	75	90	100	2
1	110	130	150	85	100	115	1
1/0	125	150	170	100	120	135	1/0
2/0	145	175	195	115	135	150	2/0
3/0	165	200	225	130	155	175	3/0
4/0	195	230	260	150	180	205	4/0
250	215	255	290	170	205	230	250
300	240	285	320	190	230	255	300
350	260	310	350	210	250	280	350
400	280	335	380	225	270	305	400
500	320	380	430	260	310	350	500
600	355	420	475	285	340	385	600
700	385	460	520	310	375	420	700
750	400	475	535	320	385	435	750
800	410	490	555	330	395	450	800
900	435	520	585	355	425	480	900
1,000	455	545	615	375	445	500	1,000
1,250	495	590	665	405	485	545	1,250
1,500	520	625	705	435	520	585	1,500
1,750	545	650	735	455	545	615	1,750
2,000	560	665	750	470	560	630	2,000

(*Continued*)

TABLE 6.1 (*Continued*)
Ampacities for Insulated Conductors Rated 0–2,000 V in Free Air

	Temperature Rating of Conductor					
	60°C (140°F)	**75°C (167°F)**	**90°C (194°F)**	**60°C (140°F)**	**75°C (167°F)**	**90°C (194°F)**
	Correction Factors					
Ambient Temp. (°C)	For Ambient Temperatures Other Than 30°C (86°F), Multiply the Allowable Ampacities Shown Above by the Appropriate Factor Shown Below					Ambient Temp. (°F)
21–25	1.08	1.05	1.04	1.08	1.05	1.04	70–77
26–30	1.00	1.00	1.00	1.00	1.00	1.00	78–86
31–35	0.91	0.94	0.96	0.91	0.94	0.96	87–95
36–40	0.82	0.88	0.91	0.82	0.88	0.91	96–104
41–45	0.71	0.82	0.87	0.71	0.82	0.87	105–113
46–50	0.58	0.75	0.82	0.58	0.75	0.82	114–122
51–55	0.41	0.67	0.76	0.41	0.67	0.76	123–131
56–60	—	0.58	0.71	—	0.58	0.71	132–140
61–70	—	0.33	0.58	—	0.33	0.58	141–158
71–80	—	—	0.41	—	—	0.41	159–176

Note: This table is adapted from NEC-2014 Table 310.15(B)(16) (NEC, 2014).

[a] For branch circuits rated 15-A, 20-A, and 30-A, conductors of at least respectively, 14, 12, and 10 AWG should be used.

To better understand various factors that affect the ampacity of a conductor, a simplified steady-state thermal model is considered in Figure 6.3. In particular, the model accounts for the mechanism of heat transfer from a conductor to the ambient air.

The heat generated by the flowing current, I, through the conductor of resistance, R, is determined through Joule effect:

$$\text{Heat} = RI^2 \tag{6.1}$$

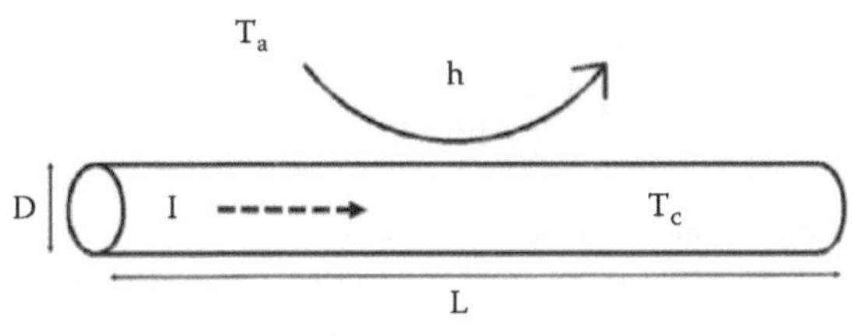

FIGURE 6.3 Heat transfer from a conductor to ambient air.

As shown in Chapter 2, the electrical resistance R can be determined from the length L, and the cross-sectional area, $A = \pi D^2/4$, of the conductor:

$$R = \rho \frac{L}{A} = \rho \frac{4L}{\pi D^2} \tag{6.2}$$

where ρ is the resistivity of the conductor material. Thus, the heat generated by the conductor

$$\text{Heat} = \rho \frac{4L}{\pi D^2} I^2 \tag{6.3}$$

Through mainly convection and conduction heat transfer mechanisms, the generated heat is dissipated to the ambient air through the surface of the conductor. The heat dissipation can be expressed as follows:

$$\text{Heat} = hA_o(T_c - T_a) \tag{6.4}$$

where

- h is the combined convective–conductive heat transfer at the surface of the conductor
- T_c is the conductor temperature assumed to be uniform along the conductor length
- T_a is the ambient air temperature
- A_o is the conductor's outer surface area exposed to the ambient air which can be expressed as a function of L and D:

$$A_o = \pi DL \tag{6.5}$$

Setting steady-state heat balance between heat generation (i.e., Equation 6.3) and heat dissipation (i.e., Equation 6.4), the following relationship can be obtained by using Equation 6.5:

$$\rho \frac{4L}{\pi D^2} I^2 = \pi DLh(T_c - T_a) \tag{6.6}$$

By simplifying Equation 6.6, the current can be determined as a function of T_a, T_c, D, h, and ρ:

$$I = \sqrt{\frac{\pi^2 D^3}{4\rho} h(T_c - T_a)} \tag{6.7}$$

Equation 6.7 indicates that the ampacity or the current that can flow through the conductor is affected by several parameters. Figure 6.4 illustrates the variation of the ampacity, I, as a function of three among five parameters outlined in Equation 6.7. In particular, the following observations can be made using the simplified thermal model of Equation 6.7:

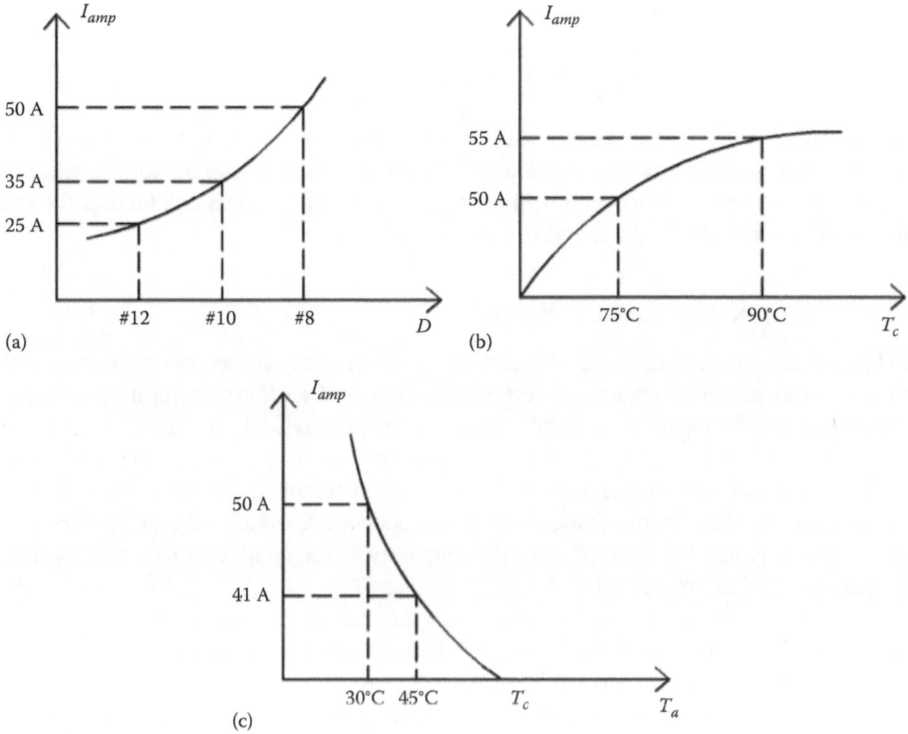

FIGURE 6.4 (a) Ampacity variations with diameter, (b) conductor temperature, and (c) ambient air temperature.

- The higher is the diameter of the conductor, the higher is the ampacity.
- The higher is the conductor temperature, the higher is the ampacity.
- The higher is the ambient air temperature, the lower is the ampacity.
- The higher is the conductor resistivity, the lower is the ampacity.
- The lower is the coefficient h (through lower air movement or higher conductor insulation), the lower is the ampacity.

When selecting the size of the conductor, not only the initial costs but also the operating costs should be considered. As it will be illustrated in Chapter 10, higher operating costs including the cost of energy encourages the installation of larger conductors than those required by the loads. Unfortunately, most designers do not consider the operating costs in their design due to the desire to reduce first costs and to the uncertainty in electricity prices.

6.3 DESIGN OF CONDUCTORS

To determine the correct size of conductors to be used for feeders and branch circuits for building electrical distribution systems, generally three criteria need to be considered:

1. Ampacity or continuous current rating
2. Short-circuit current rating
3. Maximum allowable voltage drop

In the following sections, each requirement is discussed with specific calculation procedures and examples to illustrate the selection process of conductors. Then, the three criteria are combined and utilized to select the size of the conductors for the design of feeders, subfeeders, and branch circuits.

6.3.1 Continuous Current Rating

Table 6.1 can be used to select the conductor size based on the continuous current rating of the electrical load to be served. As discussed earlier, the main parameters that affect the ampacity of a conductor include the physical characteristics of the wire such as its cross-sectional area (or size) and its material as well as the conditions under which the wire operates such as the ambient temperature and the number of conductors installed in the same conduit or raceway. Various derating factors and correction may need to be applied to the ampacity of the conductor to select its size, as illustrated in Examples 6.1–6.3.

It should be noted that when several conductors are housed in the same raceway or conduit as illustrated in Figure 6.5, the ambient temperature, T_a, is increased due to the heat dissipated by all the conductors. As shown in the thermal model of Equation 6.7, the ampacity of each of the conductors is reduced due to this higher ambient temperature. The ampacity percent reduction can be estimated from Table 6.2 using derating factors that depend on the number of conductors located in the same raceway or conduit.

Example 6.1

Problem

Determine the ampacity of a feeder made up of three conductors rated 500 MCM THW (Cu), placed in a conduit where the temperature is 35°C.

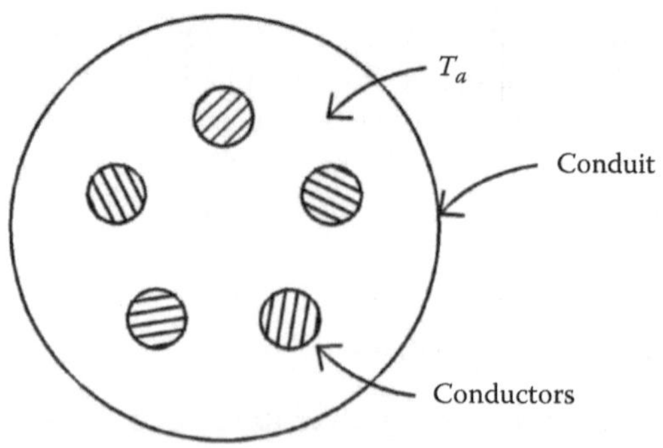

FIGURE 6.5 Several conductors or cables housed in the same conduit or raceway.

TABLE 6.2
Derating Factors for Several Conductors Housed in the Same Raceway or Cable

Adjustment Factors for More Than Three Current-Carrying Conductors in a Raceway or Cable	
Number of Current-Carrying Conductors	**Percent of Values in Table 6.1 as Adjusted for Ambient Temperature**
4–6	80
7–9	70
10–20	50
21–30	45
31–40	40
41 and above	35

Note: This table is adapted from NEC-2014 Table 310.15(B)(3)(a) (NEC, 2014).

Solution

From Table 6.1, the ampacity of 500 MCM THW (Cu) is 380 A. From Table 6.1, a correction factor of 0.94 should be used when the ambient air temperature is 35°C. Thus, the ampacity of each 500 MCM THW (Cu) conductor is

$$I_{\text{amp}} = 380 \times 0.94 = 357 \text{ A}$$

Example 6.2

Problem

Determine the ampacity of 6 #10 THW (Cu) conductors placed in the same conduit.

Solution

From Table 6.1, the ampacity of #10 THW (Cu) is 35 A. From Table 6.2, a derating factor of 80% should be used. Thus, the ampacity for each of the 6 #10 THW (Cu) conductors is

$$I_{\text{amp}} = 35 \times 0.8 = 28 \text{ A}$$

Example 6.3

Problem

Determine the ampacity of 8 #1/0 THHN (Cu) conductors placed in the same conduit where the ambient temperature is 40°C.

Solution

From Table 6.1, the ampacity of #1/0 THHN (Cu) is 170 A. In addition and from Table 6.1, a correction factor of 0.91 should be used when the ambient air temperature is 40°C. Moreover and from Table 6.2, a derating factor of 70% should be used for eight conductors located in the same conduit. Thus, the ampacity for each of the 8 #1/0 THHN (Cu) conductors is

$$I_{amp} = 170 \times 0.91 \times 0.7 = 108 \text{ A}$$

6.3.2 Short-Circuit Current Rating

As discussed in Chapter 5, high short-circuit currents can impose significant thermal or magnetic stresses not only on the conductor but also on all the components of the electrical distribution system. The conductor has to withstand these relatively high short-circuit currents since the protective device requires some finite time before detecting and interrupting the fault current. The size of the conductor should be selected so it can handle the significant heat generated at the start of the fault by the asymmetrical currents, which are higher than the symmetrical currents, as noted by Equation 6.8:

$$I_{asym} = K(\theta) I_{sym} \tag{6.8}$$

where K is a function of the total clearing time, θ, of the protection device and varies between 1 and 1.732 as noted in Figure 6.6. Table 6.3 lists the K factors and clearing times, θ, expressed in cycles, for common protection devices used in building electrical systems.

The amount of asymmetrical fault current that a conductor can safely handle can be estimated using the following correlation:

$$\left(\frac{I_{asym}}{A}\right)^2 t = c \times \ln\left(\frac{234 + T_2}{234 + T_1}\right) \tag{6.9}$$

where

A is the conductor cross-sectional area expressed in CM which reflects directly the conductor size

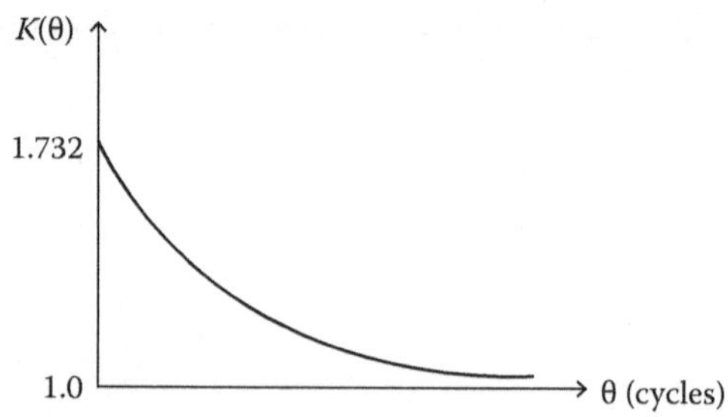

FIGURE 6.6 Variation of K factor as a function of the clearing time θ.

TABLE 6.3
K Factors and Clearing Times for Selected Protection Devices

Protection Device	System Voltage (V)	*K* Factor	Total Clearing Time (Cycles)
Circuit breaker	0–1,000V	1.30	2
Current-limiting fuse	0–1,000V	1.40	0.5
Air circuit breaker	Above 1,000V	1.15	5
Oil circuit breaker	Above 1,000V	1.10	8
Power fuse	Above 1,000V	1.60	1
Current-limiting fuse	Above 1,000V	1.60	0.5

TABLE 6.4
Values for T_1, T_2, and c for Selected Conductor Types

Conductor Type	T_1 (°C)	T_2 (°C)	c (A^2s/CM^2)
THW (Cu)	75	150	0.0297
XHHW (Cu)	90	250	0.0297
XHHW (Al)	90	250	0.0125

t is the duration of the fault current and is expressed in seconds (note that 1 cycle = 1/60 second = 0.01667 second)
T_1 is the maximum operating temperature for the conductor (typically 75°C or 90°C)
T_2 is the maximum fault current temperature (typically 150°C or 250°C)
c is a coefficient that depends on the type of conductor material and its insulation

Table 6.4 summarizes, for selected conductor types, the values for T_1, T_2, and c. Figures 6.3–6.5 illustrate charts to estimate the value of I_{asym} for various conductor sizes and fault duration times for three types of conductor and insulation. Specifically, Figure 6.7 applies to the case of THW copper conductors while Figures 6.8 and 6.9 are specific to XHHW copper and XHHW aluminum conductors.

Examples 6.4–6.6 illustrate the procedure of using the charts outlined in Figures 6.7–6.9 to estimate the symmetrical and asymmetrical fault current for conductors using Equation 6.8.

Example 6.4

Problem

Determine the symmetrical short-circuit current for a #3/0 THW (Cu) conductor protected by 150 A circuit breaker rated at 480V.

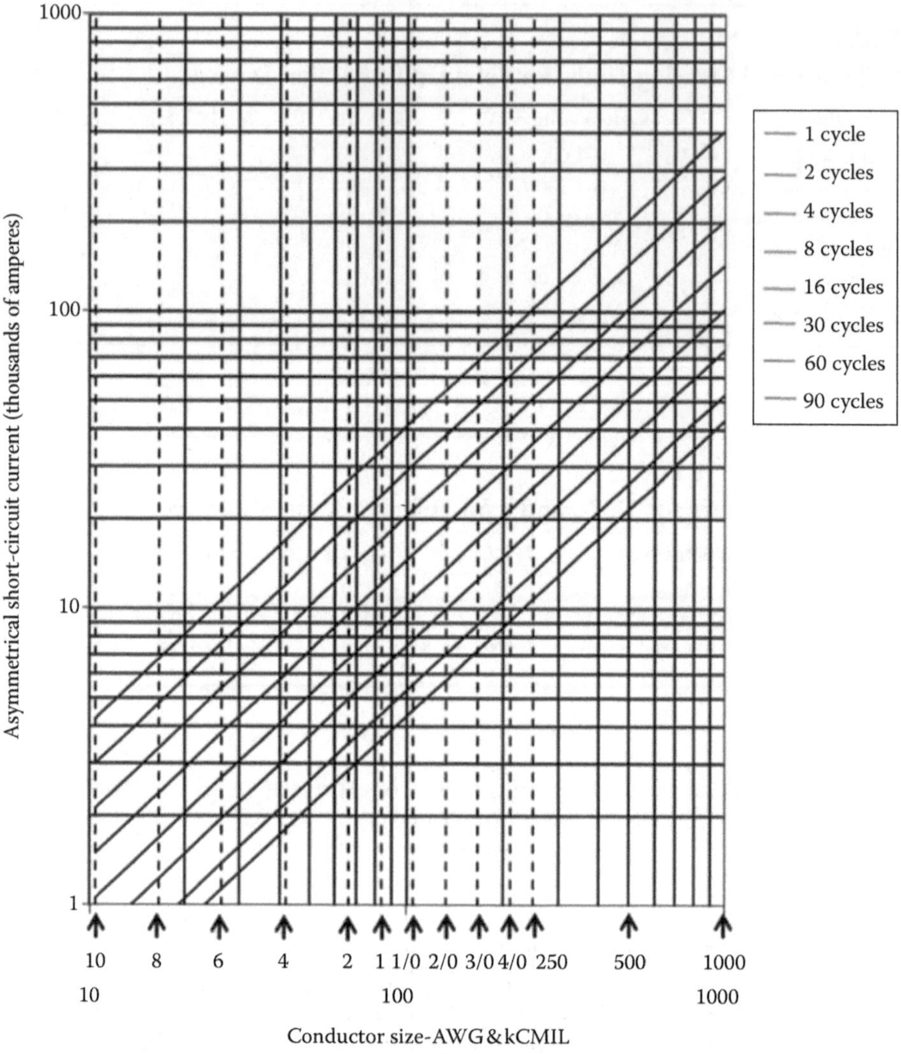

FIGURE 6.7 Allowable short-circuit currents for THW copper conductors.

Solution

From Table 6.3, the clearing time θ and the K factor for the circuit breaker can be determined: $\theta = 2$ cycles, $K = 1.3$.

From Figure 6.7, the value of asymmetrical short-circuit current can be estimated knowing θ and conductor size of #3/0: $I_{asym} = 50{,}000$ A. Thus and using Equation 6.8, the symmetrical short-circuit current can be calculated:

$$I_{sym} = \frac{50{,}000 \text{ A}}{1.3} = 38{,}000 \text{ A}$$

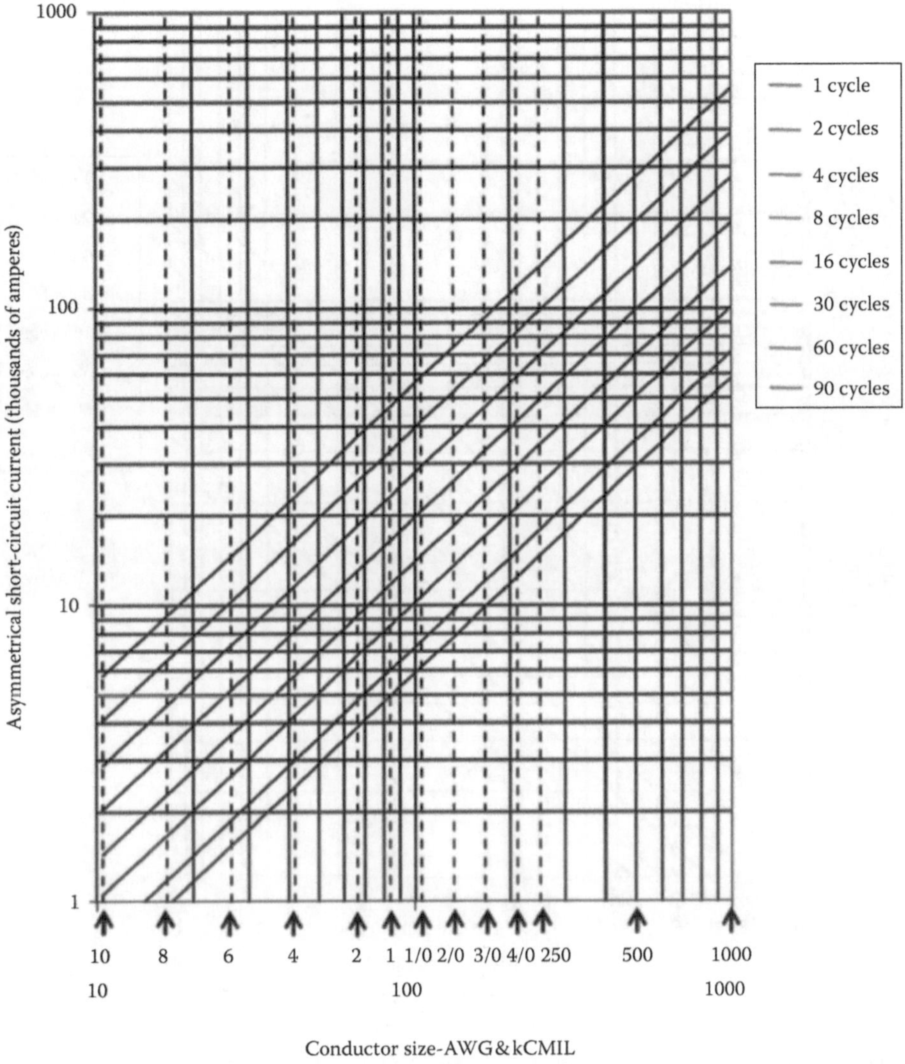

FIGURE 6.8 Allowable short-circuit currents for XHHW copper conductors.

Example 6.5

Problem

Determine the size of XHHW (Cu) feeder protected by a 480 V rated circuit breaker that can handle 40,000 A symmetrical short-circuit current.

Solution

From Table 6.3, the clearing time θ and the K factor for the circuit breaker can be determined: $\theta = 2$ cycles, $K = 1.3$.

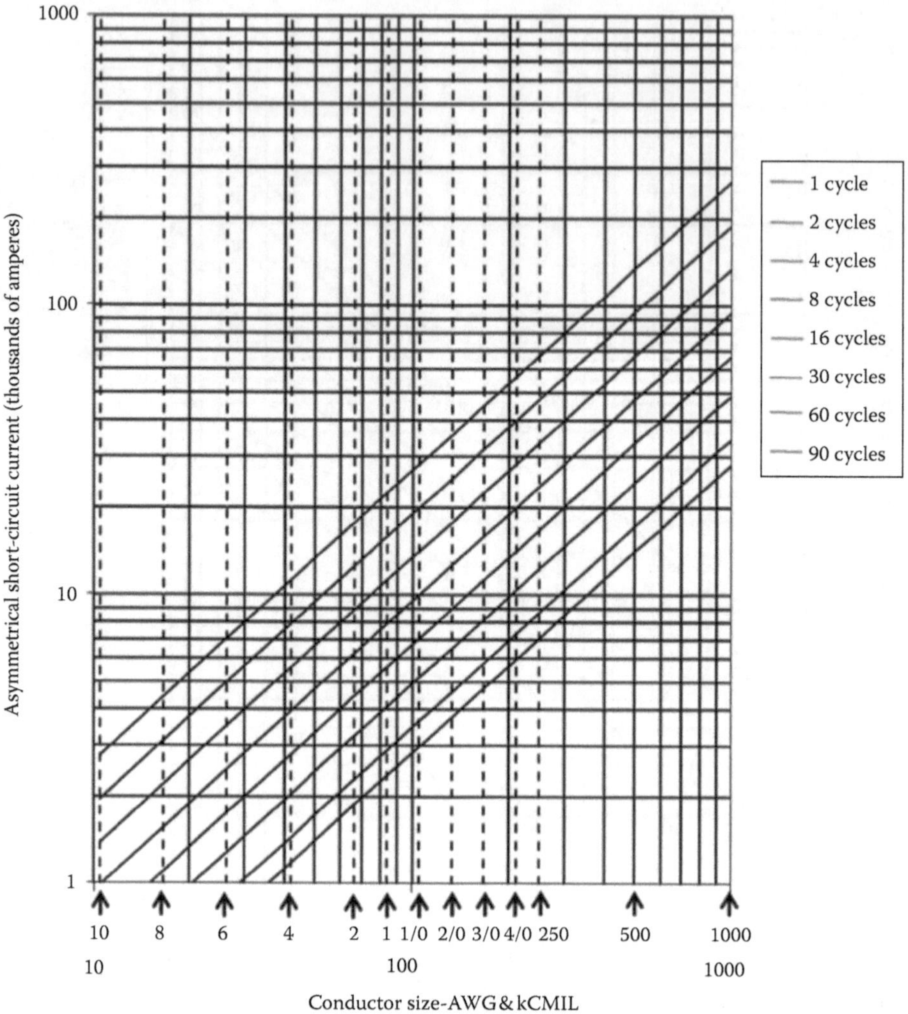

FIGURE 6.9 Allowable short-circuit currents for XHHW aluminum conductors.

From Figure 6.8, the safe size of XHHW (Cu) conductor can be estimated knowing θ and $I_{asym} = 40{,}000 \times 1.3$ A $= 52{,}000$ A: 2/0 XHHW (Cu).

Example 6.6

Problem

Determine the symmetrical short-circuit current a # 250 MCM XHHW (Al) feeder protected by a power fuse rated at 15 kV.

Solution

From Table 6.3, the clearing time θ and the K factor for the high voltage power fuse can be determined: $\theta = 1$ cycle, $K = 1.6$.

From Figure 6.9, the asymmetrical short-circuit current value can be estimated knowing θ and conductor size of 250 MCM: $I_{asym} = 67{,}000$ A. Thus and using Equation 6.8, the symmetrical short-circuit current can be calculated:

$$I_{sym} = \frac{67{,}000 \text{ A}}{1.6} = 42{,}000 \text{ A}$$

6.3.3 Maximum Allowable Voltage Drop

The third requirement for sizing conductors is the maximum allowable voltage drop across the length of branch circuits and feeders. Most electrical loads require a specific voltage level for normal operation. It is, therefore, important to reduce the voltage drop that occurs across the feeders and the branch circuits. The NEC recommends a maximum voltage drop of 5% from the service entrance to the utilization outlets with 3% limit for branch circuits and 2% limit for feeders and subfeeders, as illustrated in Figure 6.10.

The voltage drop across a conductor depends on its impedance and the current flow, as shown in Figure 6.11. Using Ohm's law, the voltage drop due to the conductor can be estimated:

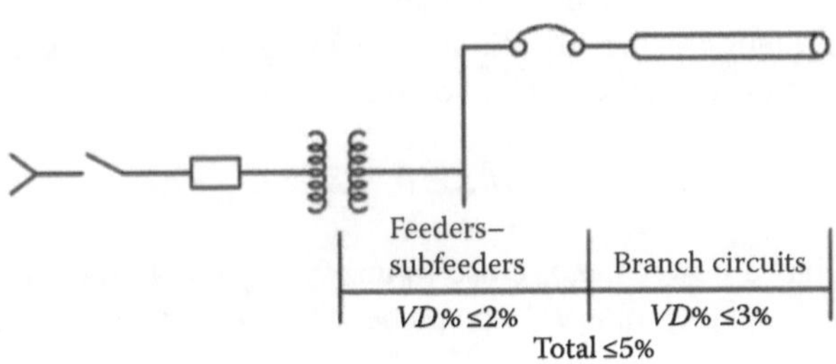

FIGURE 6.10 Typical voltage drop allowances across feeders and branch circuits.

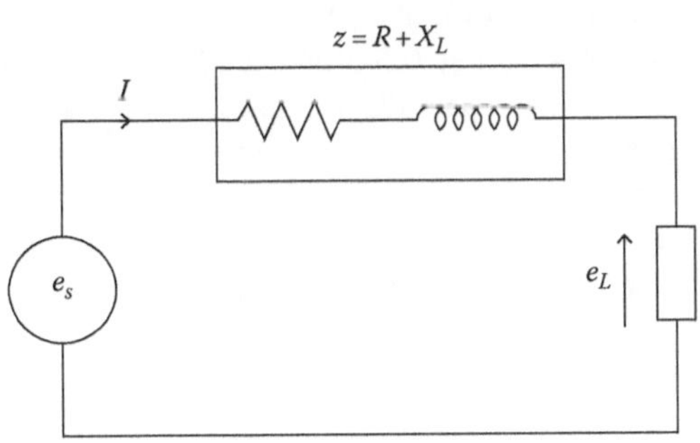

FIGURE 6.11 Voltage drop across a wire connecting the source to the load.

$$VD = Z \times I = e_S - e_L \tag{6.10}$$

The voltage drop in percent can be determined by dividing VD estimated by Equation 6.10 by the source voltage (either phase or line voltage depending on how VD is calculated):

$$VD(\%) = \frac{Z \times I}{e_S} \tag{6.11}$$

Using the concept of phasor diagram, as outlined in Chapter 2, it can be shown that the phase voltage drop can be estimated using the following expression:

$$VD = E_s + IR\cos\theta + IX\sin\theta - \sqrt{(E_s^2 + (IX\cos\theta + IR\sin\theta)^2} \tag{6.12}$$

where

- E_s is the root mean square or rms source phase voltage (V);
- I is the load current flowing through the conductor (A);
- R is the resistance of the conductor (Ω);
- X is the reactance of the conductor (Ω);
- θ is the phase angle for the load (rad or degree).

It should be noted that the AC resistance and the reactance of a conductor depends on its length as well as the types of insulation and conduit. A simplification of the voltage drop estimation using Equation 6.12 can be made in most cases when VD remains small compared to the source voltage:

$$VD = IR\cos\theta + IX\sin\theta \tag{6.13}$$

When the load has a power factor close to unity (i.e., $\theta = 0$), the calculation can be further simplified and the voltage drop can be determined using the basic formula:

$$VD = IR \tag{6.14}$$

For general design analysis and to make the calculations of the voltage drop easier for branch circuits and feeders, Table 6.5 is provided to estimate the voltage drop per 1,000 A-ft (or 305 A-m) for various conductor sizes, power factors, and conduit types (magnetic and nonmagnetic). Example 6.7 illustrates how Table 6.5 can be used to determine the voltage drop across a feeder. General rules of the thumb can be utilized to help in the preliminary design phase of the electrical distribution systems for buildings. Some of these rules of thumb include the following:

- For 120 and 208 V branch circuits, the wire length should be limited to 100 ft (30 m) so that the voltage drop does not exceed 3%.
- For 277 V branch circuits, the wire length should be limited to 200 ft (61 m) so that the voltage drop does not exceed 3%.

TABLE 6.5
Voltage Drop Estimation to Account for Load Power Factor and Type of Conduit

Copper Wire Size (AWG/MCM)	Magnetic Conduit			Nonmagnetic Conduit		
	pf=80%	*pf*=90%	*pf*=100%	*pf*=80%	*pf*=90%	*pf*=100%
12	1.570	1.749	1.917	1.565	1.749	1.923
10	0.993	1.103	1.200	0.987	1.103	1.201
8	0.635	0.699	0.750	0.629	0.693	0.751
6	0.421	0.462	0.485	0.461	0.456	0.485
4	0.277	0.300	0.306	0.271	0.294	0.306
2	0.185	0.196	0.196	0.179	0.191	0.191
1	0.150	0.162	0.150	0.150	0.156	0.150
1/0	0.127	0.133	0.121	0.121	0.127	0.121
2/0	0.109	0.110	0.098	0.098	0.104	0.092
3/0	0.092	0.092	0.081	0.081	0.087	0.075
4/0	0.081	0.075	0.064	0.069	0.069	0.057
250	0.070	0.070	0.054	0.064	0.064	0.051
300	0.064	0.064	0.045	0.056	0.055	0.042
350	0.058	0.055	0.039	0.051	0.049	0.036
400	0.055	0.051	0.035	0.047	0.044	0.032
500	0.049	0.045	0.029	0.042	0.039	0.026
600	0.046	0.041	0.024	0.038	0.034	0.022
700	0.043	0.038	0.021	0.036	0.032	0.019
750	0.042	0.037	0.020	0.034	0.031	0.017
1,000	0.038	0.032	0.016	0.029	0.025	0.013

Notes: Values in the table in volt drop for line-to-neutral per 1,000 A-ft (305 A-m).

- For 208 V feeders, the wire length should be limited to 100 ft (30 m) so that the voltage drop does not exceed 2%.
- For 480 V feeders, the wire length should be limited to 400 ft (122 m) so that the voltage drop does not exceed 2%.

Values are specific to conductors rated at 75°C. For 90°C-rated conductors, use a correction factor of 1.1.

Example 6.7

Problem

Determine the voltage drop in percent for a 200 ft (61 m) long 300-MCM THW (Cu) feeder serving a 3-phase load rated at 300A, 200 V, and *pf*=0.9. The feeder is housed in a nonmagnetic conduit.

Solution

From Table 6.5,

$$\frac{VD}{1{,}000\ \text{A}-\text{ft}} = 0.055(\text{nonmagnetic conduit})$$

For the load and length of the feeder, the A-ft can be calculated:

$$\text{A}-\text{ft} = 300\ \text{A} \times 200 = 60 \times 10^3$$

Thus, the voltage drop line-to-neutral, VD_{NL}, can be estimated:

$$VD_{NL} = 0.0605 \times 60 = 3.63\ \text{V}$$

Therefore, the voltage drop line-to-line, VD_{LL}, is

$$VD_{LL} = \sqrt{3} \times 3.63 = 6.29\ \text{V}$$

Finally, the voltage drop expressed in percent is found to be

$$VD(\%) = \frac{VD_{LL}}{E_L} = \frac{6.29}{208} = 3.0\%$$

When the overall voltage drop is found to exceed the limit of 5% from the source to the load, a few design options can be considered:

- Increase the wire size by at least one level (for instance increase of branch circuits from #12 to #10). This option has the additional benefit of reducing the energy losses through the wire. However, additional costs are incurred due to larger wires.
- Reduce the load (in Amps) for the branch circuit or the feeder. This option may result in additional circuits and therefore costs.
- Place the distribution panels, including lighting and power panels, in more central locations to reduce the length for branch circuits and/or subfeeders.

6.3.4 Design Approach for Conductors

Table 6.6 summarizes the design criteria and the various tables and figures that need to be used to size conductors for both feeders and branch circuits. For each requirement, a specific conductor size is determined (S_1 using the ampacity, S_2 for the short-circuit current, and S_3 for the voltage drop allowance). The size of the conductor, S, has to be selected to meet all the three requirements, that is,

$$S = \max(S_1, S_2, S_3) \tag{6.15}$$

The design approach summarized in Table 6.6 is applied in Example 6.8 to size the subfeeders serving two panels, as defined in Figure 6.12.

TABLE 6.6
Summary of Tables and Figures Used to Size Conductors

Criterion	Tables/Figures	Size
Ampacity	Table 6.1 (ampacity in free air) and Table 6.2 (derating factors)	S_1
Short-circuit current	Table 6.3 (*K* factors) and Figure 6.7 (THW-Cu), Figure 6.8 (XHHW-Cu), and Figure 6.9 (XHHW-Al)	S_2
Voltage drop	Table 6.5	S_3

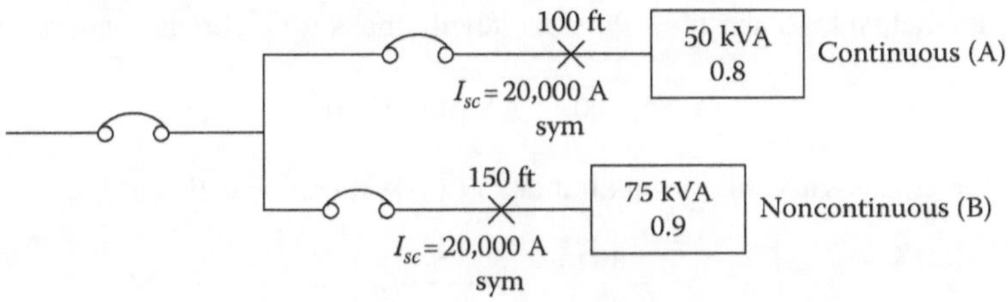

FIGURE 6.12 Feeder and subfeeders serving two panels used in Example 6.8.

Example 6.8

Problem

Size the THW (Cu) conductors for both subfeeders A and B shown in Figure 6.12 considering all the three design requirements. The panels are rated at 480 V, 3-phase, and are protected by circuit breakers. Use a voltage drop allowance of 2% for both subfeeders. IMC steel conduits are used to house the conductors.

Solution

Size of Subfeeder A:

First, the load is expressed in amperes:

$$I_L = \frac{P_T}{\sqrt{3}E_L} = \frac{50,000 \text{ VA}}{\sqrt{3} \times 480 \text{ V}} = 60.14 \text{ A}$$

Ampacity: Since the load is continuous (the conductors have to handle 80% of its rating)

$$I_{\text{conductor}} = 1.25 \times I_L = 75 \text{ A}$$

From Table 6.1

$$S_1 = \#4 \text{ THW(Cu)}$$

Short-circuit current: From Table 6.3, the clearing time θ and the K factor for the circuit breaker can be determined: $\theta = 2$ cycles, $K = 1.3$.

From Figure 6.7, the safe size of THW (Cu) conductor can be estimated knowing θ and $I_{asym} = 20{,}000 \times 1.3\ \text{A} = 26{,}000\ \text{A}$:

$$S_2 = \frac{1}{0\ \text{THW}}(\text{Cu})$$

Voltage drop allowance: The 2% voltage drop allowance is converted using line-to-neutral voltage:

$$VD_{LN} = 2\% \times 277 = 5.5\ \text{V}$$

Using the actual load and the subfeeder length, the A-ft can be calculated:

$$\text{A} - \text{ft} = 60.14\ \text{A} \times 100 = 6 \times 10^3$$

Thus, the voltage drop line-to-neutral per 1,000 A-ft can be estimated:

$$\frac{VD_{LN}}{1000\ \text{A} - \text{ft}} = \frac{5.5}{6} \approx 0.9$$

From Table 6.5, with a $pf = 0.80$ and a magnetic conduit, the size of the copper conductor can be selected:

$$S_3 = \#8\ \text{THW}(\text{Cu})$$

Conductor size: Considering the sizes found for all three criteria, the size of the conductors for the subfeeder A is defined by the short-circuit current requirement:

$$S = \max(S_1, S_2, S_3) = \frac{1}{0\ \text{THW}}(\text{Cu})$$

Size of Subfeeder B:

First the load is expressed in amperes:

$$I_L = \frac{P_T}{\sqrt{3}E_L} = \frac{75{,}000\ \text{VA}}{\sqrt{3} \times 480\ \text{V}} = 90.21\ \text{A}$$

Ampacity: Since the load is noncontinuous (the conductors can handle 100% of its rating)

$$I_{\text{conductor}} = 1.00 \times I_L = 90.2\ \text{A}$$

From Table 6.1

$$S_1 = \#3\text{THW}(\text{Cu})$$

Short-circuit current: From Table 6.3, the clearing time θ and the K factor for the circuit breaker can be determined: $\theta = 2$ cycles, $K = 1.3$.

From Figure 6.8, the safe size of THW (Cu) conductor can be estimated knowing θ and $I_{asym} = 20{,}000 \times 1.3$ A $= 26{,}000$ A:

$$S_2 = \frac{1}{0}\ \text{THW(Cu)}$$

Voltage drop allowance: The 2% voltage drop allowance is converted in line-to-neutral voltage drop:

$$VD_{LN} = 2\% \times 277 = 5.5\ \text{V}$$

For the load and the length of the subfeeder B, the A-ft can be calculated:

$$\text{A} - \text{ft} = 90.2\ \text{A} \times 200 = 18 \times 10^3$$

Thus, the voltage drop line-to-neutral per 1,000 A-ft can be estimated:

$$\frac{VD_{LN}}{1000\ \text{A} - \text{ft}} = \frac{5.5}{18} \approx 0.3$$

From Table 6.5 with a $pf = 0.90$ and a magnetic conduit, the size of the copper conductor can be selected:

$$S_3 = \#4\ \text{THW(Cu)}$$

Conductor size: Considering all the sizes, the size of the conductors for the subfeeder B, like subfeeder A, is defined by the short-circuit current requirement:

$$S = \max(S_1, S_2, S_3) = \frac{1}{0\ \text{THW}}(\text{Cu})$$

6.4 SELECTION OF CONDUITS

In commercial and large residential buildings, electrical conductors and cables are housed inside conduits and raceways for additional protection from any potential damages. The types of conduits commonly used in buildings are listed here:

- Rigid metal conduit (RMC) can be made of either steel or aluminum and has the thickest wall of all types of conduits. Rigid metal conduits with galvanized steel are typically used in hazardous locations that have high exposure to chemicals.
- Intermediate metal conduit (IMC) has a thinner wall than the rigid metal conduit but can be used for the same applications.
- Electrical metallic tubing (EMT) is a metal conduit but with a very thin wall. The NEC restricts the use of EMT to locations where it is not subjected to severe physical damages during and after installation.
- Electrical nonmetallic conduit (ENC) is made of nonmetallic materials such as fiber or rigid polyvinyl chloride (PVC). In general, a rigid nonmetallic

conduit cannot be used in locations where it can be subject to physical damages.

- Electrical nonmetallic tubing (ENT) is a pliable corrugated conduit that can be easily bent by hand. Electrical nonmetallic tubing can be concealed within walls, floors, and ceilings.
- Flexible conduit can be readily flexed and thus is not affected by vibration. Common applications of the flexible conduits are for the final connections to motors or recessed lighting fixtures.
- PVC conduit is a nonmetallic lightweight tubing and is relatively inexpensive. The main advantages of PVC are the ease of installation and the ability to withstand a wide range of environmental conditions. In particular, it can be used in underground installations. However, PVC is susceptible to corrosion caused by chemical products.

The number of electrical conductors that can be installed in one conduit is restricted to avoid any damage to the conductors and cables (especially when the conductors or cables are pulled through the conduit). The NEC restricts the percentage fill to 40% for three or more conductors. For conduits with only one and two conductors, the percentage fill is set to be 53% and 31%, respectively (NEC, 2014). The percentage fill is defined as the fraction of the total cross-sectional area of the conductors including the insulation and protection coating over the inner cross-sectional area of the conduit. Table 6.7 shows the specific areas for the 40% percent fill associated with standard sizes of EMT, IMC, and PVC conduits.

The cross-sectional areas of conductors and cables depend on their diameter as well as their insulation type. Table 6.8 lists the cross sections for standard sizes of THW (Cu), XHHW (Cu), and THHN (Cu) conductors. Using the data such as those provided by Tables 6.7 and 6.8, the maximum number of conductors that can be housed in specific conduit type and size can be determined. Table 6.9 can be utilized to size the

TABLE 6.7
Cross-Sectional Areas for Conduits Associated with 40% Fill

Trade Conduit Size	EMT Conduit in.^{2} (mm^2)	IMT Conduit in.^{2} (mm^2)	PVC in.^{2} (mm^2)
½	0.122 (78)	0.137 (89)	0.087 (56)
¾	0.213 (137)	0.235 (151)	0.164 (105)
1	0.346 (222)	0.384 (248)	0.275 (178)
1 1/4	0.598 (387)	0.659 (425)	0.495 (320)
1½	0.814 (526)	0.890 (573)	0.684 (442)
2	1.342 (866)	1.452 (937)	1.150 (742)
2½	2.343 (1,513)	2.054 (1,323)	1.647 (1,064)
3	3.538 (2,280)	3.169 (2,046)	2.577 (1,660)
3½	4.618 (2,980)	4.234 (2,729)	3.475 (2,243)
4	5.901 (3,808)	5.452 (3,490)	4.503 (2,907)

Note: This table is adapted from NEC-2014 Table 4, Chapter 9 (NEC, 2014).

TABLE 6.8
Cross-Sectional Areas for THW, XHHW, and THHN Coppers Conductors

Conductor Size (AWG, MCM)	THW (Cu)[a] in.^{2} (mm^2)	XHHW (Cu)[b] in.^{2} (mm^2)	THHN (Cu)[c] in.^{2} (mm^2)
14	0.0293 (18.9)	0.0139 (9.0)	0.0097 (6.3)
12	0.0353 (22.8)	0.0181 (11.7)	0.0133 (8.6)
10	0.0437 (28.2)	0.0243 (15.7)	0.0211 (13.6)
8	0.0835 (53.9)	0.0437 (28.2)	0.0366 (23.6)
6	0.1041 (67.2)	0.0590 (38.1)	0.0507 (32.7)
4	0.1333 (86.0)	0.0814 (52.5)	0.0824 (53.2)
3	0.1521 (98.1)	0.0962 (62.1)	0.0973 (62.8)
2	0.1750 (112.9)	0.1146 (73.9)	0.1158 (74.7)
1	0.2260 (171.6)	0.1534 (99.0)	0.1562 (100.8)
1/0	0.3039 (196.1)	0.1825 (117.7)	0.1855 (119.7)
2/0	0.3505 (226.1)	0.2190 (141.3)	0.2223 (143.4)
3/0	0.4072 (262.7)	0.2642 (170.5)	0.2679 (172.8)
4/0	0.4754 (306.7)	0.3197 (206.3)	0.3237 (208.8)
250	0.6291 (405.9)	0.3904 (251.9)	0.3970 (256.1)
300	0.7088 (457.3)	0.4536 (292.6)	0.4608 (297.3)
350	0.7870 (507.7)	0.5166 (333.3)	0.5242 (338.2)
400	0.8626 (556.5)	0.5782 (373.0)	0.5863 (378.3)
500	1.0082 (650.5)	0.6984 (450.6)	0.7073 (456.3)

Notes: This table is adapted from NEC-2014 Table 5, Chapter 9 (NEC, 2014).
[a] The listed cross-sectional areas for THW (Cu) are applicable to TW, THHW, THHW-2.
[b] The listed cross-sectional areas for XHHW (Cu) are applicable to XHH, XHHW-2.
[c] The listed cross-sectional areas for THHN (Cu) are applicable to THWN, THWN-2.

EMT conduits for THW (Cu) and THHW (Cu) conductors while Table 6.10 is suitable to select the conduit size for THHN (Cu) when using EMT, IMC, and RMC conduits. Example 6.9 illustrates the procedure to design conduits for feeders.

Example 6.9

Problem

Determine the conduit EMT size for four #3/0 THHN (Cu) conductors. All four are current-carrying conductors.

Solution

From Table 6.10, EMT conduit of 2 in. should be selected since it can handle 5 #3/0 THHN (Cu) conductors. It should be noted that the ampacity obtained from Table 6.1 of the 3/0 THHN (Cu) conductors has to be derated by 80%, as indicated by Table 6.2:

$$I_{amp} = 225 \times 0.80 = 180 \text{ A}$$

TABLE 6.9
Maximum Number of THW and THHW Conductors in EMT Conduits

Trade Size		Wire Size (THW, THHW) Conductor Size AWG/MCM																				
(in.)		14	12	10	8	6	4	3	2	1	1/0	2/0	3/0	4/0	250	300	350	400	500	600	700	750
½	EMT	6	4	3	1	1	1	1	1	1	—	—	—	—	—	—	—	—	—	—	—	—
¾	EMT	10	8	6	4	3	1	1	1	1	1	1	1	—	—	—	—	—	—	—	—	—
1	EMT	16	13	10	6	4	3	3	2	1	1	1	1	1	1	1	—	—	—	—	—	—
1¼	EMT	28	23	18	10	8	6	5	4	3	2	1	1	1	1	1	1	1	1	1	—	—
1½	EMT	39	31	24	14	11	8	7	6	4	3	3	2	1	1	1	1	1	1	1	1	1
2	EMT	64	51	40	24	18	13	12	10	7	6	5	4	3	3	2	1	1	1	1	1	1
2½	EMT	112	90	70	42	32	24	20	17	12	10	9	7	6	5	4	4	3	3	2	1	1
3	EMT	169	136	106	63	48	36	31	26	18	16	13	11	9	7	6	6	5	4	3	3	3
3½	EMT	221	177	138	83	63	47	40	34	24	20	17	15	12	10	8	7	7	6	4	4	4
4	EMT	282	227	177	106	81	60	52	44	31	26	22	19	16	13	11	10	9	7	6	5	4

Note: This table is adapted from NEC-2014 Table C.1 (NEC, 2014).

TABLE 6.10
Maximum Number of THWN and THHN Conductors in EMT, IMC, and RMC Conduits

Trade Size (in.)		Wire Size (THWN, THHN) Conductor Size AWG/MCM																				
		14	12	10	8	6	4	3	2	1	1/0	2/0	3/0	4/0	250	300	350	400	500	600	700	750
½	EMT	12	9	5	3	2	1	1	1	1	1	—	—	—	—	—	—	—	—	—	—	—
	IMC	14	10	6	3	2	1	1	1	1	1	1	—	—	—	—	—	—	—	—	—	—
	RMC	13	9	6	3	2	1	1	1	1	1	—	—	—	—	—	—	—	—	—	—	—
¾	EMT	22	16	10	6	4	2	1	1	1	1	1	1	1	—	—	—	—	—	—	—	—
	IMC	24	17	11	6	4	3	2	1	1	1	1	1	1	—	—	—	—	—	—	—	—
	RMC	22	16	10	6	4	2	1	1	1	1	1	1	1	—	—	—	—	—	—	—	—
1	EMT	35	26	16	9	7	4	3	3	1	1	1	1	1	1	1	1	—	—	—	—	—
	IMC	39	29	18	10	7	4	4	3	2	1	1	1	1	1	1	1	1	—	—	—	—
	RMC	36	26	17	9	7	4	3	3	1	1	1	1	1	1	1	1	1	—	—	—	—
1¼	EMT	61	45	28	16	12	7	6	5	4	3	2	1	1	1	1	1	1	1	1	1	—
	IMC	68	49	31	18	13	8	6	5	4	3	3	2	1	1	1	1	1	1	1	1	1
	RMC	63	46	29	16	12	7	6	5	4	3	2	1	1	1	1	1	1	1	1	1	—
1½	EMT	84	61	38	22	16	10	8	7	5	4	3	2	1	1	1	1	1	1	1	1	1
	IMC	91	67	42	24	17	10	9	7	5	4	4	3	2	1	1	1	1	1	1	1	1
	RMC	85	62	39	22	16	10	8	7	5	4	3	3	2	1	1	1	1	1	1	1	1
2	EMT	138	101	63	36	26	16	13	11	8	7	6	5	4	3	2	1	1	1	1	1	1
	IMC	149	109	68	39	38	17	15	12	9	8	6	5	4	3	3	2	2	1	1	1	1
	RMC	140	102	64	37	27	16	14	11	8	7	6	5	4	3	3	2	2	1	1	1	1

(Continued)

TABLE 6.10 (*Continued*)
Maximum Number of THWN and THHN Conductors in EMT, IMC, and RMC Conduits

Trade Size		Wire Size (THWN, THHN) Conductor Size AWG/MCM																				
(in.)		14	12	10	8	6	4	3	2	1	1/0	2/0	3/0	4/0	250	300	350	400	500	600	700	750
2½	EMT	241	176	111	64	46	28	24	20	15	12	10	8	7	6	5	4	4	3	2	2	1
	IMC	211	154	97	56	40	25	21	17	13	11	9	7	6	5	4	4	3	3	2	1	1
	RMC	200	146	92	53	38	23	20	17	12	10	8	7	6	5	4	3	3	2	1	1	1
3	EMT	364	226	167	96	69	43	36	30	22	19	16	13	11	9	7	6	6	5	4	3	3
	IMC	362	238	150	86	62	38	32	27	20	17	14	12	9	8	7	6	5	4	3	3	3
	RMC	309	225	142	82	59	36	31	26	19	16	13	11	9	7	6	5	5	4	3	3	3
3½	EMT	476	347	219	126	91	56	47	40	29	25	20	17	14	11	10	9	8	6	5	4	4
	IMC	436	318	200	115	83	51	43	36	27	23	19	16	13	10	9	8	7	6	5	4	4
	RMC	412	301	189	109	79	48	41	34	25	21	18	15	12	10	8	7	7	5	4	4	4
4	EMT	608	443	279	161	116	71	60	51	37	32	26	22	18	15	13	11	10	8	7	6	5
	IMC	562	410	258	149	107	66	56	47	35	29	24	20	17	13	12	10	9	7	6	5	5
	RMC	531	387	244	140	101	62	53	44	33	27	23	19	16	13	11	10	8	7	6	5	5

Note: This table is adapted from NEC-2014 Table C.1 (NEC, 2014).

6.5 BRANCH CIRCUITS AND FEEDERS FOR NONMOTOR LOADS

6.5.1 Design of Branch Circuits

Branch circuit is a conductor or a set of conductors that extends typically from a panel (such as a power or lighting panel) to the utilization equipment such as a receptacle, a motor, or a lighting fixture. Branch circuits are generally not rated based on the ampacity of the conductors but by the size of the overcurrent protection devices connected to them with ratings of 15, 20, 30, 40, and 50 A. Therefore, the ampacity of the conductors is typically higher than the branch circuit rating. For instance and based on a voltage drop criterion, a 20 A branch circuit can have #10 AWG THW (Cu) conductors with a rating of 35 A instead of #12 AWG THW(Cu) conductors with a rating of 25 A. It should be noted that the minimum size of copper conductors that can be used for branch circuits in residential buildings is #14 AWG. For commercial buildings and industrial facilities, #12 conductors are typically the smallest conductor size used for branch circuits. To protect their insulation, branch circuit conductors can be installed in conduits, as discussed in the previous section. In many cases, and especially for residential buildings with wood frame construction, however, the branch circuits are placed inside the walls without any conduit.

According to the NEC, the conductors should have an ampacity that is higher than the maximum load served by a branch circuit (NEC, 2014). In the case when the branch circuits serve loads that are totally or partially continuous, the conductors should have an ampacity that is not less than the noncontinuous loads plus 125% of the continuous loads. The NEC defines continuous loads as electrical loads that operate without interruption for at least 3 hours. Thus, the ampacity of any branch circuit conductor, I_{cond}, should be determined according to the following expressions:

For noncontinuous loads

$$I_{cond} = 100\%\ I_L \tag{6.16}$$

For continuous loads

$$I_{cond} = 125\%\ I_L \tag{6.17}$$

In the United States, the NEC provides guidelines to estimate the minimum lighting loads for specific spaces defined by their occupancy type. These minimum loads are expressed in volt-amperes per unit floor area, as summarized in Table 6.11, for selected occupancy types. Note that the actual loads should be used to design branch circuits and feeders for nonmotor loads. Example 6.10 shows how to determine the number of branch circuits for lighting load in an office space.

TABLE 6.11
NEC Minimum Loads for General Lighting for Select Occupancy Types

Occupancy Type	General Lighting Load	
	VA/m²	VA/ft²
Auditoriums	11.0	1.0
Banks[a]	39.0	3.5
Churches	11.0	1.0
Court rooms	22.0	2.0
Dwelling units[b]	33.0	3.0
Hospitals	22.0	2.0
Hotels and motels (rooms without cooking appliances)	22.0	2.0
Lodge rooms	17.0	1.5
Office buildings[a]	39.0	3.5
Restaurants	22.0	2.0
Schools	33.0	3.0
Stores[c]	33.0	3.0
Warehouses for storage	3.0	0.25
Other spaces such as halls, corridors, stairways, and closets	6.0	0.50

Note: This table is adapted from NEC-2014 Table 220.12 (NEC, 2014).

[a] For banks and office buildings, the larger loads estimated the values in this table and those estimated values based on actual connected lighting fixtures and receptacles should be utilized in the design specifications.

[b] For dwellings, refer to Chapter 5 for more details on the design specifications for branch circuits and service entrance feeders.

[c] For stores, additional show windows loads can be estimated using 200 VA/0.30 m (1 ft).

Example 6.10

Problem

To meet the lighting requirements for an office space, 105 luminaires are specified and are placed along three rows. Each luminaire is rated at 252 VA and requires 120 V (from a 208Y/120 V source). Determine the number of 20 A branch circuits required to serve all the luminaires.

Solution

First the load for each luminaire is estimated:

$$I_{\frac{L}{1u}} = \frac{252 \text{ VA}}{120 \text{ V}} = \frac{2.1 \text{ A}}{1 \text{ u}}$$

One 20 A branch circuit can handle 16 A (80% of 20 A) since the lighting load is continuous for office spaces, the number of luminaires that can be connected to one circuit can be found:

$$\frac{\frac{16\text{ A}}{\text{circuit}}}{\frac{2.1\text{ A}}{1\text{ u}}} \approx \frac{7\text{l u}}{\text{circuit}}$$

Then, the total number of branch circuits needed to serve all the 105 luminaires is estimated to be

$$\frac{105}{7} = 15 \text{ circuits}$$

6.5.2 Design of Feeders

For sizing the feeders and subfeeders, demand loads rather than the actual connected loads should be considered to account for the fact that not all the loads would be utilized simultaneously. The demand loads can be calculated using demand factors as summarized in Table 6.12 for lighting loads and in Table 6.13 for the receptacle loads

TABLE 6.12
Demand Factors for Lighting Loads for Select Occupancy Types

Occupancy Type	Connected Loads Range (VA)	Demand Factors (%)
Dwelling units	0–3,000	100
	3,001–120,000	35
	Over 120,000	25
Hospitals[a]	0–50,000	40
	Over 50,000	20
Hotels and motels[a] (rooms without cooking appliances)	0–20,000	50
	20,001–100,000	40
	Over 100,000	30
Warehouses for storage	0–12,500	100
	Over 12,500	50
All other spaces	All	100

Note: This table is adapted from NEC-2014 Table 220.42 (NEC, 2014).

[a] Do not use these demand factors for feeders that are serving lighting loads that can be operated simultaneously (such as in operating rooms or dining rooms).

TABLE 6.13
Demand Factors for Nondwelling Receptacle Loads

Connected Loads Range (VA)	Demand Factors (%)
0–10,000	100
Over 10,000	50

Note: This table is adapted from NEC-2014 Table 220.44 (NEC, 2014).

specific to nonresidential buildings. Example 6.11 illustrates the design procedure for feeders serving a 480 V panel as part of a commercial building electrical distribution system.

Example 6.11

Problem

A 480Y/277 V panel has the following loads:

- *Lighting*: Four circuits at 277 V with load = 16 kVA, *pf* = 0.95 (continuous)
- *Appliances*: Two circuits at 277 V with load = 9 kVA, *pf* = 0.9 (noncontinuous)
- *A/C system*: One circuit at 480 V with load = 15 kVA, *pf* = 0.85 (continuous)

Determine the size of the panel feeder using THW (Cu) conductors, EMT conduit, and circuit breaker as a protection device for the panel.

Solution

First, the demand load for the panel is calculated using the demand factors for various loads as detailed in Table 6.14.

Thus, the total demand load for the panel is $P_T = \sqrt{(43.0)^2 + (20.1)^2} = 47.4$ kVA.

The current based on the demand load can then be estimated to size both the feeder conductors and the circuit breaker:

$$I_L = \frac{P_T}{\sqrt{3}E_L} = \frac{47{,}400}{(1.732)(480)} = 57.0 \text{ A}$$

Table 5.4, 60 A/3P can be used for the trip rating for the circuit breaker protection the panel.

Table 6.1 #6THW (Cu) conductors should be considered for the panel feeder based only on the ampacity requirement.

Table 6.9, 1 in. EMT conduit can be used so it can handle 4 #6THW (Cu) conductors.

TABLE 6.14
Demand Load for Panel of Example 6.11

	Connected Load			Demand Load	
Load Type	P_R **(kW)**	P_X **(kVAR)**	**DF**	P_R **(kW)**	P_X **(kVAR)**
Lighting	15.2	5.0	1.25	19.0	6.3
Appliances	8.1	3.9	1.0	8.1	3.9
A/C	12.7	7.9	1.25	15.9	9.9
Total				43.0	20.1

6.6 BRANCH CIRCUITS AND FEEDERS FOR MOTORS

As discussed in Chapter 4, motors have inherent special characteristics that nonmotor loads do not have, including the following:

- Variable current requirements during the operation of the motors. In particular, the starting current of induction is generally high ranging from 4 to 10 times the normal rated current reached during normal operation. The high starting current can last for 3–10 seconds and should be accounted for when designing the protection device for motors. Moreover and when overloaded, the motor can overheat and should be protected.
- The motor has to be controlled to operate under various conditions. For instance, motors for air-conditioning systems may have to change their speed (rpm) using variable frequency drives to meet different heating and cooling thermal loads. Specific controls have to be specified when designing branch circuits and feeders for motors.

In this section, the design procedures for both branch circuits and feeders for motors are described with a focus on induction motors that are commonly used in buildings.

6.6.1 Motor Branch Circuits

A branch circuit for a motor has to include various components to provide a starting mechanism for the motor as well as protection against both overloading conditions and short-circuit faults. Figure 6.13 illustrates typical components that are part of a three-phase motor branch circuit. In particular, the motor branch circuit includes a protection device, an overload relay, and a controller as well as a series of switches or disconnects depending on the location of the voltage source, the starter, and the motor. The equivalent one-line diagram for the same motor branch circuit of Figure 6.13 is shown in Figure 6.14.

Several branch circuit configurations can be considered depending on whether or not the motor is out of sight of the controller and whether or not the controller is out of sight of the voltage source. Figure 6.15 represents the one-line diagrams for three motor branch circuit configurations.

Unlike other building electrical loads such as lighting and receptacles, motors have controllers to achieve several functions: start manually or automatically, change

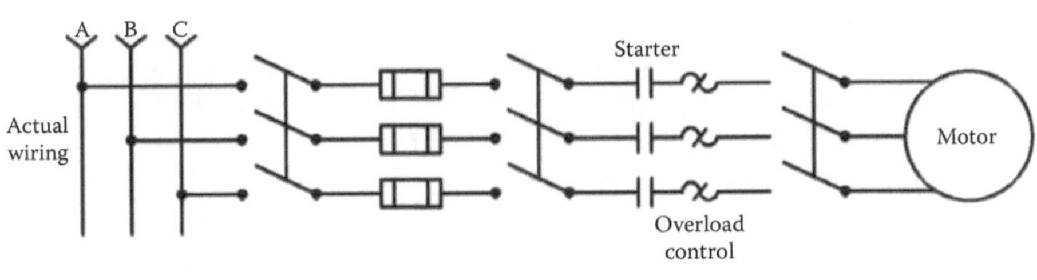

FIGURE 6.13 Typical branch circuit for a three-phase motor.

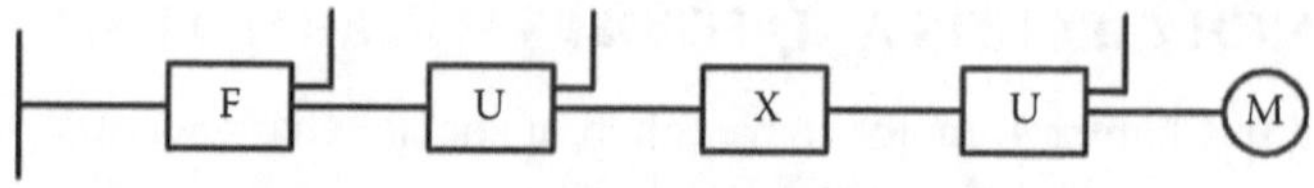

FIGURE 6.14 Equivalent one-line diagram for a motor branch circuit.

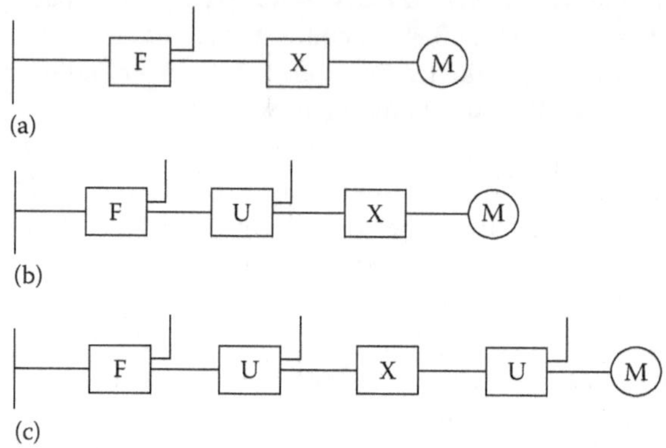

FIGURE 6.15 One-line diagrams for three motor branch circuit configurations. (a) All components are within the sight of each other. (b) Only motor and controller are within the sight of each other. (c) All components are out of sight of each other.

the direction and the speed of rotation, and provide protection from either overload or short-circuit currents. There are several motor controllers:

1. The simplest motor controller types include manual starters used for small motors (less than 1 horsepower) with 1, 2, or 3 poles.
2. Magnetic starters are also simple motor controllers and include three main types: open type, combination type, and group mounted.
3. Full-voltage nonreversing (FVNR) magnetic starters are used for full-voltage starting and stopping of AC squirrel cage motors.
4. Full-voltage reversing (FVR) magnetic starters are used for full-voltage starting, stopping, and reversing of AC squirrel cage motors. They are available in either horizontal or vertical arrangements.

It should be noted that for both the FVNR and FVR starters, motor overload protection is provided using melting alloy type thermal overload relays. These starters are available in NEMA sizes 00–7 and are designed for operation at 600 V, 50–60 Hz. They have also solid-state overload relay protection with available sizes 00–7. Figure 6.16 shows the control diagram for FVNR and FVR starters. In particular, the FVR starter is based on switching the connections between two phases in order to reverse the rotor magnetic field rotation and ultimately to reverse the rotation of the motor.

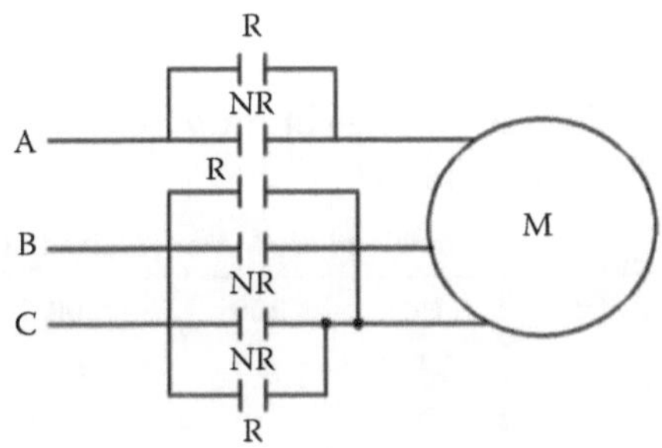

FIGURE 6.16 Control diagram for both FVNR and FVR for a motor.

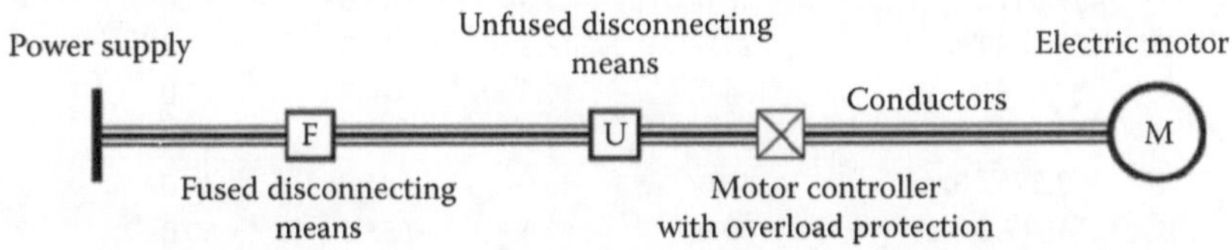

FIGURE 6.17 Typical branch circuit components for a single motor.

There are other starter types that utilize lower voltage to start the operation of the motors including

- Reduced-voltage nonreversing (RVNR) magnetic starters
- Autotransformer starters
- Wye–delta starters

However, these types of starters are rapidly being made obsolete by the variable frequency drive systems, as discussed in Chapter 4.

To design the branch circuit for a single motor, four essential components need to be sized: the conductor, the protective device with the disconnecting means, the overload protection, and the unfused disconnect means, as shown in Figure 6.17. The following steps are suggested to size all the components of a motor branch circuit:

1. *Determine motor load*: The rated load current for a motor can be determined using the manufacturer data, calculations (based on the specifications of horsepower, voltage, efficiency, and power factor, as outlined in Chapter 4), or simply from Table 6.15 for single-phase motors, Table 6.16 for three-phase induction-type motors, or Table 6.17 for three-phase synchronous-type motors.
2. *Conductor size*: Using the NEC requirements for motor branch circuits, the minimum ampacity of the conductor should be at least 125% of the rated load current. Using Table 6.1, the size of conductors can be determined. Tables 6.9 and 6.10 are then considered to obtain the conduit size.

TABLE 6.15
Full-Load Current Ratings in Amperes for Single-Phase Motors

Horsepower HP (kW)	Single-Phase Motor Voltage Rating (V) 115	200	208	230
1/6 (0.12)	4.4	2.5	2.4	2.2
1/4 (0.19)	5.8	3.3	3.2	2.9
1/3 (0.25)	7.2	4.1	4.0	3.6
1/2 (0.37)	9.8	5.6	5.4	4.9
3/4 (0.56)	13.8	7.9	7.6	6.9
1 (0.75)	16.0	9.2	8.8	8.0
1.5 (1.12)	20.0	11.5	11.0	10.0
2 (1.49)	24.0	13.8	13.2	12.0
3 (2.24)	34.0	19.6	18.7	17.0
5 (3.73)	56.0	32.2	30.8	28.0
7.5 (5.60)	80.0	46.0	44.0	40.0
10 (7.46)	100.0	57.5	55.0	50.0

Note: This table has been adapted from NEC-2014 Table 430-248 (NEC, 2014).

3. *Protective device with a disconnecting means*: The protection device for a motor is sized using demand factors provided in Table 6.18 depending on the type of the overcurrent protection device. To better understand the demand factor requirements, Figure 6.18 illustrates the response times of various protection devices relative to the performance curve of an induction motor that has a starting current equivalent to six times the rated current. As noted in Figure 6.18, the time-delay fuse has a response curve that matches closely the operation of the induction motor and thus its demand factor is the lowest. However, the instantaneous circuit breaker does not match the motor operation curve and has to be sized significantly higher than the motor-rated current to avoid tripping the device whenever the motor starts. With the fuses, disconnect means or switches have also to be specified. The standard sizes for the disconnecting means connected to fuses (often referred to as fused switches) include 30, 60, 100, 200, 400, 600, 800, and 1,200 A. According to the NEC, the size of the disconnecting means should be rated no less than 115% of the motor full-load current rating (NEC, 2014). Table 6.19 provides a guide to select three-pole fused switches depending on the fuse types, system voltages, and the horsepower ratings of three-phase induction motors.
4. *Overload protection and motor controller*: The rating of the overload protection depends on the motor service factor (SF). A rating factor of 115% is used for $SF = 1.0$ and of 125% for $SF = 1.15$.

TABLE 6.16
Full-Load Current Ratings in Amperes for Three-Phase Induction Motors

Horsepower HP (kW)	Three-Phase Motor Voltage Rating (V) 115	200	208	230	460	575	2,300
1/2 (0.37)	4.4	2.5	2.4	2.2	1.1	0.9	—
3/4 (0.56)	6.4	3.7	3.5	3.2	1.6	1.3	—
1 (0.75)	8.4	4.8	4.6	4.2	2.1	1.7	—
1.5 (1.12)	12.0	6.9	6.6	6.0	3.0	2.4	—
2 (1.49)	13.0	7.8	7.5	6.8	3.4	2.7	—
3 (2.24)	—	11.0	10.6	9.6	4.8	3.9	—
5 (3.73)	—	17.5	16.7	15.2	7.6	6.1	—
7.5 (5.60)	—	25.3	24.2	22.0	11.0	9.0	—
10 (7.46)	—	32.2	30.8	28.0	14.0	11.0	—
15 (11.19)	—	48.3	46.2	42.0	21.0	17.0	—
20 (14.92)	—	62.1	59.4	54.0	27.0	22.0	—
25 (18.65)	—	78.2	74.8	68.0	34.0	27.0	—
30 (22.38)	—	92.0	88.0	80.0	40.0	32.0	—
40 (29.84)	—	120.0	114.0	104.0	52.0	41.0	—
50 (37.30)	—	150.0	143.0	130.0	65.0	52.0	—
60 (44.76)	—	177.0	169.0	154.0	77.0	62.0	16.0
75 (55.95)	—	221.0	211.0	192.0	96.0	77.0	20.0
100 (74.60)	—	285.0	273.0	248.0	124.0	99.0	26.0
125 (93.25)	—	359.0	343.0	312.0	156.0	125.0	31.0
150 (111.90)	—	414.0	396.0	360.0	180.0	144.0	37.0
200 (149.20)	—	552.0	528.0	480.0	240.0	192.0	49.0

Note: This table has been adapted from NEC-2014 Table 430-250 (NEC, 2014).

5. As noted earlier, several types of controllers do exist to operate motors including manual starters, FVNR starters, FVR starters, and RVNR starters. The NEMA size of the starter can be selected based on Table 6.20 depending on the voltage and the horsepower rating of one-phase and three-phase motors.
6. *Unprotected disconnecting means*: The rating of unprotected disconnected means, also known as unfused switches, depends on the size of the motor. For systems with 600 V or less, the NEC specifies that the size of the disconnecting means should be no less than 115% of the full-load current rating of the motor. However, lower ratings are permitted for unfused switches. Table 6.21 provides a guide for selecting unfused switches for different system voltages and horsepower ratings.

A typical one-line diagram for a motor branch circuit is shown in Figure 6.19. Example 6.12 uses the outlined step-by-step approach to design a branch circuit for a three-phase induction motor.

TABLE 6.17
Full-Load Current Ratings in Amperes for Three-Phase Synchronous Motors

Horsepower HP (kW)	Three-Phase Motor Voltage Rating (V) 230	460	575	2,300
25 (18.65)	53	26	21	—
30 (22.38)	63	32	26	—
40 (29.84)	83	41	33	—
50 (37.30)	104	52	42	—
60 (44.76)	123	61	49	12
75 (55.95)	155	78	62	15
100 (74.60)	202	101	81	20
125 (93.25)	253	126	101	25
150 (111.90)	302	151	121	30
200 (149.20)	400	201	161	40

Note: This table has been adapted from NEC-2014 Table 430-250 (NEC, 2014). The values are specific to the case of power factor of 1.0. If the motors have a power factor of 0.90 and 0.80, the values should be multiplied by respectively, 1.1 and 1.25.

TABLE 6.18
Maximum Rating or Setting of Motor Branch Circuit Short-Circuit and Ground-Fault Protective Devices

Type of Motor	Percentage of Full-Load Current: Non-Time-Delay Fuse	Dual Element (Time-Delay) Fuse	Instantaneous Trip Breaker	Inverse Time Breaker
Single-phase motors	300	175	800	250
AC polyphase motors other than wound-rotor/squirrel cage other than design/B energy efficient	300	175	800	250
Design B energy efficient	300	175	1,100	250
Synchronous	300	175	800	250
Wound rotor	150	150	800	150
Direct current (constant voltage)	150	150	250	150

Note: This table is adapted from NEC-2014 Table 430-52 (NEC, 2014).

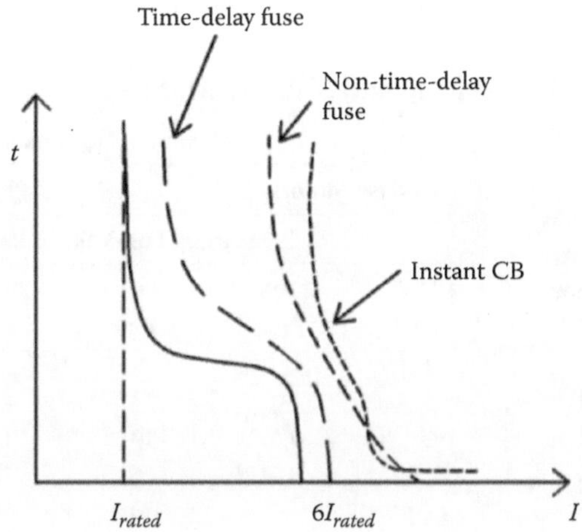

FIGURE 6.18 Induction motor operation compared to response curve of various protection devices.

TABLE 6.19
Ratings for Three-Pole Fused Switches for Motors Based on Maximum Horsepower Size

	System Voltage					
	240V		480V		600V	
Switch Rating (A)	Non-Time-Delay Fuse	Time-Delay Fuse	Non-Time-Delay Fuse	Time-Delay Fuse	Non-Time-Delay Fuse	Time-Delay Fuse
30	3	7.5	7.5	15	10	20
60	7.5	15	15	30	20	50
100	15	30	25	60	30	75
200	25	60	50	100	60	100
400	50	100	—	100	—	100

Example 6.12

Problem

Design the branch circuit for a motor with the following characteristics: 30 HP, 460 V, $SF = 115\%$, non-time-delay fuse, FVR, and with nameplate full-load current of 35 A. the controller and the motor are out of sight of the protection device. Use THW (Cu) conductor and EMT conduit for the branch circuit.

TABLE 6.20
NEMA Starter Size for 1-Phase and 3-Phase Motors

		Maximum Horsepower				
		1-Phase Motors		3-Phase Motors		
		Starting Full-Voltage Rating				
NEMA Size	Continuous Current Rating (A)	120V	230V	200V	230V	460 V/575V
00	9	1/3	1	1.5	1.5	2
0	18	1	2	3	3	5
1	27	2	3	7.5	7.5	10
2	45	3	7.5	10	15	25
3	90	7.5	15	25	30	50
4	135	—	—	40	50	100
5	270	—	—	75	100	200
6	540	—	—	150	200	400
7	810	—	—	200	300	600
8	1,215	—	—	400	450	900
9	2,250	—	—	—	800	1,600

TABLE 6.21
Ratings for Three-Pole Unfused Switches for Motors Based on Maximum Horsepower Size

	System Voltage		
Switch Rating (A)	240V	480V	600V
30	7.5	15	20
60	15	30	50
100	30	60	75
200	60	100	100
400	100	—	—

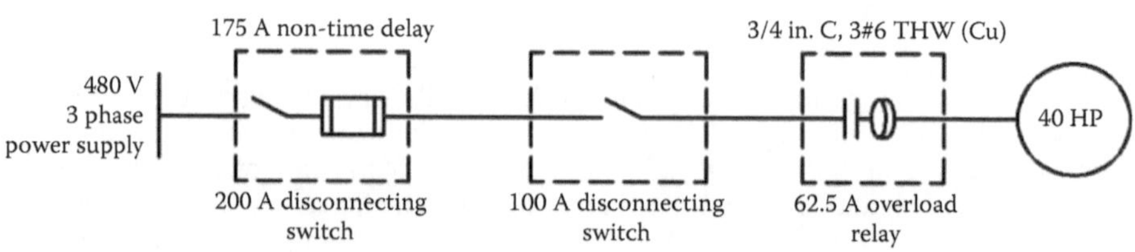

FIGURE 6.19 The branch circuit specifications for a 40 HP motor.

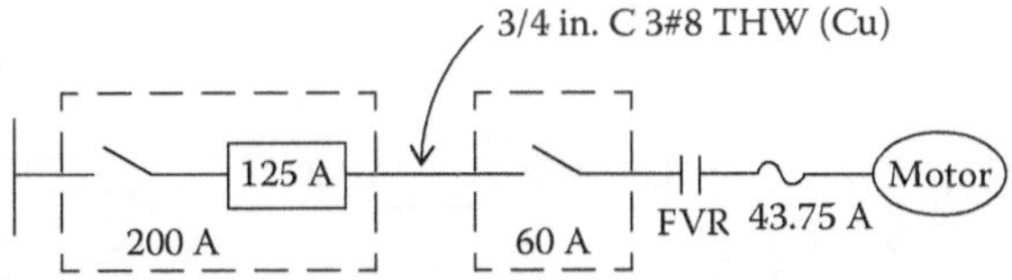

FIGURE 6.20 One-line diagram for the motor branch circuit specifications for Example 6.12.

Solution

Using the procedure outlined earlier, the following design specifications can be made for the components of a motor branch circuit:

1. Table 6.16: load $I_L = 40$ A for 30 HP, 460 V, three-phase induction motor
2. *Conductor size*: $I_{cond} = 1.25\ I_L = 50$A Table 6.1: #8 THW (Cu) Table 6.9 (using EMT conduit): 3/4 in. C assuming only three wires for the motor
3. *Fuses*: Table 6.18: $DF = 300\%$. Thus, $I_{fuse} = 300\% \times 40$ A $= 120$ A Table 5.2 (fuse ratings): Use 125 A fuses with fused switches rated at 200 A
4. *Overload protection*: Since $SF = 115\%$, $DF = 1.25$

$$I_{OL} = 1.25 \times 35 = 43.75 \text{ A since } I_{FLC} = 35 \text{ A}$$

Table 6.20: The FVR controller has a 3 NEMA starter.

Unfused switches: Table 6.21: For the unfused switches, use 60 A.

Figure 6.20 illustrates the one-line diagram for the motor branch circuit specifications defined in Example 6.12.

6.6.2 Feeder for Several Motors

In general, two components need to be sized to design a motor feeder that serves several motors: the conductor and the protective device with its disconnecting means. Typically, overload protection is not provided to the feeder since individual motors are cleared by their own overload relays in case any excessive currents occur.

1. *Motor load*: The rated load currents can be calculated as outlined in Chapter 4 for both single-phase and three-phase motors. However and in the cases when complete specifications of the motors are not available, the load currents for motors can be determined from Tables 6.15–6.17 depending on their types, sizes, and voltages.
2. *Conductor size*: Using the NEC specifications for motor branch circuit, the minimum ampacity of the conductor is 125% of the largest rated motor load plus the sum of the remainder rated motor loads:

$$I_{\text{cond}-f} = 1.25 \times I_{\text{largest}} + \sum I_{\text{remaider}} \tag{6.18}$$

3. Using Table 6.1, the size of conductors for the feeder can be determined. Tables 6.9 and 6.10 can then be used to obtain the feeder conduit size depending on the types of insulated conductors and conduits.

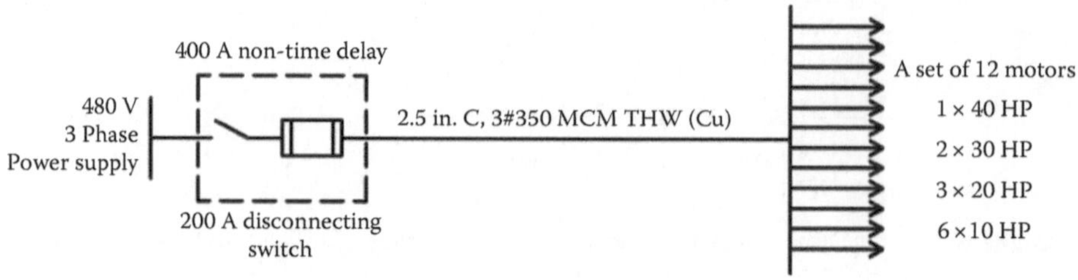

FIGURE 6.21 The feeder specifications to supply several motors.

4. *Protective device with a disconnecting means*: According to the NEC, the overcurrent protection device for a feeder (serving several motors) should be set at a rating no greater than the value calculated by taking the largest rating of the largest motor current multiplied by the appropriate demand factor, DF, provided by Table 6.18, plus the sum of the full-load currents of all the remaining motors.

$$I_{OC-f} = DF \times I_{\text{largest}} + \sum I_{\text{remaider}} \tag{6.19}$$

Example 6.13 shows the design procedure for a feeder that serves several three-phase induction motors. A one-line diagram for a feeder serving several motors is shown in Figure 6.21. Typically, the protection devices and controllers are housed in a motor control center or MCC when there are several large motors to be supplied by the electrical distribution system. If there are only a few large motors, the MCC may be avoided by supplying power using branch circuits directly from the main distribution panels or unit substations, as will be discussed in Chapter 8.

Example 6.13

Problem

Design the feeder serving 12 three-phase, 460 V, induction motors, as outlined in Figure 6.21. Consider non-time-delay fuse as a protection device and use THW (Cu) conductor and IMC conduit for the feeder.

Solution

Using the procedure outlined by Equations 6.18 and 6.19, the feeder serving the 12 motors can be specified as follows:

Table 6.16: The loads for various motors are $I_L=52$ A for 40 HP, $I_L=40$ A for 30 HP, $I_L=27$ A for 20 HP, $I_L=14$ A for 10 HP, and all induction motors are rated 460 V.

Conductor size: $I_{\text{cond}-f}=1.25\times52+2\times40+3\times 27+6\times14=310$ A Table 6.1: 350 MCM THW (Cu) Table 6.9 (using IMC conduit): 2.5 in. C using three wires for the feeder

Fuses: Table 6.13: $DF=300\%$. Thus, $I_{OC\text{-}f}=300\%\times52$ A$+2\times40+3\times27+6\times14=401$ A Table 5.2 (fuse ratings): Use 400 A fuses with fused switches rated at 200 A.

6.7 SUMMARY

In this chapter, conductors used for branch circuits and feeders are described including their sizes and their insulation types. Three requirements are typically used to select the size of conductors including ampacity, short-circuit current tolerance, and voltage drop allowance. To protect insulated conductors in large commercial buildings, conduits and raceways are used to house all the phases and neutral wires. The specific design approaches for branch circuits and feeders are outlined for both non-motor and motor loads using the NEC guidelines. Specific examples are provided throughout this chapter to illustrate the design guidelines for branch circuits and feeders.

6.8 PROBLEMS

1. Determine the ampacity rating of each of the 12 #4 THW (Cu) conductors placed in the same conduit.
2. Determine the ampacity of each of 5 #8 THHN (Cu) conductors installed in an area with a surrounding ambient temperature is 45°C.
3. Check if a #12 THW (Cu) conductor can supply power to 10 fluorescent lighting fixtures each rated at 1.8 A.
4. Determine the size of THHN (Cu) conductors needed to supply eight rows of fluorescent lighting fixtures with seven fixtures in each row with each fixture rated at 1.5 A. If only 20 A circuits are used, estimate the number of the circuits needed to serve all the lighting fixtures and the size of the THHN (Cu) recommended for each circuit.
5. Determine the symmetrical short-circuit current a #1/0 XHHW (Cu) conductor protected by a circuit breaker rated at 480 V.
6. Estimate the maximum asymmetrical short-circuit current that #1/0 THW (Cu) conductors can handle if circuit breakers are used as overcurrent protection devices.
7. Calculate the voltage drop in percent for a feeder that serves a load of 100 kW, 480 V, three-phase, power factor of 85%, and length of 300 ft. The feeder is #1 XHHW copper conductors in an aluminum conduit.
8. Calculate the percent voltage drop of a 240 V, 1-phase feeder consisting of 150 ft one way of #10 THW copper conductors in steel EMT conduit carrying a load of 20 A, 90% PF.
9. Calculate the percent voltage drop of a 120 V, 1-phase branch circuit consisting of 25 m one way of 2 #12 THHN copper conductors in EMT conduit carrying a load of 16 A, 85% PF.
10. Calculate the percent voltage drop of a 240 V, 3-phase feeder consisting of 190 ft one way of 2 #4 THW copper conductors in steel EMT conduit carrying a load of 40 A, 85% PF.
11. A building is supplied by 750 ft, three-wire, 480 V, three-phase feeder. The total load of the building is 250 kVA with a power factor of 0.8. Determine
    a. The size of the feeder and conduit using THW copper conductors
    b. The voltage drop in the feeder (assume a steel conduit)
    c. The size of the feeder and conduit that limit the voltage drop to 1%

12. Three 120/240 V panelboards are supplied with three subfeeders: *A*, *B*, and *C*. *A* is 50 ft, *B* is 100 ft, and *C* is 50 ft in length. The load on *A* is 15 kVA with a *pf* of 1.0; load *B* is 25 kVA with a *pf* of 0.95; and load *C* is 35 kVA with *pf* of 0.9. The three subfeeders are supplied with a 75 ft long main feeder. Determine the size of the main feeder and the three subfeeders. The allowance for voltage drop (line to neutral) is 1% for all conductors. Assume THW copper conductors installed in a steel conduit. All the loads are non-continuous and are protected by molded-case circuit breakers. The available symmetrical short-circuit current for all subfeeders and the main feeder was found to be 20,000 A.
13. Select the size of conductors and conduit for a feeder that services a lighting panel with the following features:
    - Load is 275 kW with a power factor of 90% and is continuous.
    - More than 50% of the load has high third-harmonic components.
    - The panel is supplied by 480Y/277 V, three-phase, and four-wire system.
    - The length of the feeder is 250 ft.
    - The available fault current is 42,000 A symmetrical.
    - The feeder overcurrent protection is a molded-case breaker.
    - The maximum allowable voltage drop is 2%.
    - Conductors to be type THW copper in steel conduit.
14. A 208Y/120 V panelboard for an office building has the following loads:
    - Receptacles with a total load of 30.00 kVA at $pf = 0.90$
    - Lighting with a total load of 20.00 kVA at $pf = 0.95$
    - One Motor with a load of 5 kVA at $pf = 0.92$
    - Heaters with a total load 12.00 kVA at $pf = 1.00$
    - Determine the size of THW-feeder for the panelboard. Size the conduit.
15. A panelboard for a commercial building is supplied by 250 ft, four-wire, 480Y/277 V, three-phase feeder. The panel has the following loads:
    - A lighting system of a total power 30 kVA with a power factor of 0.93
    - A 15 kVA resistive heater with a power factor of 1.0
    - A motor rated at 8 kVA and a power factor of 0.82

    Determine
    a. The size of the feeder and conduit using THW copper conductors (use only the ampacity criterion and the appropriate demand loads).
    b. The voltage drop in the feeder (assume a steel conduit).
    c. The size of the feeder and conduit if a limit for the voltage drop is set to be 1% and the feeder should be able to handle a symmetrical short-circuit current is 35,000 A. A circuit breaker is used to protect the feeder.
16. Calculate the motor circuit conductors for a 30 HP, 460 V, 3-phase squirrel cage motor; use copper, 75°C THW.
17. Determine the size of the disconnecting means for a 50 HP, 230 V, 3-phase squirrel cage motor.
18. Size the branch circuit for an induction motor rated at 50 HP, three-phase, 60 Hz, 460 V, 1,175 rpm, a service factor of 1.15, and a power factor of 90%. The motor is protected by time-delay fuses. Assume that the motor is out of

sight of both the power supply and the controller (but the controller is within sight from the source). The nameplate full-load current rating of each motor is 70 A. Use THW (Cu) and EMT conduit.

19. Determine the size of THHN copper branch circuit conductors required to supply a 20 kVA, single-phase, 240 V heating unit (continuous duty) with a 10 A blower motor.
20. Estimate the size of THW copper conductors that are required to supply a group of three-phase, 230 V motors 230 V rated at 7.5, 15, and 25 HP.
21. An MCC serves nine identical motors, each motor is rated 50 HP, three-phase, 60 Hz, 460 V, 1,175 rpm, a service factor of 1.00, and a power factor of 90%. All the motors are protected by time-delay fuses and have FVNR starter type. Determine the size of the conductor, conduit, time-delay fuse, and switch for the feeder supplying the MCC.
22. A motor control center has three induction motors each rated at 40 HP and 480 V and four induction motors each rated at 30 HP and 480 V. Each 40 HP motor has an efficiency of 0.92, 1,772 rpm, and a power factor of 0.85. Each 30 HP motor has an efficiency of 0.90, and a power factor of 0.90, a nameplate rating of 30 A, a service factor of 1.15, and a FVR controller.
    a. Determine the number of poles, the slip (%), and the torque delivered by each 40 HP motor.
    b. Size the branch circuit for each 30 HP motor if the source, controller, and motor are all out of sight of each other. Each motor is protected by non-time-delay fuse.
    c. Size the main feeder (conductor and conduit) as well as the non-time-delay fuse and associated switch for the motor control center.

23 A motor control center has four induction motors each rated at 20 HP and 208-V and three induction motors each rated at 30 HP and 208-V. Each 20-HP motor has an efficiency of 0.90, 1,760 rpm, a power factor of 0.80, a nameplate rating of 55 A, a service factor of 1.00, and a FVR controller. Each 30-HP motor has an efficiency of 0.91, a power factor of 0.92, and FVNR controller.
    a. Determine the number of poles, the slip (%), and the torque delivered (in lb-ft) by each 20-HP motor.
    b. Size the conductor (THW, Cu), the fuse, the switch, and the overload protection of each 20-HP motor if a time-delay fuse is used. Assume that all the motors are within sight of both the fuse and the controller.
    c. Design and show the layout of the motor control center.

24 A motor control center has the following loads: (i) 10-HP, 3-phase motor, 460-V, a power-factor of 0.80, and an FVR controller; (ii) 15-HP, 3-phase motor, 460-V, a power-factor of 0.90, and an FVNR controller; (iii) 20-HP, 3-phase motor, 460-V, a power factor of 0.95, and an FVNR controller; and (iv) 15-kW, 480-V, 3-phase electrical heaters.
    a. Size the conductor (THW, Cu), the non-time delay fuse, and the disconnect for the 10-HP motor. Assume that the 10-HP motor is within sight of both the fuse and the controller.

b. Size the conductor (THW, Cu) of the main feeder of the MCC.
c. Size the time-delay fuse, and the disconnect of the main feeder of the MCC.
d. Determine which motor must be replaced with an equivalent synchronous motor (i.e., with same HP, efficiency, and power factor) to maximize the overall power factor of the MCC.

7 Electrical Systems for Dwellings

This chapter provides a general procedure for designing electrical systems for residential buildings based on the National Electrical Code (NEC) requirements. Step-by-step calculation procedures are outlined with specific examples. Some relevant NEC requirements are highlighted and discussed throughout this chapter. Supporting tables and examples of circuiting diagrams are provided to further show the design approach of safe electrical systems for residential buildings including detached homes, apartment units, or multifamily buildings. The required design specifications for electrified homes are compared to those using fuel-based heating systems.

7.1 INTRODUCTION

The electrical distribution systems for dwellings have to be properly designed and installed in order to provide electrical energy to the utilization equipment as safely, reliably, and economically as possible. In the United States, the NEC provides specific guidelines and requirements for the safe design of electrical systems for residential buildings. Figure 7.1 shows a typical one-line diagram of an electrical system for a residential building. A service entrance feeder is typically used to connect the utility distribution station to the main panel of the building. The main panel includes the protection devices (typically circuit breakers) for the branch circuits serving various electrical devices within the building.

Figure 7.2 illustrates some electrical devices that are served by the main panel through a set of branch circuits. A panelboard for a house is shown in Figure 7.3.

A typical U.S. home or an apartment unit is served by a two-phase service entrance (240 V/120 V). If possible, the two phases of the panel should be balanced within 10% to ensure that the neutral wire does not carry significant currents. Current transformers are typically used to measure the current flowing through any electrical device, including the main service entrance, a branch circuit, or a receptacle.

7.2 GENERAL DESIGN APPROACH

To design electrical systems for dwellings, the following assumptions are often made:

1. All loads have a power factor of 1 (i.e., $PT = PR$).
2. Independently of the receptacle type (duplex, triplex, or fourplex), the load for each receptacle outlet is 180 VA.
3. The main panel rating is the same as the main circuit breaker rating.

DOI: 10.1201/9781003276999-7

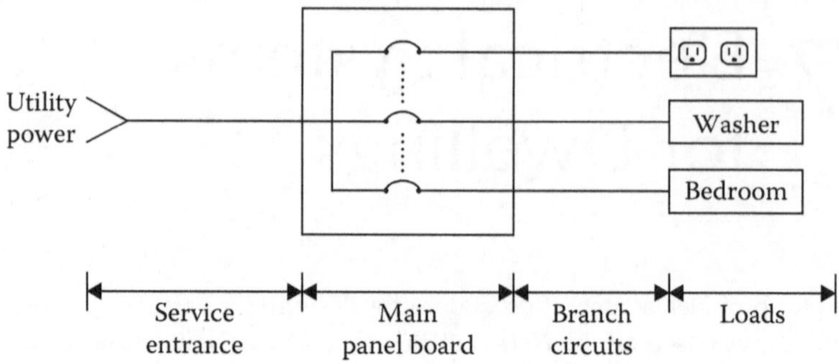

FIGURE 7.1 A schematic one-line diagram for a basic electrical distribution system within a home.

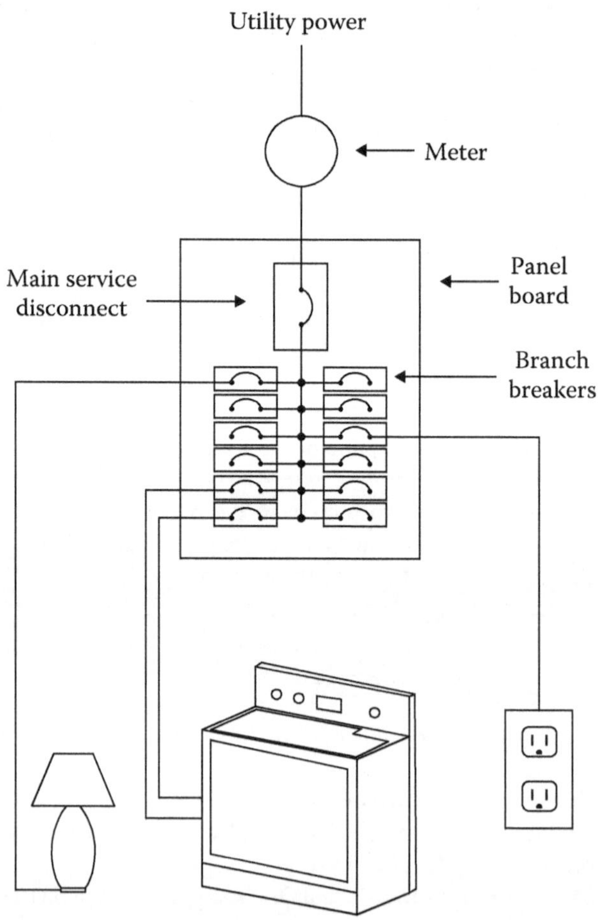

FIGURE 7.2 Common electrical loads served by a distribution panel within a home.

FIGURE 7.3 Layout of the circuit breakers in a house distribution panel.

By designing electrical systems for dwellings, the following specifications should be provided:

1. Main feeder conductor main circuit breaker for the panelboard (often referred to as panel)
2. All branch circuits to service the various electrical loads, including
   a. Lighting branch circuits (receptacles, lighting, and small appliances)
   b. Fixed appliances branch circuits (ranges/stoves, water heater, fans, dryer, disposal)
   c. Special appliances branch circuits (AC, electric heating)

Typically, an entire zone within a house is served by a lighting branch circuit. The number of the branch circuit should be identified both in the circuiting layout and in the panel schedule as shown in Figure 7.4.

As noted in Chapter 5, each dwelling should be grounded, that is electrically connected to the ground through the panelboard, using either a rod, as shown in Figure 7.5, the steel bars in foundation footing, or embedded water pipes. The size of the wire connecting the panelboard to the ground is called grounding conductor and should be determined based on the service entrance conductor size, as will be discussed in this chapter.

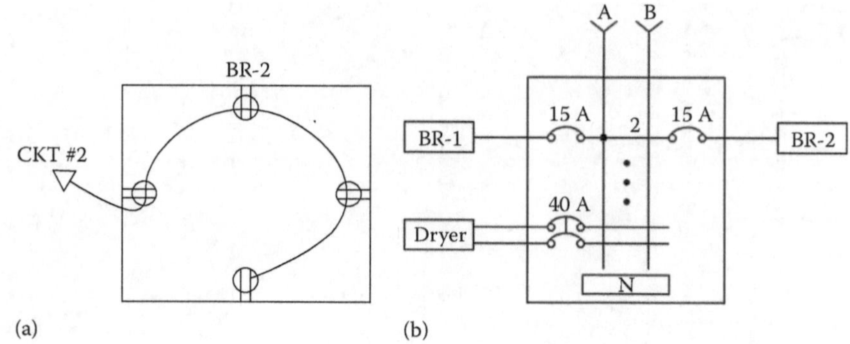

FIGURE 7.4 Identification of branch circuit serving a bedroom (BR-2) both (a) in a circuiting layout and (b) in a panelboard schedule.

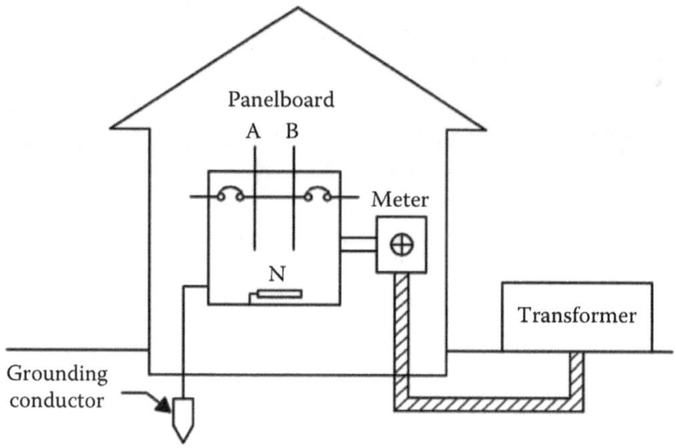

FIGURE 7.5 Typical grounding configuration for a dwelling.

7.3 MAIN SERVICE ENTRANCE DESIGN

Service entrance feeder is a conductor or a set of conductors that serves the entire dwelling or the residential building. The size of the service entrance feeder can be determined using either the standard method or the optional method outlined in the NEC. The circuit breaker protecting the panelboard is called the main circuit breaker (main CB) and is often located within the panel and defines the rating of the panelboard. Designing the service entrance includes determining the size of both the conductor and the main CB, as illustrated in Figure 7.6.

7.3.1 Standard Method for Sizing Service Entrance

To determine the service entrance size, all electrical loads for the dwelling should be obtained. These loads include lighting, receptacles, fixed appliances, special

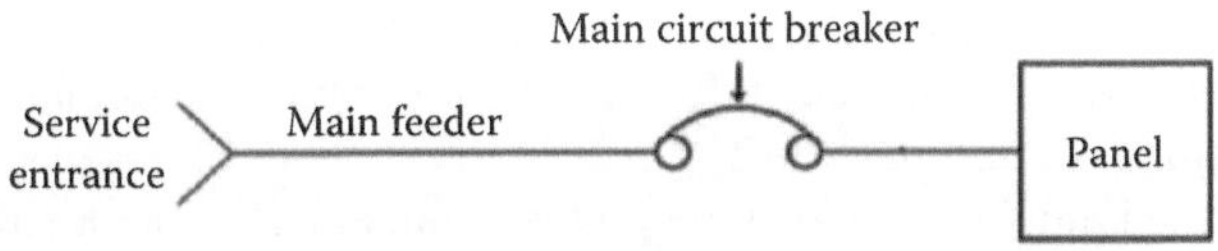

FIGURE 7.6 Sizing the main feeder conductor and main circuit breaker are part of the design of service entrance for dwellings.

appliances, and heating/cooling systems. According to NEC, two calculation methods can be utilized to determine the phases (A and B for one-phase systems and A, B, and C for three-phase systems) and the neutral (N):

- Standard method: can be used to size any phase and neural wires (A, B, N for one-phase systems and A, B, C, N for three-phase systems).
- Optional method: can be used to size only the phase wires (A, B for one-phase systems and A, B, C for three-phase systems) in the case of the main CB $\geq$ 100 A.

In this section, the standard method calculation details for estimating the demand loads for the service entrance feeder are described.

7.3.1.1 Estimation of the Demand Load for the Phases (A, B, and, if Needed, C)

1. General loads (lighting/general receptacles)
    a. General load = 3 VA/ft^2 (or 33 VA/m^2) $\times$ total livable building floor area
    b. Kitchen load = 2 branch circuits $\times$ 1,500 VA/circuit
    c. Each branch circuit has to be rated at 20 A.
    d. Laundry load = 1 branch circuit $\times$ 1,500 VA/circuit
    e. The branch circuit has to be rated at 20 A.

Then, the demand load associated with general lighting and general-use receptacles, DL_{GL}, can be estimated as follows:

$$
\begin{aligned}
DL_{GL} &= 100\% \text{of first } 3{,}000 \text{ VA} + 35\% \text{ of remainder upto } 120{,}000 \text{ VA} \\
&+ 25\% \text{ of the remainder over} 120{,}000 \text{ VA}
\end{aligned}
\tag{7.1}
$$

2. Fixed appliances load: apply demand factors (DFs) to the actual connected load (CL) for all the fixed appliances that served in the building using the following equation:

$$DL_{FA} = DF \times CL_{FA} \tag{7.2}$$

with

a. $DF = 100\%$ if the total number of fixed appliances ≤ 3
b. $DF = 75\%$ if the total number of fixed appliances ≥ 4

3. Special appliances load: Typically, dryers, ranges, space heating, and cooling systems are considered special appliances and their demand loads can be estimated according to NEC guidelines.
   a. Table 7.1 can be used to estimate the demand loads for electrical dryers.
   b. Table 7.2 can be used to estimate the demand loads for electrical ranges and ovens.
   c. A demand factor of 100% should be applied to only the largest of the space heating or cooling system serving the dwelling to determine the demand loads associated with air conditioning systems for residential buildings.
   d. If motor is present, a demand factor of 25% should be applied to the largest motor load associated typically with one of the fixed appliances.
4. Total demand load: the total demand load, DL_{tot}, in VA can be simply estimated by summing the demand loads estimated in the previous three steps according to the following equation:

$$DL_{tot} = DL_{GL} + DL_{FA} + DL_{SA} \tag{7.3}$$

Then, the total load can be expressed in terms of amperage using the following expression:

$$\begin{aligned} I_{tot} &= \frac{DL_{tot}}{E_L} \text{ for one} - \text{phase systems} \left(\text{such as} \frac{120}{240} \text{V} \right) \\ I_{tot} &= \frac{DL_{tot}}{\sqrt{3}E_L} \text{ for three} - \text{phase sytems} \left(\text{such as} \frac{208}{120} \text{V} \right) \end{aligned} \tag{7.4}$$

1. For ranges individually rated more than 12 kW but no more than 27 kW, the maximum demand in Column C shall be increased by 5% for each additional kW of rating or major fraction thereof by which the rating of individual ranges exceeds 12 kW.
2. For ranges individually rated more than $8\frac{3}{4}$ kW and of different ratings, but none exceeding 27 kW, an average value of rating shall be calculated by adding together the ratings of all ranges to obtain the total connected load (using 12 kW for any range rated less than 12 kW) and dividing by the total number of ranges. Then, the maximum demand in Column C shall be increased by 5% for each additional kW of rating or major fraction thereof by which this average value exceeds 12 kW.
3. For ranges individually rated more than $1\frac{3}{4}$ kW but no more than $8\frac{3}{4}$ kW, it is permissible instead of using the method provided in Column C to add the nameplate ratings of all household cooking appliances rated more than $1\frac{3}{4}$

TABLE 7.1
Demand Factors for Electrical Dryers

Number of Dryers	Demand Factor (%)
Less than 4	100.0
5	85.0
6	75.0
7	65.0
8	60.0
9	55.0
10	50.0
11	47.0
12	46.0
13	45.0
14	44.0
15	43.0
16	42.0
17	41.0
18	40.0
19	39.0
20	38.0
21	37.0
22	36.0
23	35.0
24	34.5
25	34.0
26	33.5
27	33.0
28	32.5
29	32.0
30	31.5
31	31.0
32	30.5
33	30.0
34	29.5
35	29.0
36	28.5
37	28.0
38	27.5
39	27.0
40	26.5
41	26.0
42	25.5
More than 43	25.0

Note: The demand load for one dryer is the largest of its nameplate rating and 5,000 VA. This table is adapted from NEC-2014 Table 220.54 (NEC, 2014).

TABLE 7.2
Demand Loads and Demand Factors for Electrical Ranges and Ovens

Number of Appliances	Demand Factors (%) Column A (Less than 3.4 kW Rating)	Demand Factors (%) Column B (for 3.5–8.75 kW Ratings)	Column C Maximum Demand (kW) (for 8.75–12 kW Ratings)
1	80	80	8
2	75	65	11
3	70	55	14
4	66	50	17
5	62	45	20
6	59	43	21
7	56	40	22
8	53	36	23
9	51	35	24
10	49	34	25
11	47	32	26
12	45	32	27
13	43	32	28
14	41	32	29
15	40	32	30
16	39	28	31
17	38	28	32
18	37	28	33
19	36	28	34
20	35	28	35
21	34	26	36
22	33	26	37
23	32	26	38
24	31	26	39
25	30	26	40
26–30	30	24	15 kW + 1 kW for each range
31–40	30	22	15 kW + 1 kW for each range
41–50	30	20	25 kW + 1 kW for each range
51–60	30	18	25 kW + 1 kW for each range
61 and over	30	16	25 kW + 1 kW for each range

Notes: This table is adapted from NEC-2014 Table 220.55 (NEC, 2014).

kW but no more than $8\frac{3}{4}$ kW and multiply the sum by the demand factors specified in Column A or Column B for the given number of appliances. Where the rating of the cooking appliances falls between Column A and Column B, the demand factors for each column should be applied to the appliances of that column, and the results added together.

7.3.1.2 Estimation of the Demand Load for the Neutral (N)

The same procedure outlined for estimating the demand load for the phases should be utilized, but only the one-phase loads should be considered, except for the loads associated with ranges and dryers. Indeed, these two loads estimated using Tables 7.1 (dryers) and 7.2 (ranges) should be considered with an additional derating factor of 70%.

Example 7.1 illustrates the application of Tables 7.1 and 7.2 to determine the loads associated with dryers and ranges, while Example 7.2 utilizes for a specific house the standard method to estimate the demand load for both the phases and neutral.

Example 7.1

Problem

Determine the demand loads associated with electrical dryers and ranges for the following cases:

six dryers, each rated at 5 kVA
one range rated at 12 kVA
one range rated at 9 kVA
one range rated at 2.5 kVA
one range rated at 27 kVA
three ranges rated at 10, 14, and 19 kVA

Solution

Using Tables 7.1 and 7.2, the demand loads for all the cases can be estimated as summarized here:

1. $DL = 0.75 \times 6 \times 5$ kVA $= 22.5$ kVA (based on Table 7.1 demand factor for six dryers)
2. $DL = 8$ kVA (based on Table 7.2, Column C)
3. $DL = 8$ kVA (based on Table 7.2, Column C)
4. $DL = 0.80 \times 2.5$ kVA $= 2$ kVA (based on Table 7.2, Column A)

Using Note 1 and Column C of Table 7.2, first the demand factor, *DF*, is estimated:

$$DF = 1 + (27 - 12) \times 0.05 = 1.75$$

Then, the demand load can be determined:

$$DL = 1.75 \times 8 = 14 \text{ kVA}$$

Using Note 2 of Table 7.2, the following steps are outlined to estimate the demand load of three ranges with different ratings:

1. Increase 10–12 kVA.
2. Take average of all three ratings = (12 + 14 + 19)/3 = 15 kVA.
3. Estimate the demand factor: $DF = 1 + (15-12) \times 0.05 = 1.15$.
4. Calculate the demand load: $DL = 1.15 \times 14 = 16.1$ kVA.

Example 7.2

Problem

Determine the demand load for both phases A and B as well as the neutral for a service entrance feeder serving 1,600 ft^2 (150 m^2) house that has the following electrical loads:

400 VA disposable (120 V), 800 VA compactor (120 V), 4.5 A blower motor (120 V), 1,400 VA dishwasher (120 V), 4,800 VA water heater (240 V), 5,000 VA dryer (240 V), 12,000 VA range (240 V), and 20,000 VA space heaters (240). There are three separately controlled heaters in the dwelling.

Use the NEC standard method to determine the demand loads for A, B, and N.

Solution

First, the load for phases A and B are estimated considering all the loads. Then, the neutral is sized using only phase-to-neutral loads (i.e., 240) except as noted by NEC the loads associated with the range and the dryer.

Phases A and B Demand Load

1. General lighting and receptacles load:
    - General load = 3 VA/ft^2 × 1,600 ft^2 = 4,800 VA
    - Kitchen load = 2 circuits × 1,500 VA/circuit = 3,000 VA
    - Laundry load = 1 circuit × 1,500 VA/circuit = 1,500 VA
    - Thus, the total connected load = 9,300 VA.

    The demand load for general lighting and receptacles load:

$$DL_{GL} = 100\% \times 3{,}000 \text{ VA} + 35\%(9{,}300 - 3{,}000) \text{ VA} = 5{,}205 \text{ VA}$$

2. Fixed appliances load:
    - Disposal (use rated VA) = 400 VA
    - Compactor (use rated VA) = 800 VA
    - Blower motor = 4.5 A × 120 V = 540 VA
    - Dishwasher (use rated VA) = 1,400 VA
    - Water heater (use rated VA) = 4,800 VA
    - Thus, the total connected load = 7,940 VA.

    The demand load for the fixed appliances load:

$$DL_{FA} = 75\% \times 7{,}940 \text{ VA} = 5{,}955 \text{ VA}$$

3. Special appliances load:
   - Range (use Table 7.2) = 8,000 VA
   - Dryer (use Table 7.1) = 5,000 VA
   - Space heaters (use rated VA) = 20,000 VA
   - Largest motor (25% × compactor VA) = 200 VA.

   Thus, the demand load for the special appliances load:

$$DL_{SA} = 33,200 \text{ VA}$$

The total demand load for phases A and B:

$$DL_{tot,A-B} = 5,205 \text{ VA} + 5,955 \text{ VA} + 33,200 \text{ VA} = 44,360 \text{ VA}$$

Therefore, the demand load in terms of current for phases A and B:

$$I_{tot,A-B} = \frac{44,360 \text{ VA}}{240 \text{ V}} = 184.8 \text{ A}$$

Neutral N demand load

1. General lighting and receptacles load:
   - General load = 3 VA/ft² × 1,600 ft² = 4,800 VA
   - Kitchen load = 2 circuits × 1,500 VA/circuit = 3,000 VA
   - Laundry load = 1 circuit × 1,500 VA/circuit = 1,500 VA
   - Thus, the total connected load = 9,300 VA.

   The demand load for general lighting and receptacles load:

$$DL_{GL} = 100\% \times 3,000 \text{ VA} + 35\%(9,300 - 3,000)\text{VA} = 5,205 \text{ VA}$$

2. Fixed appliances load: only 120 V appliances are considered:
   - Disposal (use rated VA) = 400 VA
   - Compactor (use rated VA) = 800 VA
   - Blower motor = 4.5 A × 120 V = 540 VA
   - Dishwasher (use rated VA) = 1,400 VA
   - Thus, the total connected load = 3,140 VA.

   The demand load for the fixed appliances load:

$$DL_{FA} = 75\% \times 3,140 \text{ VA} = 2,355 \text{ VA}$$

3. Special appliances load: only 120 V appliances except range and dryer loads
   - Range (use Table 7.2 and a derating factor of 70%) = 70% × 8,000 VA = 5,600 VA
   - Dryer (use Table 7.1 and a derating factor of 70%) = 70% × 5,000 VA = 3,500 VA
   - Largest motor (25% * compactor VA) = 200 VA.

Thus, the demand load for the special appliances load:

$$DL_{SA} = 9,300 \text{ VA}$$

The total demand load for the neutral N:

$$DL_{tot,N} = 5,205 \text{ VA} + 2,355 \text{ VA} + 9,300 \text{ VA} = 16,860 \text{ VA}$$

Therefore, the demand load in terms of current for the neutral N:

$$I_{tot,N} = \frac{16,860 \text{ VA}}{240 \text{ V}} = 70.3 \text{ A}$$

7.3.2 Optional Method for Sizing Service Entrance

As noted earlier, the size of phases for a dwelling service entrance size can be determined using the optional method. Indeed and when the ampacity of the service feeder conductors is larger than 100 A, the NEC allows the design to utilize the optional method for residential buildings served by 240/120 V or 208Y/120 V systems. It should be noted that the size of the neutral (N) needs to be still determined using the standard method. First, the general calculation procedure for the optional method is outlined. Then, an example is provided to clarify the application of the method to size the phases of a dwelling service entrance feeder.

The optional method to size the phase wires (i.e., A and B for one-phase systems and A, B, and C for three-phase systems) for the service entrance consists of the following steps:

1. Demand load for general loads:
   a. The demand load, $DL_{GL,O}$, for general loads for one or two-unit dwellings can be estimated as follows:
   b. $DL_{GL,O} = 100\% \times \text{first } 10,000 \text{ VA} + 40\%\left(CL_{GL,O} - 10,000 \text{ VA}\right)$
   c. where $CL_{GL,O}$ are the connected loads associated with lighting, general-use receptacles, and appliances and are calculated as follows:
   d. $CL_{GL,O} = CL_{GL} + CL_{AP}$ (7.6)
   e. where CL_{GL} is the connected general load associated with lighting and general-use receptacles and CL_{AP} is the connected load for all appliances except for space heating and cooling systems. The two connected loads are estimated as noted in the following sections.
2. General loads (lighting/general receptacles)
   - General load = 3 VA/ft^2 (or 33 VA/m^2) total livable building floor area
   - Kitchen load = 2 branch circuits × 1,500 VA/circuit
   - Each branch circuit has to be rated at 20 A.
   - Laundry load = 1 branch circuit × 1,500 VA/circuit
   - The branch circuit has to be rated at 20 A.

By adding these three loads, the connected load associated with general lighting and general-use receptacles, CL_{GL}, can be determined.

Appliances load:

The connected load, CL_{AP}, for all appliances including fixed and special appliances, except space heating and cooling systems, considered in the standard method is determined by simply adding all the nameplate ratings for the appliances.

3. Demand load for space heating and cooling systems:

   The demand load, $DL_{HC,\,O}$, for the heating and air conditioning systems is the largest of the demand loads estimated using the demand factors and connect loads specific to various systems serving the dwelling, as noted in Table 7.3.
4. *Total demand load*: The total demand load, $DL_{tot\text{-}O}$, can be estimated by simply summing up the demand loads estimated in the previous three steps according to the following equation:

$$DL_{tot-O} = DL_{GL-O} + DL_{HC-O} \tag{7.7}$$

Then, the total load can be expressed in terms of amperage depending on the line voltage for the service:

$$\begin{aligned} I_{tot} &= \frac{DL_{tot-O}}{E_L} \text{ for one} - \text{phase systems}\left(\text{such as } \frac{120}{240}\text{ V}\right) \\ I_{tot} &= \frac{DL_{tot-O}}{\sqrt{3}E_L} \text{ for three} - \text{phase systems}\left(\text{such as } \frac{208\text{ Y}}{120\text{ V}}\right) \end{aligned} \tag{7.8}$$

Example 7.3 illustrates the application of the optional method to size the service feeder for the same dwelling considered in Example 7.2.

TABLE 7.3
Demand Factors for Heating and Air Conditioning Systems for Dwellings Specific to Optional Method

Type of Space Heating and Cooling Systems	Demand Factor (%)
Air conditioning and cooling system	100
Heat pump system without supplemental electric heating	100
Heat pump (HP) system with supplemental electric heating (H)	100 for HP65 for H
Electrical heating system with less than four separately controlled units	65
Electrical heating system with four or more separately controlled units	40
Electrical thermal storage system	100

Example 7.3

Problem

For the dwelling considered in Example 7.2, estimate the demand loads for phases A and B using the optional method instead of the standard method.

Solution

Under specific conditions, the NEC allows the use of the optional method to estimate the demand load for phases A and B.

First, all the rated loads (except for the space heaters) are added:

- General load = 3 VA/ft^2 × 1,600 ft^2 = 4,800 VA
- Kitchen load = 2 circuits × 1,500 VA/circuit = 3,000 VA
- Laundry load = 1 circuit × 1,500 VA/circuit = 1,500 VA
- Disposal = 400 VA
- Compactor = 800 VA
- Blower motor = 4.5 A × 120 V = 540 VA
- Dishwasher = 1,400 VA
- Water heater = 4,800 VA
- Range = 12,000 VA
- Dryer = 5,000 VA.

Thus, the total connect load for the dwelling without the space heaters:

$$CL_{\text{total no of heaters}} = 34,240 \text{ VA}$$

The total demand load for phases A and B using the optional method for all loads except the heaters:

$$DL_{\text{total no of heaters}} = 100\% \times 10,000 \text{ VA} + 40\%(34,240 - 10,000) \text{ VA} = 19,696 \text{ VA}$$

The additional load due to the space heaters (refer to Table 7.3 for the demand factor) is

$$DL_{\text{heaters}} = 20,000 \text{ VA} \times 65\% = 13,000 \text{ VA}$$

Then, the total demand load for phases A and B using the optional method for all loads:

$$DL_{\text{total no of heaters}} = 19,696 \text{ VA} + 13,000 \text{ VA} = 32,696 \text{ VA}$$

Therefore, the demand load in terms of current for phases A and B using the optional method:

$$I_{tot,A-B} = \frac{32,696 \text{ VA}}{240 \text{ V}} = 136.2 \text{ A}$$

TABLE 7.4
Optional Method Demand Factors for Apartment Buildings

Number of Dwelling Units	Demand Factor
1–2	100% for first 10 kVA 40% for remainder
3–5	45
6–7	44
8–10	43
11	42
12–13	41
14–15	40
16–17	39
18–20	38
21	37
22–23	36
24–25	35
26–27	34
28–30	33
31	32
32–33	31
34–36	30
37–38	29
39–42	28
43–45	27
46–50	26
51–55	25
56–61	24
Over 62	23

Note: This table is adapted from NEC-2014 Table 220.84 (NEC, 2014).

In this example, the optional method results in lower demand loads than the standard method and thus can be used to justify reduced size for the service entrance for dwellings.

For dwellings with several units such as the case for apartment buildings, the demand factors provided in Table 7.4 should be considered to determine the size of the service entrance using the optional method as outlined in detail in Section 7.6.

7.3.3 Sizing Service Entrance

Based on the demand loads for the phases and the neutral, the size for the conductors as well as the main circuit breaker for the service entrance can be selected depending on the type of cable to be used (i.e., insulation layer and conductor temperature) as

TABLE 7.5
Single-Phase Service Entrance Ratings and Feeder Conductor Sizes for Dwellings

	Service Entrance Conductor Size (AWG or MCM)	
Main Service Rating (A)	**Copper Wires**	**Aluminum or Copper-Clad Aluminum Wires**
100	2 AWG	1/0 AWG
125	1 AWG	2/0 AWG
150	1/0 AWG	3/0 AWG
175	2/0 AWG	4/0 AWG
200	3/0 AWG	250 MCM
225	4/0 AWG	300 MCM
250	250 MCM	350 MCM
300	350 MCM	500 MCM
350	500 MCM	700 MCM0
400	600 MCM	Two parallel 4/0 AWG or two parallel 3/0 AWG
	900 MCM	Two parallel 250 MCM

discussed in both Chapters 5 and 6. Table 7.5 can be used to determine the appropriate conductor sizes and circuit breaker ratings for dwellings with single-phase services. It should be noted that Section 310.15 of the NEC (2014) allows to adjust by a factor of 0.83 the reduce the size of 120/240 V service entrance ungrounded conductors for dwellings rated between 100–400 A. The selection for both the conductor sizes and the main circuit breaker ratings are based on the demand load (in amps) for both the phases and neutral using either the standard or the optional method. It should be noted that for single-family residential buildings, the panelboard ratings (i.e., the main circuit breaker ratings) range from 100 to 200 A. For all residential buildings served by one-phase or three-phase systems, the sizes of conductors used should be based on the ampacity noted in Table 6.1 and the circuit breaker ratings listed in Table 5.4. Moreover and to enhance the safety of the electrical systems in residential buildings, the main panel has to be connected to the ground through a grounding conductor, as indicated in Figure 7.5. The size of the grounding conductor depends on the size of the main service entrance feeder, as discussed in Chapter 6 and as summarized in Table 7.6. It should be noted that the size of the neutral should be larger than the grounding conductor. If the size of the grounding conductor is found to be larger than that of the neutral, the size of the neutral conductor should be increased to be the same as that of the grounding wire. More discussions on the grounding options are outlined in Section 250 of the NEC (2014).

Example 7.4 illustrates the selection approach for the service entrance conductors, main circuit breaker rating, and the grounding conductor.

TABLE 7.6
Service Conductor Sizes for Dwellings

Service Entrance Conductor Size (AWG or MCM)		Service Grounding Conductor Size (AWG)	
Copper Wires	**Aluminum or Copper-Clad Aluminum Wires**	**Copper Wires**	**Aluminum or Copper-Clad Aluminum Wires**
3 or 2 AWG	1 or 1/0 AWG	8 AWG	6 AWG
1 or 1/0 AWG	2/0 or 3/0 AWG	6 AWG	4 AWG
2/0 or 3/0 AWG	4/0 AWG or 250 MCM	4 AWG	2 AWG
4/0 AWG–350 MCM	300–500 MCM	2 AWG	1/0 AWG
400–600 MCM	600–800 MCM	1/0 AWG	3/0 AWG
700–1,100 MCM	1,000–1,750 MCM	2/0 AWG	4/0 AWG

Example 7.4

Problem

For the dwelling considered in Examples 7.2 and 7.3, size the service entrance conductor sizes as well as the main circuit breaker rating. Compare the service entrance sizes using the optional method instead of the standard method. Consider copper THW conductors for the service entrance.

Solution

Standard Method:

Using the demand loads for the phases A and B as well as the neutral N found in Example 7.2 using the standard method, the rating of the panel is 200 A according to Table 7.5 (since the demand load is 184.8 A) and the size of the service entrance conductors for A and B is 3/0 THW (Cu) based on Table 7.5. Moreover, the neutral conductor size is 4 THW (Cu) based on Table 6.1 (since the demand load is 70.3 A). The grounding conductor size according to Table 7.6 is 4 AWG (Cu), which is the same as that of the neutral conductor.

Optional Method:

Using the demand loads for phases A and B found in Example 7.3 applying the optional method, the rating of the panel can be reduced to 150 A according to Table 7.5 (since the demand load is 136.2 A) and the size of conductors for A and B is 1 (based on Table 7.4). The grounding conductor size according to Table 7.6 is 6 AWG (Cu), which is lower than that of the neutral conductor found to be 4 THW (Cu) using the standard method calculations.

7.4 BRANCH CIRCUITS FOR RESIDENTIAL BUILDINGS

The NEC has specific requirements for the number of receptacles and lighting outlets to be placed in residential buildings. The basic guidelines are summarized in the following sections.

7.4.1 General Guidelines for Receptacles

NEC has specific requirements for placing receptacles in various spaces inside and outside residential buildings. Some of these receptacles have to be of the GFCI type, especially those close to any water source. Some of these requirements are outlined briefly in this section. However, the designer should refer to the NEC to ensure that the electrical system meets all the NEC requirements.

1. *Wall receptacles*: No point along an unbroken wall should be more than 1.8 m (6 ft) from a receptacle outlet [210.52(A)(1)]. Figure 7.7 illustrates the placement of receptacles along the walls of a room inside a residential building. It is better to start from reference location such as a door or location A in Figure 7.7 before placing a receptacle outlet after 6 ft at point C. Then, another receptacle outlet can be placed at point D, 12 ft from point C along the unbroken walls. Since the wall is broken after point E, receptacles need to be placed only in wall sections at least 2 ft wide. Thus, receptacles

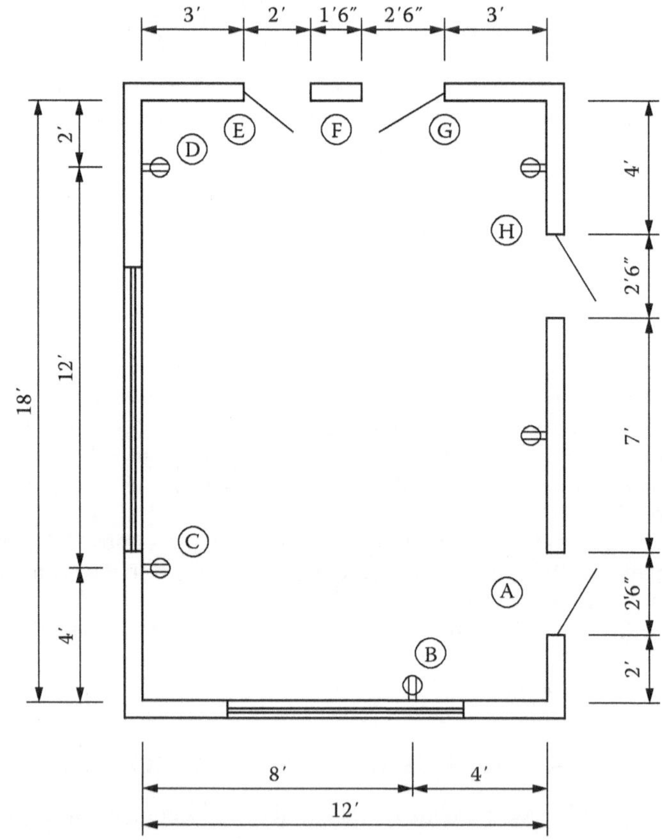

FIGURE 7.7 Locations of general-purpose receptacles inside a room within a dwelling.

are placed in sections G–H and H–A (of about 7 ft width each). However, no receptacle is needed at section F since it has a width of only 1.5 ft.

2. *Small-appliance circuits*: A minimum of two 20 A, 1,500 VA small-appliance circuits must be installed in the kitchen, dining room, pantry, and breakfast room [NEC 220.52(B)].
3. *Receptacle outlets over countertops*: Receptacle outlets are required at each counter space wider than 12 in.; no point along the wall shall be more than 0.6 m (2 ft) from a plug [NEC 210.52(C)(1) through (5)]. Figure 7.8 shows placement options for receptacles for a kitchen countertop. As discussed in item (11), the receptacles close to the sink have to have ground-fault current interrupter (GFCI) protection.
4. *Bathroom wall receptacle outlets*: A wall receptacle outlet-rated GFCI shall be installed within 0.9 m (3 ft) of the basin [NEC 210.52(D)].

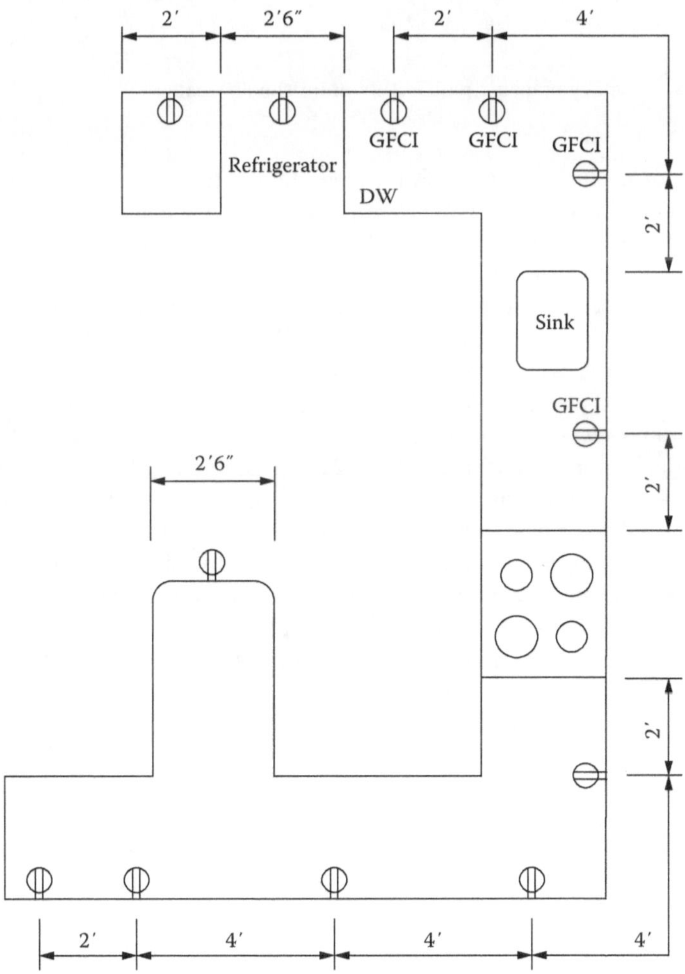

FIGURE 7.8 Locations of kitchen countertop receptacles within a dwelling.

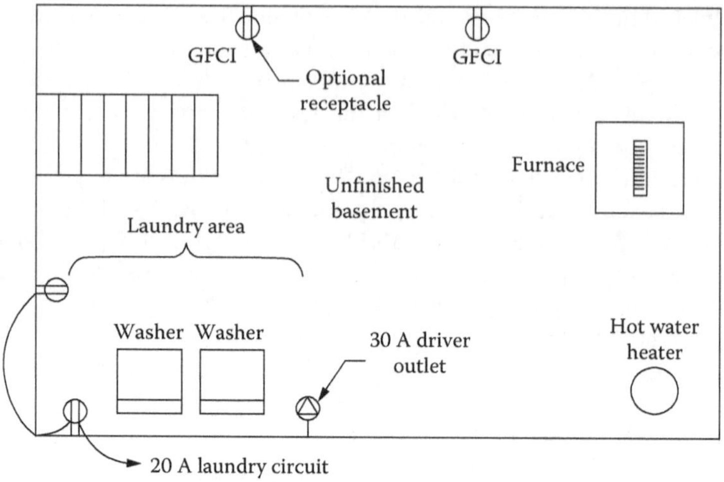

FIGURE 7.9 Laundry branch circuit within unfinished basement.

5. *Outdoor receptacle outlets*: For a single-family dwelling, at least one receptacle outlet, accessible while standing at grade level, and located not more than 2 m (6.5 ft) above grade, shall be installed at the front and back of the dwelling [NEC 210.52(E)(1)].
6. Balconies, decks, and porches that are accessible from inside the dwelling unit shall have at least one receptacle outlet installed within the perimeter of the balcony, deck, or porch. The receptacle shall not be located more than 2 m (6.5 ft) above the balcony, deck, or porch surface [NEC 210.52(E)(3)].
7. *Laundry area receptacle outlets*: At least one 20 A, 1,500 VA circuit must be installed in the laundry area [NEC 220.52(B) and 210.11(C)(2)]. Figure 7.9 illustrates a 20 A branch circuit serving a laundry area located within an unfinished basement. Note the receptacle outlet for the dryer is not served by this 20 A branch circuit. As noted in the following requirement, at least one GFCI receptacle outlet is required in the unfinished basement.
8. *Basement, crawl space, and garage receptacle outlets*: At least one GFCI-protected receptacle shall be installed in the basement, crawl space, and in the attached garage of a dwelling unit [NEC 210.52(G)(1)].
9. Hallways: In dwelling units, hallways of 10 ft or more in length shall have at least one receptacle outlet [NEC 210.52(H)].
10. Foyers that are not part of a hallway in accordance with 210.52(H) and that have an area that is greater than 60 ft^2 shall have a receptacle(s) located in each wall space of 3 ft or more in width and unbroken by doorways, floor-to-ceiling windows, and similar openings [NEC 210.52(I)].
11. All 125 V, single-phase, 15 and 20 A receptacles installed in the locations listed shall have GFCI protection for personnel: bathrooms, garages, accessory buildings, outdoors, crawl spaces, unfinished basements, kitchens, and within 6 ft of sinks.

7.4.2 General Guidelines for Lighting Outlets

NEC has specific requirements for lighting locations and related to switches as noted here:

1. At least one wall switch-controlled lighting outlet is required in all habitable rooms and bathrooms [NEC 210.70].
2. At least one wall switch-controlled lighting outlet shall be installed in hallways, stairways, attached garages, and detached garages with electric power.
3. For dwelling units, attached garages, and detached garages with electric power, at least one wall switch–controlled lighting outlet shall be installed to provide illumination on the exterior side of outdoor entrances or exits with grade-level access. A vehicle door in a garage shall not be considered as an outdoor entrance or exit.
4. Where one or more lighting outlet(s) are installed for interior stairways, there shall be a wall switch at each floor level, and landing level that includes an entryway, to control the lighting outlet(s) where the stairway between floor levels having six risers or more.
5. For attics, underfloor spaces, utility rooms, and basements, at least one lighting outlet containing a switch or controlled by a wall switch shall be installed where these spaces are used for storage or contain equipment requiring servicing. At least one point of control shall be at the usual point of entry to these spaces. The lighting outlet shall be provided at or near the equipment requiring servicing.

7.4.3 General Sizing Procedure for Branch Circuits

The rating of a branch circuit is associated with the size of the circuit breaker that protects the loads served by the circuit. The connected load should be used to estimate the size of the circuit breaker. It should be noted that NEC states that for continuous loads (i.e., operating continuously for more than 3 h), a branch circuit can only be loaded at 80% of its rating. For residential buildings, most of the loads are not continuous, and thus branch circuits can be designed to be 100% loaded. To size both circuit breaker and the conductors associated with a branch circuit, the connected and demand loads are expressed in amperes as indicated by Equations 7.9 and 7.10:

$$\begin{aligned} I_{CL-BC} &= \frac{VA-Rating}{E}\left(\text{for single}-\text{phase loads}\right) \\ I_{CL-BC} &= \frac{VA-Rating}{\sqrt{3}E}\left(\text{for three}-\text{phase loads}\right) \end{aligned} \tag{7.9}$$

where

VA-Rating is the connected load total power in VA

E is the voltage served by the load (typically 120 or 240 V for single-phase loads and 208 V for three-phase loads).

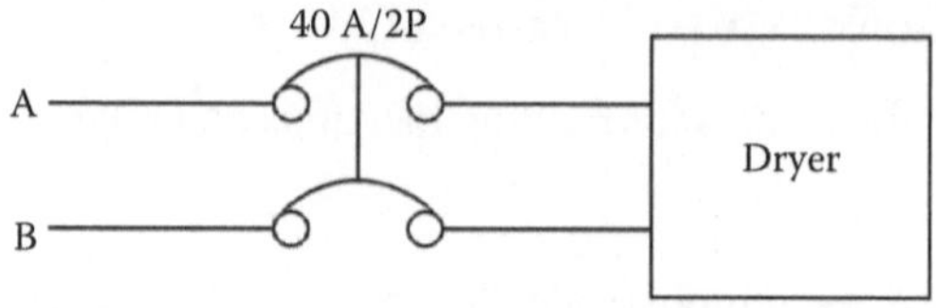

FIGURE 7.10 A two-pole circuit breaker for a branch circuit serving a dryer.

If the load is served by three-phase voltage, the

$$I_{DL-BC} = DF \times I_{CL-BC} \tag{7.10}$$

with $DF = 100\%$ for noncontinuous load and $DF = 125\%$ for continuous load.

Example 7.5 illustrates a basic calculation procedure to determine the branch circuit rating for a dryer. Figure 7.10 shows a common representation of the circuit breaker and its rating.

Example 7.5

Problem

Determine the rating of the branch circuit serving a 7,500 VA dryer using 240 V. Specify the size of the copper conductors (THW type) serving the dryer.

Solution

The load expressed in amperes for the dryer is first estimated:

$$I_L = \frac{7{,}500 \text{ A}}{240 \text{ V}} = 31.25 \text{ A}$$

From Table 5.4, the circuit breaker rating to protect branch circuit serving the dryer is 40 A having two poles (since the dryer is supplied by 240 V). The size of the copper conductors for the branch circuit is # 8 THW.

7.5 GENERAL DESIGN PROCEDURE

The design of electrical systems for dwellings is a rather straightforward procedure since it is well outlined by the NEC requirements. The approach can be summarized in the following steps:

1. Estimate the connected loads for lighting fixtures, receptacles, appliances, and heating/cooling systems. The estimation can be carried out using different sources:
   - Layout receptacles (180 VA).

Panelboard Schedule									
Description	Load A	Load B	Breaker trip (A)	A	B	Breaker trip (A)	Load A	Load B	Description
Range	40.74		50			30	24.12		Water heater
		40.74						24.12	
Dishwasher	9.1		15			15	7		Disposal
Wahing machine		12	15			30		24	Dryer
							24		
Laundry	16		20			20	16		Kitchen
Kitchen		16	20			15		12	Receptacle
Receptacle	12		15			15	12		Receptacle
Receptacle		12	15			15		12	Receptacle
Lighting		16	20			20		16	Lighting
Lighting	16		20						
Totals	93.84	96.74	190			160	83.12	88.12	

	A	B
Balance:	176.96	184.86
		4%

FIGURE 7.11 Panel schedule for a dwelling balanced within 10%.

- Lighting fixtures (rating of lamps up to 100 VA).
- Obtain manufacturing rating for any desired fixed and special appliances.

2. Size the service entrance feeder, main circuit breaker, and the grounding conductor, as outlined in Section 7.3.
3. Determine the branch circuits needed to serve all the loads including lighting, receptacles, and appliances. Typically, 15 and 20 A branch circuits are used for lighting/receptacles. Fixed and special appliances are served by dedicated branch circuits. Section 7.4 outlines the guidelines for sizing branch circuits.
4. Develop a panel schedule to ensure that the connected loads are balanced between the phases. The unbalances between phases should be kept below 10%. Different formats can be used to present the panel schedules. Figure 7.11 presents an example for a panel schedule.
5. Provide circuiting and switching layouts. Floor layouts should be provided to indicate the branch circuits serving all the loads, as well as the locations of the switches, main panels, receptacles, lighting fixtures, and special receptacles. The branch circuits should be labeled with the same numbers used in the panel schedule. Figures 7.12 and 7.13 illustrate examples for circuiting and switching layouts, respectively. Both switching and circuiting details can be provided in the same layout.

Example 7.6 illustrates a complete calculation approach to design the major elements for the electrical system for a single-family house. Figure 7.14 shows a combined circuiting and switching layout.

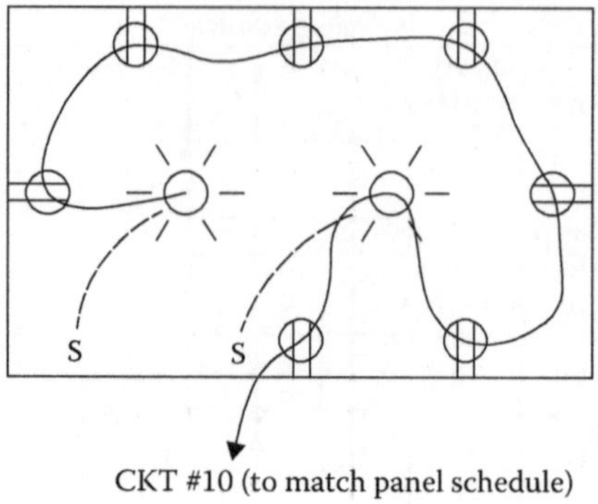

FIGURE 7.12 A branch circuit labeled in a circuiting layout.

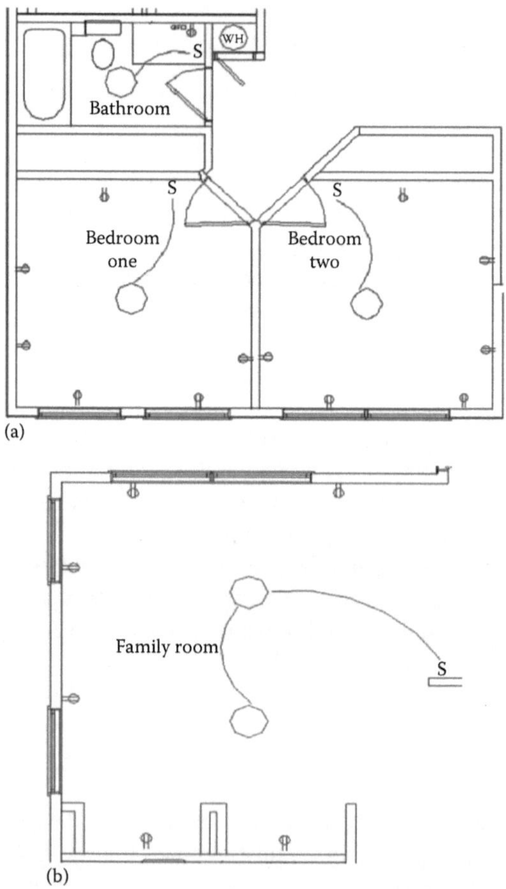

FIGURE 7.13 Location of (a) switches and (b) their associated loads in a switching layout.

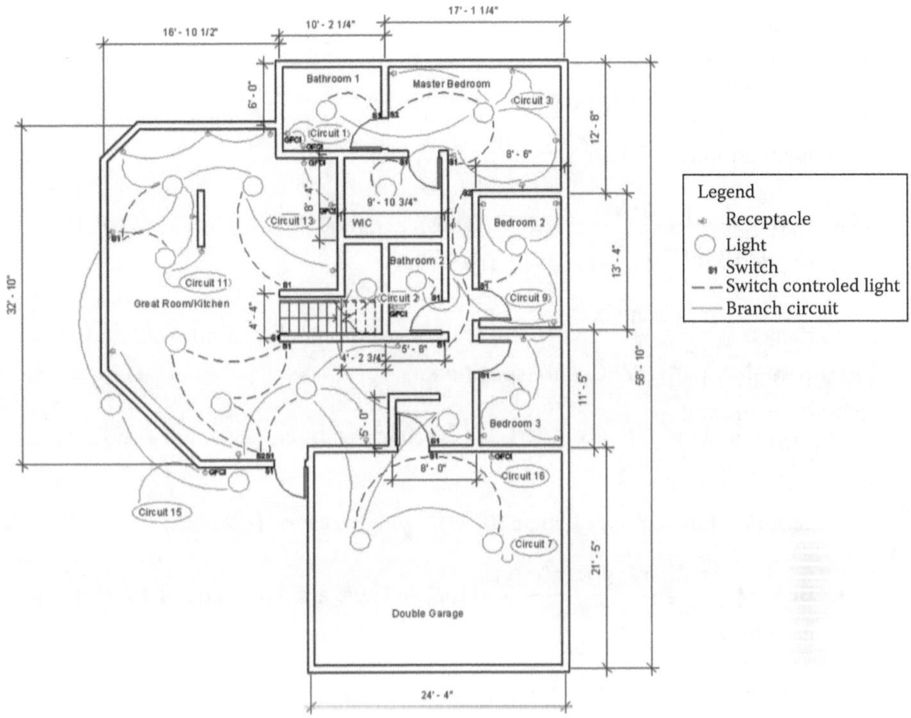

FIGURE 7.14 Combined circuiting and switching layout for a dwelling.

Example 7.6

Problem

Design of one dwelling's electrical system

- House: 40 ft × 25 ft ranch house
- One branch circuit: for bedrooms (six receptacles, two lamps each rated at 100 VA)
- One branch circuit: for living area (seven receptacles, two lamps each rated at 100 VA)
- Kitchen + laundry
- Appliances: water heater (5,000 kVA, 240 V), dryer (4,500 VA, 240 V), range (8,000 VA, 240 V)
- Size main service (panel rating, grounding, wires A/B/N)
- Show panel schedule
- Size all circuits, and estimate the number of the required branch circuits.

Solution

Branch Circuits:

- Water heater : $I_L = \dfrac{5{,}000\ \text{VA}}{240\ \text{V}} = 20.8\ \ \text{A}$. Thus, a circuit breaker of 30-A/2P is required with a wire size of 10 THW(Cu).

- Dryer : $I_L = \dfrac{4{,}500\ \text{VA}}{240\ \text{V}} = 18.8\ \ \text{A}$ Thus, a circuit breaker of 20-A/2P is needed with a wire size of 12 THW(Cu).

- Range : $I_L = \dfrac{8{,}000\ \text{VA}}{240\ \text{V}} = 33.3\ \text{A}$ Thus, a circuit breaker of 40-A/2P and a wire size of 8 THW(Cu) are specified.

- *Kitchen*: Two circuits rated at 20-A/1P with an assumed load of $I_L = 80\% \times 20\text{-A} = 16\text{-A}$.

- *Laundry*: One circuit rated at 20-A/1P with a load of 16-A.

- Bedroom : $I_L = \dfrac{6 \times 180 + 2 \times 100}{120\ \text{V}} = 10.7\ \text{A}$ Thus, a circuit rated at 15-A/1P is sufficient.

- Living : $I_L = \dfrac{7 \times 180 + 2 \times 100}{120\text{V}} = 12.2\ \text{A}$ Therefore. A circuit rated at 15-A is specified.

Thus,

$$\text{Number of branch circuits} = 2 \times 3 + 5 = 11$$

Service Entrance: Standard Method

Size phases, grounding, and main panel breaker

1. General lighting/receptacles

$$\text{Actual load} = 13 \times 180 + 4 \times 100 = 2{,}740\ \text{VA}$$

$$\text{Minimum NEC load} = \frac{3\ \text{VA}}{\text{ft}^2} \times 1{,}000\ \text{ft}^2 = 3{,}000\ \text{VA}$$

$$\frac{\text{Kitchen}}{\text{Laundry}} = 3 \times 1{,}500 = 4{,}500\ \ \text{VA}$$

$$\text{Using the minimum NEC load : CL} = 7{,}500\ \text{VA}$$

$DL = 100\%\ \ 3{,}000 + 35\,\% \times 4{,}500 = 4{,}575\ \text{VA}$

2. Fixed appliances

$$\text{Water heater} = 5{,}000\ \text{VA}$$

3. Special appliances
   - Dryer (4,500 VA) = 5,000 VA (refer to the note of Table 7.1)
   - Range (8,800 VA) = 80% × 8,000 VA = 6,400 VA (refer to Table 7.2)
   - DL_{Tot} = 20,975 VA

$$I_{L,A+B} = \frac{20,975 \text{ VA}}{240 \text{ V}} = 87.4 \text{ A}$$

Size neutral

1. General lighting/receptacles
   DL = 4,575 VA
2. Fixed appliances
   DL = 0
3. Special appliances
   - Dryer = 5,000 × 0.7 = 3,500 VA
   - Range = 6,400 × 0.7 = 4,480 VA
   - $DL_{Tot,N}$ = 12,555 VA
   - I_{LN} = 12,555 VA/240 V = 52.3 A

Panel Schedule

$$I_{L,A} = 64 + 47 = 111 \text{ A} \qquad \% \text{ unbalance} = \frac{111 \text{ A} - 105 \text{ A}}{111 \text{ A}} = 5\%$$
$$I_{L,B} = 70 + 35 = 105 \text{ A}$$

7.6 ELECTRICAL SYSTEMS FOR APARTMENT BUILDINGS

The design approach and calculation procedures outlined for single-family homes can be applied to any residential building including high-rise apartments. Both standard and optional methods can be used to determine the size of the service entrance for multifamily residential buildings, as illustrated in Figure 7.15, for an apartment with two units.

The standard method, as outlined in Section 7.3.1, can be applied to size the feeder for an apartment building. Example 7.7 illustrates the calculation details using the standard method for an apartment building with several units.

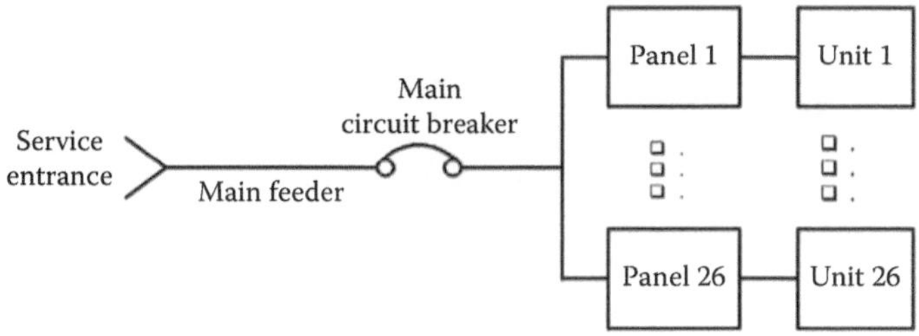

FIGURE 7.15 A feeder serving a 26-unit apartment building.

Example 7.7

Problem

Size the service entrance for an apartment building with 26 units served by 120/240 V.

Each unit has a floor area of 850 ft^2 and the following electrical loads: 1.4 kVA dishwasher, 920 VA disposal, 5 kVA dryer, 8 kVA range, and 480 VA blower motor.

Solution

The calculation is carried out using the standard method outlined in Section 7.3.1:

Size of phases:

1. General lighting
   - *Receptacles*: 3 VA/ft^2 × 850 ft^2 × 26 = 66,300 VA
   - *Kitchen*: 2 CKT × 1,500 VA × 26 = 78,000 VA
   - *Laundry*: 1,500 VA × 26 = 39,000 VA
   - *Connect load*: 66,300 + 78,000 + 39,000 = 183,300 VA
   - *Demand load*: 100% × 3,000 + 35% × 117,000 + 25% × 63,300 = 59,775 VA
2. Fixed appliances
   - *Dishwasher*: 26 × 1,400 × 75% = 27,300 VA
   - *Disposal*: 26 × 920 × 75% = 17,940 VA
   - *Blower motor*: 26 × 480 × 75% = 9,360 VA
3. Special appliances
   - *Dryer*: 26 × 32.5% × 5,000 VA = 42,250 VA (refer to Table 7.1)
   - *Range*: 26 × 24% × 8,000 VA = 49,920 VA (refer to Table 7.2)
   - *Disposal*: 25% × 920 = 230 VA (the largest and only motor load)
   - DL_{Total} A + B = 208,000 VA
   - IL A + B = 860 A

Since it is desirable to utilize 500 MCM (THW-Cu) conductors or smaller and since an ampacity of 500 MCM conductor is 380 A: The number of parallel wires = 860/380 ≈ 3

It is recommended to use three parallel 350 MCM THW (Cu) conductors with each conductor handling 286 A load.

Size of the Neutral

The calculations are the same as those outlined for phases A and B, but use only 70% of dryer range load:

$$D_{L,N} = 179,200 \text{ VA}$$

$$I_{L,N} = \frac{179,200 \text{ VA}}{(240 \text{ V} \times 3)} = 249 \text{ A}$$

Thus, three parallel 250 MCM (THW Cu) conductors can be used for the neutral of the main service entrance feeder supplying the apartment building.

The optional method can be also applied to size the conductors for the phases of the service entrance feeder for an apartment building using the same procedure outlined in Section 7.3.2. However, the demand load is estimated using the demand factors, DF_O, listed in Table 7.4 and the total building connected load, $CL_{tot\text{-}O}$, including the largest of the air conditioning and heating systems as expressed by Equation 7.11:

$$DL_{tot-O} = DF_O \times CL_{tot-O} \tag{7.11}$$

It should be noted that NEC allows for two-unit apartments to use lower of the demand load obtained using Equation 7.8 with two-unit dwellings and that found using Equation 7.11 for three-unit dwellings with the same unit floor area and number of appliances. Example 7.8 provides the application of the optional method to size the phases for a feeder of multifamily residential building considered in Example 7.7.

Example 7.8

Problem

Size the service entrance for the 26-unit apartment building of Example 7.7 using the optional method.

Solution

First, All the connected loads for the 26 units are added:

$$\text{General lighting / receptacles}: 3\ \text{VA / ft}^2 \times 850\ \text{ft}^2 \times 26 = 66{,}300\ \text{VA}$$

$$\text{Kitchens}: 2\ \text{CKTs} \times 1{,}500\ \text{VA} \times 26 = 78{,}000\ \text{VA}$$

$$\text{Laundry loads}: 1{,}500\ \text{VA} \times 26 = 39{,}000\ \text{VA}$$

$$\text{Dishwashers}: 26 \times 1{,}400 = 36{,}400\ \text{VA}$$

$$\text{Disposals}: 26 \times 920 = 23{,}920\ \text{VA}$$

$$\text{Blower motors}: 26 \times 480 = 12{,}480\ \text{VA}$$

$$\text{Dryers}: 26 \times 5{,}000\ \text{VA} = 130{,}000\ \text{VA}$$

$$\text{Range}: 26 \times 8{,}000\ \text{VA} = 108{,}000\ \text{VA}$$

The total connected load is then

$$CL_{tot-O} = 594,600 \text{ VA}$$

The DF_O from Table 7.4 for 26-unit apartment building is 34%; thus, the demand load is

$$DL_{tot-O} = 0.34 \times 594{,}600 \text{ VA} = 202{,}164 \text{ VA}$$

Thus, the load for the phases A and B expressed in Amperes is very close to the load found using the standard method:

$$I_{L,O} = \frac{202{,}164 \text{ VA}}{240 \text{ V}} = 842 \text{ A}$$

Thus, it is recommended to use three parallel 350 MCM THW (Cu) conductors, each can handle 281 A load.

7.7 CASE STUDY: ANALYSIS OF ELECTRICAL SYSTEMS FOR A RANCH HOUSE

7.7.1 Building Description

The electrical system for a residential building is evaluated for compliance with the NEC requirements in this case study. The building is a single-family, one-story, detached home located in Denver, Colorado. The original construction was in the 1880s, but additions and improvements have been made since its construction. The home consists of one ground floor, one unfinished basement, and one unfinished attic. The ground floor, as shown in Figure 7.16, has two bedrooms, one and a half bathrooms, one kitchen, and one living room. The basement, as indicated in Figure 7.17, has one laundry area. There is approximately 85 m^2 (916 ft^2) of living and conditioned space. The electrical system in the home is relatively new as it was rewired when a small 12 m^2 (130 ft^2) addition was added to the back of the home in the early 2000s. When the addition was constructed, the entire electrical wiring of the home was retrofitted and brought up to code at the time.

7.7.2 Description of the Electrical System

The power to the home is supplied by main feeders from an overhead service entrance to the main panel. The main panel is located outside the building with an electrical meter attached to the top of the box. The main panel is grounded with a rod located just below the breaker.

1. *Panel and main feeder sizes*: The main circuit breaker of the panel is rated at 125 A, as indicated in Figure 7.18, and is powered by the main feeder, which is assumed to be #1 AWG THW (Cu). This is taken from Table

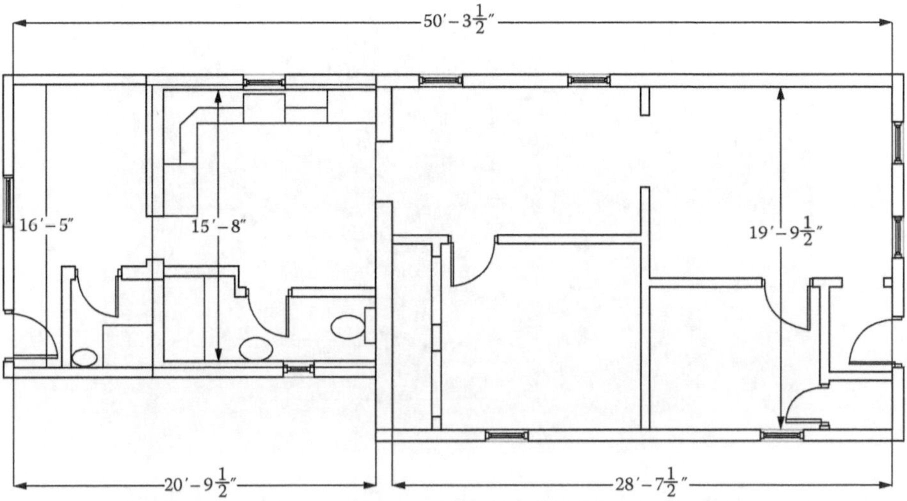

FIGURE 7.16 Floor plan of the ground floor of the audited ranch house.

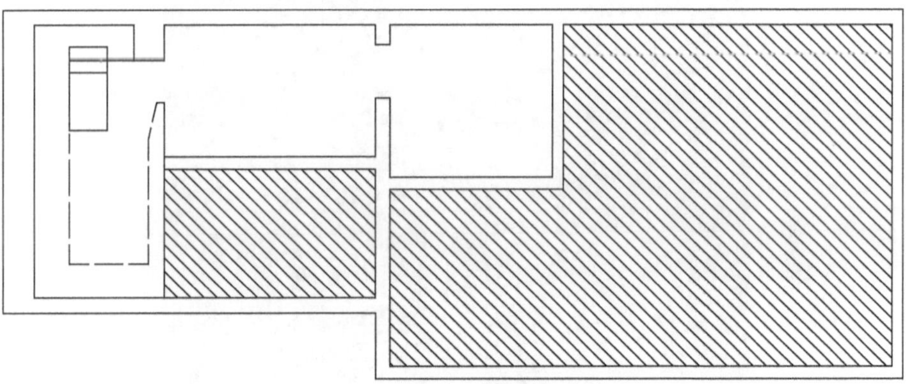

FIGURE 7.17 Floor plan of the basement floor of the audited ranch house.

310.15(B)(16) from NEC (2014), knowing the main circuit is rated at 125 A. The neutral wire is assumed to be of the same size as the main feeders since electricians typically use the same size wire.

2. *Grounding*: As previously mentioned, the home is grounded by a metal rod that is staked into the ground, as indicated in Figure 7.19. It is assumed that the grounding wire is #6 AWG (Al), as outlined from NEC Table 250.66 (NEC, 2014).
3. *Number of circuits*: The panel includes 21 circuits. When the wiring within the home was redone in the early 2000s, most of the circuit breakers were replaced with tandem circuit breakers to create more space without replacing

FIGURE 7.18 Main circuit breaker with sizing.

FIGURE 7.19 Grounding wire and rod.

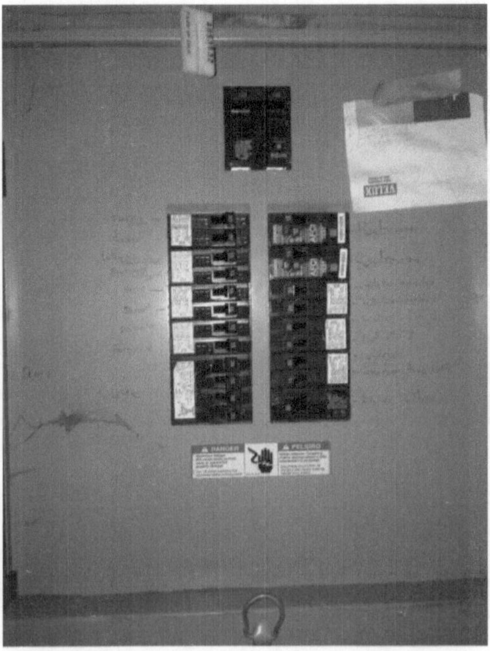

FIGURE 7.20 Main circuit breaker.

the entire encasement. There are currently no extra circuits available, so the flexibility of the panel is limited. However, if the three single-pole circuits were replaced with tandem circuits, then there would be three spare circuits. The circuit breakers protects all the home branch circuits included within the panel as shown in Figure 7.20.

4. *Receptacles, switches, and lights*: The audited house has a total of 28 double receptacles and three special double receptacles. The three receptacles in the bathrooms are GFCI receptacles along with all four receptacles above the countertop of the kitchen. There are 16 switches throughout the home, with at least one in each room. All switches are typical on/off switches except for the switch in the main bathroom, which has a motion sensor. The home has a total of 17 light fixtures, each with a varying number of bulbs. There are a total of seven 9.5 W LED bulbs, 14 numbers of 15 W CFL bulbs, six 40 W incandescent bulbs, and two 60 W incandescent bulbs. Figures 7.21 and 7.22 show the layouts for the receptacles, switches, and lights for, respectively, the ground floor and the basement of the house.
5. *Fixed appliances*: The home has the following fixed appliances:
   - Ceiling fan
   - Dishwasher
   - Garbage disposal
   - Vent fan (with 40 W light)
   - Washer

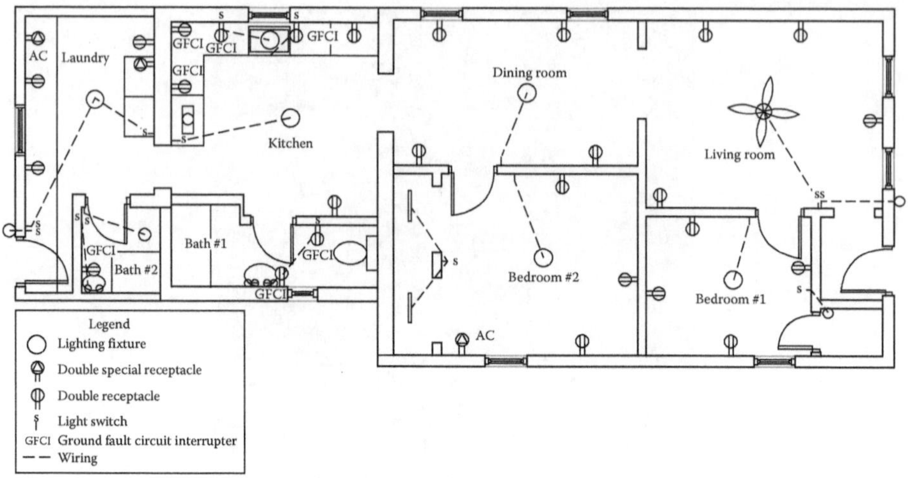

FIGURE 7.21 Receptacles, lighting, and switches of the ground floor of the audited ranch house.

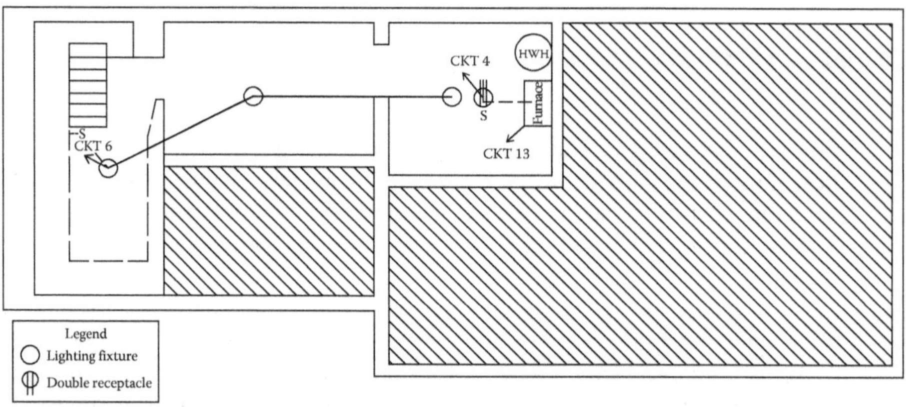

FIGURE 7.22 Receptacles, lighting, and switches of the ground floor of the audited ranch house.

6. *Special appliances*: The home has the following special appliances:
   - Electric range
   - Clothes dryer
   - Swamp cooler
   - Air conditioner
   - Furnace (gas)

The loads for each of these appliances are described in more details in the following sections.

7.7.3 Load Estimation

1. *Receptacles*: According to NEC 220.14 (I), each receptacle load can be assumed to be 180 VA. There are a total of 28 nonspecial receptacles throughout the house for a total load of 5,040 VA. All receptacles in the bathroom and in the kitchen above the countertop are GFCI for a total of seven. The locations and wattages of the receptacles are listed in Table 7.6.
2. *Lighting*: The lighting fixtures and associated wattage loads distributed throughout the home are provided in Table 7.7.
3. *Fixed appliances*: A summary of the fixed appliances and associated loads available throughout the home is shown in Table 7.8.

TABLE 7.7
List of Lighting Fixtures

Room	Type of Bulb	Watt per Bulb	# of Bulbs	Total Wattage
Living room	CFL	15 (60 W equiv.)	1	15
	LED	9.5 (60 W equiv.)	3	28.5
Bedroom #1	CFL	15 (60 W equiv.)	3	45
Bedroom #2	CFL	15 (60 W equiv.)	2	30
	CFL tube	15 (60 W equiv.)	2	30
Dining room	CFL	15 (60 W equiv.)	1	15
	LED	9.5 (60 W equiv.)	1	9.5
Bathroom #1	Incandescent	40	4	160
Bathroom #2	CFL	15 (60 W equiv.)	1	15
	Incandescent	40	2	80
Kitchen	CFL	15 (60 W equiv.)	2	30
	LED	9.5 (60 W equiv.)	1	9.5
Laundry area	CFL	15 (60 W equiv.)	1	15
	LED	9.5 (60 W equiv.)	2	19
Basement	CFL	15 (60 W equiv.)	1	15
	Incandescent	60	2	120
			Total	636.5

TABLE 7.8
Summary of the Ratings for the Fixed Appliances

Room	Appliance	Voltage	Wattage
Living room	Ceiling fan	120	55
Kitchen	Dishwasher	120	1,440
	Disposal	120	828
	Vent fan (with 40 W light)	120	256
Laundry area	Washer	120	1,200

TABLE 7.9
Summary of the Ratings of the Special Appliances

Room	Appliance	Voltage	Wattage
Kitchen	Electric range	240	12,500
Laundry area	Electric dryer	240	5,600
	Swamp cooler	120	460
Bedroom #1	Air conditioner	120	820
Basement	Furnace (gas)	120	1,200

4. *Special appliances*: The fixed appliances and their wattage loads are indicated in Table 7.9. All loads were defined using nameplate specifications.

7.7.4 Electrical Layout

The electrical layout of receptacles, switches, and lighting for specific rooms is given in Figure 7.23. Note that branch circuits include both receptacles and lighting fixtures, a common practice for residential buildings since the same voltage is required for both loads.

7.7.5 Zoning and Branch Circuit Loads

A summary of all the circuit breaker zones is provided in Table 7.10. An explanation of improperly sized circuit breakers can be found in the next section. This section simply discusses the general layout of the circuits along with an analysis on the logic of the zones and the flexibility of the system in its current state.

After reviewing the loads for all the branch circuits, the electrical distribution system is not effectively zoned. Each area of the home is not served necessarily by one circuit. For example, the outlets of bedroom #1 and bedroom #2 are on the same branch circuit, and the lighting fixtures for the two bedrooms are on a separate branch circuit. It would be more convenient to have all of the electrical components (lighting and outlets) for bedroom #1 on one branch circuit and the electrical components (lighting and outlets) for bedroom #2 on a different branch circuit. This issue has been noticed throughout the entire house. None of the branch circuits serves both the lighting and receptacle loads. This circuiting method has resulted in unbalances between low loads on the lighting branch circuits and high loads on the outlet branch circuits.

The low loads on the lighting circuits provide good flexibility for future expansion. However, no extra loads can be considered for the outlet branch circuits. Circuit #5 is overloaded with seven receptacle outlets and one air conditioning unit. If the lighting and receptacle outlets in bedroom #2 were on one circuit instead of the current circuiting scheme, the load would only be 1,420 W and the amperage would be 11.8 A, which is acceptable for a branch circuit protected by a 15 A/1P rated circuit breaker. In addition, other branch circuits seem to be undersized, primarily all the

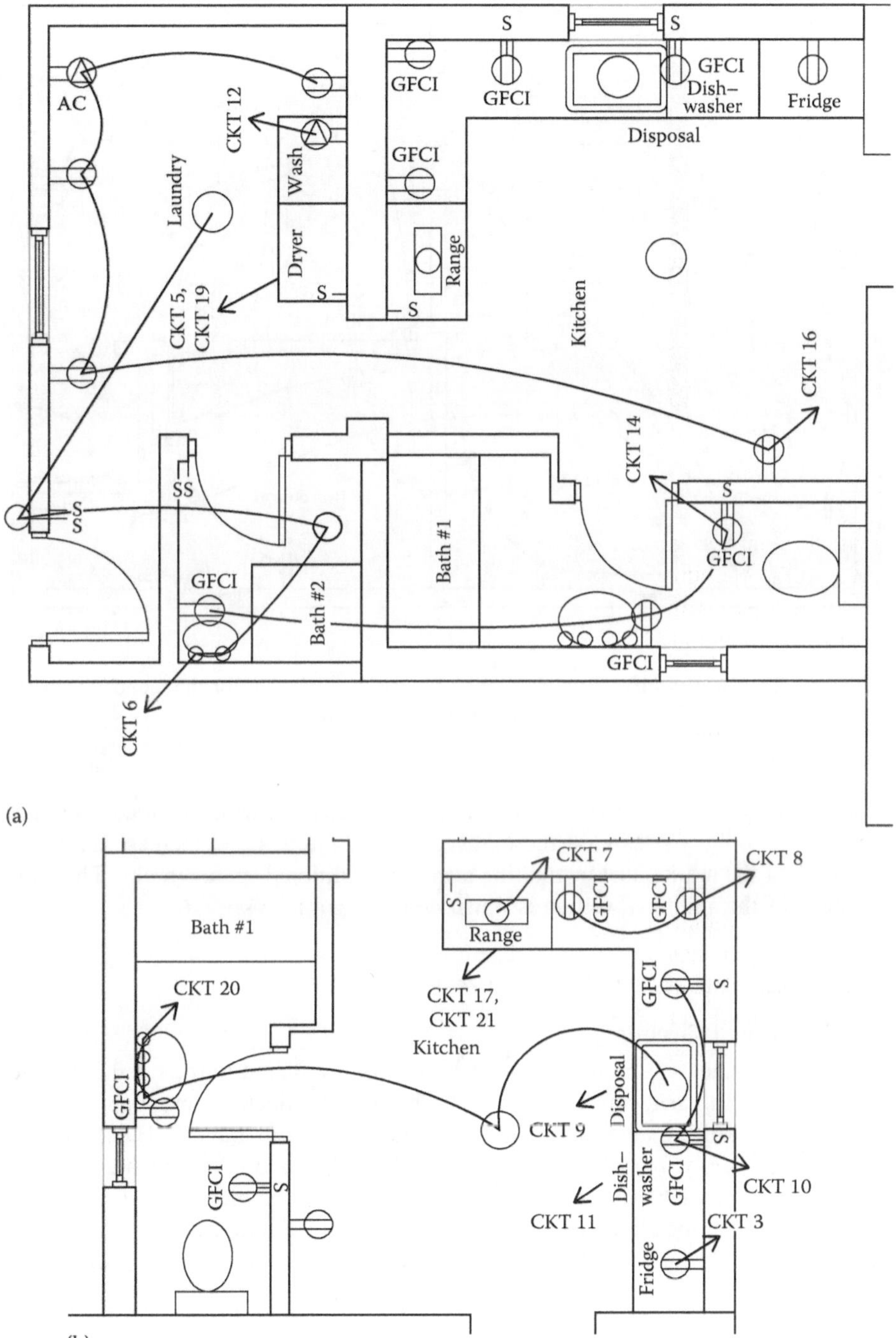

FIGURE 7.23 Branch circuits for the dwelling: (a) for the laundry area and two bathrooms as well as (b) for the kitchen. (c) for remaining rooms.

(Continued)

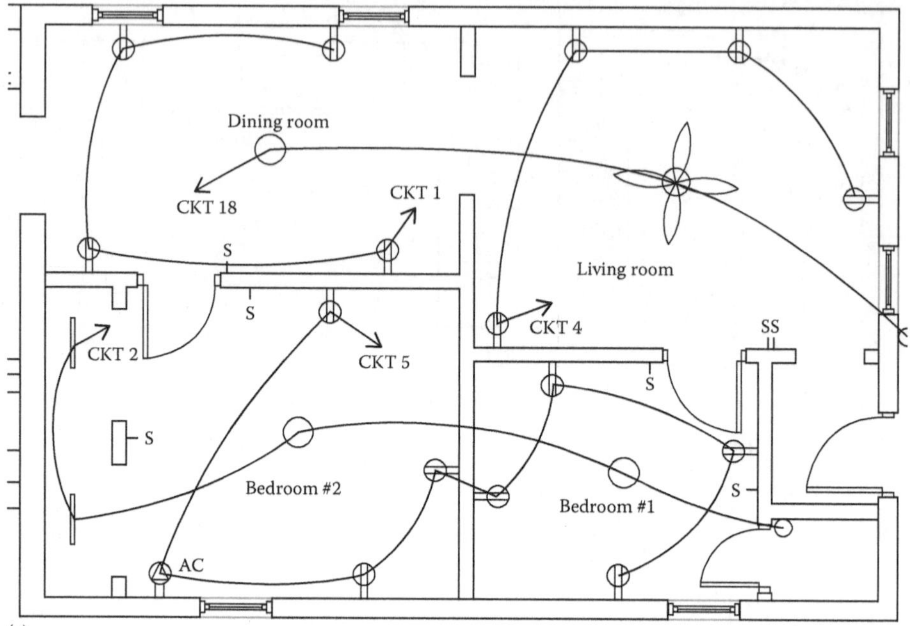

(c)

FIGURE 7.23 (*Continued*) Branch circuits for the dwelling: (a) for the laundry area and two bathrooms as well as (b) for the kitchen. (c) for remaining rooms.

lighting circuits. As a corrective measure, branch circuits with low loads could be combined in order to decrease the number of total circuits and increase the future flexibility of the main panel by making available additional spare circuits. The panel schedule for the audited dwelling is illustrated in Figure 7.24.

7.7.6 Calculations and Sizing Checks

7.7.6.1 Branch Circuits

Table 7.11 provides a summary of the actual loads compared to the rated capacities of the circuits. In particular, there are three branch circuits that do not appear to be sized properly since their actual loads (identified in bold in Table 7.11) exceed their ratings. Specifically, branch circuit #5 is rated at 15 A and is not intended to have an outlet designated to an air conditioning unit. It is recommended to increase the circuit breaker for the circuit to a 20 A breaker. The other branch circuits that are not sized properly are associated with the range circuits. The nameplate specifications show that the range is rated at 12,500 W, which indicates that each circuit should be rated at 60 A and not 50 A (since 12,500/240 = 52 A). This is most likely because standard home ranges are typically sized at 12,000 W, and so a 50 A breaker would have been adequate. All the other circuits are sized properly.

TABLE 7.10
Summary for Branch Circuits and Associated Loads

Circuit Breaker	Location or Appliance	Size	Number of Receptacles	Number of Lamps	Total Power (W)	Total Current (A)
1	Dining room (outlets)	20 A/1P	4	0	720	6
2	Bedroom #1 and bedroom #2 (lights)	15 A/1P	0	7	105	0.875
3	Kitchen (fridge)	20 A/1P	1	0	780	6.5
4	Living room and basement (outlets)	15 A/1P	5	0	900	7.5
5	Bedroom #1 and bedroom #2 (outlets and AC)	15 A/1P	7	0	2,080	17.3
6	Basement-bath #2-laundry (lights)	15 A/1P	0	9	264	2.2
7	Hood	15 A/1P	0	1	256	2.1
8	Kitchen counter (GFCI)	20 A/1P	2	0	360	3
9	Disposal	15 A/1P	0	0	828	6.9
10	Kitchen counter (GFCI)	20 A/1P	2	0	360	3
11	Dishwasher	15 A/1P	0	0	1,440	12
12	Washer	20 A/1P	0	0	1,200	10
13	Furnace	15 A/1P	0	0	1,200	10
14	Bathroom #1 and bathroom #2 (GFCI)	20 A/1P	3	0	540	4.5
15 and 19	Dryer	30 A/2P	0	0	5,600	23.3
16	Laundry outlets (AC) and 1 Kitchen outlets	20 A/1P	4	0	1,180	9.8
17 and 21	Range	50 A/2P	0	0	12,500	52
18	Living and dining room lights and ceiling fan	15/1P	0	6	123	1.0
20	Kitchen and bath #1 (lights)	20/1P	0	7	200	1.7

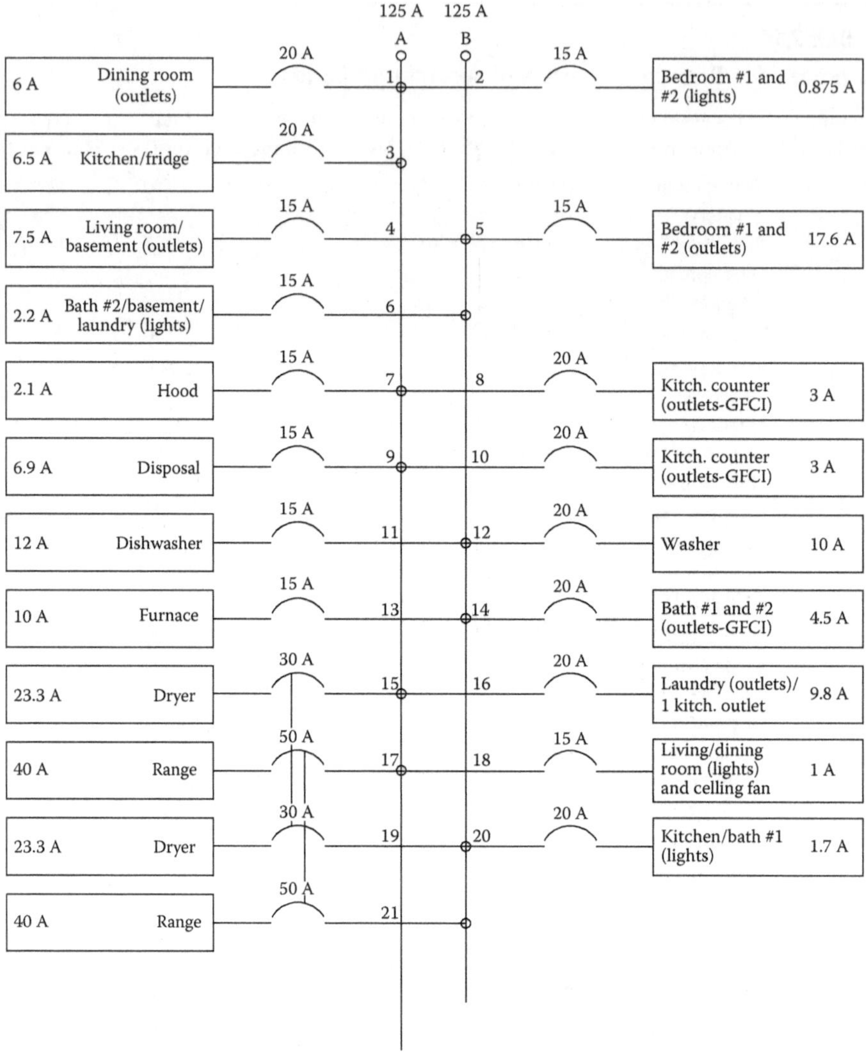

FIGURE 7.24 Electrical panel schedule for the ranch house.

7.7.6.2 Main Service Entrance

From Article 220 of the NEC, the standard method can be used to size the main circuit breaker, main feeders, and the neutral wire. The calculations and procedure are outlined here:

1. General loads (lighting and *small appliances*)

$$\text{Existing general loads in the home} = \text{Lighting loads} + \text{Receptacle loads}$$

$$= 636\ \text{VA} + 5{,}040\ \text{VA} = 5{,}677\ \text{VA}$$

TABLE 7.11
Summary of Branch Circuit Ratings and Their Connected Loads

Circuit Breaker	Location or Appliance	Actual Load (A)	Rating (A)
1	Dining room (outlets)	6	20
2	Bedroom #1 and bedroom #2 (lights)	0.88	15
3	Kitchen (fridge)	6.5	20
4	Living room and basement (outlets)	7.5	15
5	Bedroom #1 and bedroom #2 (outlets and AC)	17.3	15
6	Basement-bath #2-laundry (lights)	2.2	15
7	Hood	2.1	15
8	Kitchen counter (GFCI)	3	20
9	Disposal	6.9	15
10	Kitchen counter (GFCI)	3	20
11	Dishwasher	12	15
12	Washer	10	20
13	Furnace	10	15
14	Bathroom #1 and bathroom #2 (GFCI)	4.5	20
15	Dryer	23.3	30
16	Laundry outlets (AC) and 1 kitchen outlets	9.8	20
17	Range	52	50
18	Living and dining room lights and ceiling fan	1	15
19	Dryer	23.3	30
20	Kitchen and bath #1 (lights)	1.7	20
21	Range	52	50

If existing general loads are greater than the loads below, existing loads are used to size the main feeder.

2. NEC requirements

$$\text{Lighting load and receptacle outlets}: \text{Load} = 3\frac{\text{VA}}{\text{sf}}\text{A} = 3\frac{\text{VA}}{\text{sf}}1{,}000\text{ sf} = 3{,}000\text{VA}$$

$$\text{Small appliance receptacles}: \text{Load} = 2\times 1{,}500\text{ VA} = 3{,}000\text{ VA}$$

$$\text{Laundry receptacles}: \text{Load} = 1\times 1{,}500\text{ VA} = 1{,}500\text{ VA}$$

$$\text{Total general load} = 3{,}000 + 3{,}000 + 1{,}500 = 7{,}500\text{ VA} > 5{,}677\text{ VA}$$

$$\therefore \text{Use } 7{,}500\text{ VA for demand load}$$

$$\text{Demand load} = 3{,}000\times 100\% + (7{,}500 - 3{,}000)35\% = 4{,}575\text{ VA} = 4.58\text{ kVA}$$

3. Fixed appliance load

$$\text{Ceilingfan} = 0.055 \text{ kVA}$$

$$\text{Dishwasher} = 1.44 \text{ kVA}$$

$$\text{Disposal} = 0.828 \text{ kVA}$$

$$\text{Vent fan} = 0.256 \text{ kVA}$$

$$\text{Washer} = 1.2 \text{ kVA}$$

$$\text{Sum} = 3.78 \text{ kVA}$$

$$\text{Demand load} = 3.78 \text{ kVA} \times 75\% = 2.83 \text{ kVA}$$

4. Special appliances

$$\text{Electric dryer : Demand load} = 5.6 \text{ kVA} \times 100\% = 5.6 \text{ kVA}$$

$$\text{Electric range: Actual load} = 12.5 \text{ kVA}$$

$$\text{Demand load} = 8 \text{ kVA}(1 + 0.5 \times 5) = 8.2 \text{ kVA}$$

$$\text{Heat/AC load:}$$

$$\text{Total AC load} = 820 + 640 = 1{,}280 \text{ VA}$$

$$\text{Total heat load} = 1{,}200 \text{ VA}$$

$$\therefore \text{Use AC load to size}$$

$$\text{Demand load} = 1.28 \text{ kVA} \times 100\% = 1.28 \text{ kVA}$$

$$\text{Motors :}$$

$$\text{Ceiling fan} = 0.055 \text{ kVA}$$

$$\text{Disposal} = 0.828 \text{ kVA}$$

$$\text{Demand load} = 0.828 \times 25\% = 0.207 \text{ kVA}$$

$$\text{Demand : load total} = 4.58 + 2.83 + 5.6 + 8.2 + 1.28 + 0.207 = 22.7 \text{ kVA}$$

5. Main circuit breaker

$$I = \frac{\text{Demand load}}{240 \text{ V}} = \frac{22{,}700}{240} = 94.6 \text{ A}$$

Select a main circuit breaker of 100 A

6. Main feeder size
   From Table 7.5, the main feeders can be sized to #2 AWG THW (Cu). According to the textbook, aluminum conductors are used as service entrance and feeder conductors that carry electricity from the transformer to the panelboard. Therefore, a 1/0 AWG THW (Al) wire has been chosen to make a comparison to the existing main feeder size.
7. Neutral wire
   The demand load for the neutral wire is the sum of all of the one-pole demand loads given plus 70% of the dryer and range loads.

$$\text{Demand load}_N = DL_{Gen} + DL_{\text{fixed}} + DL_{AC} + DL_{\text{Motor}}$$

$$+DL_{\text{Dryer}} \times 70\% + DL_{\text{Range}} \times 70\%$$

$$\text{Demand load}_N = 4.58 + 2.83 + 0.70 \times 5.6 + 0.70 \times 8.2 + 1.28 + 0.207$$

$$= 18.6 \text{ kVA}$$

$$I = \frac{\text{Demand load}_N}{240 \text{ V}} = \frac{18{,}600}{240} = 77.3 \text{ A}$$

From Table 7.5, the neutral wire can be sized to #4 AWG THW (Cu).

Since the existing service feeder is oversized, there is no need to consider the optional method. It should be noted, however, that the optional method would provide a load current of less than 94 A, which would be sized up to a 100 A circuit breaker anyways.

It is clear from the summary outlined in Table 7.12 that the current service equipment is adequate for the loads of the home, but is oversized. The main circuit breaker could be brought down to a 100 A breaker, the main feeders can be resized to a #2 AWG THW (Cu) wire, and the required neutral wire size is #4 AWG THW (Cu). Since the service entrance is oversized, it does have room for flexibility and future additions.

7.7.6.3 Grounding Check

The size of the required ground wire was found using Table 7.6 as listed in Table 7.13.

The existing grounding wire is assumed to be sized properly according to the existing main circuit breaker size. If the system is sized to the minimum required size according to NEC, then the grounding wire size could also be reduced to #8 (Cu).

TABLE 7.12
Summary of Entrance Service Sizes

Service Entrance System	Existing	Required by NEC
Main circuit breaker	125 A	100 A
Main feeder wire	#1 AWG THW (Cu)	#2 AWG THW (Cu)
Neutral wire	#1 AWG THW (Cu)	#4 AWG THW (Cu)

TABLE 7.13
Grounding Wire Summary

Grounding	Existing	Required by NEC
Wire	#6 (Cu)	#8 (Cu)

TABLE 7.14
Total Loads for Phases A and B for the Ranch House

Circuit Breaker	Location or Appliance	Phase A (A)	Phase B (A)
1	Dining room (outlets)	6	
2	Bedroom #1 and bedroom #2 (Lights)	0.88	
3	Kitchen (fridge)	6.5	
4	Living room and basement (outlets)		7.5
5	Bedroom #1 and bedroom #2 (outlets and AC)		17.3
6	Basement-bath #2-laundry (lights)		2.2
7	Hood	2.1	
8	Kitchen counter (GFCI)	3	
9	Disposal	6.9	
10	Kitchen counter (GFCI)	3	
11	Dishwasher		12
12	Washer		10
13	Furnace		10
14	Bathroom #1 and bathroom #2 (GFCI)		4.5
15	Dryer	23.3	
16	Laundry outlets (AC) and 1 kitchen outlet	9.8	
17	Range	52	
18	Living and dining room lights and ceiling fan	1	
19	Dryer		23.3
20	Kitchen and bath #1 (lights)		1.7
21	Range		52
	Total	114.48	140.5

7.7.6.4 Panelboard Balance

Table 7.14 shows the total loads for both phases A and B due to various connected loads. Table 7.14 also indicates the level of unbalance between the phases.

$$\% \text{ Unbalanced}(B, A) = \frac{B - A}{B} = \frac{140.5 - 114.8}{140.5} = 18.5\% \text{ unbalanced}$$

The current layout of the panel is not balanced. The owners could easily rectify this problem by switching circuit #2 to phase A and circuit #4 to phase B. This would create an unbalance of only 5%.

7.7.6 Additional NEC Requirements

In addition to the NEC calculation checks, other NEC requirements should be checked within the home.

1. Branch circuit ratings
   NEC 210.3 states that individual branch circuits should be rated at 15, 20, 30, 40, and 50 A. All the circuit breakers at this home are rated at 15, 20, 30, or 50 A, and therefore meet the NEC requirements.
2. GFCI receptacles
   NEC 210.8 states that for dwelling units the following areas should have GFCI protections: bathrooms, unfinished basements, kitchens (above countertops), 6 ft within the outside edge of a bathtub or shower, and in the laundry areas. The audited residence meets most of these requirements except two items. Indeed, the bathroom, and kitchen countertop outlets are all GFCI, which also satisfies the "within 6′ of and bathtubs and showers" requirement. However, the unfinished basement outlet and the laundry area outlets are not GFCI. To meet the current code, the owners should consider switching the outlets to GFCI.
3. Small-appliance branch circuits
   NEC 210.11 (C)(1) states that two or more 20 A small-appliance branch circuits shall be provided for receptacle outlets in the kitchen. NEC 210.52 (B)(2) states that the small-appliance branch circuits shall have no other outlets. NEC 210.11(C)(2) states that at least one additional 20 A small-appliance branch circuit shall be provided to supply the laundry receptacle outlets and it should have no other outlets. The home meets the specific requirement that the laundry small-appliance branch circuit is rated at 20 A. However, only one small-appliance branch circuit of 20 A is available in the kitchen.
4. Lighting outlets
   NEC 210.70 states that for a dwelling unit there shall be at least one wall switch-controlled lighting outlet in every habitable room and bathroom, as well as in the basement. As can be seen in the electrical layout plan, this requirement is met.
5. Receptacle outlets
   NEC 210.52 has several general requirements for receptacle outlets for a dwelling unit. Here are some of the major requirements:
   - Receptacles should be installed so no point measured horizontally along any wall space is more than 6 ft from a receptacle outlet.
   - Any wall space 2 ft or more in width shall have a receptacle.
   - Wall countertop receptacles should be installed so no point along the wall line is more than 2 ft from a receptacle outlet.

- At least one receptacle outlet shall be installed in areas designated for laundry.
- At least one receptacle outlet shall be installed in the unfinished portion of the basement.

The home meets all of these requirements as can be seen in Figure 7.25.

7.7.7 Recommendations and Conclusions

It is recommended that the circuit branches be evaluated and possibly combined. Several branch circuits are very oversized while three circuits are undersized. This could be rectified by takings some loads from the oversized circuit and adding them to the undersized circuits. If a redefinition of the circuits is not feasible, then circuit #5 rating can be increased from 15 A/1 P to 20 A/1 P and the range circuit rating can be replaced so its rating increases from 50 A/2 P to 60 A/2 P.

In addition, the layout of the circuits is not as logical as it could be. Since the circuits are already established, it does not economically make sense to completely reestablish the zones. However, if this were a new construction home, I would recommend that the zones be laid out by room/area instead of by outlets and lighting fixtures.

To rectify the unbalanced panelboard, moving circuit #5 to phase A and moving circuit # 2 to phase B are recommended. This action reduces the percent unbalanced from 18% to 5%.

As established in the calculations section of this report, the existing main circuit breaker, feeder size, neutral size, and ground wire are currently oversized. Since the system is oversized, additional loads and circuits can be added safely. In fact, the home only has one single 20 A circuit outlet in the kitchen area. It is recommended that an additional 20 A, single-outlet—small-appliance—circuit be installed in order to meet the code for the kitchen area. Since the panelboard is already full, it is recommended that one of the single-phase breakers be changed to a tandem breaker so that there is room for the additional circuit. Since the main circuit breaker is already oversized, there is no need to replace the panel.

In conclusion, there are only a few items for which the electrical system of the home does not meet the current NEC requirements, but the system is generally in a good condition and the number of changes and retrofits the electrical system has probably endured over the years.

7.7.8 Heating Systems for Building Electrification

7.7.8.1 Background

It is estimated that space heating, domestic hot water, and cooking equipment in commercial and residential buildings are responsible for 6% of the global CO2 emissions associated mostly to on-side use of oil and natural gas products (McKinsey, 2022). In the US, 13% of the national annual greenhouse emissions are attributed to space heating, hot water heating, and cooking (ACEEE, 2022).

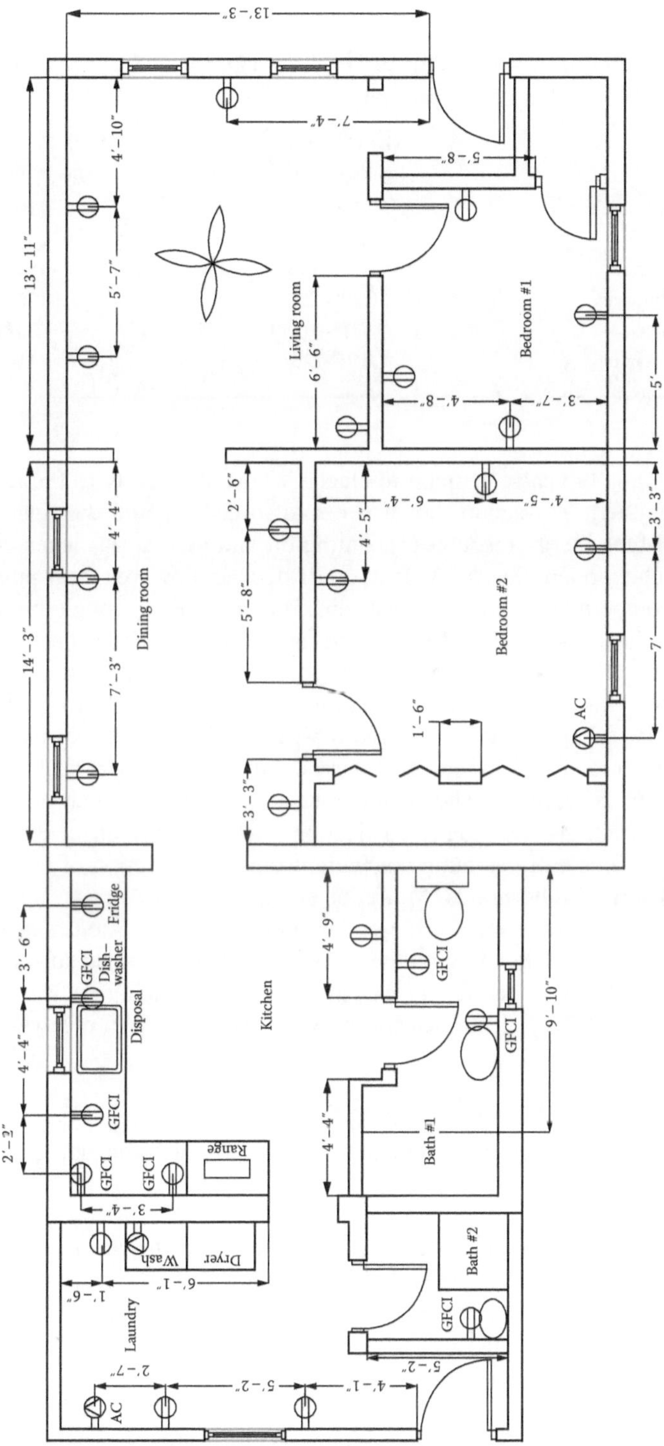

FIGURE 7.25 Distances between two adjacent receptacles along the ground floor walls.

TABLE 7.15
Range of Carbon Emission Per Unit of Heat Energy (lb of CO_2e/MMBtu) Delivered for Various Heating Systems

Heating System	Grid Generation Electricity	With 70% produced from renewable
Electrical resistance baseboards	250–310	110–130
Oil-fired boilers	230–275	230–275
Steam (district heating)	205–270	185–220
Gas-fired boilers/furnaces	200–250	170–190
Air-to-air heat pumps	115–160	30–40
Ground-source heat pumps	75–120	15–25

Space heating is typically difficult to electrify cost-effectively to the reliance of most existing systems on steam, hot water, and forced air and the relatively low prices of fossil-fuel energy resources including oil and natural gas when compared to those of grid-supplied electricity. In the last decade, however, heat pumps have become cost-competitive and energy-efficient equipment that provide both heating and cooling with the benefits of decarbonizing buildings. Indeed, heat pumps are on average more efficient that furnaces for space heating by at least a factor of 4. According to McKinsey predictions, heat pumps could represent 90% of new heating units sold in 2,050 compared to 35% currently with the ability to reduce carbon emissions globally by three gigatons annually. Already in some countries such as Norway, more than 25% of existing households utilize heat pumps to heat their homes.

For new constructions, building electrification can be cost-effective when accounting for the avoided cost of gas main connections and the adoption of meters in all-electric homes and neighborhoods (McKenna et al., 2020).

Table 7.15 provides the range of the carbon emissions for various space heating systems used in the US depending on the electricity source options: current US grid generation, and 70% produced from renewable energy technologies. It is estimated that currently only 20% of electricity generated in the US is from renewables (EIA, 2021).

7.7.8.2 Overview of Electrical Heating Systems

The common heating systems used for electrification programs include the following technologies:

- Air source heat pumps include both air-to-air and air-to-water heat pumps and can heat and cool buildings. These heat pumps rely on a vapor compression cycle with both condenser and evaporator to provide air conditioning through ducted or ductless distribution systems. The ducted systems are typical to provide conditioned air through a centralized unit. On the other hand, ductless systems also referred to as min-split units, deliver heating and cooling directly to individual spaces. Air-to-water heat pumps can be coupled with radiant panels for added energy efficiency and improved thermal comfort.

- Cold-Climate heat pumps: These systems provide heating even in cold climates typically through auxiliary electric heating elements. Current technologies can provide 100% heating at -15°C (5°F) and up to 75% heating at –25°C (–13°F) without any auxiliary systems due to improvements in inverter-driven compressors.
- Ground source heat pumps or also referred to as geothermal heat pumps are based on the use of deep ground or water as heat source and sink to provide heating and cooling to buildings. Indeed, deep ground is characterized by temperatures that remain constant through the year and thus can be a source to both reject and extract heat. The energy efficiency performance ratings of these systems exceed typically than those for air source heat pumps, but their installation costs are significantly higher.
- Water-source heat pumps are like air-source heat pumps but use water instead air to provide heating and cooling through water distribution loops. They are generally more energy efficient than the air source heat pumps due to the higher heat capacity of water compared to air. They can also provide simultaneous heating and cooling for different spaces in the same building (Halbhavi, 2016).

As noted above, heat pumps represent the most common and energy-efficient alternatives to fossil-based systems as well as electrical resistance heaters.

7.7.8.3 Building Electrification Programs

In the US, several states and cities have implemented incentive programs for building electrification including encouraging the deployment of heat pumps, converting to induction cooking and electrical domestic hot water heating as summarized in Table 7.16 based on an ACEEE evaluation of 38 surveyed programs (ACEEE, 2022). All these programs target the replacement of fossil-fuel-based heating and cooking equipment with electrical systems. Some of these programs intend to also replace the inefficient electrical resistance heating with heat pumps. In addition, all the surveyed programs are intended for residential buildings with only 11 encompassing small commercial buildings (ACEEE, 2022).

In the US, the EIA reports that only 12.1 million households use heat pumps out of the 139 million households representing only 8.7% adoption rate (EIA, 2017). Currently, there are no building code requirements for adopting heat pumps, so unless builders seek to achieve certificate requirements (i.e., LEED, electrification, or other labelling standards), there is no motivation to specify these systems. Heat pumps have higher initial costs compared to gas or oil heating systems and require more specialized knowledge to install. In addition, heat pumps are not suitable for all climates.

However, they offer large potential for energy savings. According to the US DOE, heat pumps can reduce electricity by approximately 50%, in comparison to their electric heating systems = (DOE, 2022). Geothermal heat pumps could reduce heating

TABLE 7.16
Summary of Program Types Enforced in Various US Cities and States for Building Electrification Efforts

Program Category	Technologies	Number of Programs
Space heating	All heating types	38
	Ductless air source heat pumps	21
	Ground source heat pumps	9
	Ducted air source heat pumps	5
	Cold Climate heat pumps	5
Water heating	All system types	30
Cooking	Induction Equipment	12
Others	On-site Solar Power Generation	8
	EV Chargers	4
	Batteries	2

energy usage by 30%–60% in US homes, if they have access to geothermal sources (DOE, 2022). Wu and Skye (2018) indicate that Ground Source Heat Pumps (GSHPs) can save between 24.3% and 39.2% of energy in heating-dominated regions, depending on specific climate zone. A case study in California showed a 50% reduction in household carbon emissions when switching from natural gas to heat pumps for both domestic hot water and space heating (Brockway and Delforge, 2018). Another case study in the United Kingdom residences showed a savings of 37% in both cost and emissions when switching from a natural gas boiler to a heat pump (Gaur et al., 2021). When heat pumps are used for domestic hot water heating, they are 2–3 times more efficient than standard electric water heaters (DOE, 2022).

7.7.8.4 Design of Electrified Dwellings

The design approach outlined in Section 7.5 is valid for electrified dwellings including detached homes and apartment buildings. The required size of the service entrance of electrified dwellings may increase compared to non-electrified buildings due to the added electrical loads associated with space heating, water heating, and cooking systems. Example 7.9 estimates the service entrance size when the space heating system of the ranch house considered in Example 7.6 is electrified using a heat pump.

Example 7.9

Problem

The heating system for the ranch house used in Example 7.6 is converted to a 6,500-VA heat pump instead of natural gas furnace. Determine the new service entrance required for the house using the standard method.

Solution

Size Phases, Grounding, and Main Panel Breaker

1. General lighting/receptacles

   This load remains unchanged and is estimated at DL = 4,575 VA (Refer to Example 7.6).

2. Fixed appliances

   This load specific to only the water heater is also the same as noted in Example 7.6: DL = 5,000 VA

3. Special appliances

   In addition to the dryer (5,000 VA) and the range (6,400 VA) loads, the load of the heat pump (6,500 VA) should be included.

   Thus, the total demand load for the electrified ranch house becomes:

$$DL_{Tot} = 27{,}475 \text{ VA}$$

Therefore, the current that the phases have to serve is increased to:

$$I_{L,A+B} = \frac{27{,}475 \text{ VA}}{240 \text{ V}} = 114.5 \text{ A}$$

Size Neutral

Since the heat pump is rated at 240-V, the size of the neutral remains unchanged with and without electrified space heating with # 6 TWH(Cu), which is the same as the new grounding conductor.

7.8 SUMMARY

In this chapter, the general design procedure for sizing electrical distribution systems for residential buildings is described. Several examples are presented to illustrate the application of various steps in designing the service entrance feeder and main circuit breaker as well as the branch circuits supplying all the electrical loads within a dwelling. In particular, the examples and the case studies cover the design of main service feeders as well as branch circuits for detached homes as well as apartment buildings with and without electrified heating systems. While some of the major NEC requirements are outlined in this chapter, the designer should refer to the applicable standards and codes to ensure that the design specifications meet all the requirements to ensure the safety, reliability, and flexibility of the electrical distribution system.

7.9 PROBLEMS

1. Determine the size of the service entrance THW (Cu) conductors required for phases A and B and the neutral of a 310 m^2 (3,200 ft^2) dwelling with the following loads: 12.0 kVA range; 6.0 kVA dryer; 30.0 kVA heating system; 6.0 kVA water heater; 1.5 kVA dishwasher; 0.9 kVA compactor no. 1; 0.8 kVA compactor no. 2; 1.6 kVA attic fan; 0.4 kVA vent fan; and 6.0 kVA A/C unit. Use the standard method for sizing the service conductors.

2. A dwelling unit is supplied by a single-phase, 120/240 V service. The electric load of the unit includes: (1) 7,000 VA water heater, 240 V; (2) 500 VA vent fan, 120 V; (3) 1,200 VA compactor, 120 V; (4) 7,500 VA dryer, 240 V. Each appliance is served by one branch protected by a circuit breaker.
   a. Determine the rating and the number of poles of each circuit breaker.
   b. Assuming that all the appliances have the same power factor, estimate the minimum percent unbalanced current between the two phases.
3. Determine the size of the service entrance THW (Cu) conductors required for phases A and B using the optional method of a 310 m^2 (3,200 ft^2) dwelling with the following loads: 12.0 kVA range; 6.0 kVA dryer; 30.0 kVA heating system; 6.0 kVA water heater; 1.5 kVA dishwasher; 0.9 kVA compactor no. 1; 0.8 kVA compactor no. 2; 1.6 kVA attic fan; 0.4 kVA vent fan; and 6.0 kVA A/C unit. Compare the size of phases A and B to those obtained with the standard method. Conclude.
4. A 274 m^2 (2,950 ft^2) dwelling is supplied by a single-phase, 120/240 V service and has the following loads: 1.8 kVA dishwasher, 120 V; 1.2 kVA compactor, 120 V; 1.0 kVA disposal, 120 V; 0.5 kVA vent fan, 120 V; 16 kVA range, 240 V; 4.5 kVA dryer, 240 V; 15 kVA heating unit, 240 V; 9 kVA air conditioning unit, 240 V; and 6.5 kVA water heater, 240 V.
   a. Using the standard method, size the THW service entrance conductors (A, B, and N).
   b. Size the main breaker and the grounding conductor.
   c. Determine the minimum number of 15 A branch circuits required for general lighting and receptacles.
5. Determine the size of the service entrance conductors required for phases A and B and the neutral of a 186 m^2 (2,000 ft^2) residence having the following load: 15 kVA space heating (five electric baseboards separately controlled); 10 kVA range; 4.5 kVA dryer; 1.5 kVA dishwasher; 5 kVA air conditioning unit; and 3 kVA water heater.
   a. Use the standard to size the conductors (A, B, and N) as well as the main circuit breaker.
   b. Use the optional method for sizing the service conductors (A and B).
   c. Show a possible panel schedule that serves the house (that meets NEC).
6. A 348 m^2 (3,750 ft^2) house has the following loads: 12 kVA range, 240 V; 15 kVA air conditioning unit, 240 V; and 8 kVA water heater, 240 V.
   a. Using the standard method, size the THW service entrance conductors.
   b. Determine the minimum number of 15 A circuits for general lighting and receptacles. Estimate the total number of circuits required for the entire house.
   c. Size the main breaker and the grounding conductor.
7. A 316 m^2 (3,400 ft^2) house has the following loads: 1.5 kVA compactor, 120 V; 1.2 kVA ceiling fan, 120 V; 14 kVA range, 240 V; 30 kVA heating unit, 240 V; 40 A air conditioning unit, 240 V; and 10 kVA water heater, 240 V.
   a. Using the standard method, size the THW service entrance conductors.
   b. Determine the minimum number of 15 A circuits for general lighting and receptacles. Estimate the total number of circuits required for the entire house.

   c. Size the main breaker and the grounding conductor.
   d. Size the circuit breaker for the heating unit and the A/C unit.
8. Determine the size of the service entrance conductors required for phases A and B and the neutral for $20 \times 83.5\,m^2$ ($900\,ft^2$) units in a multifamily dwelling. The units have the following loads:
   - General lighting and receptacle loads.
   - Two small kitchen appliance loads and one laundry load per unit.
   - Special appliance loads. Each unit has 10 kVA space heating unit, 12 kVA range, 5 kVA water heater, 1.2 kVA dishwasher, and 1 kVA disposal.
   - Use both the standard and the optional methods to determine the service entrance conductor size for the phases. Assume 120/240 V service.
9. A $300\,m^2$ ($3{,}230\,ft^2$) dwelling unit is supplied by a single-phase, 120/240 V service. The electric load of the unit includes (1) 5,000 VA water heater, 240 V; (2) 900 VA vent fan, 120 V; (3) 1,000 VA compactor, 120 V; (4) 4,000 VA dryer, 240 V, (5) 6,500 VA range, 240 V, and (5) a 6,000-VA heat pump, 240 V. Use the standard and optional methods to size the service entrance feeder including the main circuit breaker and the grounding conductor.
10. For the house define in Problem 7.5, determine the new size of the service entrance conductors required for phases A and B and the neutral if space heating and air conditioning system is provided by a heat pump rated at 24-A and 240-V.
11. Determine the change in the size of the service entrance including the main feeder and panel rating required to electrify an 800-ft^2 ($74\,m^2$) apartment unit served by 208Y/120-V with the following electrical loads: 8 kVA range (208-V), 3 kVA water heater (208-V), 0.8 kVA dishwasher (120-V), and 0.7 kVA disposal (120-V). For the electrified apartment unit, assume that a 5-kVA heat pump (208-V) is used to replace an existing natural gas fireplace.
12. A $2{,}950\,ft^2$ ($274\,m^2$) dwelling is supplied by a single-phase, 120/240-volt service and has the following loads: 1.5-kVA dishwasher, 120 volts; 1.2-kVA compactor, 120-volts; 0.5-kVA vent fan, 120 volts; 15-kVA range, 240 volts; 7.5-kVA dryer, 240 volts; and 12-kVA air conditioning unit, 240 volts; and 5.5-kVA water heater, 240 volts.
   a. Using the standard method, size the THW service-entrance conductors (A, B, N, and G) and the main circuit breaker.
   b. Determine the minimum number of 15-A circuits to serve the general lighting and receptacles.

8 Electrical Systems for Commercial Buildings

This chapter provides a general procedure for designing electrical systems for commercial buildings. In particular, general design guidelines are outlined for lighting and power panels, motor control centers, main distribution panels, and unit substations. Moreover, short-circuit currents at various locations within the electrical distribution system are estimated using simplified analysis methods. Throughout this chapter, step-by-step calculation procedures are outlined with specific examples. Tables and diagrams are provided to further assist the readers in designing safe electrical systems for commercial buildings. A case study for designing a radial electrical distribution system for an office building is described with detailed calculations and specifications.

8.1 INTRODUCTION

The electrical distribution systems for commercial buildings can be arranged in different configurations depending on the level of desired reliability, flexibility, and ease of maintenance. Moreover, and especially for large buildings or a complex of several buildings, electrical distribution systems include a wide range of transformers, unit substations, switchboards, lighting and power panels, and motor control centers (MCCs). All these systems have to be adequately designed and installed in order to safely supply electricity to the building loads. Common electrical equipment in buildings can be divided into three main categories: lighting, receptacles, and motors for heating and air conditioning systems. All electrical distribution system components have to be strategically located throughout the building to minimize energy losses, conductor lengths, and voltage drops associated with feeders and branch circuits. Ideally, the layout of the electrical distribution system for a large building would be designed using a set of networks such as that illustrated in Figure 8.1 with a centrally located unit substation feeding several distribution panels that provide power to low-voltage transformers and panels located throughout the building to serve the loads using short branch circuits.

It should be noted that only one main electrical service entrance is allowed for one building. However, there are a few exceptions when separate services are allowed for special loads such as fire pumps, emergency or standby loads, and large loads (i.e., exceeding 2,000 A). Typically, three categories of arrangements for electrical distribution systems can be specified for commercial buildings:

1. *Radial distribution systems*: These systems provide the simplest arrangements for supplying electricity to buildings using one primary feeder and one transformer through which all the loads are served. Figure 8.2 shows a

DOI: 10.1201/9781003276999-8

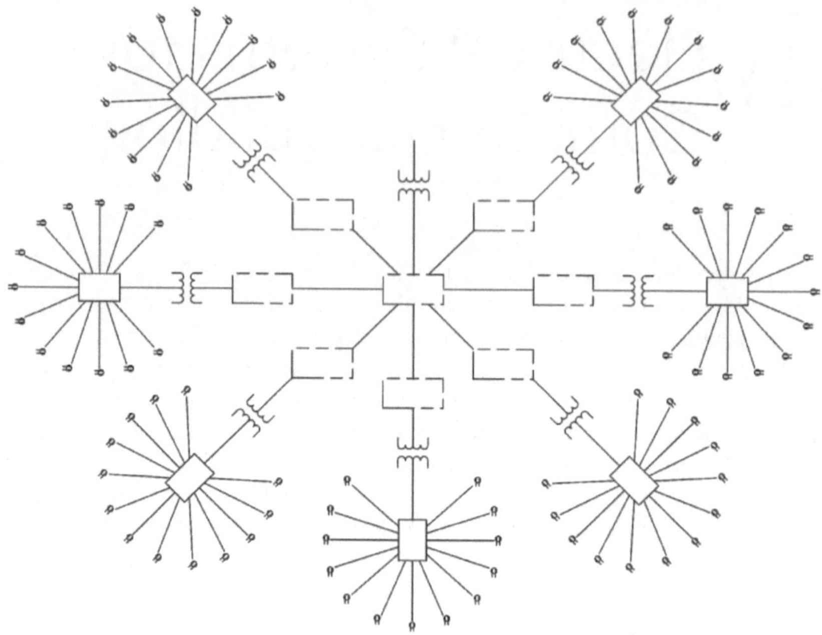

FIGURE 8.1 Ideal arrangement of a commercial building electrical distribution system with a main unit substation, transformers, and panels.

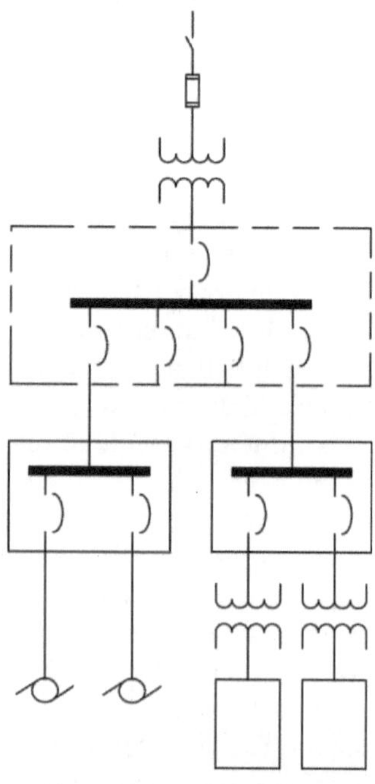

FIGURE 8.2 One-line diagram for a simple radial distribution system.

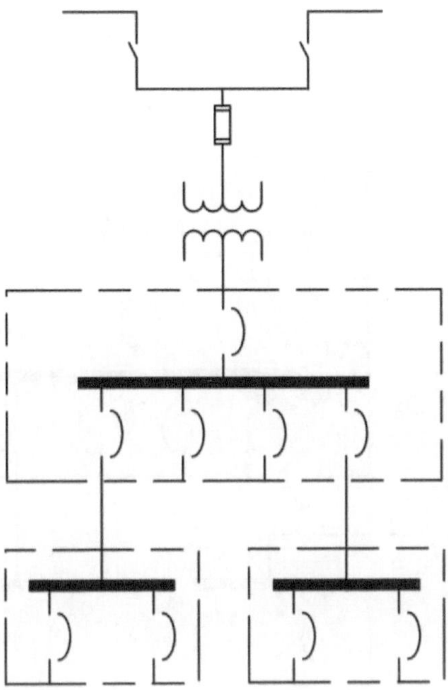

FIGURE 8.3 One-line diagram for a radial distribution system with loop utility feed.

one-line diagram for a simple radial distribution system arrangement. The main advantages of these types of distribution systems include ease of design and operation as well as low implementation cost. However, this system has low reliability and is generally difficult to maintain. Indeed, any faults or any maintenance needs within the primary feeder or the main transformer result in loss of power for the entire distribution system. A variation of the simple radial system, a utility loop feed option can be considered to increase the liability of the distribution system as illustrated in Figure 8.3. When the building has significant critical, emergency, and/or standby loads, a generator can be added as outlined in Figure 8.4.

2. *Secondary selective distribution systems*: These systems have two primary feeders and transformers. Each transformer serves part of the building with some potential duplication depending on the size of the primary transformers. Several options are available for these types of electrical distribution systems. Figure 8.5 shows one simple arrangement of secondary selective distribution system with two primary feeders and transformers serving two separate radial systems. Figure 8.6 illustrates a secondary selective system with a tie breaker that connects the two radial systems. Under normal operation, the tie breaker is open so the two radial systems operate independently. When the tie breaker is closed due to any power outage in one radial system, the other transformer can energize the entire secondary bus and serve loads in both radial systems. Figure 8.7 provides a one-line diagram

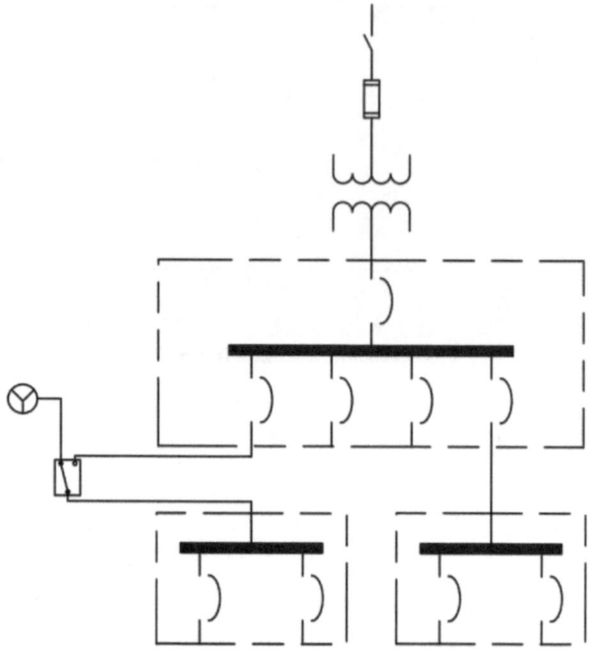

FIGURE 8.4 One-line diagram for a radial distribution system with a generator.

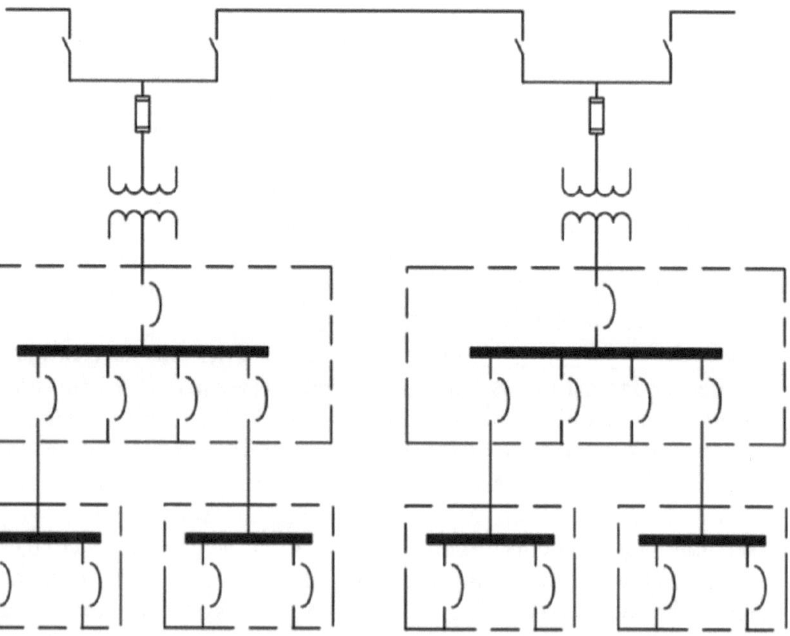

FIGURE 8.5 One-line diagram for a secondary selective distribution system without a tie breaker.

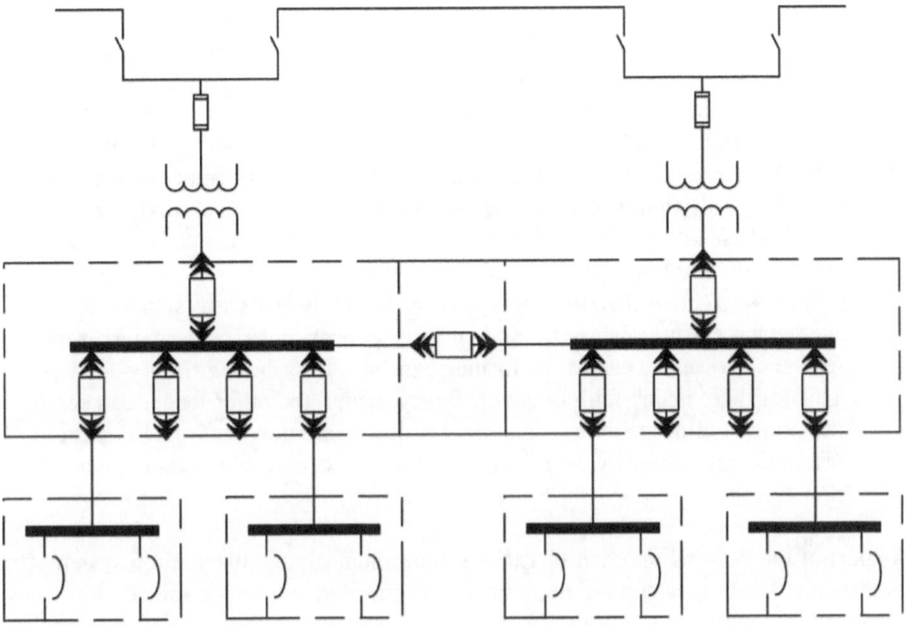

FIGURE 8.6 One-line diagram for a secondary selective distribution system with a tie breaker.

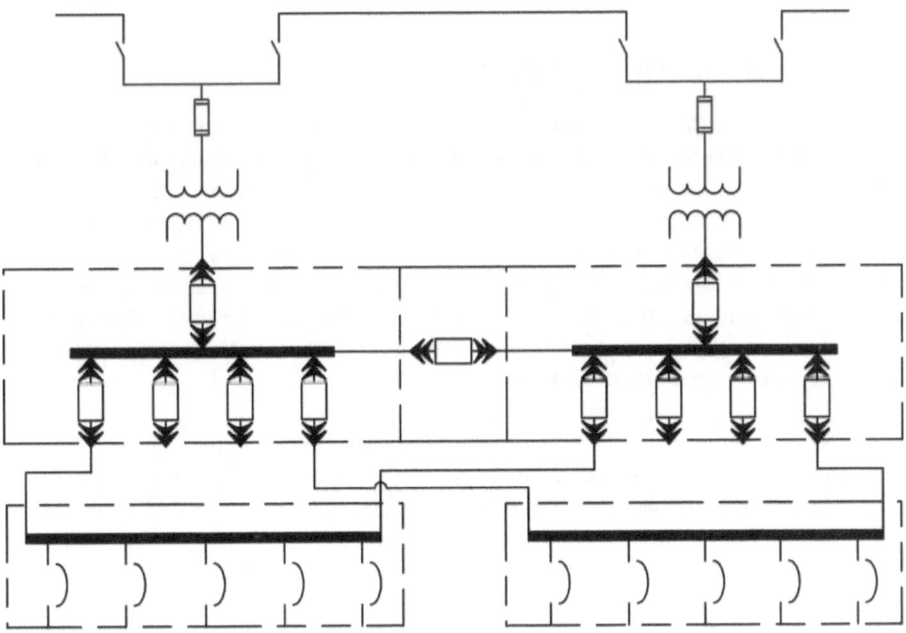

FIGURE 8.7 One-line diagram for a double-ended secondary selective distribution system with a tie breaker.

of a double-ended secondary selective distribution system with a tie breaker that allows one transformer to serve some of the loads in the other radial system even under normal operation. The secondary selective distribution arrangements provide more reliability, ease of maintenance, and flexibility compared to the radial distribution systems since whenever one of the transformers is subject to power outage, the other transformer is capable of providing sufficient power to operate typically at least 75% of all building loads. These advantages, however, come with the implementation of higher costs compared to the simple radial systems.

3. *Primary selective distribution systems*: These systems are similar to the secondary selective systems and include two primary feeders and transformers. However, each transformer can be fed by both primary feeders. Thus, if there is any fault on any primary feeder, the other feeder can serve the entire building if the transformers are adequately sized. This system offers higher flexibility when compared to the secondary selective distribution systems.

As part of the design approach of various equipment of any distribution system, the estimation of fault (i.e., short-circuit) currents is needed at least for specific locations. Some of the common methods for calculating short-circuit currents for electrical distribution systems are outlined first. Then, the design for the general approach is provided to select sizes and ratings of lighting and panels as well as to design the layouts of MCCs, switchboards, and unit substations.

8.2 SHORT-CIRCUIT CURRENTS

The calculation of short-circuit currents at specific points within the distribution system is an important step in designing safe electrical systems for buildings. Indeed, as noted in the previous chapters, the design of the main components of the electrical systems is significantly affected by the magnitude of the short-circuit currents that these components can potentially experience. In particular, the size of protection devices and conductors for both feeders and branch circuits depends on the levels of short-circuit currents. There are several methods to estimate the magnitude of short-circuit currents available at specific points within an electrical building distribution system. These methods include the following:

- Per-unit method that relies on dimensionless analysis to determine the magnitude of short-circuit currents relative to the normal rated current. Moreover, all the parameters of the electrical systems such as voltages and impedances are also normalized using the rated conditions as references.
- Mega-Volt-Ampere (MVA) method that utilizes the total power expressed in MVA (i.e., 1,000 kVA) at various locations along the distribution system to estimate the short-circuit currents using the rated voltages and impedances.
- Ohmic method that directly uses Ohm's law and impedances at various points within the distribution systems to compute short-circuit currents.

- In this section, the first two methods are introduced and applied to estimate short-circuit currents at various locations along a typical radial electrical distribution system commonly used for buildings. It should be noted that the Ohmic method, a more direct method, utilizes a similar analysis approach as the per-unit method since it relies on the estimation of the total impedances but without any normalization.

8.2.1 Per-Unit Calculation Method

To estimate the short-circuit currents at various locations within the electrical distribution system, some data are needed related to the impedances of the main components of the system including the utility, transformer, and feeders. The main procedure of short-circuit calculations is summarized in five steps.

1. Collect data including
   - Utility short-circuit capability expressed in kVA, $P_{SC,U}$
   - Transformer parameters including the kVA rating, P_T, impedance Z_T, the ratio X/R, the primary and secondary voltages, as outlined in Figure 8.8
   - Specifications of feeders and subfeeders (size, conduit type, and length).
   - Potential sources of electricity including generators and motors. In case of faults, motors can actually generate currents and contribute to the short-circuit currents.
2. Identify strategic locations for estimating short-circuit currents.
3. In this step, the locations along the distribution system have to be selected where the short-circuit current, I_{SC}, needs to be calculated. These locations depend on the position and number of the protection devices for feeders and subfeeders, as noted in Figure 8.9.
4. Estimate the short-circuit current at the secondary side of the main transformer.
5. Point A noted in Figure 8.10 is the first location within an electrical distribution system where short-circuit current should be calculated. Indeed, this point affects the short-circuit conditions for all the other points, as noted in step 4. It is also where the magnitude of the fault current is typically the highest.

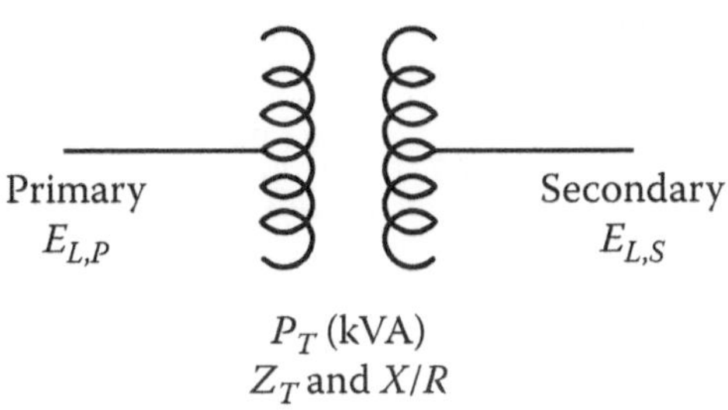

FIGURE 8.8 Main parameters of a transformer.

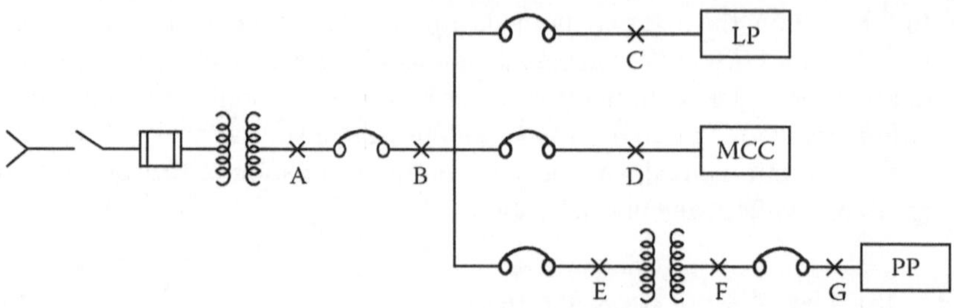

FIGURE 8.9 Typical points along a radial distribution system where short-circuit currents are typically estimated.

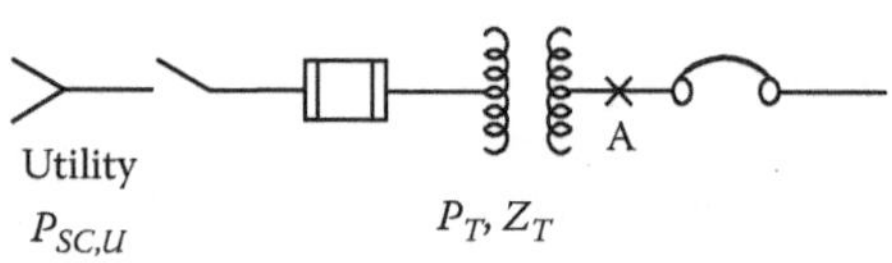

FIGURE 8.10 Parameters needed to estimate short-circuit current at point A.

The per-unit impedance at point A depends on both the utility and transformer impedance, as outlined in Equations 8.1 and 8.2:

$$Z_{UT} = Z_T + Z_U \tag{8.1}$$

with

$$Z_U(pu) = \frac{P_T}{P_{SC,U}} \tag{8.2}$$

The short-circuit current due to the utility-transformer can be estimated using Ohm's law using the per-unit values, as noted in Equation 8.3:

$$I_{SC}(pu) = \frac{E(pu)}{Z_{UT}(pu)} \tag{8.3}$$

The short-circuit at point A in Amperes without motor contribution can then be calculated using the rated current at the secondary side of the transformer:

$$I_{SC}(A) = I_{SC}(pu) \times I_{L,S}(A) \tag{8.4}$$

where the secondary line current $I_{L,S}$ is calculated from the transformer rated total power, P_T, and the secondary line voltage, $E_{L,S}$, as shown in Equation 8.5:

$$I_{L,S}(A) = \frac{P_T}{\sqrt{3}E_{L,S}} \tag{8.5}$$

It should be noted that to be conservative, the value of $P_{SC,U}$ can be assumed to be infinite (i.e., ∞). This assumption provides higher short-circuit estimation at point A than the actual value.

Example 8.1

Problem

Estimate the short-circuit at point A for a system with the following specifications for the main transformer and utility:

$$P_T = 1{,}000 \text{ kVA}, \quad Z_T = 4\%, \quad P_{SC,U}$$

Solution

First, the combined utility-transformer per-unit impedance is calculated:

$$Z_{UT} = Z_U + Z_T$$

with

$$Z_u = \frac{P_T}{P_{SC,u}} = \frac{1{,}000 \text{ kVA}}{100{,}000 \text{ kVA}} = 0.01$$

Thus,

$$Z_{UT}(pu) = 4\% + 1\% - 0.05$$

The application of Ohm's law using the per-unit value provides

$$I_{SC}(pu) = \frac{E(pu)}{Z_{UT}(pu)} = \frac{1}{0.05}$$

or

$$I_{SC}(pu) = 20$$

Without any motor contribution, the short-circuit at point A can be estimated:

$$I_{SC}(A) = I_{SC}(pu) \times I_{L,S}(A)$$

with

$$I_{L,S}(A) = \frac{1{,}000 \text{ kVA}}{0.48 \text{ V}\sqrt{3}} = 1{,}202 \text{ A}$$

Thus, the short-circuit current at point A without any motor contribution is

$$I_{SC} = 24{,}056 \text{ A}$$

When the electrical distribution system includes one or several motors, then the short-circuit current at point A can be higher than the value estimated from Equation 8.4. Indeed, when short-circuit occurs, the motors continue to rotate and actually can act as generators and thus may increase the magnitude of the short-circuit current that the system experiences at various locations including point A. The detailed calculation of the additional current that the motors can generate under short-circuit conditions is rather complex. However, a simple rule of thumb can be used to estimate the contribution of all motors to the short-circuit current at point A, $I_{SC, M}$, based on the secondary line voltage value, $E_{L,S}$, as summarized by Equations 8.6 and 8.7:

$$I_{SC,M} = 4 \times \sum I_M \tag{8.6}$$

with

$$\sum I_M = \begin{matrix} 100\% \text{ of } I_{L,S} \text{ if } E_{L,S} = 480 \text{ V} \\ 50\% \text{ of } I_{L,S} \text{ if } E_{L,S} = 208 \text{ V} \end{matrix} \tag{8.7}$$

Example 8.2

Problem

For the system considered in Example 8.1, estimate the motor contribution and determine the total short-circuit current at Point A.

Solution

Using Equation 8.6 and the fact that the voltage of the system is 480 A

$$I_{SC,M} = 4I_{L,S} = 4 \times 1{,}202 = 4{,}808 \text{ A}$$

Therefore, the final short-circuit current at point A, $I_{SC, A}$, can be calculated:

$$I_{SC,A} = 24{,}056 + 4{,}808 = 28{,}800 \text{ A}$$

Estimate short-circuit for any point connected by only conductors to point A.

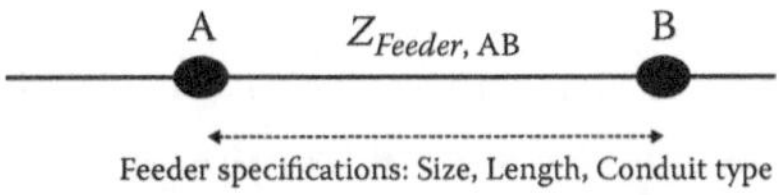

FIGURE 8.11 Feeder separating two points A and B where the short-circuit currents are to be estimated.

TABLE 8.1
Resistance and Reactance of Feeders Made Up of Three Copper Conductors in a Conduit for Three-Phase Systems

	Resistance (Ω/1,000 ft [Ω /1,000 m])			Reactance (Ω/1,000 ft [Ω /1,000 m])	
Wire Size in AWG or MCM	**PVC Conduit**	**Aluminum Conduit**	**Steel Conduit**	**PVC, Aluminum Conduit**	**Steel Conduit**
14	3.1 (10.2)	3.1 (10.2)	3.1 (10.2)	0.058 (0.190)	0.073 (0.240)
12	2.0 (6.6)	2.0 (6.6)	2.0 (6.6)	0.054 (0.177)	0.068 (0.223)
10	1.2 (3.9)	1.2 (3.9)	1.2 (3.9)	0.050 (0.164)	0.063 (0.207)
8	0.78 (2.56)	0.78 (2.56)	0.78 (2.56)	0.052 (0.171)	0.065 (0.213)
6	0.49 (1.61)	0.49 (1.61)	0.49 (1.61)	0.051 (0.167)	0.064 (0.210)
4	0.31 (1.02)	0.31 (1.02)	0.31 (1.02)	0.048 (0.157)	0.060 (0.197)
3	0.25 (0.82)	0.25 (0.82)	0.25 (0.82)	0.047 (0.154)	0.059 (0.194)
2	0.19 (0.62)	0.20 (0.66)	0.20 (0.66)	0.045 (0.148)	0.057 (0.187)
1	0.15 (0.49)	0.16 (0.52)	0.16 (0.52)	0.046 (0.151)	0.057 (0.187)
1/0	0.12 (0.39)	0.13 (0.43)	0.12 (0.39)	0.044 (0.144)	0.055 (0.180)
2/0	0.10 (0.33)	0.10 (0.33)	0.10 (0.33)	0.043 (0.141)	0.054 (0.177)
3/0	0.077 (0.253)	0.082 (0.269)	0.079 (0.259)	0.042 (0.138)	0.052 (0.171)
4/0	0.062 (0.203)	0.067 (0.220)	0.063 (0.207)	0.041 (0.135)	0.051 (0.167)
250 MCM	0.052 (0.171)	0.057 (0.187)	0.054 (0.177)	0.041 (0.135)	0.052 (0.171)
300 MCM	0.044 (0.144)	0.049 (0.161)	0.045 (0.148)	0.041 (0.135)	0.051 (0.167)
350 MCM	0.038 (0.125)	0.043 (0.141)	0.039 (0.128)	0.040 (0.131)	0.050 (0.164)
400 MCM	0.033 (0.108)	0.038 (0.125)	0.035 (0.115)	0.040 (0.131)	0.049 (0.161)
500 MCM	0.027 (0.089)	0.032 (0.105)	0.029 (0.095)	0.039 (0.128)	0.048 (0.157)

Note: This table is adapted from NEC-2014 Table 9, Chapter 9 (NEC, 2014).

Once the impedance in point A is estimated using Ohm's law using the short-circuit current calculated in step 3, the impedance at any other point B as shown in Figure 8.11 can be estimated using the impedance of the feeder that separate the two points A and B as expressed by Equation 8.8:

$$Z_B = Z_A + Z_{\text{Feeder},AB} \tag{8.8}$$

The feeder impedance depends on the size of the conductor as well as its length and the type of conduit. Table 8.1 lists the resistance and the reactance per 1,000 m and per 1,000 ft length for select conductor sizes and conduit types.

To determine the per-unit value for the feeder impedance, a reference value for impedances, Z_{ref}, is considered using the per phase values for voltage, $E_{\Phi,S}$, and current, $I_{\Phi,S}$, at the secondary side of the main transformer, as defined by Equation 8.9:

$$Z_{ref}(\Omega) = \frac{E_{\Phi,S}}{I_{\Phi,S}} \tag{8.9}$$

Then, the per-unit feeder impedance is determined using Equation 8.10:

$$Z_{\text{feeder}}(pu) = \frac{Z_{\text{feeder}}(\Omega)}{Z_{ref}(\Omega)} \tag{8.10}$$

Example 8.3

Problem

For Examples 8.1 and 8.2, point B is connected to point A with a 250 MCM feeder housed in a steel conduit. If the feeder length is estimated to be 500 ft, determine the per-unit feeder impedance for the feeder and the short-circuit current at point B. The feeder is part of a 480Y/277-V system.

Solution

For a 250 MCM feeder housed in a steel conduit, the resistance and the reactance can be obtained:

$$\text{From Table 8.1} \quad \begin{aligned} R &= \frac{0.054\ \Omega}{1{,}000\ \text{ft}} \\ X &= \frac{0.052\ \Omega}{1{,}000\ \text{ft}} \end{aligned}$$

Since the length of the feeder is 500 ft, the actual resistance and reactance of the feeder can be estimated:

$$R = \frac{0.054\ \Omega \times 500\ \text{ft}}{1{,}000\ \text{ft}} \rightarrow R_F = 0.027$$

$$X = \frac{0.052\ \Omega \times 500\ \text{ft}}{1{,}000\ \text{ft}} \rightarrow R_F = 0.026$$

The reference impedance can be estimated using the rated phase values for voltage and current, as illustrated in Figure 8.12 for this example.

Using Equation 8.9, the reference impedance can be calculated:

$$Z_{ref} = \frac{277\ \text{V}}{1{,}202\ \text{A}} = 0.23\ \Omega$$

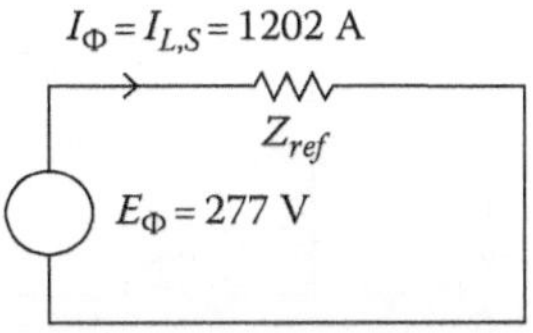

FIGURE 8.12 Per phase circuit to estimate the reference impedance for Example 8.3.

Thus, the per-unit resistance and reactance of the 250 MCM feeders are

$$R_F(pu) = \frac{0.027}{0.23} = 0.0117$$

$$X_F(pu) = \frac{0.026}{0.23} = 0.0113$$

Using Equation 8.8, the per-unit impedance at point B can be obtained from the impedance at point A and the feeder impedance:

$$Z_B = Z_A + Z_F$$

with

$$Z_A = R_A + jX_A$$

To estimate the resistance and the reactance of the impedance at point A, it is typically assumed that at this point $X/R = 4$ or that the phase angle for point A is $\theta = 76°$. Thus, the split of the magnitude of the impedance at point A to resistance and reactance can be obtained:

$$Z_A(pu) = \frac{I_{L,S}}{I_{SC,A}} = \frac{1{,}202}{28{,}800} = 0.042$$

$$Z_A(pu) = 0.042(\cos 76° + j\sin 76°) = 0.0101 + j0.0408$$

Therefore, the per-unit impedance at point B can be calculated:

$$Z_B(pu) = (0.0101 + 0.0117) + j(0.0408 + 0.0113)$$

or

$$Z_B(pu) = 0.0218 + j0.0521$$

The magnitude of the impedance at point B is

$$|Z_B(pu)| = \sqrt{0.0218^2 + 0.0521^2} = 0.0565$$

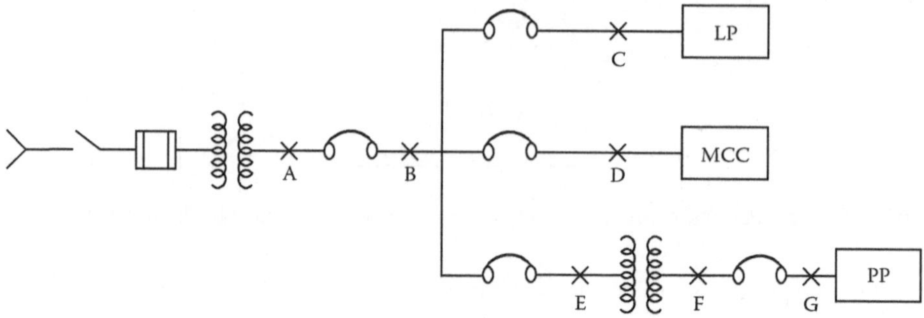

FIGURE 8.13 Analysis for estimating short-circuit at point F from point E.

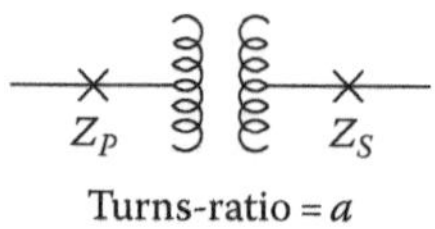

FIGURE 8.14 Transformation of impedance for a transformer.

The application of Ohm's law using the per-unit value provides

$$I_{SC,B}(pu) = \frac{E(pu)}{Z_B(pu)} = \frac{1}{0.0565} = 17.71$$

Thus, the short-circuit at point B can be estimated in Amperes:

$$I_{SC,B}(A) = 17.71 \times 1{,}202 = 21{,}300\text{A}$$

Estimate short-circuit for a point located after a second transformer.

For any point separated from points A and B by a feeder or a set of feeders, the analysis outlined in Step 4 can be used. However, when the points are separated by a transformer as shown in Figure 8.13 for points E and F, the calculation has to be adjusted as shown in this step.

First, the impedance of the primary side has to be transferred to the secondary side using the transformer turns-ratio, as indicated in Figure 8.14 and expressed in Equation 8.11:

$$Z_S = \frac{1}{a^2} Z_P \tag{8.11}$$

It should be noted that Equation 8.11 is applicable only when impedances are expressed in Ohms and is not valid when the per-unit impedances are used.

The impedance at point F, Z_F, can be obtained by adding the impedance at point E, Z_E, to the second transformer impedance, $Z_{T,2}$, as shown in Equation 8.12:

$$Z_F = \frac{1}{a^2} Z_E + Z_{T,2} \tag{8.12}$$

This expression for Z_F can be used with Equation 8.8 to estimate the short-circuit currents at any point located after point F along the secondary side of the second transformer

Example 8.4

Problem

For a simple radial distribution system shown in Figure 8.13 and considered for Examples 8.1–8.3, determine the short-circuit at points E and F if the system specifications are

- Second Transformer: 200 kVA, 480Y/277–208Y/120
- $Z_{T,2} = 5\%, \frac{X}{R} = 1.5$
- Subfeeder BE: 250 MCM THW (Cu), 50-ft length, steel conduit
- Impedance at point B : $Z_{B,pu} = 0.0565, \frac{X}{R} = 2.4$

Solution

Using Equation 8.8, the impedance at point E can be estimated:

$$Z_E(pu) = Z_{B,pu} + Z_{\text{Feeder},BE}(pu)$$

with

$$Z_B(pu) \text{given with} \frac{X}{R} = 2.4$$

Thus,

$$Z_B(pu) = 0.0565[\cos 67° + j\sin 67°] = 0.0218 + j0.0528$$

For the feeder BE, the impedance can be obtained from Table 8.1:

$$R = \frac{0.054}{1{,}000 \text{ ft}} \rightarrow R = 0.0027$$

$$X = \frac{0.052}{1{,}000 \text{ ft}} \rightarrow X = 0.0026$$

The reference impedance, $Z_{ref,1}$, for the feeder BE is defined by Equation 8.9 using the first transformer rated at 1,000 kVA and secondary phase voltage of $E_{\phi,S,1} = 277\,V$ and phase current, $I_{\phi,S,1} = I_{L,S,1}$:

$$Z_{ref,1} = \frac{E_{\phi,S,1}}{I_{\phi,S,1}} = \frac{277}{1,202} = 0.23\Omega$$

with the secondary current for the first transformer, $I_{L,S,1}$, as calculated in Example 8.1:

$$I_{L,S,1} = \frac{1,000 \text{ kVA}}{\sqrt{3}(0.48 \text{ kV})} = 1,202 \text{ A}$$

Thus, the per-unit impedance for feeder BE is

$$Z_{\text{Feeder},BE}(pu) = \frac{0.0027}{0.23} + j\frac{0.0026}{0.23}$$

or

$$Z_{\text{Feeder},BE}(pu) = 0.0117 + j0.0113$$

Thus, the per-unit impedance at point E can be determined using Equation 8.8:

$$Z_E(pu) = (0.0218 + 0.0117) + j(0.0521 + 0.0113) = 0.0335 + j0.0634$$

The magnitude of the per-unit impedance at point E, Z_E, is then

$$|Z_E(pu)| = \sqrt{0.0335^2 + 0.0634^2} = 0.0718$$

It should be noted that the X/R value of the impedance Z_E can be calculated as $X/R = 0.0634/0.0335 = 1.9$, which corresponds to a phase angle of 62.1°.

The short-circuit current expressed in per-unit at point E is calculated to be

$$I_{SC,E}(pu) = \frac{1}{0.0718} = 13.94$$

The short-circuit current in Amperes at point E is then

$$I_{SC,E} = 13.94 \times 1,202 = 16,750 \text{ A}$$

The per-unit impedance at point F (at the secondary side of the second transformer) is estimated using Equation 8.12:

$$Z_F(pu) = Z_{T,2}(pu) + Z_{S,E}(pu)$$

with

$$Z_{S,E} = \frac{1}{a^2} Z_E$$

Since the equation here is only applicable when the impedances are expressed in Ohms, the impedance at point E is converted to be expressed in Ohms:

$$Z_E(\Omega) = Z_{ref,1}(\Omega) \times Z_E(pu)$$

or

$$Z_E(\Omega) = (0.23\ \Omega) \times (0.0718) = 0.0165\ \Omega$$

The turn ratio of the second transformer can be calculated as

$$a = \frac{480}{208} = 2.3$$

Thus,

$$Z_{S,E} = \frac{1}{a^2} Z_E = 0.00312\ \Omega$$

The impedance of the second transformer, $Z_{T,2}$, can be obtained from the values of X/R and percent transformer impedance:

$$Z_{T,2}(pu) = \left\| Z_{T,2}(pu) \right\| \cos(\theta) + j\sin(\theta)$$

with

$$\sin(\theta) \rightarrow \tan^{-1}\left(\frac{X}{R}\right)$$

or

$$\tan^{-1}(1.5) = 56.3°$$

Thus, the per-unit transformer impedance is

$$Z_{T,2}(pu) = 0.05 \times (\cos 56.3° + j\sin 56.3°)$$

or

$$Z_{T,2}(pu) = 0.0278 + j0.0416$$

It should be noted that for points F and G, the reference impedance is related to second transformer characteristics and is defined as

$$Z_{ref,2} = \frac{E_{\phi,s,2}}{I_{\phi,S,2}} = \frac{120 \text{ V}}{555 \text{ A}} = 0.216\ \Omega$$

since the reference voltage is the secondary phase voltage for the second transformer, that is, $E_{\phi,\,S,2} = 120\,\text{V}$, and the reference current is

$$I_{L,S,2} = \frac{200 \text{ kVA}}{\sqrt{3}(0.208 \text{ kV})} = 555 \text{ A}$$

Thus,

$$Z_{E,S}(pu) = \frac{Z_E(\Omega)}{Z_{ref,\,2}}$$

$$Z_{E,S}(pu) = \frac{0.00312}{0.216}\left[\cos 62.1° + j\sin 62.1°\right] = 0.00675 + j0.01276$$

Therefore, the per-unit impedance at point F can be calculated:

$$Z_F(pu) = 0.0345 + j0.0544$$

The magnitude of the per-unit impedance at point F is then

$$\|Z_F(pu)\| = \sqrt{0.0345^2 + 0.0544^2} = 0.06438$$

Finally, the per-unit and the actual short-circuit current at point F can be estimated:

$$I_{SC,F}(pu) = \frac{1}{0.06438} = 15.53$$

Therefore and since the reference current is 555 A,

$$I_{SC,F}(A) = 15.53 \times 555 \text{ A} = 8{,}600 \text{ A}$$

8.2.2 Short-Circuit Calculations Using MVA Method

The MVA method is a simplified method for calculating short-circuit currents at various locations within an electrical distribution system. The basic principle of the method is to express all the loads within the distribution system in terms of total power using MVA units for all components within the distribution system. The components include typically the utility source, transformers, motors, generators, and feeders. Typically, the fault currents obtained from the MVA method are within 3%–5% of those obtained from the per-unit method as long as the condition $X/R > 10$ is met for the steady-state impedance values for all the components. In addition to its ease of application, the MVA method deals with large numbers unlike the per-unit or Ohmic methods that consider rather small numbers and thus it is easier to make computational errors.

The general calculation approach for the MVA method is provided in the following four steps:

1. Estimate the admittance, Y, for all components of the distribution system as follows:

$$Y = \frac{100}{Z(\%)} \tag{8.13}$$

2. Calculate the short-circuit MVA values, MVA_{SC}, for all the components based on their MVA ratings:

$$MVA_{SC} = \frac{MVA}{Y} \tag{8.14}$$

3. Estimate the combined $MVA_{SC,tot}$ depending on the parallel or series connections of various components using the following rules:

For parallel connections:

$$MVA_{SC,tot} = MVA_{SC,1} + MVA_{SC,2} + \cdots + MVA_{SC,n} \tag{8.15}$$

For series connections:

$$\frac{1}{MVA_{SC,tot}} = \frac{1}{MVA_{SC,1}} + \frac{1}{MVA_{SC,2}} + \cdots + \frac{1}{MVA_{SC,n}} \tag{8.16}$$

4. Convert the $MVA_{SC,tot}$ to symmetrical short-circuit current, I_{SC}, depending on the system line voltage, E_L, for the three-phase system:

$$I_{SC} = \frac{MVA_{SC,tot}}{\sqrt{3}E_L} \tag{8.17}$$

It should be noted that the MVA_{SC} for a feeder within the distribution system can be obtained from the impedance of the feeder conductors and the system line voltage using the following expression:

$$MVA_{SC} = \frac{E_L^2}{Z(\Omega)} \tag{8.18}$$

Example 5 illustrates the application of the MVA method to find the short-circuit at the secondary side of the main transformer for a simple radial distribution system.

Example 8.5

Problem

Estimate the symmetrical short-circuit current at the secondary side of a 13.8 kV-208Y/120-V transformer rated at 1,000 kVA with $Z=5\%$ and supplied by a utility having 15 kVA fault current capabilities.

Solution

First, the MVA_{SC} values for both the utility and the transformer are estimated.
For the utility (using the utility distribution voltage and the utility fault current),

$$MVA_{SC,U} = \sqrt{3} \times 13.8 \text{ kV} \times 15 \text{ kVA} = 358.52 \text{ MVA}$$

For the transformer using Equations 8.13 and 8.14,

$$MVA_{SC,T} = \frac{1{,}000 \text{ kVA}}{0.05} = 20 \text{ MVA}$$

Since the utility source and the transformer are in series, the total $MVA_{SC\text{-}tot}$ value can be estimated using Equation 8.16:

$$\frac{1}{MVA_{SC-tot}} = \frac{1}{MVA_{SC,U}} + \frac{1}{MVA_{SC,T}} = \frac{1}{358.52} + \frac{1}{20} = 0.05279$$

Thus,

$$MVA_{SC-tot} = \frac{1}{0.05279} = 18.94 \text{ MVA}$$

Therefore, the symmetrical short-circuit current at the secondary side of the transformer can be calculated using Equation 8.17 with a line voltage of $E_L = 208$ V:

$$I_{SC} = \frac{18.94 \text{ MVA}}{\sqrt{3}(208 \text{ V})} = 52{,}583 \text{ A}$$

8.3 LIGHTING AND POWER PANELS

To determine the electrical loads for different building types, rough estimates can be used in the preliminary phases of the design depending on the building type. Specifically, power densities for common electrical loads such as those provided in Table 8.2 for various building types can be used initially in the preliminary design. However, specific connected loads using actual loads are determined in later design phases, especially when better estimates are available based on the final design specifications of lighting and mechanical systems.

As indicated in Chapter 6, demand loads, and not the connected loads, are used to size both the conductors and protection devices for power and lighting panels. These demand loads are obtained from the connected loads using the demand factors specified for lighting and receptacles as well as any motor and nonmotor loads for commercial buildings. For instance, the demand factor for the lighting loads in a commercial building is 1.25 since it is a continuous load. However, the demand factor for the receptacle loads is 1.0 for the first 10,000 kVA and 50% for the remainder kVA. As outlined in Chapter 6, Table 6.1 can be used to select the conductors, while Tables 6.11 and 6.12 can be considered for selecting the size of conduits needed for subfeeders serving the lighting and power panels. Example 8.6 indicates the design of the feeder for a panel serving mostly the lighting loads in a commercial building.

Example 8.6

Problem

A panelboard for a commercial building is supplied by a 250 ft, four-wire, 480Y/277 V, three-phase feeder. The panel has the following loads:

- A lighting system with a total power of 55 kVA with a power factor of 0.90
- A 5 kVA resistive heater with a power factor of 1.0
- Determine:
  a. The size of the feeder and conduit using THW copper conductors (use the ampacity criterion only).
  b. The voltage drop in the feeder (assume a steel conduit.)
  c. The size of the feeder and conduit if a limit for the voltage drop is set to be 1% and the feeder should be able to handle a symmetrical short-circuit current is 35,000 A. A circuit breaker is used to protect the feeder.

Solution

a. Size of feeder based on ampacity:

The calculations are summarized in Table 8.3 using the complex power (power triangle) calculation method to account for the effect of power factors for all the loads.

The total demand load is then 73.29 kVA with a power factor of 0.913. The demand current is calculated:

$$I_L = \frac{73.290\text{ VA}}{\left(1.732 \times 480\text{ V}\right)} = 88.2\text{ A}$$

so use #3 THW (Cu) (refer to Table 6.1).

For the conduit, use $1^{1/4}$-in (refer to Table 6.11).

TABLE 8.2
Power Densities for Preliminary Estimation of Electrical Loads for Various Building Types

Building/Space Type	Lighting W/ft² (W/m²)	Receptacles W/ft² (W/m²)	HVAC W/ft² (W/m²)	Computer W/ft² (W/m²)	Others W/ft² (W/m²)	Total W/ft² (W/m²)
Apartments						
High rise Low rise	0.7 (7.5)0.7 (7.5)	2.2 (23.7)2.0 (21.5)	3.0 (32.3)3.0 (32.3)	0.00.0	1.0 (10.8)1.0 (10.8)	6.9 (74.3)6.7 (72.1)
Auditoriums						
General Stage	1.8 (19.4)1.6 (17.2)	0.5 (5.4)0.5 (5.4)	3.5 (37.7)15.0 (161.5)	0.00.0	0.8 (8.6)0.5 (5.4)	6.6 (71.0)17.6 (189.5)
Banks						
Branch offices main offices	1.5 (16.1)1.5 (16.1)	2.1 (22.6)1.5 (16.1)	5.7 (61.4)5.7 (61.4)	1.6 (17.2)1.4 (15.1)	1.4 (15.1)1.4 (15.1)	12.3 (132.4)11.3 (121.6)
Computer rooms	1.2 (12.9)	1.2 (12.9)	8.5 (91.5)	11.5 (123.8)	2.2 (23.7)	24.6 (264.8)
Corridors	0.9 (9.7)	0.3 (3.2)	0.0	0.0	0.0	1.2 (12.9)
Hospitals	1.2 (12.9)	1.0 (10.8)	5.0 (53.8)	0.0	1.0 (10.8)	8.2 (88.3)
Hotels	1.0 (10.8)	1.0 (10.8)	5.0 (53.8)	0.0	1.3 (14.0)	8.3 (89.3)
Libraries	1.3 (14.0)	1.0 (10.8)	5.5 (59.2)	0.0	0.5 (5.4)	8.3 (89.3)
Motels	1.0 (10.8)	0.5 (5.4)	3.2 (34.4)	0.0	0.5 (5.4)	5.2 (56.0)
Office buildings						
High rise low rise	1.0 (10.8)1.0 (10.8)	2.0 (21.5)1.9 (20.5)	4.7 (50.6)5.3 (57.1)	1.7 (18.3)1.7 (18.3)	1.2 (12.9)2.0 (21.5)	10.6 (114.1)11.9 (128.1)
Restaurants	1.3 (14.0)	0.3 (3.2)	6.0 (64.6)	0.0	0.5 (5.4)	8.1 (87.2)
Retail stores	1.5 (16.1)	0.9 (9.7)	5.5 (59.2)	0.0	1.4 (15.1)	9.3 (100.1)
Schools						
Elementary middle high	1.2 (12.9)1.2 (12.9)1.2 (12.9)	1.7 (18.3)1.6 (17.2)1.5 (16.1)	6.1 (65.7)5.3 (57.1)4.6 (49.5)	0.00.6 (6.5)0.5 (5.4)	3.7 (39.8)1.5 (16.1)2.3 (24.8)	12.7 (136.7)10.2 (109.8)10.1 (108.7)
Warehouses						
Grocery Storage	0.8 (8.6)0.8 (8.6)	0.6 (6.5)0.3 (3.2)	0.00.0	0.00.0	0.5 (5.4)0.5 (5.4)	1.9 (20.5)1.6 (17.2)

TABLE 8.3
Calculation of the Demand Load for Panel Feeder of Example 8.6

Loads	kVA	PF	Connected Load	DF	Demand Load
Lighting	55	0.90	49.5 + *j*23.97	1.25	61.88 + *j*29.97
Heater	5.0	1.00	5.0 + *j*0.0	1.0	5.0 + *j*0.0
Total			54.5 + *j*23.97		66.88 + *j*29.97

b. From Table 6.7, the $VD_{L\text{-}N}=(0.300+0.196)/2$ V/1,000 A-ft for # 3 THW (Cu), $pf=0.90$, and steel conduit. Thus, $VD_{L\text{-}N}=0.248\times I_L\times 250$ ft/1,000. The actual load is $I_L=59{,}538/(1.732\times 480)=$ 71.6 A$VD_{L\text{-}N}=0.248\times 71.6\times 250$ ft/1,000 = 4.44 V or $VD(\%)=4.44/277=1.6\%$
c. Size of feeder taking into account the three criteria (ampacity, short-circuit, and voltage drop):
   Ampacity: Based on (a) $I=88.2$ A, the size of the feeder should be #3 THW (Cu), as noted in Table 6.1
   Table 6.11 indicates, for four wires, the size of the conduit is 1.25″
- *Short-circuit*: Figure 6.7 indicates for I_{sc}, $a_{sym}=1.3\times 35{,}000=45{,}500$ A and $\theta=2$ cycles, the size of the feeder should be: #3/0 THW (Cu)

Table 6.11 indicates, for four wires, the size of the conduit is 2″
- *Voltage drop*: For $VD=1\%$, VDL-N = 0.01 × 277-V = 2.77 V

Since A-ft = 71.6 × 250 = 17.9 × 10^3 A-ft, VD/1,000 A-ft = 2.77/17.9 = 0.155
From Table 6.7, for the steel conduit, $pf=0.9$, and copper wires, the size of the feeder should be #1/0 THW (Cu)
Table 6.11 indicates, for 4 wires, the size of the conduit is 2″
Thus, the size of the feeder is 2″C-4 # 3/0 THW (Cu)

8.4 MOTOR CONTROL CENTERS DESIGN

MCCs are recommended when several motors need to be controlled from one location. The centralized location of all the control and protection units can offer several benefits including the convenience of operation and ease of maintenance. For commercial buildings, MCCs are typically located in the mechanical rooms that house the fans, pumps, and the heating, ventilating, and air conditioning (HVAC) equipment.

8.4.1 Types of MCCs

There are two main classes defined by the National Electrical Manufacturers Association (NEMA) for MCCs:

- **Class 1** —*Independent units*: These units include mechanical groupings and combinations of motor control units, branch feeders, other units, and electrical devices arranged in a convenient assembly. For these centers, the

MCC manufacturers complete all the required wiring between unit components so that field wiring is not needed.

- **Class 2** —*Interconnected units*: These units are the same as Class 1 with the addition of manufacturer-furnished electrical interlocking and wiring between units as specified in the control system diagrams supplied by the client.

In addition, different types of wiring schemes can be considered for the MCCs:

Type A. For these units, field wiring connects directly to device terminals internal to the unit. Type A wiring is provided only for Class 1 MCCs.

Type B. In these units, field control wiring connects to unit terminal blocks within or adjacent to each motor control unit. There are two sizes:

1. The load power wiring for Size 3 or a smaller combination of motor control units connects as follows:
   a. *Type B–D*: Load power wiring connects directly to the device terminals, located immediately adjacent to the vertical busway.
   b. *Type B–T*: Load power wiring connects to factory-wired power terminal blocks located within or adjacent to each unit.
2. The load power wiring for Size 4 or larger combination motor control units and branch feeder units connects directly to the unit device terminals.

Type C. Two sizes are considered for this type of wiring:

1. *Size 3 and smaller*: Field control wiring and load wiring on motor control units connect to master terminal blocks mounted at the top or bottom of the vertical sections containing control units. The unit control wiring and load wiring are factory wired from unit-located terminal blocks to the master terminal blocks.
2. *Size 4 and larger*: The user load power wiring connects directly to the device terminals. The MCCs are made of modules or sections. The typical module dimensions are
   - 38 and 50 cm (i.e., 15, and 20 in.) deep
   - 50, 63, and 76 cm (i.e., 20, 25, and 30 in.) wide

Manufacturer specifications should be consulted before designing and selecting MCCs. In addition to the layout of various modules, both vertical and horizontal buses should be defined when designing MCCs. The standard aluminum or copper bus ratings are 400, 600, 800, 1,200, 1,600, 2,000, and 2,500 A. Figure 8.15 shows the typical modules of an MCC with the available space that houses the motor controllers (typically 182 cm or 72 in height) with spaces for horizontal and vertical buses or wire ways (which are typically 10 cm or 4 wide for 50 cm or 20 in. sections and 23 cm or 9 in. wide for 63 cm 25 in. sections).

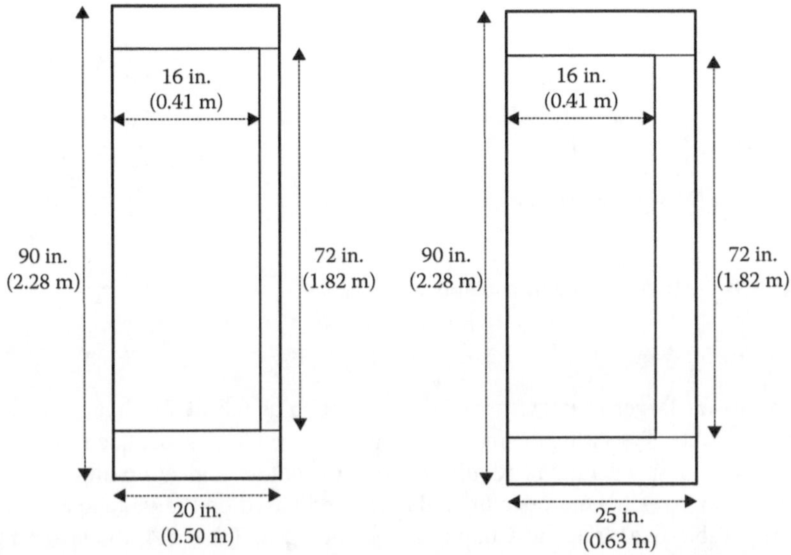

FIGURE 8.15 Typical dimensions for an MCC section.

8.4.2 Layout of MCCs

When planning to design the MCC layout for each section, it is important to determine the space requirements often expressed in space factors for each motor controller. One space factor corresponds to 15 cm (6 in.) vertical subsection within each MCC module. Since each MCC section has a total available height of 182 cm (72 in.), as outlined in Figure 8.15, there are 12 space factors per section. The designer typically is responsible for specifying and planning out the MCC layout based on the motor controller units and heights. However, manufacturer representatives can also assist in the specifications of the layout for all the MCC sections.

8.4.2.1 Design Approach

The overall size of MCCs depends on several factors, including the number of motors, the type of the motor starters, and the rating of the protective devices. In the US, MCCs are typically arranged in modules of 50 cm (20 in.) wide by 50 cm (20 in.) deep and 228 cm (90 in.) high. The procedure for determining the space requirements of MCCs is illustrated in this section using an MCC that feeds the motors defined in Figure 8.16 (refer to Example 6.13).

8.4.2.1.1 Space Requirements for Motor Starters

Each module (i.e., a section of 20 in. × 20 in. × 90 in.) of an MCC can hold a variable number of drawout control units. Each drawout control unit houses the starter, the protection device (circuit breaker or fused switch), and a disconnecting means for one motor. The drawout control units have standard incremental dimensions. Commonly, the smallest unit has a dimension of 12 in. (30 cm) with an increment of

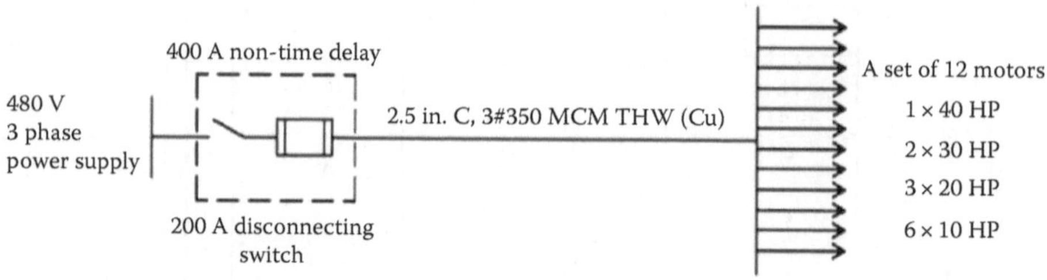

FIGURE 8.16 Motors served by a motor control center.

6in. (15cm) for larger units. Other increments such as 6.5 or 7in. (16.5 or 17.7cm) are also available. Typically, the increments are referred to as space factors. Table 8.4 provides typical space factors required for both motors and nonmotor loads. The NEMA starter sizes for motors can be determined based on the voltage and HP ratings using Table 6.20 (refer to Chapter 6). As noted in Table 8.4, the space factors depend on the type of protection devices and starter types for motors and on the rating of the protection devices for the nonmotor loads.

Table 8.5 provides the space factor requirements for various drawout control units used in the MCC that serves the motors considered in Figure 8.16.

Therefore, the total number of space factors required for all the motor control units is 31. In addition, one space factor is typically allocated to the incoming feeder cables. Thus, 32 space factors are needed for the MCC that serves 40 HP, 2×30 HP, 3×20 HP, and 6×10 HP motors. Since each module can hold 12 space factors, the number of modules needed for the MCC is 32/12 or three modules with four spare space factors.

8.4.2.1.2 Layout of the MCC

Figure 8.17 shows a possible layout for the MCC including the position of the drawout control units for all the motors.

It should be noted that the MCC can be free standing or mounted against a wall. To allow for easy access to the drawout control units, sufficient working space in front of the MCC should be available. For an MCC rated at 480V, the U.S. National Electrical Code (NEC) requires a minimum clearance of 3.5ft (1.07m) from the front face of the MCC to the nearest grounded surface such as a wall.

8.5 SWITCHBOARDS AND UNIT SUBSTATIONS

8.5.1 Overview

Unit Substations and switchboards typically house the large over-current protection devices (typically, power circuit breakers) for the distribution systems for buildings. They are made of one or several metallic free-standing modules or sections that are custom or field assembled by the manufacturers. The main difference between the switchboards and unit substations is related to the location of the main service

TABLE 8.4
Typical Space Factors Required for MCCs (Based on a Space Factor = 6 in [15 cm])

NEMA Starter Size (Refer to Table 6.20)	Starter Type: FVNR	FVR	RV – Autotransformer	RV – Wye-Delta
	Circuit Breaker Type Protection			
1	2	3	–	–
2	2	3	6	5
3	4	4	9	6
4	4	4	9	7
5	6	10	12	12
6	12	12	12	–
	Fusible Switch Type Protection			
1	2	3	–	–
2	2	3	6	5
3	4	5	10	8
4	6	6	10	12
5	7	12	12	12
6	12	12	12	–

Trip Rating (A)	Circuit Breakers	Fusible Switches
30	–	2
60	–	2
100	2	2
150	2	4
200	3	4
225	3	4
400	4	6

Note: Manufacturers' specifications should be consulted for a final selection.
(a) Space factors for various NEMA starter sizes and protection devices associated with motors.
(b) Space factors for various protection device ratings associated with nonmotor loads.

entrance transformer. Indeed, the unit substations are integrated main distribution panels and include both the switchgear and the transformer in addition to the power circuit breakers for the main secondary feeder and the subfeeders serving the various panels. Therefore, the main sections of a unit substation include, as illustrated in Figure 8.18, the following:

TABLE 8.5
Number of Space Factors Required by the Motor Control Units for MCC of Figure 8.16

Motor HP	Type of Starter	Space Factor per Motor	Number of Motors	Number of Space Factors
40	FVR	5	1	5
30	FVNR	4	2	8
20	FVNR	2	3	6
10	FVNR	2	6	12

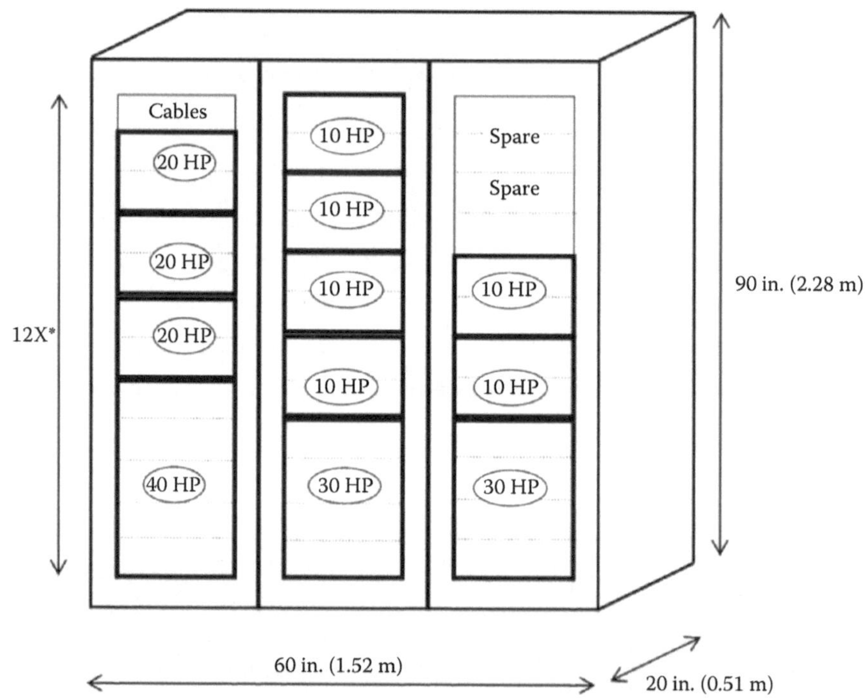

FIGURE 8.17 Layout for the MCC of Figure 8.16. The symbol *X* is used to designate one space factor (that is 6 in. [15 cm]).

- *Primary section*: Housing the primary feeder cables, primary switchgear, and often lightning protection. The switchgear consists of fused or infused load-interrupter switches or drawout circuit breakers.
- *Transformer section*: In this section, a typically ventilated dry-type step-down transformer is installed to reduce the primary high-voltage to secondary low-voltage for distribution to various building loads.

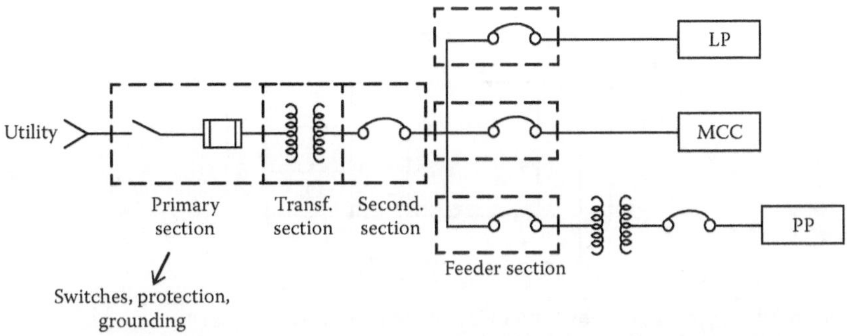

FIGURE 8.18 Typical distribution system and associated sections for a unit substation.

- *Secondary section*: This section provides the protection and control for low-voltage for the secondary feeder. It may include fusible switches, molded-case circuit breakers, or power circuit breakers.
- *Feeder section*: This section includes protection and control devices for sub-feeders serving lighting panels, power panels, MCCs, or branch circuits for single motors.
- Switchboards, on the other hand, include only the switchgears and act as the primary distribution points of the electrical systems. They typically include the following:
- *Cable pull section*: This section includes cables or busways for low-voltage distribution.
- *Main switchgear section*: This section provides protection and control for low voltage for the main feeder or busway. It may include fusible switches, molded-case circuit breakers, or power circuit breakers. A ground fault detection system may also be included in this section.
- *Feeder section*: This section includes protection and control devices for sub-feeders serving lighting panels, power panels, MCCs, or branch circuits.

Switchboards are available in different voltages and include 120/240 V, 208Y/120 V, and 277Y/480 V and ratings of 1,200, 1,600, 2,000, 2,500, 3,000, and 4,000 A. Figure 8.19 illustrates typical sections for both unit substations and switchboards.

8.5.2 Design Procedure for Unit Substations

The design procedure for unit substations includes the selection of the size and rating of the primary, transformer, main secondary, and feeders sections. The following steps provide the general design approach for unit substations:

1. *Primary section*: First, the primary current is estimated based on the size of the transformer:

$$I_{L,P} = \frac{P_T}{\sqrt{3}E_L} \tag{8.19}$$

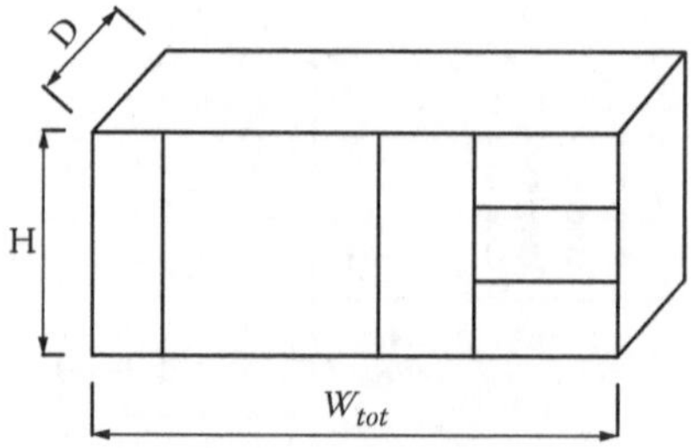

FIGURE 8.19 Typical sections for a unit substation or a switchboard.

TABLE 8.6
Ratings and Dimensions of Dry-Type Transformer Sections Suitable for Unit Substations

kVA Rating of the Transformer	Dimensions of the Transformer Section[a]					
	5 kV Systems			15 kV Systems		
	H	W	D	H	W	D
111.5	90	30	54	90	30	54
150	90	30	54	90	30	54
225	90	30	54	90	42	54
300	90	42	54	90	90	54
500	90	42	54	90	90	54
750	90	90	54	90	90	54
1,000	90	90	54	90	100	54
1,500	90	100	54	90	100	54
2,000	90	100	54	100	100	54
2,500	90	100	54	100	100	54

[a] The dimensions are expressed in inches and should be multiplied by 2.54 to express them in cm.

Based on this primary current increased by 25%, the primary section is selected using a current rating of 600 A or 1,200 A and a voltage rating of 5 or 15 kV. A standard section is available with a width of 36 in. (91 cm), height of 90 in. (228 cm) or 100 in. (253 cm), and depth of 54 in. (137 cm).

2. *Transformer section*: The selection and dimensions of this section depend on the manufacturers' specifications and are typically a function of the transformer kVA and primary voltage ratings, as summarized in Table 8.6, for typical transformer section dimensions available from some U.S. manufacturers.
3. *Main secondary section*: The secondary current of the transformer is calculated using Equation 8.20. Then, the size of the protection device (typically a power circuit breaker) is selected using the demand load obtained by Equation 8.21:

TABLE 8.7
Frame Sizes and Typical Ratings for Low-Voltage Power Circuit Breakers

		Instantaneous Trip Interrupting Symmetrical Current Ratings (A)			Short-Time (0.5 seconds Delay) Trip Interrupting Symmetrical Ratings (A)		
Frame Size (A)	Trip Ratings[a] (A)	208/240 V	480 V	600 V	240 V	480 V	600 V
600	50–600	42,000	30,000	30,000	30,000	30,000	30,000
600	50–800	42,000	30,000	30,000	30,000	30,000	30,000
1,600	50–1,600	65,000	50,000	42,000	50,000	50,000	42,000
2,000	50–2,000	65,000	50,000	50,000	50,000	50,000	50,000
3,000	1,200–3,000	85,000	65,000	65,000	65,000	65,000	65,000
4,000	1,600–4,000	130,000	85,000	85,000	85,000	85,000	85,000

[a] Standard trip ratings include 50, 100, 150, 200, 225, 300, 400, 600, 800, 1,200, 2,000, 2,400, 3,000, and 4,000 A.

TABLE 8.8
Ratings and Dimensions of Main Feeder and Subfeeder Sections

Unit No.	Width [in. (cm)]	Maximum Current Ratings (A)	Number of Compartments	Applications
Unit 1	30 (76)	3,000	1	Main secondary sections
Unit 2	18 (46)	600	4	Feeder sections
Unit 3	26 (66)	1,600	3	Feeder sections
Unit 4	30 (76)	3,000	3	Feeder sections
Unit 5	34 (86)	4,000	2	Main secondary or feeder sections
Unit 6	26 (66)	600–1,600	4	Feeder sections (with different feeder ratings)

$$I_{L,S} = \frac{P_T}{\sqrt{3}E_{L,S}} \tag{8.20}$$

$$I_{CB} = 1.25 \times I_L \tag{8.21}$$

The frame size of the power Main CB can be obtained based on the I_{CB} value as well as the short-circuit current value obtained at the main secondary feeder (refer to Table 8.7 for selecting the frame size and the trip rating of power circuit breakers). Once the frame size of the circuit breaker is selected, Table 8.8 is used to select the unit number and associated dimensions that meet the requirements of the main secondary section.

4. *Subfeeder section*: The selection of the subfeeder sections is similar to the selection of the main secondary section using Table 8.8.

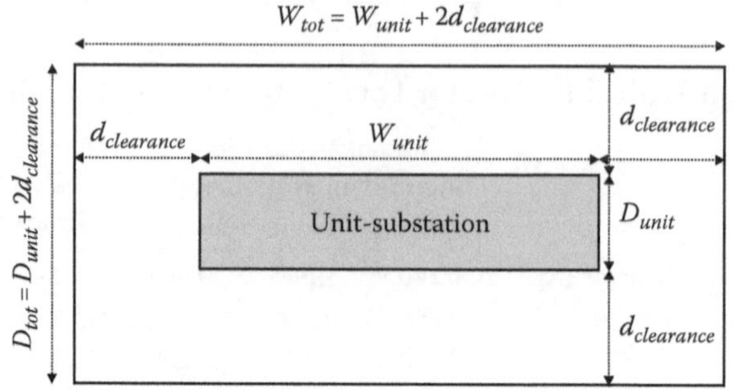

FIGURE 8.20 Typical clearance and electrical room size for 15 kV systems.

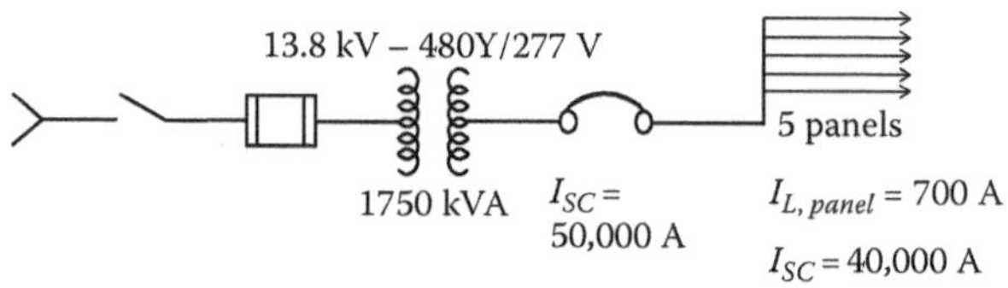

FIGURE 8.21 One-line distribution system for Example 8.7.

5. *Electrical room size*: Once all the sections for the unit substation or the switchboard are selected, the width and depth of the main distribution panel can be determined. Then, the size of the electrical room can be estimated using NEC clearance requirements, which call for a specific clearance distance, $d_{clearance}$, around the main distribution panels assumed centered within the electrical room. The clearance distance is set to 6 ft (1.83 m) for high-voltage systems (i.e., with 15 kV rating) and 5 ft (1.52 m) for low-voltage systems (i.e., with 5 kV rating) when a unit substation is considered. Figure 8.20 illustrates the clearance distances and dimensions of the electrical room that houses a unit substation.

Example illustrates the design approach outlined here to design a unit substation layout for a large commercial building.

Example 8.7

Problem

Determine all the required sections of a unit substation for a radial distribution system shown in Figure 8.21. Then, size the electrical room assuming the unit substation is placed in the center of the room.

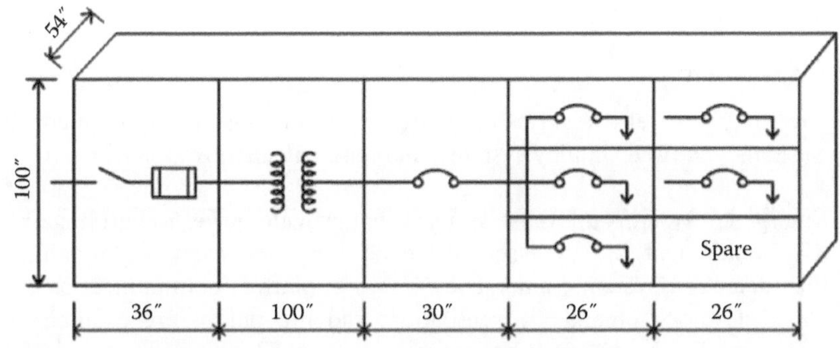

FIGURE 8.22 Layout for the unit substation of Example 8.7.

Solution

Using the different steps outlined in the general design procedure, the unit substation sections can be selected for the system of Figure 8.21.

1. *Primary section*:

$$I_L = \frac{1{,}750 \text{ kVA}}{(1.732)(13.8 \text{ kV})} = 73.2 \text{ A}$$

   Use 600 A, 15 kV, W = 36 in., H = 90 in., D = 54 in. primary section
2. *Transformer section*:
   Table 8.6 → W = 100 in., H = 100 in., D = 54 in.
3. *Main secondary section*:

$$I_L = \frac{1{,}750 \text{ kVA}}{\sqrt{3}(0.48 \text{ kV})} = 2{,}105 \text{ A}$$

$$I_{CB} = 1.25 \times I_L = 2{,}631 \text{ A}$$

   Table 8.7 → use 3,000 A CB (ISC = 65,000 A < 50,000 A)
   Table 8.8 → unit #1, W = 30 in.
4. *Feeder section*:
   Table 8.7 → 1,600 A CB (ISC = 50,000 A < 40,000 A)
   Table 8.8 → use unit #3, W = 26 in.
   Each can handle three panels → two feeder sections are needed
5. *Electrical room size*: Figure 8.22 shaunit substation based on the selection of all the sections. Since the total width of the unit substation is 218 in. or 18.16 ft and its depth is 54 in., the size of the room with a clearance distance of 6 ft in all sides can be determined:

$$\text{Electrical room size} = 30.16 \text{ ft (width)} \times 16.5 \text{ ft (depth)}$$

8.6 EMERGENCY SYSTEMS

8.6.1 Overview

There are three general categories of emergency power systems and include emergency systems, required standby systems, and optional standby systems:

- Emergency systems are those systems that provide power to building electrical equipment that are essential for the immediate safety of human life. Fire detection systems, emergency lighting, alarms, communication systems, elevators, emergency ventilation, and fire pumps are examples of emergency loads that must be served in order to allow occupants to safely egress the building. These loads must be powered almost immediately after an emergency event occurs. The National Fire Protection Agency (NFPA) requires that emergency systems be operational within 10 seconds after the main supply power fails.
- Required standby systems are those systems that are necessary for building safety, but do not immediately endanger building inhabitants in their absence. Systems for fire fighting, control of health hazards, long-term rescue operations, and industrial hazard prevention are examples of electrical loads that would be powered by the required standby systems. Because these systems do not need to be operational immediately, required standby systems must be ready to serve the loads within 60 seconds of power loss.
- Optional standby systems are intended to minimize the economic loss due to a building emergency. These types of loads can range widely depending on the building owner's requirements.

8.6.2 Types of Emergency Systems

An emergency power system is made up of all the components required to provide power to the necessary building electrical systems. Both the power source and the distribution system must be completely separate from the primary system components. This requirement is to ensure that an internal power failure within the building will not jeopardize the functionality of the emergency power components. NFPA requirements state that emergency systems, required standby systems, and optional standby systems can all be powered by one source. However, an automatic load pickup and a load shedding equipment must be integrated into the system to ensure that power can be distributed to the individual system components based on well-defined priority.

There are several options to generate backup power for emergency loads. These options include the following:

- *Batteries*: When power failure occurs, a connection from the emergency loads to a battery system is made through a transfer switch. This option is especially easy to design if the emergency loads can be operated using direct current (DC). In case alternating current (AC) is needed to serve the emergency loads, an inverter must be installed into the system to change the battery DC output into AC power. Small loads like emergency lighting, alarms, and communication systems are best powered by a battery backup. For larger loads, the most practical application is a power generator.

- *Generators*: These systems burn fossil fuels to convert chemically stored energy into mechanical and then electrical power. Generators are available in a variety of voltage ratings and can be used to power both single- and three-phase loads that make them a versatile backup power source for a wide range of load types. Typically, lights and fire pumps are powered by a generator system. Chapter 13 discusses in more details the design specifications and operation strategies for generators and cogenerators.
- *Secondary utility services*: For large buildings or building complexes, two separate utility services can be utilized: primary and secondary. If the primary utility service is subject to any failure, the secondary (or standby) utility service can provide the necessary supplemental power. The two service entrances have to be connected to the building separately at different locations. Moreover, the power sources of the two services have to be independent so that any failure from one service does not affect the other.

While most generator systems require the use of fossil fuels to provide electrical energy, there are several backup systems that run off from natural power systems. Solar-, wind-, and water-powered systems can replace those systems when the emergency loads are small. These systems gather power over an extended period of time and send that power to a battery system for storage. Then, in the event of an emergency power failure, these batteries can supply the required loads as a typical battery system would. Such naturally powered systems are ideal for locations with high natural resources or with limited fossil fuel resources. Chapter 14 describes design and analysis approaches to integrate distributed power generation systems including renewable energy technologies to meet the electrical needs of buildings.

It should be noted that the main reason to install optional standby systems is to reduce economical impacts rather than enhance human safety due to any loss of power. Indeed and for special building applications, economic losses due to power failure can be substantial even if the power outage lasts a few seconds. Power losses can damage important and sensitive equipment and products that are vulnerable even within the typical 10 seconds response time needed to activate conventional emergency systems. For these loads, uninterruptible power supply (UPS) systems can offer a good alternative to the typical emergency systems. A UPS is a combination of batteries and ballasts designed to continue power to a load until a backup power system can be connected. Usually, the UPS can run at full load only for a few seconds, but this time is sufficient to avoid significant damages associated with power loss.

8.6.3 Emergency Lighting

Emergency lighting systems are triggered in case of failure of normal lighting systems in order to provide sufficient illuminance for occupants who should be able to see the path to the building exits. An emergency lighting system consists of area lighting along exit routes, and directional exit signs. According to codes, both must be fully operational at the required levels within ten seconds of failure and be able to provide the required illuminance for 90 minutes. The normal lighting system can fail in three ways: a utility power outage, a building electrical system failure, or a lighting fixture failure. To respond to all three cases, the emergency lighting system must sense failure locally (either by voltage sensor or photo sensor) and be powered independently.

- *Area lighting*: There are commonly two approaches to light exit routes—special emergency lighting fixtures or integrated lighting fixtures. Emergency lighting fixtures are easier to specify and install since they are specifically designed to light egress paths. However, they are add-ons to the normal lighting systems and may not be appealing esthetically and their light beam may cause glare. These fixtures are available in a wide variety of designs and forms and can use almost any light source. Alternatively, emergency area lighting can be supplied by select fixtures among the normal lighting system by the addition of integrated battery system. The main benefit of this integrated system is it is part of the normal lighting system and provides more uniform illuminance in case of emergency.
- *Exit lighting*: Exit signs are also essential in case of emergency events and are required by most codes. Indeed, building occupants have to be guided by clear directional signs from any location in the building until they reach safety, typically outdoors. These can be externally or internally illuminated and use several light sources, such as fluorescent or LED. It should be noted that the exit signs need be visible even through smoke.

8.7 FIRE ALARMS

Fire alarm systems enhance the safety of people within buildings since they assist in timely detection and provide early warnings as soon as any smoke and/or fire occurs. When designing fire alarm systems for buildings, it is important to meet the requirements set by the local authorities and applicable codes. In the US, most local codes follow the specifications outlined by the National Fire Protection Associate (NFPA) code 72 (NFPA, 2022). The main scope of NFPA 72 is to cover design, installation, testing, and maintenance considerations of various fire alarm systems as well as supervising station alarm systems, public emergency alarm reporting systems, fire and carbon monoxide detection and warning equipment, and emergency communications systems. The design of fire alarm systems is based primarily on the main goal of saving people as well as materials from any fire hazards inside buildings. Other considerations should be, however, accounted for when specifying a fire alarm system including ease of maintenance, flexibility to future change, and longevity.

A typical fire alarm system includes several components including detection devices, notification devices, communication equipment, and fire suppression (i.e., sprinkler) systems. The type and number of various devices required for a given building depend on several factors such as the number of floors, the construction materials, the number of exits, and the heat sources.

8.8 CASE STUDY

8.8.1 Overview

In this case study, the specifications of an electrical distribution system for an office building are determined using the design approach outlined in this chapter. Specifically, the case study considers a three-story medium-sized office building with a total floor area of 4,200 m^2 (45,000 ft^2) with a basement, open office plan, and three conference rooms. The building electrical loads are supplied by a radial distribution system from a 13.8 kV utility line through a step-down transformer to reduce the voltage to 480Y/277 V in order to serve air conditioning and lighting as well as a 208Y/120-V transformer connected to panels to serve receptacles.

8.8.2 Building Description

Figure 8.23 illustrates the first-floor layout for the office building. Other floors have a similar layout. Figure 8.24 shows the 3-D rendering of the building. In particular, the building has the following spaces:

- Mechanical in basement
- Hydraulic elevator
- Vestibule
- Naturally lit stairwell
- Open reception area
- Open space for play area
- Kitchen
- Shower stalls in the bathrooms
- Three large conference rooms
- Huddle room
- Electrical room

8.8.3 Electrical Loads

8.8.3.1 Lighting

The building's general lighting system is made up of fluorescent fixtures operating at 277 V. Specifically, 2×4 troffers are used for ambient and task lighting in most areas. In the restrooms, the vestibule, and the kitchen area, downlights are utilized. Table 8.9 provides a summary of the number of fixtures and downlight units used in the entire building as well as the total power lighting load. Based on the magnitude of this load, one lighting panel is deemed sufficient for the entire building.

The line current for the connected load of the lighting panel is estimated to be

$$I_L = \frac{P_T}{\sqrt{3} \times E_L} = \frac{31{,}800\ \text{VA}}{(1.732)(480\ \text{V})} = 37.3\ \text{A} \tag{8.22}$$

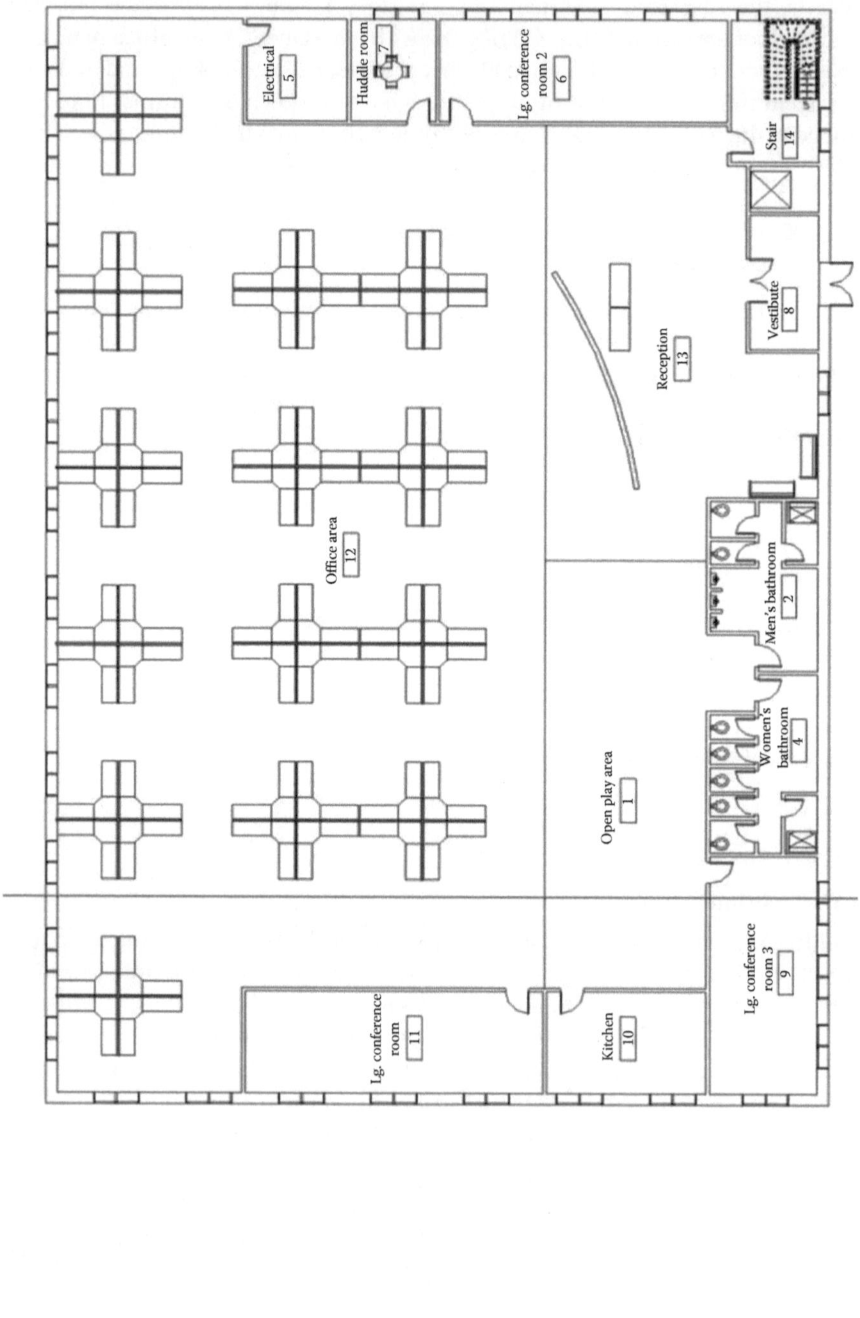

FIGURE 8.23 Typical floor layout for the first-floor plan for the medium office building.

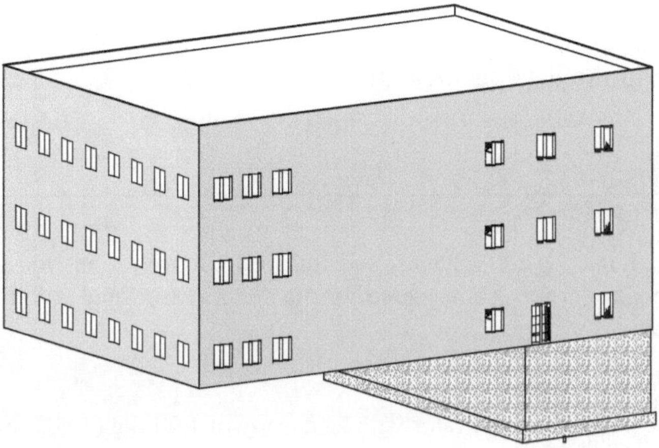

FIGURE 8.24 3-D rendering of the office building.

TABLE 8.9
Summary of Lighting Loads

Fixture ID	Description	Quantity	Load (VA)	Total Load (kVA)
D1	6″ Downlight	57	60	3.42
D2	8″ Downlight	78	100	7.80
R1	2 × 4 Troffer	258	80	20.6
Total				31.8

TABLE 8.10
Summary of Motor Specifications for the Air Conditioning System

Equipment	Motor Size (HP)	Type	Voltage (V)	Full Load Current (A)
Supply fan	75	FVNR	460	96.0
Return fan	40	FVNR	460	52.0
Exhaust fan	15	FVNR	460	21.0
Chilled water pump	15	FVNR	460	21.0
Hot water pump	20	FVNR	460	27.0
Condenser	75	RVNR	460	96.0

8.8.3.2 Mechanical System

To heat and cool the building, an air handling unit (AHU) is used. In particular, supply air is heated through a hot water coil served by a boiler and cooled using a chilled water coil fed by a chiller. The conditioned air is moved through both supply and return ducts using fans. Pumps drive hot and chilled water flows. All the fans and pumps are powered by electric motors. Table 8.10 lists the mechanical power output

TABLE 8.11
Elevator Motor Rating Specifications

Equipment	Motor Size (HP)	Type	Voltage (V)	Full Load Current (A)
Elevator motor	25	FVR	460	34.0

(expressed in horsepower), voltages, and full load current ratings for all motors as well as the controller types considered for the motors associated with the air conditioning system.

8.8.3.3 Elevator Motor

For the office building, an elevator with a capacity of 1,600 kg (3,500 lb) is specified (Allen and Jano, 2007) since a low-to-medium speed is acceptable for a three-story building. With a speed of 61 m/min (200 ft/min), the elevator would travel from the basement to the third floor (about 13 m or 42 ft) in approximately 13 seconds. The elevator motor specifications are selected as summarized in Table 8.11 (Baldor, 2015).

8.8.4 Electrical System Design

8.8.4.1 Distribution System

For this office building, a simple radial distribution system is considered since there are no critical loads and the owner has a budget constraint. The one-line and riser diagram for the electrical distribution system is shown in Figure 8.25. A unit substation is specified for this building and is located in the electrical room. The main distribution panel serves one MCC as well as one lighting panel and three power panels.

8.8.4.2 Electrical Panels

The electrical room could be located in the center of the building to allow for shorter runs to most of the loads. However, in order to keep the center area available for open offices as well as to minimize the main feeder connecting the utility line with the building distribution system, an electrical room is placed on the east side of each floor.

In the basement, a large electrical room houses the unit substation that includes the main transformer (T-SVC) lowering the utility distribution voltage of 13.8 kV to 480/277 V. The main transformer feeds a lighting panel (LP-1), an MCC, as well as a secondary transformer (T-1), which feeds the low-voltage panels. All lighting fixtures are served by one 480Y/277-V lighting panel (LP-1) located in the first-floor electrical room. All of the large motors are supplied by the MCC, which is located in the basement electrical room. The convenience receptacles and all other 120 V loads are circuited, by floor, to one of three low-voltage power panels (P-1, P-2, P-3). The power for these panels is fed from the unit substation via a secondary transformer (T-1). To reduce the length of branch circuits, power panel P-1 is located in the first-floor electrical room, P-2 is located in the second-floor electrical room, and P-3 is located in the third-floor electrical room. Figures 8.26 and 8.27 illustrate the branch circuits for lighting and general-purpose receptacles for selected floors, respectively.

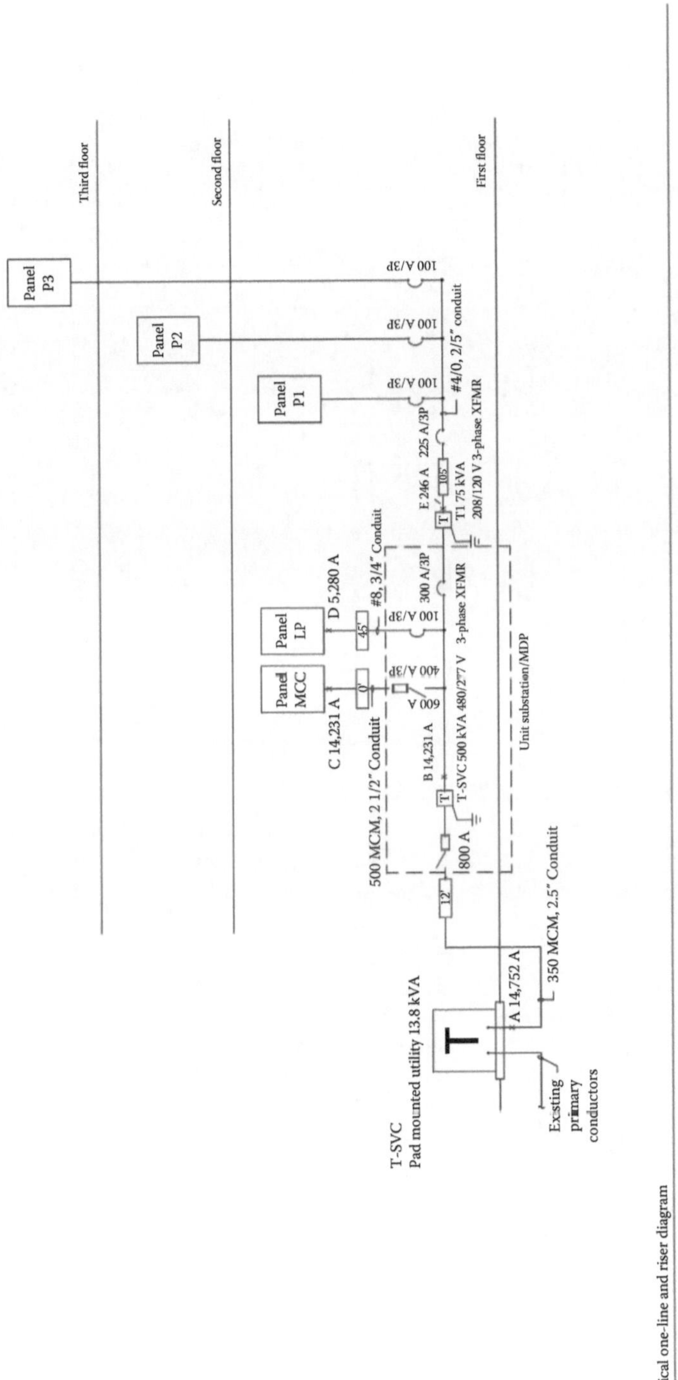

FIGURE 8.25 One-line and riser diagram for the radial system serving the office building.

FIGURE 8.26 Lighting branch circuits associated with the first floor.

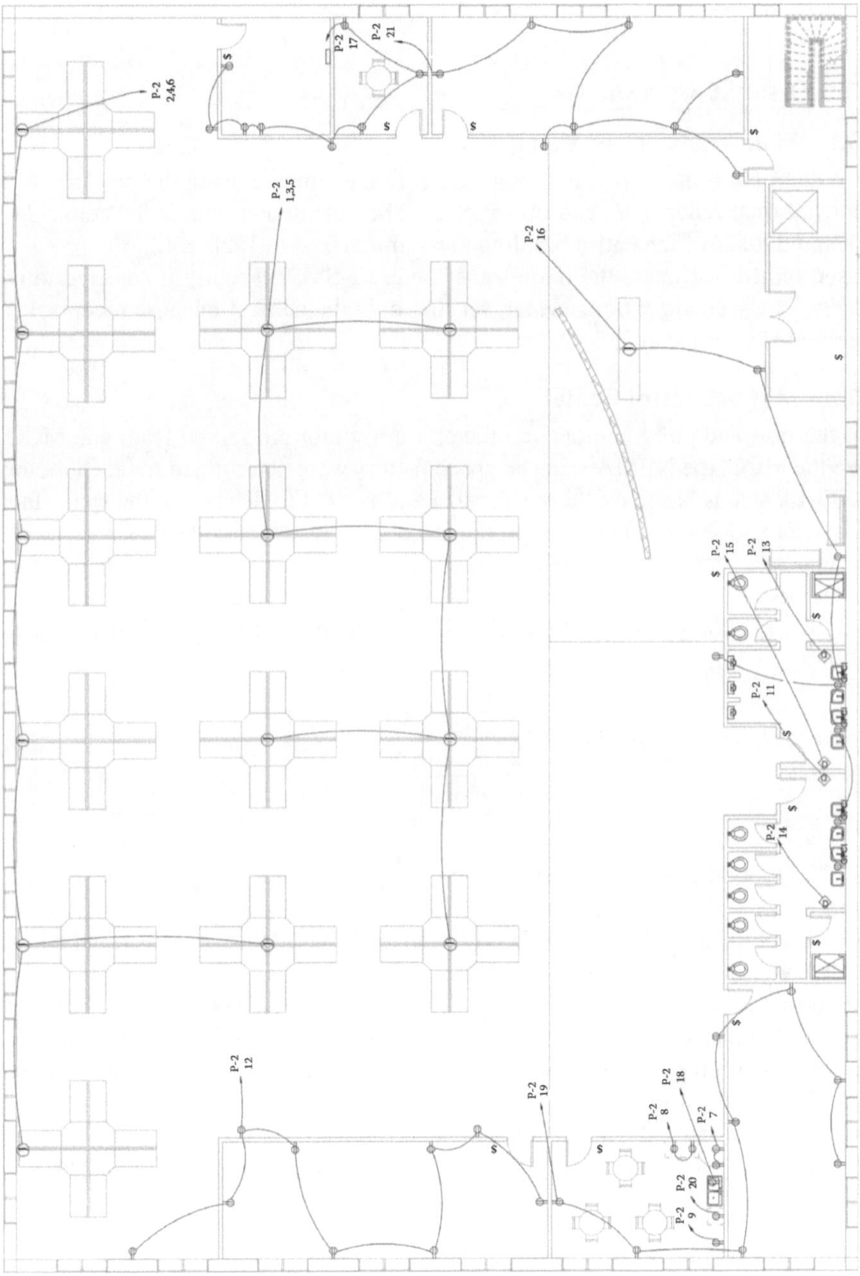

FIGURE 8.27 Power branch circuits and power panel associated with the second floor.

All the panels are specified to ensure that all the branch circuits are protected with molded case circuit breakers. Moreover, the three phases for the panels are generally balanced within 10%. Figure 8.28 shows the lighting panel schedule while Figure 8.29 presents the schedule for power panel P-2.

8.8.5 Design Calculations

8.8.5.1 Main Transformer Sizing

To size the main transformer, the total load is first estimated using the power densities for lighting, receptacle, and other loads. The calculation details for connected and demand loads of the entire building are summarized in Table 8.12.

Based on this estimate, the main transformer (T-SVC) capacity is selected to be 500 kVA. The secondary transformer, serving only the general-purpose receptacles, is specified to be rated at 75 kVA.

8.8.5.2 Motor Control Center

All of the fans and pump motors and the elevator motor are served from one MCC. To size the MCC, the NEMA size and space factors were determined for each motor. The NEMA size is based on the starter size and is used to determine the space factors. Space factor requirements for all the motors are listed in Table 8.13.

Adding one space factor for the main fuse and cabling, the required number of space factors is 31. Based on 15 cm (6 in.) of height for each space factor and a total of 2.28 m (90 in.) of height for each module, the total number of modules is determined as follows:

$$\text{Number of modules} = \frac{31\ \text{space factors}}{\dfrac{12\ \text{space factors}}{\text{module}}} = 2.58 \rightarrow 3\ \text{modules} \tag{8.23}$$

The MCC layout including the horsepower and controller for each motor is provided in Table 8.14.

For each of the modules, the total connected load is used to size the vertical buses. A summary of these loads is listed in Table 8.15.

Based on the largest line current of 148 A (module 1), the vertical buses selected shall be 300 A each. The total demand load for the MCC is 371 A (refer to the calculation details in the next section), thus the rating for the main horizontal bus is set to be 600 A. The overall dimensions of the MCC are 1.52 m (60 in.) wide × 2.28 m (90 in.) tall × 0.51 m (20 in.) deep. For this MCC rated at 480 V, NEC requires a minimum clearance of 1.1 m (3.5 ft) from the front face to the nearest grounded surface, such as a wall.

8.8.5.3 Size of Feeder and Subfeeders

To size the main feeder and subfeeders for both motor and nonmotor panels, the connected loads are estimated first and demand factors are then applied to calculate the demand loads. The demand loads are utilized to size the feeders and associated

Branch Panel: P-2

Location: Electrical 203	Volts: 206/120 V	A.I.C. Rating:
Supply From: T-1	Phases: 3	Mains Type:
Mounting: Surface	Wires: 4	Mains Rating: 100 A
Enclosure:		

Notes:

CKT	Circuit Description	Trip	Poles	A		B		C		Poles	Trip	Circuit Description	CKT
1	Workstations	20 A	3	1167...	1167...					3	20 A	Workstations	2
3	—	—	—			1167...	1167...			—	—	—	6
5	—	—	—					1167...	1167...	—	—	—	6
7	Receptacles—Kitchen counter	20 A	1	360 VA	360 VA					1	20 A	Receptacles—Kitchen receptacles	8
9	Kitchen refrigerator	20 A	1			1200...	0 VA			1	0 A	Spare	10
11	Hand dryer	20 A	1					1725...	1440...	1	20 A	Receptacles—Large conference room 2	12
13	Hand dryer	20 A	1	1725...	1725...					1	20 A	Hand dryer	14
15	Hand dryer	20 A	1			1725...	1760...			1	20 A	Receptacles—Restrooms, reception area	16
17	**Receptacles—Private office, electrical room**	20 A	1					1440...	1176...	1	20 A	Motor—Garbage disposal	18
19	Receptacles—Kitchen convenience, Lg Conf Rm 3	20 A	1	1440...	900 VA					1	20 A	Kitchen dishwasher	20
21	Receptacles—Large conference Rm 2	20 A	1			1440...	0 VA			1	0 A	Spare	22
23	Spare	0 A	1					0 VA	0 VA	1	0 A	Spare	24
25	Spare	0 A	1	0 VA	0 VA					1	0 A	Spare	26
27	Spare	0 A	1			0 VA	0 VA			1	0 A	Spare	28
29	Spare	0 A	1					0 VA	0 VA	1	0 A	Spare	30
31					0 VA					—	—	Space	32
33													34
35													36
37													38
39													40
41													42
			Total Load:	8743 VA		8458 VA		8114 VA					
			Total Amps:	73 A		71 A		68 A					

Legend:

Load Classification	Connected Load	Demand Factor	Estimated Demand	Panel Totals	
Motor	1176 VA	100.00%	1176 VA		
Power	7500 VA	100.00%	7500 VA	Total Conn. Load:	25316 VA
Receptacle	9740 VA	100.00%	9740 VA	Total Est. Demand:	25316 VA
Equipment	6900 VA	100.00%	6900 VA	Total Conn. Current:	70 A
				Total Est. Demand Current:	70 A

Notes:

FIGURE 8.28 Schedule for lighting panel LP.

Branch Panel: LP-1

Location: Electrical 103	Volts: 480 V/277 V	A.I.C. Rating:
Supply From: T-SVC	Phases: 3	Mains Type:
Mounting: Surface	Wires: 4	Mains Rating: 100 A
Enclosure: Type 1		MCB Rating: 250 A

Notes:

CKT	Circuit Description	Trip	Poles	A		B		C		Poles	Trip	Circuit Description	CKT
1	Level 1—Restrooms/vestibule lighting	20 A	1	1340...	960 VA					1	20 A	Level 1—Row 1 lighting	2
3	Level 1—Conference room 3 lighting	20 A	1			600 VA	840 VA			1	20 A	Level 1—Huddle/conference room 2 lighting	4
5	Level 1—Reception area lighting	20 A	1					720 VA	800 VA	1	20 A	Level 1—Open area lighting	6
7	Level 1—Kitchen, conference room 1 lighting	20 A	1	1200...	800 VA					1	20 A	Level 1—Row 4 lighting	8
9	Level 1—Row 2 lighting	20 A	1			960 VA	800 VA			1	20 A	Level 1—Row 5 lighting	10
11	Level 1—Row 3 lighting	20 A	1					800 VA	800 VA	1	20 A	Level 1—Row 6 lighting	12
13	Level 2—Row 1 lighting	20 A	1	960 VA	960 VA					1	20 A	Level 2—Row 2 lighting	14
15	Level 2—Row 3 lighting	20 A	1			800 VA	800 VA			1	20 A	Level 2—Row 6 lighting	16
17	Level 2—Row 4 lighting	20 A	1					800 VA	800 VA	1	20 A	Level 2—Row 5 lighting	18
19	Spare	0 A	1	0 VA	1140...					1	20 A	Level 2—Restroom lighting	20
21	Level 2—Conference rooms/kitchen lighting	20 A	1			1,800...	960 VA			1	20 A	Level 2—Open area lighting	22
23	Level 3—Row 1 lighting	20 A	1					960 VA	960 VA	1	20 A	Level 3—Row 2 lighting	24
25	Level 3—Row 3 lighting	20 A	1	800 VA	800 VA					1	20 A	Level 3—Row 4 lighting	26
27	Level 3—Row 5 lighting	20 A	1			800 VA	800 VA			1	20 A	Level 3—Row 6 lighting	28
29	Level 3—Conference rooms/kitchen lighting	20 A	1					1800...	1080...	1	20 A	Level 3—Restroom lighting	30
31					960 VA					1	20 A	Level 3—Open area lighting	32
33	Level 3—Reception, huddle, vestibule lighting	20 A	1			1520...	1600...			1	20 A	Level 2—Reception, huddle, vestibule lighting	34
35													36
37													38
39													40
41													42
			Total Load:	9920 VA		12,280 VA		9520 VA					
			Total Amps:	36 A		45 A		34 A					

Legend:

Load Classification	Connected Load	Demand Factor	Estimated Demand	Panel Totals	
Lighting	20,560 VA	125.00%	25,700 VA		
Other	11,160 VA	100.00%	11,160 VA	Total Conn. Load:	31,720 VA
				Total Est. Demand:	36,860 VA
				Total Conn. Current:	38 A
				Total Est. Demand Current:	44 A

Notes:

FIGURE 8.29 Schedule for power panel P-2.

TABLE 8.12
Total Building Electrical Load Estimations

			Connected Load				Demand Load	
Load	PF	VA/ft² (VA/m²)	P_T (kVA)	P_R (kW)	P_X (kVAR)	DF	P_R (kW)	P_X (kVAR)
Lighting	0.95	1 (10.8)	45	42.75	14.05	1.25	53.44	17.56
Receptacles first 10 Kva remainder	0.90	1.9 (20.5)	85.5	967.95	42,532.91	1,000.50	90,033.98	43,616.45
Computers	0.90	1.7 (18.3)	76.5	68.85	33.35	1.00	68.85	33.35
Elevator	0.90	0.3 (3.2)	13.5	12.15	5.88	1.25	15.19	7.36
Air conditioning	0.90	5 (53.8)	225	202.5	98.08	1.00	202.50	98.08
Miscellaneous	1.00	2 (21.5)	90	90	0	1.00	90.00	0.00
Total							472.95	177.15
							505.04 kVA	

TABLE 8.13
MCC Space Factor Estimations

Motor	Size (HP)	Type	Voltage (V)	Full Load Current (A)	NEMA Size	Space Factor
Supply fan	75	FVNR	460	96.0	4	6
Return fan	40	FVNR	460	52.0	3	4
Exhaust fan	15	FVNR	460	21.0	2	2
Chilled water pump	15	FVNR	460	21.0	2	2
Hot water pump	20	FVNR	460	27.0	2	2
Condenser	75	RVNR	460	96.0	4	10
Elevator	25	FVR	460	34.0	2	4
Total						30

TABLE 8.14
MCC Layout

	Module	
1	2	3
Fuse	15 HP FVNR (2)	75 HP RVNR (10)
Spare		
40 HP FVNR (4)	15 HP FVNR (2)	
	20 HP FVNR (2)	
75 HP FVNR (6)	25 HP FVR (4)	

TABLE 8.15
Line Current by MCC Module

Module	Line Current, I_L (A)
1	148
2	103
3	96

TABLE 8.16
Demand Factors for Sizing Feeders and Circuit Breakers for Panels

Load Type	Demand Factor
Lighting	1.25
Receptacles (first 10 kVA)	1.00
Receptacles (remainder)	0.50
Equipment	1.00
Other	1.00

protection devices. The actual loads are used to estimate the voltage drop in various wires.

For the lighting panel and the power panels, the connected loads are calculated based on power ratings of lighting fixtures and receptacles served by the branch circuits. In particular and for each duplex receptacle, 180 VA is considered as the connected power load (NEC, 2014). The connected load for each lighting fixture is estimated based on the type of fixture as noted previously in the load description section. To calculate the demand loads, the demand factors outlined in Table 8.16 are defined based on the NEC requirements (refer Table 220.42 [Lighting] and Table 220.44 [Receptacles]).

For motors (all air conditioning equipment and elevators), the following equation was used to find the demand load (in Amps) and size of the branch circuit feeder:

$$I_L = 1.25\left(I_{\text{largest}}\right) + \sum I_{\text{remainder}} \tag{8.24}$$

The demand factor of 1.25 is applied to the largest motor. Using the loads listed for each motor, the demand load for MCC can be determined:

$$I_L = 1.25(96 \text{ A}) + (21 \text{ A} + 21 \text{ A} + 27 \text{ A} + 52 \text{ A} + 34 \text{ A} + 96 \text{ A}) = 371 \text{ A}$$

The conductors for the main feeder and subfeeders are sized by using Table 6.1. After the wire size is determined, the voltage drop is calculated to ensure it is no more than

TABLE 8.17
Summary of the Calculations Performed for the Subfeeders

Panel	Voltage	Connected Load (VA)	Demand Load (VA)	Wire Size	Length (ft)	Voltage Drop	Conduit
LP-1	480/277	31,620	36,760	#8	45	0.40%	¾″
P-1	208/120	25,496	23,771	#6	20	0.54%	1″
P-2	208/120	25,496	23,771	#6	35	0.95%	1″
P-3	208/120	25,496	23,771	#6	50	0.88%	1″
MCC	480/277	288,482	308,435	500 MCM	10	0.06%	2 ½″

2% of the phase voltage. The following equation is used to calculate the voltage drop for all subfeeders:

$$VD(\%) = \frac{\text{Length(ft)} \times \text{Actual Load(A)} \times \text{Voltage Drop}\left(\dfrac{\text{V}}{1{,}000\ \text{A}-\text{ft}}\right)}{1{,}000 E_{\phi}} \tag{8.25}$$

A summary of the subfeeders' sizing calculations, conductor sizes, and conduit sizes is shown in Table 8.17.

The main feeder is sized based on the kVA rating of the main transformer:

$$I_L = \frac{500{,}000}{\sqrt{3} \times 480} = 601\ \text{A} \tag{8.26}$$

Two parallel main conductors are chosen for the main feeder and are denoted as 2×350 MCM THW (Cu) housed in steel conduit size of 2½″. For the unit substation considered for the building, the length of the main feeder conductors is only 3.70 m (12 ft).

The subfeeder serving the transformer T1 has an amperage load of

$$I_L = \frac{75{,}000}{\sqrt{3} \times 208} = 208\ \text{A} \tag{8.27}$$

The subfeeder is thus sized at 4/0 THW (Cu) with a conduit size of 2.5 in.

8.8.5.4 Protection Devices

The lighting panel and the three power panels are protected by power circuit breakers. The size of each circuit breaker is determined based on the demand load associated with the panel. A summary of the rating for the power circuit breakers is provided in Table 8.18. All power circuit breakers have three poles.

The motor loads are protected with dual-element (time-delay) fuses. For this type of fuse protecting squirrel-cage induction-type motor, the demand factor is 175% (refer to Table 6.19 or NEC Table 430.52). Since the fuses are protecting the MCC

TABLE 8.18
Summary of Power Circuit Breaker Ratings

Panel	Voltage	Demand Load (kVA)	Demand Load (A)	Circuit Breaker
LP-1	480/277	36,760	44	600 AF/50 AT
P-1	208/120	23,591	65	600 AF/100 AT
P-2	208/120	23,591	65	600 AF/100 AT
P-3	208/120	23,771	66	600 AF/100 AT

panel that serves several motors, this demand factor is applied only to the largest motor:

$$I_{\text{fuse}} = DF\left(I_{\text{largest}}\right) + \sum I_{\text{remainder}} \tag{8.28}$$

Thus, the demand load used to size the fuses protecting the entire MCC is

$$I_{\text{fuse}} = 1.75(96\text{ A}) + (21\text{ A} + 21\text{ A} + 27\text{ A} + 52\text{ A} + 34\text{ A} + 96\text{ A}) = 419\text{ A}$$

As noted in Section 6.6.2 and according to NEC 430.52 (NEC, 2014), a protective device (i.e., a fuse in this case) with the nearest standard size that does not exceed this amount shall be selected. Therefore, a 400 A fuse is required to protect the MCC based on the standard ratings provided in Table 5.2. A switch of 600 A is selected. A similar fuse and switch selection is performed as part of the design of each motor branch circuit.

8.8.5.5 Short-Circuit Calculations

The short-circuit calculations are carried out for various points in the building electrical distribution system, as shown in Figure 8.30. Only power panel is shown in Figure 8.30 since the calculations for the other two panels are similar.

The 500 kVA transformer is selected with an impedance of 4%. The utility has a short-circuit power capacity of 100,000 kVA. The per-unit method calculation procedures, outlined in Section 8.2, are considered for all the points of Figure 8.30. The calculation details for the short-circuit currents at each point are provided.

8.8.5.5.1 Short-circuit at point A

$Z_u = P_T/P_{SC,U} = 500\text{ kVA}/100{,}000\text{ kVA} = 0.005$
$Z_{UT}\,(pu) = Z_U + Z_T = 0.045$
$I_{SC,UT}\,(pu) = 1/0.045 = 22.2$
$I_{L,S} = 500\text{ kVA}/(1.73 * 0.48\text{ kV}) = 602\text{A}$
$I_{SC,UT} = 13{,}364\text{ A}$
$I_{sc,A}\,(\text{A}) = I_{SC,UT} + 4 \times I_{L,S} = 13{,}364 + 4 \times 602 = 15{,}772\text{ A}$
$I_{sc,A}\,(pu) = I_{sc,A}\,(\text{A})/I_{L,S} = 15{,}772/602 = 26.2$
$Z_{sc,A}\,(pu) = 1/26.2 = 0.0382$

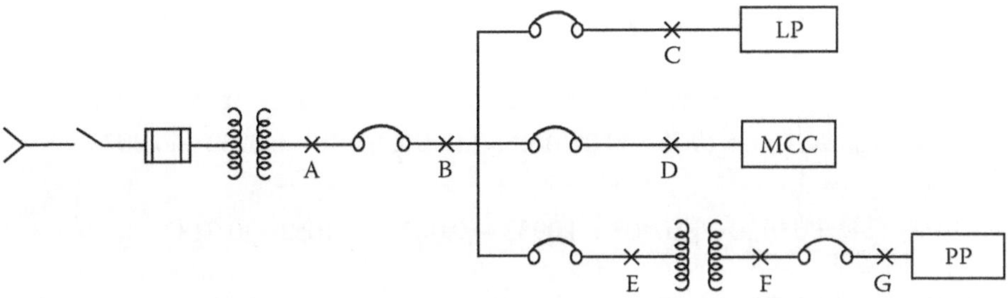

FIGURE 8.30 Locations for short-circuit calculations.

8.8.5.5.2 *Short-Circuit at B*

$$Z_B = Z_A + Z_{\text{Feeder}} \text{ AB}$$

Table 8.1, for Feeder AB, $R = 0.039\ \Omega/1{,}000\,\text{ft}$, $X = 0.050\ \Omega/1{,}000\,\text{ft}$,

Thus, with the 32-ft feeder length for AB:

Z_{Feeder}, AB (Ω) $= 0.00125 + j0.0016$
$Z_{ref,1}$ (Ω) $= 277/602 = 0.46$
Z_{Feeder}, AB (pu) $= 0.00271 + j0.00348$
$Z_{sc,A}\,(pu) = 0.0382 \times (\cos(76o) + j\sin(76o)) = 0.00924 + j0.03707$ (at point A, X/R = 4)
$Z_{sc,B}\,(pu) = 0.00924 + j0.03707 + 0.00271 + j0.00348 = 0.0120 + j0.0405$
$Z_{sc,B}\,(pu)\| = 0.0120 + j0.0405\| = 0.0423$
$I_{sc,B}\,(pu) = 23.64$
$I_{sc,B}\,(A) = 23.64 \times 602 = 14{,}231$ A

8.8.5.5.3 *Short-Circuit at C*

$$Z_C = Z_B + Z_{\text{Feeder}},\ \text{BC}$$

$$Z_B\,(pu) = 0.0120 + j0.0405$$

Based on Table 8.1, for a feeder #8 AWG serving the lighting panel (LP):

$$R = 0.78\ \Omega/1{,}000\,\text{ft and } X = 0.065\ \Omega/1{,}000\,\text{ft}$$

The feeder DC is estimated to be 55 ft long, thus using the $Z_{ref,1}$ =0.46Ω, feeder BC impedance is:

$$Z_{Feeder}, BC\ (pu) = (0.78 + j0.065) \times 55/1{,}000 \times 1/0.46 = 0.0932 + j0.0078$$

$$Z_C\ (pu) = 0.0120 + j0.0403 + 0.0932 + j0.0078 = 0.1052 + j0.0483$$

$$Z_C\ (pu) = 0.116$$

$$I_{sc,C}\ (pu) = 1/0.116 = 8.64$$

$$I_{sc,C}\ (A) = 8.64 \times 602 = 5{,}200\ \text{A}$$

8.8.5.5.4 Short-Circuit at D and E

It is estimated that the short-circuit currents and impedances at both point D and point E are the same as those at point B because the feeders BD and BE are short.

8.8.5.5.5 Short-Circuit at F

$$Z_F = Z_{T2} + Z_{E,S}$$

The impedance for the 480Y/277-208Y/120-V is $Z_{T2} = 5\%$ with $X/R = 1.5$:

$$Z_{T2}\ (pu) = 0.05\ [\cos(a\tan(1.5)) + j\sin(a\tan(1.5))] = 0.0277 + j0.0416$$

Based on the short length of BE feeder:

$$Z_E\ (pu) = Z_B\ (pu) = 0.0120 + j0.0405$$

Thus, since $Z_{ref,1}$ =0.46 Ω, the impedance Z_E in ohms can be calculated:

$$Z_E\ (\Omega) = 0.46 \times (0.0120 + j0.0405) = 0.0055 + j0.01865$$

Thus, the impedance of the transferred point E to secondary of transformer T_2 using:

$$Z_{E,S}\ (\Omega) = Z_E(\Omega)/a_2 = 0.0055 + j0.01865/(2.3)2\ (a = 480/208 = 2.3)$$

$$Z_{E,S}\ (\Omega) = 0.00104 + j0.00353$$

The reference current at the secondary side of T2 is:

$$I_{L,S,2}\,(\text{A}) = 75/(1.73 \times 0.208) = 208\ \text{A}$$

Thus, the reference impedance at the secondary side of T2 is:

$$Z_{ref,2}\,(\Omega) = 120/208 = 0.576$$

The per-unit impedance for $Z_{E,S}$ can be estimated:

$Z_{E,S}\,(\text{pu}) = Z_{E,S}\,(\Omega)/Z_{ref,2}\,(\Omega) = (0.00104 + \text{j}0.00353)/0.576 = 0.0018 + \text{j}0.0061$
$Z_F\,(pu) = Z_{T2}\,(pu) + Z_{E,S}\,(pu) = 0.0277 + \text{j}0.0416 + 0.0018 + \text{j}0.0061 = 0.0295$
$4 + \text{j}0.04772$
$Z_F\,(pu) = 0.05612$
$I_{sc,F}\,(pu) = 1/0.05612 = 17.8$
$I_{sc,F}\,(\text{A}) = I_{sc,F}\,(pu) \times I_{L,S,2}\,(\text{A}) = 17.8 \times 208\ \text{A} = 3710\ \text{A}$

8.8.5.6 Voltage Drop Calculations

For each of the feeders, the voltage drop is calculated as shown here (refer to the section entitled Feeders and Subfeeders). In addition, the voltage drop for each branch circuit is determined based on the location of various loads and the estimation of the length of the branch circuit from the panel. Based on the results of the calculations, few branch circuits have a voltage drop higher than the recommended 3%. To reduce the voltage drop for these wires without relocating any panels, the total load is reduced by shifting the devices to other circuits. Figures 8.31 and 8.32 show the calculated voltage drop for the branch circuits and subfeeders.

8.8.5.7 Unit Substation Design

As discussed in Section 8.5, the unit substation includes the main feeder conductors, the primary fused switch, the primary transformer, the secondary main power circuit breaker, and the power circuit breakers for the subfeeders servicing the various panels. To estimate the size of the primary main feeder conductors of the unit substation, the primary current is first estimated based on the size of the transformer:

$$I_{L,P} = \frac{P_T}{\sqrt{3}E_L} = \frac{500\ \text{kVA}}{(1.732)(13.8\ \text{kV})} = 20.9\ \text{A} \tag{8.29}$$

Increasing this value by 25% gives a line current of 26.1 A. A primary section with 600 A current rating and 15 kV voltage rating is selected with the following physical dimensions: 0.91 m (36 in.) W × 2.28 m (90 in.) H × 1.37 m (54 in.) D. The 500 kVA transformer section has the following dimensions: 2.28 m (90 in.) W × 2.28 m (90 in.)

Panel P-3 Branch Circuit Voltage Drop						
Panel	**Circuit Number**	**Load Name**	**Voltage (V)**	**Wire Size**	**Length**	**Voltage Drop (V)**
P-3	1,3,5	Power	208	3-#8, 1-#8, 1-#8	227′ - 0 21/32″	3
P-3	2,4,6	Power	208	3-#10, 1-#10, 1-#10	184′ - 5 21/32″	4
P-3	7	Refrigerator	120	1-#6, 1-#6, 1-#6	193′ - 3 7/16″	2
P-3	8	Receptacles—Large conference room 1	120	1-#4, 1-#4, 1-#4	244′ - 11 5/16″	2
P-3	9	Receptacles—Kitchen counter	120	1-#12, 1-#12, 1-#12	175′ - 0 5/16″	2
P-3	10	Kitchen dishwasher	120	1-#8, 1-#8, 1-#8	190′ - 1 7/1″	2
P-3	11	Motor—Garbage disposal	120	1-#6, 1-#6, 1-#6	186′ - 6 15/16″	2
P-3	12	Receptacles—Kitchen counter	120	1-#12, 1-#12, 1-#12	181′ - 11 15/16″	2
P-3	14	Hand dryer	120	1-#6, 1-#6, 1-#6	163′ - 7 5/16″	2
P-3	15	Hand dryer	120	1-#6, 1-#6, 1-#6	146′ - 9 15/16″	2
P-3	16	Hand dryer	120	1-#6, 1-#6, 1-#6	146′ - 10 5/16″	2
P-3	18	Hand dryer	120	1-#6, 1-#6, 1-#6	132′ - 0 15/16″	2
P-3	19	Receptacles—Huddle room, electrical room	120	1-#10, 1-#10, 1-#10	82′ - 7 9/16″	2
P-3	21	Receptacles—Large conference room 3	120	1-#8, 1-#8, 1-#8	124′ - 8 13/16″	2
P-3	23	Receptacles—Large conference room 2	120	1-#4, 1-#4, 1-#4	268′ - 6″	2
P-3	24	Receptacles—Restrooms, reception area	120	1-#4, 1-#4, 1-#4	212′ - 1 5/16″	2

FIGURE 8.31 Voltage drop estimations for Panel P-3 branch circuits.

H × 1.37 (54 in.) D. The main secondary section contains a circuit breaker. For the main secondary section, the total load expressed in current is estimated:

$$I_L = \frac{500 \text{ kVA}}{(1.732)(0.480 \text{ kV})} = 601 \text{ A}$$

Then, the size of the power circuit breaker based on the following demand load is

$$I_{CB} = 1.25(601 \text{ A}) = 752 \text{ A}$$

Using the next standard size, a trip rating of 800 A is selected (800AF/800AT) for the main secondary power circuit breaker. Unit 1 can house this power circuit breaker, with a maximum current rating of 3,000 A and a width of 0.76 m (30 in.).

To select the subfeeder sections, the ratings of the protection devices for the MCC as well as the lighting and power panels (refer to Table 8.18) are considered. Since the ratings for all the power circuit breakers are less than 600 A, unit 2 subfeeder section is selected. As noted in Table 8.8, the width of unit 2 is 0.46 m (18 in.).

Panel LP-1 Branch Circuit Voltage Drop						
Panel	Circuit Number	Load Name	Voltage (V)	Wire Size	Length	Voltage Drop (V)
LP-1	1	Level 1—Restrooms/vestibule lighting	277	1-#12, 1-#12, 1-#12	225′ - 10 1/2″	4
LP-1	2	Level 1—Row 1 lighting	277	1-#12, 1-#12, 1-#12	179′ - 9 26/32″	2
LP-1	3	Level 1—Conference room 3 lighting	277	1-#12, 1-#12, 1-#12	213′ - 6 5/8″	2
LP-1	4	Level 1—Huddle/conference room 2 lighting	277	1-#12, 1-#12, 1-#12	121′ - 4 5/8″	1
LP-1	5	Level 1—Reception area lighting	277	1-#12, 1-#12, 1-#12	174′ - 7 1/16″	2
LP-1	6	Level 1—Open area lighting	277	1-#12, 1-#12, 1-#12	199′ - 8 15/16″	2
LP-1	7	Level 1—Kitchen, conference room 1 lighting	277	1-#12, 1-#12, 1-#12	227′ - 10 1/8″	4
LP-1	8	Level 1—Row 4 lighting	277	1-#12, 1-#12, 1-#12	127′ - 0 5/8″	1
LP-1	9	Level 1—Row 2 lighting	277	1-#12, 1-#12, 1-#12	168′ - 3 25/32″	2
LP-1	10	Level 1—Row 5 lighting	277	1-#12, 1-#12, 1-#12	134′ - 4 27/32″	1
LP-1	11	Level 1—Row 3 lighting	277	1-#12, 1-#12, 1-#12	138′ - 2 3/4″	2
LP-1	12	Level 1—Row 6 lighting	277	1-#12, 1-#12, 1-#12	144′ - 4 27/32″	2
LP-1	13	Level 2—Row 1 lighting	277	1-#12, 1-#12, 1-#12	194′ - 0 25/32″	3
LP-1	14	Level 2—Row 2 lighting	277	1-#12, 1-#12, 1-#12	182′ - 3 25/32″	2
LP-1	15	Level 2—Row 3 lighting	277	1-#12, 1-#12, 1-#12	152′ - 2 3/4″	2
LP-1	16	Level 2—Row 6 lighting	277	1-#12, 1-#12, 1-#12	158′ - 4 27/32″	2
LP-1	17	Level 2—Row 4 lighting	277	1-#12, 1-#12, 1-#12	141′ - 0 5/8″	2
LP-1	18	Level 2—Row 5 lighting	277	1-#12, 1-#12, 1-#12	148′ - 4 27/32″	2
LP-1	20	Level 2—Restrooms lighting	277	1-#12, 1-#12, 1-#12	244′ - 10 7/16″	4
LP-1	21	Level 2—Conference rooms/kitchen lighting	277	1-#10, 1-#10, 1-#10	295′ - 11 1/8″	4
LP-1	22	Level 2—Open area lighting	277	1-#12, 1-#12, 1-#12	213′ - 8 15/16″	3
LP-1	23	Level 3—Row 1 lighting	277	1-#12, 1-#12, 1-#12	208′ - 0 25/32″	3
LP-1	24	Level 3—Row 2 lighting	277	1-#12, 1-#12, 1-#12	196′ - 3 25/32″	3
LP-1	25	Level 3—Row 3 lighting	277	1-#12, 1-#12, 1-#12	166′ - 2 3/4″	2
LP-1	26	Level 3—Row 4 lighting	277	1-#12, 1-#12, 1-#12	155′ - 0 5/8″	2
LP-1	27	Level 3—Row 5 lighting	277	1-#12, 1-#12, 1-#12	162′ - 4 27/32″	2
LP-1	28	Level 3—Row 6 lighting	277	1-#12, 1-#12, 1-#12	172′ - 4 27/32″	2
LP-1	29	Level 3—Conference rooms/kitchen lighting	277	1-#10, 1-#10, 1-#10	310′ - 11 1/8″	5
LP-1	30	Level 3—Restrooms lighting	277	1-#12, 1-#12, 1-#12	227′ - 8 15/16″	3
LP-1	33	Level 3—Receptions, huddle, vestibule lighting	277	1-#10, 1-#10, 1-#10	273′ - 10 1/8″	3
LP-1	34	Level 2—Receptions, huddle, vestibule lighting	277	1-#12, 1-#12, 1-#12	240′ - 7 19/32″	5

FIGURE 8.32 Voltage drop estimations for Panel LP branch circuits.

Figure 8.33 shows the layout of the unit substation including its various sections based on the specification outlined.

8.8.5.8 Emergency Systems

As noted in Section 8.6, NEC requires emergency lighting to be connected to either a generator or a UPS (uninterruptable power supply) system or for the fixtures to contain an integral battery back-up (NEC, 2014). In this building, separate emergency fixtures are specified and are operated using battery-powered units in case of any failure. In addition, the NFPA requires that stairs have a minimum of 1 lux (10 fc)

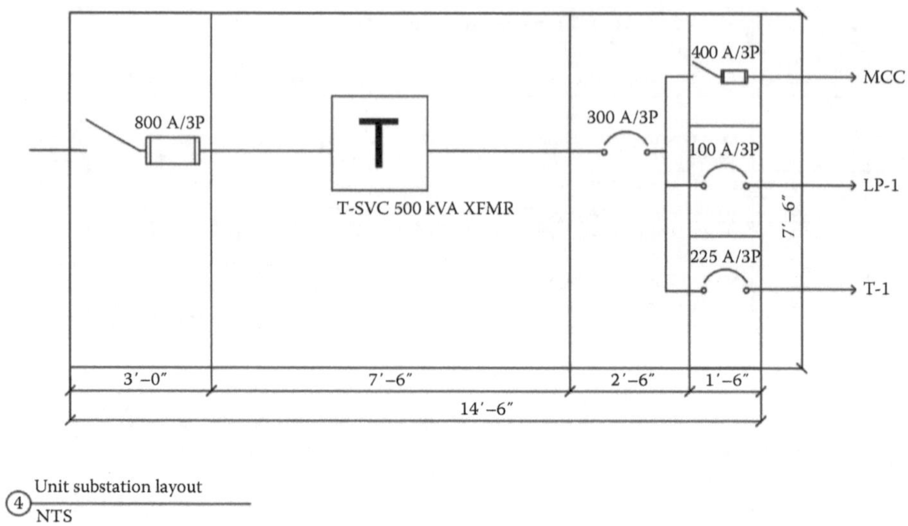

FIGURE 8.33 Layout for the unit substation.

illuminance level along the paths of egress. Figure 8.26 provides the layout of emergency lighting fixtures and exit signs. In particular, the emergency lighting units are labeled at Y1 and exit signs are labels as X1.

8.9 SUMMARY

Common arrangements as well as major components of electrical distribution systems for commercial buildings are described throughout this chapter. Moreover, simplified design approaches to select the adequate ratings for all the components of the distribution systems are presented with some illustrative examples. In particular, two calculation methods are outlined to estimate short-circuit currents at various locations of a typical distribution system. To apply the calculation and design procedures introduced in this chapter, a detailed case study is presented for selecting the ratings for various components of a radial distribution system designed to meet the electrical needs of a three-story office building.

8.10 PROBLEMS

1. Determine the circuit breaker and the feeder size using THW(Cu) for a 480Y/277 V lighting panelboard serving 100-kW load with a power factor of 0.95 for a commercial building.
2. Determine the circuit breaker and the feeder size using THW(Cu) for a 208Y/120 V power panelboard serving 100 receptacles in an office building. The power factor for the load associated with the receptacles can be assumed to be 0.90.
3. A 480Y/277 V lighting panelboard has the following loads:

- Lighting (four circuits at 277 V) with a total load of 16.0 kVA at $pf = 0.95$ (continuous load).
- Appliances (two circuits at 277 V) with a total load of 9.0 kVA at $pf = 0.90$ (noncontinuous load).
- A/C system (one circuit at 480 V) with a total load of 15.00 kVA at $pf = 0.85$ (continuous load).
  a. Determine the size of all the circuit breakers in the panelboard (provide the trip rating).
  b. Show a schedule for the panelboard with the least load unbalance between the three phases. Determine the percent load unbalance (based on total power in each phase) for the proposed schedule.
  c. Select the proper main circuit breaker for the panelboard (provide both the frame size and the trip rating).

4. A 208Y/120 V panelboard for an office building has the following loads:
   - Receptacles with a total load of 50.00 kVA and a power factor of 0.90.
   - Lighting with a total load of 80.00 kVA and a power factor of 0.95.
   - One 20-HP motor with an efficiency of 0.90 and a power factor of 0.92.
   - Heaters with a total load 25.00 kVA and a power factor of 1.00.
     a. Determine the size of the THW-feeder for the panelboard. Size the conduit.
     b. Select the proper circuit breaker for the panelboard. The motor is protected by a circuit breaker with an inverse time-delay element.
     c. Size the equipment grounding conductor (Cu) for the panelboard.
5. A panel for a commercial building is supplied by a 500 ft four-wire 208Y/120-V three-phase feeder. The panel serves three major loads:
   - A lighting system of a total power of 70 kVA with a power factor of 0.8.
   - 500 receptacles connected to loads with an average power factor of 0.90.
   - Heaters rated at 17.5 kVA and a power factor of 1.00.
   - Several motors with a total power rating of 20-kVA and a power factor of 80% with the largest motor rated at 50 kW and a power factor of 0.85.
   - Determine:
     a. The size of the feeder and conduit using THW copper conductors.
     b. The voltage drop in the feeder. (Assume a steel conduit.)
     c. The size of the feeder and conduit that limit the voltage drop to 1%.
6. A panel for an office building is supplied by a 200 ft four-wire and a 480Y/277 V feeder. The loads of the panel include
   - A lighting system of a total power of 10 kVA with a power factor of 0.8.
   - 100 receptacles with an average power factor 0.90.
   - Heaters rated at 7.5 kVA and a power factor of 1.00.
   - Several motors with a total power rating of 40 kVA and a power factor of 80% with the largest motor rated at 5 kW and a power factor of 0.85.

   Determine:
     a. The size of the feeder and conduit using THW copper conductors.
     b. The voltage drop in the feeder expressed in %. (Assume a steel conduit).

   c. The size of the feeder and conduit if the voltage drop is limited to 1%.
   d. Would the size for the feeder found in (c) change if the symmetrical short-circuit current is 20,000-A and a circuit breaker is used to protect the feeder? Show your calculations.

7. A 208Y/120 V panelboard for an office building has the following loads:
   - Receptacles with a total load of 30.00 kVA at $pf = 0.90$
   - Lighting with a total load of 20.00 kVA at $pf = 0.95$
   - One motor with a load of 5 kVA at $pf = 0.92$
   - Heaters with a total load 12.00 kVA at $pf = 1.00$
   - Determine the size of the THW-feeder for the panelboard. Size the conduit.
   - Select a proper circuit breaker for the panelboard. The motor is protected by a circuit breaker with an inverse time-delay element.
   - Size the equipment grounding conductor (Cu) for the panelboard.
8. Consider an MCC with six identical motors, each motor rated at 60 HP, three-phase, 60 Hz, 460 V, 1,760 RPM, a service factor of 1.00, and a power factor of 90%. All the motors are protected by time-delay fuses and have full voltage nonreversing starter type.
   a. Estimate the efficiency of each motor if its line current is 70 A.
   b. Determine the number of poles, the slip (%), and the torque delivered (in lb-ft) by each motor.
   c. Design a branch circuit for one motor. Assume that the motor is out of sight of both the power supply and the controller (the controller is within sight from the source). The nameplate full-load current rating of each motor is 70 A.
   d. Determine the size of the conductor, conduit, time-delay fuse, and switch for the feeder supplying the MCC.
9. Consider an MCC with nine identical motors, each motor rated at 50 HP, three-phase, 60 Hz, 460 V, 1,175 RPM, a service factor of 1.00, and a power factor of 90%. All the motors are protected by time-delay fuses and have full voltage non-reversing starter type.
   a. Design a branch circuit for one motor. Assume that the motor is out of sight of both the power supply and the controller (the controller is within sight from the source). The nameplate full-load current rating of each motor is 70 A.
   b. Determine the size of the conductor, conduit, time-delay fuse, and switch for the feeder supplying the MCC.
10. A MCC has three induction motors each rated at 40 HP and 480 V and four induction motors each rated at 30 HP and 480-V. Each 40-HP motor has an efficiency of 0.92, 1,172 rpm, a power factor of 0.85, a nameplate rating of 45 A, a service factor of 1.15, and a FVNR controller. Each 30-HP motor has an efficiency of 0.90, a power factor of 0.90, and an FVR controller.
    a. Size the conductor (THW, Cu), the fuse, and the overload protection of each 40-HP motor if a time-delay fuse is used. Assume that all the motors are within sight of both the fuse and the controller.

   b. Design and show the layout of the MCC.
   c. Determine the induction motors (by specifying their HP and their number) that need to be replaced with synchronous motors so that the overall power factor of the MCC is as close to unity as possible without changing the total kW load of the MCC nor the HP ratings of the motors. Assume that the power factor for any synchronous motor is 0.90 leading.
11. A MCC has the following loads: (1) four 10-HP, 3-phase motor, 460-V, a power-factor of 0.80, and a FVR controller; (2) two 15-HP, 3-phase motor, 460-V, a power-factor of 0.90, and a FVNR controller; (3) two 20-HP, 3-phase motor, 460-V, a power factor of 0.95, and a FVNR controller; and (4) one 20-kW, 480-V, 3-phase electrical heater.
    - Design and show the layout of the MCC. Allow one space factor for the main feeder. Assume that all control units are the fusible switch type.
    - Determine which motors have to be replaced with an equivalent synchronous motor (i.e., with the same HP, efficiency, and power factor) to maximize the overall power factor of the MCC.
12. A MCC is designed to control the following motors using the type of starters noted: Six 20 HP, FVNR; Four 25 HP, FVNR; Two 25 HP, FVR; One 30 HP, FVNR; One 60 HP, FVNR; and One 75 HP, FVR.
    - The source of supply is 208 V, three-phase, and the motors are rated at 200 V, three-phase. The control units are the circuit breaker type. Determine the overall size and draw a layout of the MCC.
    - Same questions as in (1) except the supply is 480-V and the motors are rated at 460 V.
13. Consider a resistive load with an impedance of $Z = 1.5$ supplied by 120 V. Assume that the system impedance (i.e., wires, protection devices, etc.) is resistive with $Z = 0.1$.
    - Calculate the normal load current of the circuit.
    - Determine the fault current if the load is short-circuited.
    - Provide the per unit values for system impedance and fault current.
    - In reality the system impedance is $Z = 0.08 + j\ 0.06$. What is then the fault current (both in Amps and p.u.).
14. A unit substation has a 4.16 Kv-208Y/120 V, three-phase, 1,000 kVA transformer. The available fault current at the main secondary bus is 50,800 A symmetrical. The unit substation services 10 feeders. The continuous load for each feeder is 400-A with 45,000 A symmetrical fault current.
    - Select the frame size for the main secondary circuit breaker.
    - Select the frame size of the feeder circuit breakers.
    - Design the unit substation including the dimensions for all the sections.
    - Determine the size of the electrical room that should house the unit substation.
15. For this problem, consider the single-line diagram for a building power system shown in Figure 8.34. In particular, the utility system short-circuit kVA capability is 850,000 kVA with $X/R = 6.0$. Table 8.19 shows the impedance and X/R of the transformers.

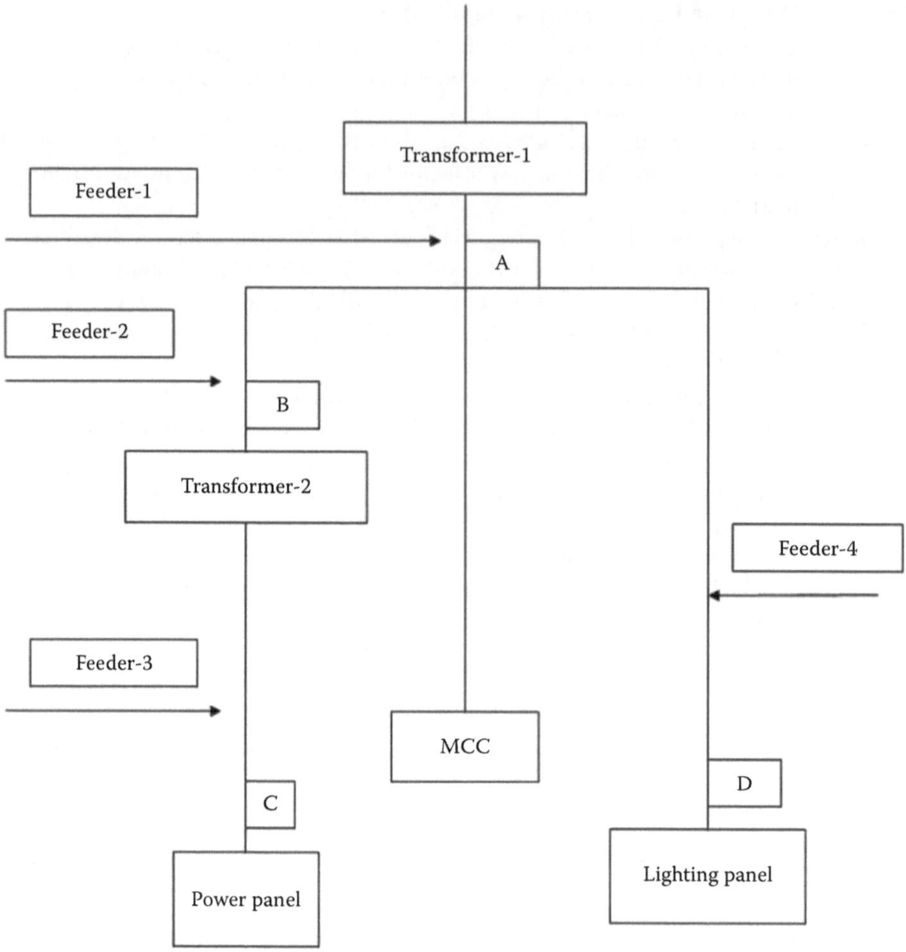

FIGURE 8.34 One-line diagram for a commercial building electrical distribution system.

TABLE 8.19
Specifications of the Transformers and Feeders

Transformer	Rating (kVA)	Z (%)	X/R
Transformer-1: 13.8 kV–480Y/277 V	2,500	2.5%	8.00
Transformer-2: 480Y/277-V–208Y/120-V (TA)	112.5	4.0%	1.50
Feeder	**Conductors**	**Length (ft)**	**Conduit**
Feeder-1	4 [3"C-4#350 MCM THW (Cu)]	20	Steel
Feeder-2	2", 3#2/0 THW (Cu)	30	Steel
Feeder-3	3.5", 4 #500 MCM THW (Cu)	40	PVC
Feeder-4	2.5", 4#4/0 THW (C)	125	PVC

The purpose of this problem is to analyze the short-circuit currents at different locations of the system. Note that roof top A/C unit is basically an MCC panel. For all feeders, steel conduits are specified.

a. Determine the short-circuit current at point A. Note that the main secondary feeder is 20 ft long with four parallel sets of 3 in. C-4#350 MCM THW (Cu).
b. Determine the short-circuit current at point B just before Transformer-2 and point C at the power panel.
c. Determine the short-circuit current at point D at the lighting panel.
d. Based on this calculation, select the proper frame size and the trip rating for the circuit breakers of the main secondary and power panel as well as the lighting panel.

16. Using the radial electrical distribution system of Figure 8.34, determine the unit substation specifications and the electrical room size. The primary feeder is protected by a fused switch.
    a. Determine the dimensions of each section of the unit substation.
    b. Draw an elevation view of the unit substation.
    c. Size the room required to house the unit substation.

9 Economic Analysis of Energy Projects

This chapter provides an overview of common economic analysis methods that can be utilized to determine the cost-effectiveness of a wide range of energy efficiency measures and distributed generation technologies suitable for buildings. The purpose of this chapter is to describe simple yet rigorous evaluation methods, including life cycle cost (LCC), analysis technique, and levelized cost of energy (LCOE) analysis approach, used to make decisions about designing and retrofitting building energy systems. The simple payback period (SPP) method is also presented since it remains widely used in the preliminary design or retrofitting stages of energy-efficient buildings and distributed power generation systems.

9.1 INTRODUCTION

In most building design and retrofit applications, initial investments are required to incorporate and implement energy efficiency measures or renewable energy systems. These initial costs must be generally justified in terms of a reduction in the operating costs (due to energy cost savings). Therefore, most improvements in the efficiency of energy systems have a delayed reward, that is, expenses come at the beginning of a construction or a retrofit project while the benefits are incurred later. For an energy efficiency or renewable energy project to be economically worthwhile, the initial expenses have to be lower than the sum of savings obtained by the reduction in the operating costs over the lifetime of the project.

The lifetime of an energy system project spans typically over several years. Therefore, it is important to properly compare savings and expenditures of various amounts of money over the lifetime of a project. Indeed, an amount of money at the beginning of a year is worth less at the end of the year and has even less buying power at the end of the second year. Consequently, the amounts of money due to expenditures or savings incurred at different times of an energy project cannot be simply added.

In engineering economics, savings and expenditures of amounts of money during a project are typically called *cash flows*. To compare various cash flows over the lifetime of a project, a life cycle cost (LCC) analysis is typically used. In this chapter, basic concepts of engineering economics are described. First, some of the fundamental principles and parameters of economic analysis are presented. In addition, data are provided to help the reader estimate relevant economic parameters. Then, the general procedure of an economic evaluation of any energy project is described. Finally, some of the advantages and disadvantages of various economic analysis methods are discussed.

DOI: 10.1201/9781003276999-9

9.2 BASIC CONCEPTS

Several economic parameters affect a decision between various investment alternatives. To perform a sound economic analysis for energy projects, it is important that the designer or the auditor be (1) familiar with the most important economic parameters and (2) aware of the basic economics concepts. The parameters and the concepts that significantly affect the economic decision-making include

- The time value of money and interest rates, including simple and compounded interest
- Inflation rate and composite interest rate
- Taxes including sales, local, state, and federal tax charges
- Depreciation rate and salvage value

In the following sections, these listed parameters are briefly described to help the reader better understand the LCC analysis procedure to be introduced and applied in this chapter. A more detailed description of these parameters and common engineering economic analysis approaches is provided by Krarti (2010).

9.2.1 Interest Rate

When money is borrowed to cover part or all the initial cost of a project, a fee is charged for the use of this borrowed money. This fee is called *interest* (I) and the amount of money borrowed is called *principal* (P). The amount of the fee depends on the value of the principal and the length of time over which the money is borrowed. The interest charges are typically normalized to be expressed as percentage of the total amount of money borrowed. This percentage is called *the interest rate* (i):

$$i = \frac{I}{P} \tag{9.1}$$

It is clear that an economy with low interest rates encourages money borrowing (to invest in projects or simply to buy goods) while an economy with high interest rates encourages money saving. Therefore, if money has to be borrowed for a project, the interest rate is a good indicator of whether or not the project can be cost-effective.

To calculate the total interest charges over the lifetime of a project, the compounded method is typically considered. In this method, the lifetime, N, is divided into smaller periods, typically called interest periods (such as 1 month or often 1 year). The interest fee is charged at the end of each interest period and is allowed to accumulate from one interest period to the next. The interest charge, I_k, for the total mount, F_k, accumulated after k periods, is estimated as follows:

$$I_k = iF_k \tag{9.2}$$

Therefore, the total payment, F_{k+1}, at the end of period k (and thus at the beginning of period $k+1$) is estimated by

TABLE 9.1
Average Long-Term Interest Rates for Selected Countries

Period/Year	France	Germany	Japan	United States
Period				
1961–1973	6.9	7.2	7.0	5.3
1974–1980	11.2	8.1	8.0	8.6
1981–1990	12.0	7.8	6.5	10.3
1990–1995	8.5	7.5	5.1	7.2
Year				
1995	7.5	6.9	3.4	6.6
2000	5.4	5.3	1.7	6.0
2005	3.4	3.4	1.4	4.8
2010	4.1	4.0	2.0	4.4

Source: OECD, Organization for Economic Cooperation and Development, Economic Statistics, http://www.ocde.org, 2011.

$$F_{k+1} = F_k + I_k = (1+i)F_k \tag{9.3}$$

If the principal is P, which is the total payment at the beginning of first period (i.e., $F_1 = P$), the total amount due at the end of lifetime N can be expressed using Equation 9.4:

$$F = F_{N+1} = (1+i)^N P \tag{9.4}$$

It is clear that the total amount of money, F, increases exponentially with N. The interest charges follow the law of compounded interest. When economic analysis is carried out for energy projects, the interest rate is typically assumed to be constant throughout their lifetime. Therefore, it is common to use an average interest rate when an economic analysis is performed for energy projects including the integration of energy efficiency and renewable energy technologies to new or existing buildings. Table 9.1 provides historical data for the long-term interest rates for selected countries. Example 9.1 illustrates the difference between simple and compound interest rates.

Example 9.1

Problem

A homeowner has $2,500 available and has the option to invest this money in either (1) a bank that has an annual interest rate of 5% or (2) buying a new air conditioning system for his home. If he decides to invest all the money in the bank, how much will the homeowner have after 15 years? Compare this amount if simple interest rather than compounded interest had been paid.

Solution

If the interest is compounded with $P = \$2{,}500$, $N = 15$, and $i = 0.05$, the investment will accumulate to the total amount F:

$$F = \$2{,}500(1 + 0.05)^{15} = \$5{,}197$$

Thus, the homeowner's original investment will have doubled over the 15-year period.

If simple interest had been paid, the total amount that would have accumulated is slightly less:

$$F = \$2{,}500(1 + 0.05 \times 15) = \$4{,}375$$

9.2.2 Inflation Rate

Inflation occurs when the cost of goods and services increases from one period to the next. While the interest rate, i, defines the cost of money, the inflation rate, λ, measures the increase in the cost of goods and services. Therefore, the future cost of a commodity, FC, is higher than the present cost, PC, of the same commodity:

$$FC = PC(1 + \lambda) \tag{9.5}$$

Table 9.2 presents historical data for inflation rates for selected countries.

If the interest and inflation rates are compounded during the same period, a composite interest rate, θ, can be defined to account for the fact that inflation decreases the buying power of money due to increases in the cost of commodities:

$$\theta = \frac{i - \lambda}{1 + \lambda} \tag{9.6}$$

TABLE 9.2
Average Inflation Rates for Selected Countries

Period/Year	France	Germany	Japan	United States
Period				
1971–1980	9.8	5.0	8.8	7.1
1981–1990	6.2	2.5	2.1	4.7
1990–1995	1.9	2.8	1.0	2.6
Year				
1995	1.8	1.2	−0.1	2.8
2000	1.8	1.4	−0.5	3.4
2005	1.9	1.9	−0.6	3.4
2010	0.7	0.4	−1.4	1.0

Source: OECD, Organization for Economic Cooperation and Development, Economic Statistics, http://www.ocde.org, 2011.

It should be mentioned that theoretically the composite interest rate can be negative. In this case, the value of money is reduced with time. Example 9.2 includes the effect of inflation rate in assessing the future value of the investment considered in Example 9.1.

Example 9.2

Problem

Determine the actual value of the \$2,500 investment for the homeowner of Example 9.1 if the economy experiences an annual inflation rate of 3%.

Solution

The composite interest rate can be determined using Equation 9.6:

$$\theta = \frac{i - \lambda}{1 + \lambda} = \frac{0.05 - 0.03}{1 + 0.03} = 0.01942$$

With $P = \$2,500$ and $N = 15$, the investment will accumulate to the total amount F:

$$F = \$2,500(1 + 0.01942)^{15} = \$3,336$$

Thus, the future value of the investment when the inflation rate is accounted for is significantly lower than the value found in Example 9.1.

9.2.3 Tax Rate

In most economies, the interest incurred from an investment is subject to taxation. If this taxation has a rate, t, over a period that coincides with the interest period, then the amount of taxes, T, to be collected from an investment, P, with an interest rate, i, is determined as follows:

$$T = tiP \tag{9.7}$$

Therefore, the composite interest rate defined by Equation 9.6 can be generalized to account for both inflation and tax rates related to present and future values:

$$\theta = \frac{(1 - t)i - \lambda}{1 + \lambda} \tag{9.8}$$

Example 9.3

Problem

If the homeowner is in the 28% tax bracket, determine the actual value of his \$2,500 investment considered in Example 9.1 if the economy experiences an annual inflation rate of 3%.

Solution

The composite interest rate can be determined using Equation 9.8:

$$\theta = \frac{(1-t)i-\lambda}{1+\lambda} = \frac{(1-0.28)\times 0.05-0.03}{1+0.03} = 0.00583$$

With $P = \$2{,}500$ and $N = 15$, the investment will accumulate to the total amount F:

$$F = \$2{,}500(1+0.00583)^{15} = \$2{,}728$$

The inclusion of the tax rate reduces further the future value of the investment compared to the cases considered in Examples 9.1 and 9.2.

9.2.4 Cash Flows

In evaluating energy projects, it is important to account for the total cash receipts and disbursements due to the implementation of an energy efficiency measure (such as the installation of premium efficiency motors) or the integration of renewable energy system (such as the installation of a photovoltaic panel) during the entire lifetime of the project. The difference between the total cash receipts (inflows) and total cash disbursements (outflows) for a given period is called a cash flow.

Over the project lifetime, an accurate accounting of all the cash flows should be performed. For energy projects, the cash flow accounting can be outlined in a tabular format as illustrated in Table 9.3 for the receipts and disbursements attributed to the installation of an energy-efficient motor. In particular, Table 9.3 accounts for the cost related to the initial cost of motor installation (counted as a disbursement for Year 0) and the cost savings due to the higher energy efficiency of the new motor (counted as

TABLE 9.3
Cash Flows for the Installation of a New Motor over a Lifetime of 10 Years

End of Year	Total Cash Receipts ($)	Total Cash Disbursements ($)	Total Cash Flows ($)	Comments
0	0	4,000	−4,000	Installation cost of a new furnace
1	400	0	+400	Net cost savings
2	380	0	+380	Net cost savings
3	360	0	+360	
4	340	0	+340	
5	330	0	+330	
6	320	0	+320	
7	310	0	+310	
8	305	0	+305	
9	300	0	+300	
10	295	0	+295	

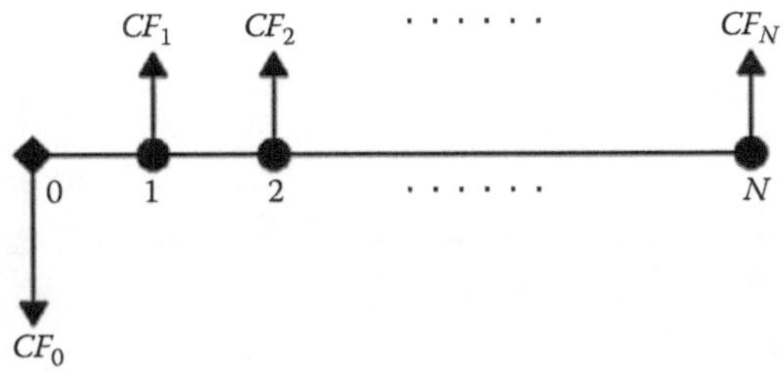

FIGURE 9.1 Typical cash flow diagram for a project.

receipts in Year 1–10). The reduction in the annual receipts is attributed to the aging of the equipment.

Note that the cash flows are positive when they represent inflows (i.e., receipts) and are negative when they are outflows (i.e., disbursements). To better visualize the evolution over time of cash flows, a cash flow diagram can be used as depicted in Figure 9.1. Note that in Figure 9.1, the initial cash flow, $C_0 = -\$4{,}000$ (disbursement), is represented by a downward pointing arrow. Meanwhile, the cash flows occurring later, C_1 through C_N with $N = 10$, are receipts and are represented by upward pointing arrows.

It should be noted again that cash flows cannot be simply added since the value of money changes from one period to the next. In the next section, various factors are defined to correlate cash flows occurring at different periods.

9.3 COMPOUNDING FACTORS

Two types of payment factors are considered in this section. These payment factors are useful in the economic evaluation of various energy projects. Without loss of generality, the interest period is assumed to be 1 year. Moreover, a nominal discount rate, *d*, is used for the analysis described in this chapter. The discount rate is an effective interest rate that includes the effects of several parameters such as inflation and taxation, as shown by Equation 9.8.

9.3.1 Single Payment

In this case, an initial payment is made to implement a project by borrowing an amount of money, *P*. If this sum of money earns interest at a discount rate, *d*, then the value of the payment, *P*, after *N* years is the future amount, *F*. The ratio *F/P* is often called single payment compound-amount factor (*SPCA*). The *SPCA* factor is a function of *d* and *N* and is defined as:

$$SPCA(d,N) = \frac{F}{P} = (1+d)^N \tag{9.9}$$

Using the cash flow diagram of Figure 9.1, the single payment represents the case where $C_0 = P$, $C_1 = \cdots = C_{N-1} = 0$, and $C_N = F$ as illustrated in Figure 9.2.

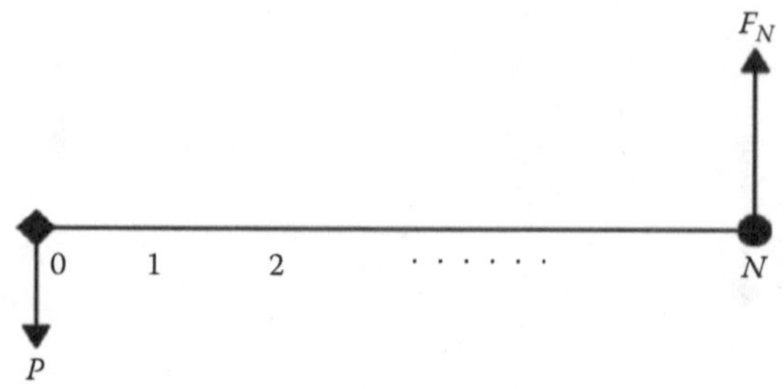

FIGURE 9.2 Cash flow diagram for single payment.

The inverse ratio *P/F* can be used to estimate the value of the cash flow, *P*, needed to attain a given amount of cash flow *F* after *N* years. The ratio *P/F* is called single payment present worth (*SPPW*) factor and is equal to:

$$SPPW(d,N) = \frac{P}{F} = (1+d)^{-N} \tag{9.10}$$

Table 9.4 provides the *SPPW* values for various discount rates, *d*, and life cycles, *N*.

9.3.2 Uniform-Series Payment

In most energy projects, energy cost savings are estimated annually after an initial investment to incorporate or implement one or several sets of energy efficiency measures and/or renewable energy systems. It is hoped that during the lifetime of the project, the sum of all the annual energy cost savings can surpass the initial investment. Consider then an amount of money, *P*, that represents the initial investment, and a receipt of an amount, *A*, that is made each year and represents the cost savings due to the energy project. To simplify the analysis, the amount A is assumed to be constant for all the years during the lifetime of the project. Therefore, the cash flows can be set to $C_0 = P$, $C_1 = \cdots = C_N = A$ as depicted in Figure 9.3.

It can be shown that the ratio *A/P*, called uniform-series capital recovery factor (*USCR*), can be expressed as a function of both *d* and *N*:

$$USCR(d,N) = \frac{A}{P} = \frac{d(1+d)^N}{(1+d)^N - 1} = \frac{d}{1-(1+d)^{-N}} \tag{9.11}$$

The uniform-series present worth factor (*USPW*), needed to calculate the value of *P* knowing the amount *A*, is the ratio *P/A* and can be expressed as follows:

$$USPW(d,N) = \frac{P}{A} = \frac{(1+d)^N - 1}{d(1+d)^N} = \frac{1-(1+d)^{-N}}{d} \tag{9.12}$$

Table 9.5 provides the *USPW* factors for various values of discount rate, *d*, and life cycle, *N*.

Example 9.4 shows the values calculated for the various compound factors using the expressions introduced in this section.

TABLE 9.4
SPPW* Values for Selected Values of Discount Rate, *d*, and Number of Years, *N

SPPW N (Years)	Discount Rate, d (%) 0	1	2	3	4	5	6	7	8	9	10
1	1.000	0.990	0.980	0.971	0.962	0.952	0.943	0.935	0.926	0.917	0.909
2	1.000	0.980	0.961	0.943	0.925	0.907	0.890	0.873	0.857	0.842	0.826
3	1.000	0.971	0.942	0.915	0.889	0.864	0.840	0.816	0.794	0.772	0.751
4	1.000	0.961	0.924	0.888	0.855	0.823	0.792	0.763	0.735	0.708	0.683
5	1.000	0.951	0.906	0.863	0.822	0.784	0.747	0.713	0.681	0.650	0.621
6	1.000	0.942	0.888	0.837	0.790	0.746	0.705	0.666	0.630	0.596	0.564
7	1.000	0.933	0.871	0.813	0.760	0.711	0.665	0.623	0.583	0.547	0.513
8	1.000	0.923	0.853	0.789	0.731	0.677	0.627	0.582	0.540	0.502	0.467
9	1.000	0.914	0.837	0.766	0.703	0.645	0.592	0.544	0.500	0.460	0.424
10	1.000	0.905	0.820	0.744	0.676	0.614	0.558	0.508	0.463	0.422	0.386
11	1.000	0.896	0.804	0.722	0.650	0.585	0.527	0.475	0.429	0.388	0.350
12	1.000	0.887	0.788	0.701	0.625	0.557	0.497	0.444	0.397	0.356	0.319
13	1.000	0.879	0.773	0.681	0.601	0.530	0.469	0.415	0.368	0.326	0.290
14	1.000	0.870	0.758	0.661	0.577	0.505	0.442	0.388	0.340	0.299	0.263
15	1.000	0.861	0.743	0.642	0.555	0.481	0.417	0.362	0.315	0.275	0.239
16	1.000	0.853	0.728	0.623	0.534	0.458	0.394	0.339	0.292	0.252	0.218
17	1.000	0.844	0.714	0.605	0.513	0.436	0.371	0.317	0.270	0.231	0.198
18	1.000	0.836	0.700	0.587	0.494	0.416	0.350	0.296	0.250	0.212	0.180
19	1.000	0.828	0.686	0.570	0.475	0.396	0.331	0.277	0.232	0.194	0.164
20	1.000	0.820	0.673	0.554	0.456	0.377	0.312	0.258	0.215	0.178	0.149
21	1.000	0.811	0.660	0.538	0.439	0.359	0.294	0.242	0.199	0.164	0.135
22	1.000	0.803	0.647	0.522	0.422	0.342	0.278	0.226	0.184	0.150	0.123
23	1.000	0.795	0.634	0.507	0.406	0.326	0.262	0.211	0.170	0.138	0.112
24	1.000	0.788	0.622	0.492	0.390	0.310	0.247	0.197	0.158	0.126	0.102
25	1.000	0.780	0.610	0.478	0.375	0.295	0.233	0.184	0.146	0.116	0.092
26	1.000	0.772	0.598	0.464	0.361	0.281	0.220	0.172	0.135	0.106	0.084
27	1.000	0.764	0.586	0.450	0.347	0.268	0.207	0.161	0.125	0.098	0.076
28	1.000	0.757	0.574	0.437	0.333	0.255	0.196	0.150	0.116	0.090	0.069
29	1.000	0.749	0.563	0.424	0.321	0.243	0.185	0.141	0.107	0.082	0.063
30	1.000	0.742	0.552	0.412	0.308	0.231	0.174	0.131	0.099	0.075	0.057

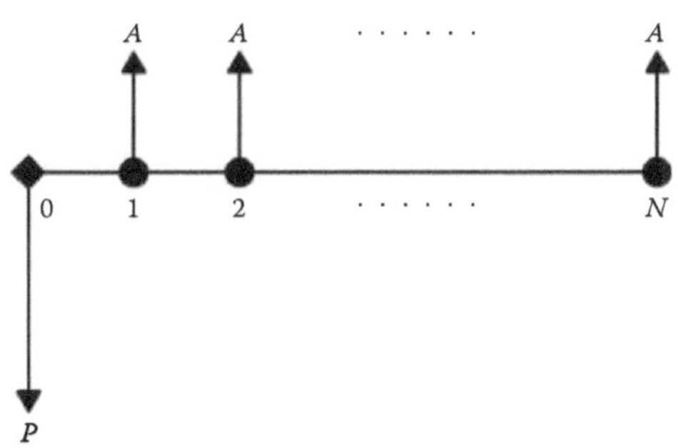

FIGURE 9.3 Cash flow diagram for uniform-series payment.

TABLE 9.5
USPW* Values for Selected Values of Discount Rate, *d*, and Number of Years, *N

USPW *N*	Discount Rate, *d* (%)										
(Years)	0	1	2	3	4	5	6	7	8	9	10
1	1.00	0.99	0.98	0.97	0.96	0.95	0.94	0.93	0.93	0.92	0.91
2	2.00	1.97	1.94	1.91	1.89	1.86	1.83	1.81	1.78	1.76	1.74
3	3.00	2.94	2.88	2.83	2.78	2.72	2.67	2.62	2.58	2.53	2.49
4	4.00	3.90	3.81	3.72	3.63	3.55	3.47	3.39	3.31	3.24	3.17
5	5.00	4.85	4.71	4.58	4.45	4.33	4.21	4.10	3.99	3.89	3.79
6	6.00	5.80	5.60	5.42	5.24	5.08	4.92	4.77	4.62	4.49	4.36
7	7.00	6.73	6.47	6.23	6.00	5.79	5.58	5.39	5.21	5.03	4.87
8	8.00	7.65	7.33	7.02	6.73	6.46	6.21	5.97	5.75	5.53	5.33
9	9.00	8.57	8.16	7.79	7.44	7.11	6.80	6.52	6.25	6.00	5.76
10	10.00	9.47	8.98	8.53	8.11	7.72	7.36	7.02	6.71	6.42	6.14
11	11.00	10.37	9.79	9.25	8.76	8.31	7.89	7.50	7.14	6.81	6.50
12	12.00	11.26	10.58	9.95	9.39	8.86	8.38	7.94	7.54	7.16	6.81
13	13.00	12.13	11.35	10.63	9.99	9.39	8.85	8.36	7.90	7.49	7.10
14	14.00	13.00	12.11	11.30	10.56	9.90	9.29	8.75	8.24	7.79	7.37
15	15.00	13.87	12.85	11.94	11.12	10.38	9.71	9.11	8.56	8.06	7.61
16	16.00	14.72	13.58	12.56	11.65	10.84	10.11	9.45	8.85	8.31	7.82
17	17.00	15.56	14.29	13.17	12.17	11.27	10.48	9.76	9.12	8.54	8.02
18	18.00	16.40	14.99	13.75	12.66	11.69	10.83	10.06	9.37	8.76	8.20
19	19.00	17.23	15.68	14.32	13.13	12.09	11.16	10.34	9.60	8.95	8.36
20	20.00	18.05	16.35	14.88	13.59	12.46	11.47	10.59	9.82	9.13	8.51
21	21.00	18.86	17.01	15.42	14.03	12.82	11.76	10.84	10.02	9.29	8.65
22	22.00	19.66	17.66	15.94	14.45	13.16	12.04	11.06	10.20	9.44	8.77
23	23.00	20.46	18.29	16.44	14.86	13.49	12.30	11.27	10.37	9.58	8.88
24	24.00	21.24	18.91	16.94	15.25	13.80	12.55	11.47	10.53	9.71	8.98
25	25.00	22.02	19.52	17.41	15.62	14.09	12.78	11.65	10.67	9.82	9.08
26	26.00	22.80	20.12	17.88	15.98	14.38	13.00	11.83	10.81	9.93	9.16
27	27.00	23.56	20.71	18.33	16.33	14.64	13.21	11.99	10.94	10.03	9.24
28	28.00	24.32	21.28	18.76	16.66	14.90	13.41	12.14	11.05	10.12	9.31
29	29.00	25.07	21.84	19.19	16.98	15.14	13.59	12.28	11.16	10.20	9.37
30	30.00	25.81	22.40	19.60	17.29	15.37	13.76	12.41	11.26	10.27	9.43

Example 9.4

Problem

Find various compounding factors for $N = 10$ years and $d = 5\%$.

Solution

The values of the compounding factors for $d = 0.05$ and $N = 10$ years are summarized here:

Compounding Factor	Equation Used	Value
SPCA	Equation 9.9	1.629
SPPW	Equation 9.10	0.614
USCR	Equation 9.11	0.129
USPW	Equation 9.12	7.722

9.4 ECONOMIC EVALUATION METHODS

To evaluate the cost-effectiveness of energy projects, several evaluation tools can be considered. The basic concept of all these tools is to compare among the design or retrofit alternatives the net cash flow that can be obtained during the lifetime of the project. As discussed earlier, a simple addition of all the cash flows, such as those represented in Figure 9.1, is not appropriate. However, by using the compound factors introduced in Section 9.3, the conversion of the cash flows from one period to another is feasible. This section provides a brief description of the common economic evaluation methods used in most engineering projects.

9.4.1 Net Present Worth

The basic principle of this method is to evaluate the present worth of the cash flows incurred during the lifetime of the project. Referring to the cash flow diagram of Figure 9.1, the sum of all the present worth of the cash flows can be obtained by using the single payment present worth factor defined in Equation 9.10:

$$NPW = -CF_0 + \sum_{k=1}^{N} CF_k \times SPPW(d,k) \tag{9.13}$$

Note that the initial cash flow is negative (a capital cost for the project) while the cash flows for the other years are generally positive (revenues). In the particular but common case of a project with constant annual revenues (due to energy operating cost savings), $CF_k = A$, the net present worth (NPW) is reduced to

$$NPW = -CF_0 + A \times USPW(d,N) \tag{9.14}$$

For a project to be economically viable, the NPW has to be positive or at worst be equal to zero (i.e., $NPW \geq 0$). Obviously, the higher the value of *NPW*, the more economically sound is the project. The NPW value method is often called the net savings method since the revenues are often due to the energy cost savings from implementing the project.

9.4.2 Rate of Return

In this method, the first step is to determine the specific value of the discount rate, d', that reduces the NPW to zero. This specific discount rate is called the rate of return

(*ROR*) or, in some cases, the internal rate of return (IRR). Depending on the case, the expression of *NPW* provided in Equation 9.13 or 9.14 can be used. For instance, in the general case of Equation 9.13, the rate of return, d', is the solution of the following equation:

$$-CF_0 + \sum_{k=1}^{N} CF_k \times SPPW(d',k) = 0 \tag{9.15}$$

To solve accurately this equation, any numerical method can be used. However, an approximate value of d' can be obtained by trial and error. This approximate value can be determined by finding the two d-values for which the *NPW* is slightly negative and slightly positive, and then interpolate linearly between the two values. It is important to remember that a solution for *ROR* may not exist in some special cases.

Once the rate of return is obtained for a given alternative of the project, the actual market discount rate or the minimum acceptable rate of return is compared to the *ROR* value. If the value of *ROR* is larger (i.e., $d' > d$), the project can be considered as cost-effective.

9.4.3 Benefit–Cost Ratio

The benefit–cost ratio (BCR) method is also called the savings-to-investment ratio (SIR) and provides a measure of the net benefits (or savings) of the project relative to its net cost. The net values of both benefits (B_k) and costs (C_k) are computed relative to a baseline case. The present worth of all the cash flows is typically used in this method. Therefore, the BCR can be computed as follows:

$$BCR = \frac{\sum_{k=0}^{N} B_k \times SPPW(d,k)}{\sum_{k=0}^{N} C_k \times SPPW(d,k)} \tag{9.16}$$

The alternative option for the project is considered economically viable relative to the base case when the BCR is greater than 1.0 (i.e., $BCR > 1.0$).

9.4.4 Cost of Energy

For power-generating projects, the cost of energy is often used to determine their cost-effectiveness compared to the baseline options such as buying power from the utility grid system. Specifically, the cost of energy is the cost of generating 1 kWh by the power-generating facility or technology accounting for its capital costs, financing costs, fuel costs, and fixed and variable operating and maintenance (O&M) costs over an assumed lifetime. Typically, the term levelized cost of energy (LCOE) is utilized when the present worth of all costs is considered. The specific calculations for LCOE depend on the generating technology, but a general method is provided by Equation 9.17:

$$LCOE = \frac{\sum_{k=0}^{N} C_k \times SPPW(d,k)}{\sum_{k=0}^{N} E_k} \tag{9.17}$$

where

C_k are the total costs incurrent in year k and include investment expenditures as well as O&M and fuel costs

E_k is the total electrical energy produced at year k

For a generating project or a technology to be economically competitive, LCOE should be lower than the baseline (typically from the grid) prices. Chapter 14 provides a detailed economic analysis of various distributed generation technologies using specific case studies.

9.4.5 Discounted and Simple Payback Periods

In this evaluation method, the period Y (expressed typically in years) required to recover an initial investment is determined. Using the cash flow diagram of Figure 9.1, the value of Y is the solution of the following equation:

$$CF_0 = \sum_{k=1}^{Y} CF_k \times SPPW(d,k) \tag{9.18}$$

If the payback period Y is less than the lifetime of the project N (i.e., $Y<N$), then the project is economically viable. The value of Y obtained using Equation 9.18 is typically called discounted payback period (*DPP*) since it includes the value of money.

In the vast majority of applications, the time value of money is neglected in the payback period analysis method. In this case, Y is called simple payback period (*SPP*) and is the solution of the following equation:

$$CF_0 = \sum_{k=1}^{Y} CF_k \tag{9.19}$$

In the case where the annual net savings are constant ($CF_k = A$), the SPP can be easily calculated as the ratio of the initial investment over the annual net savings:

$$Y = \frac{CF_0}{A} \tag{9.20}$$

The values for the SPP are shorter than for the discounted payback periods since the undiscounted net savings are greater than their discounted counterparts. Therefore, acceptable values for SPPs are typically and significantly shorter than the lifetime of the project.

9.4.6 Summary of Economic Analysis Methods

Table 9.6 summarizes the basic characteristics of the economic analysis methods used to evaluate single alternatives of an energy efficiency project. Chapter 10 presents several examples to assess the cost-effectiveness of energy efficiency measures suitable for building electrical systems.

TABLE 9.6
Summary of the Basic Criteria for the Various Economic Analysis Methods

Evaluation Method	Equation	Criterion
Net present worth (*NPW*)	$NPW = -CF_0 + \sum_{k=1}^{N} \times CF_K \times SPPW(d,k)$	$NPW > 0$
Rate of return (*ROR*)	$-CF_0 + \sum_{k=1}^{N} \times CF_K \times SPPW(d',k)$	$d' > d$
Benefit–cost ratio (*BCR*)	$BCR = \frac{\sum_{k=0}^{N} \times B_K \times SPPW(d,k)}{\sum_{k=0}^{N} \times C_K \times SPPW(d,k)}$	$BCR > 1$
Discounted payback period (*DPP*)	$CF_0 = \sum_{k=1}^{N} \times CF_K \times SPPW(d,k)$	$Y < N$
Simple payback period (*SPP*)	$Y = \frac{CF_0}{A}$	$Y \ll N$

It is important to note that the economic evaluation methods described in this section indicate whether or not a single alternative of a design and a retrofit project is cost-effective. However, these methods cannot be used or relied on to compare and rank various alternatives for a given energy efficiency project. The LCC analysis method is more appropriate and should be considered to make a sound decision.

9.5 LIFE CYCLE COST ANALYSIS METHOD

The LCC analysis method is the most commonly accepted method to assess the economic benefits of energy efficiency or distributed generation projects over their lifetime. Typically, the method is used to evaluate and provide priority for at least two alternatives of a given project (for instance, evaluate and decide between two alternatives for the same project: install an energy-efficient transformer or consider premium efficiency motors). The basic procedure of the LCC method is relatively simple since it seeks to determine the total cost incurred by various alternatives over the lifetime of the project. For all project alternative options including the baseline case, the total costs are estimated over the project lifetime including installation, operation, replacement, and maintenance costs. The cost is commonly determined using one of two approaches: the present worth or the annualized cost estimate. Then, the project alternative with the lowest total cost (or *LCC*) is typically selected.

Using the cash flow diagram of Figure 9.1, the *LCC* amount for each alternative can be computed by projecting all the costs (including costs of acquisition, installation, maintenance, and operation of the energy systems related to the project implementation) to

1. One single present value amount that can be computed as follows:

$$LCC = \sum_{k=0}^{N} CF_k \times SPPW(d,k) \tag{9.21}$$

This is the most commonly used approach in calculating *LCC* in energy projects.

Or

2. Multiple annualized costs over the lifetime of the project:

$$LCC_a = USCR(d,N)\left[\sum_{k=1}^{N} CF_k \times SPPW(d,k)\right] \tag{9.22}$$

Note that the two approaches for calculating the *LCC* values are equivalent.

In several energy efficiency projects, the annual cash flow remains the same after the initial investment. In these cases, *LCC* can be estimated based on the initial cost, *IC*, and the annual cost, *AC*, as follows:

$$LCC = IC + USPW(d,N)AC \tag{9.23}$$

Example 9.5 illustrates the application of the *LCC* analysis method to select the best design option for a transformer to serve a building's electrical system.

Example 9.5

Problem

An owner has asked the design team to provide a recommendation on the most economical among three design options for an electrical transformer. The three options are briefly described here.

a. Use a premium energy efficiency transformer with an efficiency of 99.7%. The additional cost of this transformer is $10,000 relative to a standard transformer.
b. Use a medium energy-efficiency transformer with an efficiency of 99.5%. The incremental cost of this transformer relative to a standard transformer is $4,000.
c. Use a standard energy-efficiency transformer with an efficiency of 99.3%.

Determine the best cost-effective option for the building owner. Assume that the life time of the building electrical system project is 30 years and the discount rate is 5%. The building has a peak load of 750 kVA with annual average load factor of 0.876 and power factor of 0.90. Use all the LCC analysis method to determine the best option assuming an electricity cost of $0.12/kWh.

Solution

The total cost of operating the transformer is considered for the three options. In this analysis, the salvage value of the transformer and the maintenance fee is assumed to be the same for all three design options and thus are not considered. To present the calculations for LCC analysis, it is recommended to present the results in a tabular format and proceed as shown here:

Cost Item	Option A	Option B	Option C
	Initial investment		
(a) Incremental cost ($)	10,000	4,000	0
	Annual operating costs:		
(b) Electricity use (kWh/year)	2,965,396	2,971,357	2,977,341
(c) Fuel cost ($) [$0.12×(b)]	355,848	356,363	357,281
	USPW factor		
d=5%, N=30, Equation 9.12	15.37	15.37	15.37
Present worth ($)			
(a)+USPW ∗ (c)	54,802,449	5,485,244	5,492,284

Therefore, the LCC for option A is the lowest. Thus, it is recommended for the building owner to use a premium efficiency transformer.

This conclusion is different from that obtained by using the simple payback analysis. (Indeed, the payback period for option A—relative to the base case C—is *SPP*(A)=($10,000)/($1,433/year)=7.0 years, while for option B, *SPP*(B)=($4,000)/($718/year)=5.6 years).

9.6 GENERAL PROCEDURE FOR AN ECONOMIC EVALUATION

It is important to remember that the recommendations for energy efficiency and renewable energy projects should be based on an economically sound analysis. In particular, the designer should ask several questions before making the final recommendations such as the following:

- Will project savings exceed additional costs?
- Which design solution will be most cost-effective?
- What design option will minimize overall building operating costs?
- Which combination of interrelated project options will maximize net savings?
- What priority should be given to various project alternatives if the owner has limited capital funding options?

As discussed earlier, the best suitable economic assessment method is the LCC method described in Section 9.5. Before the application of the LCC method, several data are needed to perform an appropriate and meaningful economic analysis. To help the designer collect the required information and apply the LCC method, the following systematic approach is proposed:

1. Define the problem that the proposed project is attempting to address, and state the main objective of the project. For instance, for building a constant volume heating and cooling system, the project is to use instead of variable air volume fans connected to motors with variable speed drives (VSDs) to condition all the spaces in order to reduce the overall building energy use and cost.

2. Identify the constraints related to the implementation of the project. These constraints can vary in nature and include financial limitations or space requirements. For instance, the new motors require more space to install VSDs and associated controllers.
3. Identify technically sound strategies and alternatives to meet the objective of the project. For instance, three options can be considered for the motors and VSDs: (a) standard efficiency motors and VSDs, (b) premium efficiency motors and standard VSDs, and (c) premium efficiency motors and VSDs.
4. Select a method of economic evaluation suitable for all the project alternatives. When several alternatives including the base case are considered, the LCC method is preferred to make definitive and sound decisions for energy projects. For a preliminary economic analysis mentioned earlier, the SPP method can be used. As mentioned earlier, the payback period method is not accurate and should be used with care to make decisions.
5. Compile data and establish assumptions. The data include discount rates, energy costs, installation costs, operating costs, and maintenance costs. Some of these data are difficult to acquire and some assumptions or estimations are needed. For instance, an average discount rate over the life cycle of the project may be assumed based on any available historical data.
6. Calculate indicators of economic performance. These indicators depend on the economic evaluation method selected. The indicators can be based on a wide range of analysis methods such as the SPP, NPW, IRR, LCOE, and LCC.
7. Evaluate the economic indicators obtained for various alternatives. This evaluation can be performed by simply comparing the values of LCC obtained for various alternatives.
8. Perform a detailed sensitivity analysis to assess the impact of variations in some of the unknown parameters on the results obtained from the economic analysis. Indeed and since the economic evaluation performed in Step 6 is typically based on some assumed values (for instance, the annual discount rate), it is important to determine whether or not the results of the evaluation performed in Step 7 depend on these assumptions. For this purpose, the economic evaluation is repeated for all alternatives using different but plausible assumptions.
9. Take into account unqualified effects. Some of the alternatives may have some effects that cannot be accounted properly in the economic analysis but may be a determining factor in decision-making. For instance, environmental impacts (emission of pollutants) can be important to disqualify otherwise economically sound alternatives.
10. Make specific recommendations. The final selection will be based on the findings of the three previous steps (i.e., Steps 7–9). Typically, the alternative with the lowest LCC value would be recommended.

Once the project is found to be cost-effective, it is important to decide on the financing options to actually carry out the project and implement the selected alternatives.

9.7 ELECTRICITY RATES

One of the most important factors that impact the cost-effectiveness of any energy efficiency or renewable energy projects is the cost of electricity purchased from the grid. In this section, some of the common electricity rates are presented. The rates depend mostly on the generation, transmission, and distribution systems utilized by utilities to serve their customers. Typically, most electrical utilities are faced with the following cost items:

- *Generation plant*: The cost of operating the power plant to generate electricity represents typically the highest cost category. Indeed, the power generation plants have to meet several regulatory and safety requirements. For instance, several plants are required to meet strict pollution standards, especially if they operate in highly populated areas. Meeting these regulations can increase significantly the generating costs.
- *Transmission/distribution systems*: To deliver the electricity from the generation plant, where it is produced, to areas where it is utilized, transmission lines, substations, and distribution networks have to be used. The cost of the transmission/distribution systems depends on the distances to be covered as well as transformers, capacitors, and meters to be used. Moreover, the delivery energy losses can be a significant part of the transmission/distribution costs.
- *Fuel costs*: The electricity is generated using a primary fuel source depending on the power plant. The fuel cost can be small as in the case of hydroelectric plants or significant as in conventional fuel oil or coal power plants. The cost of fuel may fluctuate depending on the world markets.
- *Administrative costs*: The salaries of management, technical, and office staff as well as insurance and maintenance costs for the power plant equipment are part of the administrative costs. These administrative costs can be large, especially for nuclear power plants.
- Other factors that affect the cost of electricity include the generating capacity of the utility, and the demand/supply conditions at a given time (i.e., on-peak and off-peak periods).
- Utilities allocate the cost of electricity differently depending on the type of customers by offering various rate schedules. Three customer types are generally considered by utilities: residential, commercial, and industrial customers. Each utility may offer several rate structures for each customer type. It is therefore important that the auditor know various rate structures that can be offered to the audited facility.

Utilities can tailor their rates to the customer's needs for electricity using several methods. Some of the common rate structures used by U.S. utilities are summarized here:

- Block pricing rates
- Seasonal pricing rates
- Time of use rates
- Real-time pricing rates

In addition to these rates, utilities provide some riders and discounts to their customers. A rider may modify the structure of a rate based on specific qualifications of the customer. For instance, utilities can change summer energy charge of their residential customers using an air conditioning rider. Moreover, utilities can reduce energy and/or demand charges using discount provisions. For instance, several utilities offer voltage discount when the customer is willing to receive voltages higher than the standard voltage level. To benefit from this discount, the customer may have to install and maintain a properly sized transformer. More detailed discussion on each of the commonly available rates is provided in the following sections with some examples to illustrate how a designer or an energy auditor can apply the utility rate provisions to check the calculation of utility bills. First, common features of all electric utility rates are discussed in detail.

9.7.1 Common Features of Utility Rates

There are several utility rate features and concepts that a designer or an energy auditor should be familiar with to be able to interpret and analyze correctly the utility billing procedure. Some of these concepts are described in the following sections.

9.7.1.1 Billing Demand

The demand that is billed by the utility is referred to as the billing demand. The billing demand is often determined from the peak demand obtained for 1 month (or any billing cycle). The peak demand, also known as actual demand, is defined as the maximum demand or maximum average measured demand in any 15-minute period in the billing cycle. To better understand the concept of billing demand, consider two different monthly load profiles: a rugged profile A and a flat profile B as illustrated in Figure 9.4. It is further assumed that the average demand for profile A coincides with that of profile B (and thus the total energy use in kWh for both profiles A and B is the same). It is not equitable for the utility to bill the same charges for the two profiles. Indeed, profile A requires that the utility supplies higher demand and thus increases its generation capacity only for a short period of time. Profile B is ideal for the utility since it does not change over time. Therefore, some utilities charge their customers for the peak demand incurred during the billing period. This demand charge may serve as an incentive to the customers to reduce or "shave" their peak demand. For some rates, the billing

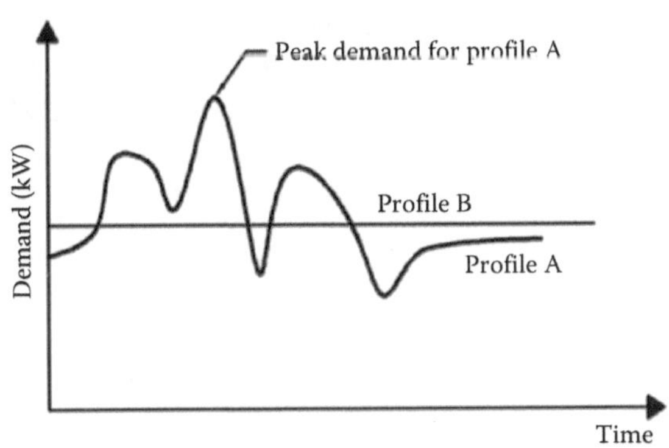

FIGURE 9.4 Peak demand for two different electric load profiles.

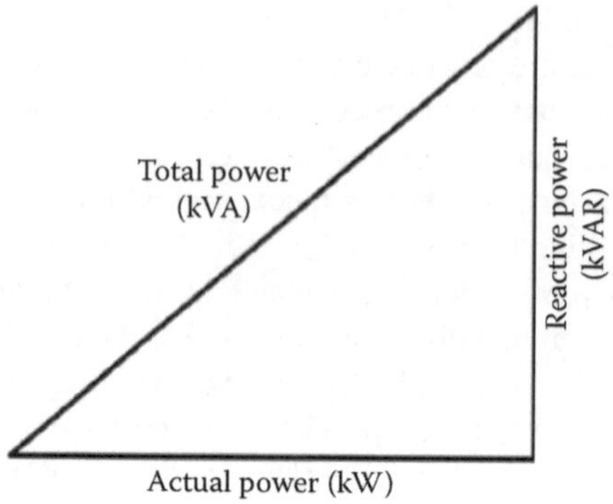

FIGURE 9.5 Power triangle diagram.

demand is determined based on the utility specifications and may be different from the maximum demand actually measured during a billing period. For instance, ratchet and power factor clauses can change the determination of the billing demand. Moreover, a minimum billing demand can be specified in a contractual agreement between the utility and the customer. This minimum demand is often called the contract demand.

9.7.1.2 Power Factor Clause

The power factor is defined as the ratio of actual power used by the consumer (expressed in kW) to the total power supplied by the utility (expressed in kVA). Figure 9.5 shows the power triangle to illustrate the concept of power factor. The reader is referred to Chapter 2 for more details on the concept of power factor. For the same actual power consumed by two customers but with different power factor values, the utility has to supply higher total power to the customer with lower power factor. To penalize customers for low power factor (generally lower than 0.85), some utilities use a power factor clause to increase the billing demand charges or to impose new charges (on total power demand or reactive power demand). For instance, the billing demand can be increased whenever the measured power factor of the customer is below a reference value or base power factor:

$$\text{Billed demand} = \text{Actual demand}\left(\frac{pf_{\text{base}}}{pf_{\text{actual}}}\right) \tag{9.24}$$

Example 9.6 illustrates how the power factor clause may affect the utility bill.

Example 9.6

Problem

A utility has the following monthly billing structure for their commercial customers:

- Customer charge: \$450/month
- Demand charge: \$20/kW
- Energy charge: \$0.03/kWh

A power factor clause states that the billing demand, upon which the demand charge is based, shall be the actual demand for the month corrected for the power factor by multiplying the measured actual demand by 80 and dividing the product by the actual average power factor expressed in percent. The power factor clause applies whenever the average monthly power factor is less than 80%.

i. Calculate the utility bill for a facility with the following energy use characteristics during a specific month:
   a. Actual demand: 300 kW
   b. Energy consumption: 50,000 kWh
   c. Average monthly power factor: 60%
ii. Determine the cost savings achieved for the month if the facility power factor is improved to be always above 80%.

Solution

i. First, the billing demand for the month is determined using the power factor clause since the average power factor is below 80%:

$$\text{Billed demand} = \text{Actual demand} \times \frac{80}{pf} = 300 \times \frac{80}{60} = 400 \text{ kW}$$

   Then, the monthly bill can be calculated using the utility rate structure:
   a. Customer charge = $450.00
   b. Demand charge = 400 kW × $20/kW = $8,000.00
   c. Energy charge = 50,000 kWh × $0.03/Wh = $1,500.00
   Total monthly charges = (a) + b + (c) = $9,950.00
   The actual bill may include other components such as fuel adjustment cost and sales taxes.
ii. The cost savings from improved power factor are due to a reduction in the billing demand. Indeed, the billing demand for the considered month will be the actual demand (i.e., 300 kW). Thus, a reduction of 100 kW in the billing demand can be achieved resulting in a reduction in demand charges of:

$$\text{Cost savings} = 100\text{kW} \times \$20 \text{ kW} = \$2{,}000 \text{ month}$$

The reader is referred to Chapter 2 for a more detailed description of power factor improvement measures.

9.7.1.3 Ratchet Clause

Typically, the utility charges are billed monthly. In particular, the demand charges are based on monthly peak demand. However, when the peak demand for 1 month is significantly higher than for the other months (such as the case for buildings with high cooling loads in the summer months), the utility has to supply the required peak demand and thus operate additional generators only for 1 or 2 months. For the rest of the year, the utility will have to maintain these additional generators. To recover some of this maintenance cost and to encourage demand shaving, some utilities use a ratchet clause in the determination of the billed demand. For instance, the billed demand for any given month is a fraction of the highest maximum demand of the previous 6 months (or 12 months) or the actual demand incurred in the month. Example 9.7 illustrates the calculation procedure of the utility bill with a ratchet clause.

Example 9.7

Problem

The utility has added a ratchet clause to the rate structure described in Example 9.6. Specifically, the ratchet clause states that no billing demand shall be considered as less than the 70% of the highest on-peak season maximum demand corrected for the power factor previously determined during the 12 months ending with the current month.

Calculate the utility bill for the facility considered in Example 9.6 taking into account the ratchet clause. Assume that the previous highest demand (during the last 12 months) is 700 kW.

Solution

First, the minimum billing demand determined by the ratchet clause is calculated:

$$\text{Minimum billing demand}(\text{ratchet clause}) = 700 \text{ kW} \times 0.70 = 490 \text{ kW}$$

Since the billing demand based on the power factor clause was found to be 400 kW (see Example 9.6), the billing demand for the month is

$$\text{Billed demand} = \max(490, 400 \text{ kW}) = 490 \text{ kW}$$

Then, the monthly bill can be calculated using the utility rate structure:

a. Customer charge = $ 450.00
b. Demand charge = 490 kW × $20/kW = $9,800.00
c. Energy charge = 50,000 kWh × $0.03/Wh = $1,500.00

Total monthly charges = (a) + b + (c) = $11,750.00

Thus, the ratchet clause increases the utility bill for the month.

9.7.1.4 Fuel Cost Adjustment

Most utilities have to purchase primary energy sources (fuel oil, natural gas, and coal) to generate electricity. Since the cost of these commodities change over time, the utilities impose an adjustment to their energy charges to account for any cost variation of their primary energy sources. Generally, utilities provide in the description of their rate structure a formula that they use to calculate the fuel cost adjustment. In addition to fuel cost adjustment, utilities may levy taxes and surcharges to recover imposts required from them by federal or state governments or agencies. In some cases, the fuel cost adjustment can be a significant proportion of the utility bill. Example 9.8 illustrates the effect of both the fuel cost adjustment and sales tax on a monthly utility bill.

Example 9.8

Problem

Calculate the utility bill for the facility considered in Example 9.7 taking into account both a fuel cost adjustment of 0.015/kWh and a sales tax of 7%.

Solution

The fuel cost adjustment should be applied to the energy use, while the sales tax should be applied to the total cost. Thus, the monthly bill for the industrial facility considered in Example 9.2 becomes

a. Customer charge = \$450.00
b. Demand charge = 490 kW × \$20/kW = \$9,800.00
c. Energy charge = 50,000 kWh × [\$0.03/kWh + \$0.015/kW] = \$2,250.00

Total monthly charges (before sales tax) = (a) + (b) + (c) = \$12,500.00

Total monthly charges (after sales tax) = \$12,500.00 × 1.07 = \$13,375.00

The cumulative effect of both fuel cost adjustment and sales tax increased the utility bill for the month by [\$13,375 − \$11,750 = \$1,625].

9.7.1.5 Service Level

Utilities typically offer several rate structures for a given customer depending on the type of service. For instance, utilities may have different rates depending on the voltage level provided to the customers. The higher the delivery voltage level, the cheaper the energy rate. In particular, utilities offer a reduced rate for demand and/or energy charges to customers that own their service transformers. This type of rate is generally described under various clauses of a utility bill, including customer-owned transformer, voltage level, or service type. The owner has to determine the cost-effectiveness of owning a transformer and thus receive a higher voltage level by comparing the utility bill savings against the cost of purchasing/leasing and maintaining the transformer. Typically, transformer energy losses are charged to the owner. In addition, the owner should either have a standby transformer or make an alternate arrangement (such as a standby utility service) in case of breakdown of the service transformer. Example 9.9 presents an estimation of savings in utility bills due to owning a service transformer.

Example 9.9

Problem

The facility of Example 9.6 has an option of owning and operating its own service transformer with the advantage of a reduced rate structure as described here:

- Customer charge: \$650/month
- Demand charge: \$15/kW
- Energy charge: \$0.025/kWh

Calculate the reduction in the utility bill during the month for which the energy use characteristics are given in Example 9.6. In the savings calculation, do not consider the power factor clause, ratchet clause, sales tax, and fuel cost adjustment.

Solution

First, the utility bill is determined in the case the facility owns its service transformer:

a. Customer charge = $650.00
b. Demand charge = 400 kW × $15/kW = $6,000.00
c. Energy charge = 50,000 kWh × $0.25/Wh = $1,250.00

Total monthly charges = (a) + (b) + (c) = $7,900.00

Thus, the savings in the utility bill for the month are

$$\text{Utility bill savings} = \$9,950 - \$7,900 = \$2,050$$

A LCC-based economic analysis may be needed to determine the cost-effectiveness of replacing or owning transformers.

9.7.2 Block Pricing Rates

In these rates, the energy price depends on the rate of electricity consumption using either inverted blocks or descending blocks. An inverted block pricing rate structure increases the energy price as the consumption increases. On the other hand, a descending (also referred to as a declining) block rate structure reduces the price as the energy consumption increases. Typically, the rate is referred to as "flat" rate when the energy price does not vary with the consumption level. Table 9.7 outlines a block pricing rate for energy charges. Example 9.10 illustrates the calculation details of utility bills for a residential building subject to a block pricing rate.

TABLE 9.7
An Electric Utility Rate with No-Demand Charges for Residential General Service (Utility Rate A)

Billing Item	Winter (Nov–Apr)	Summer (May–Oct)
Customer charge ($)	7.50	7.50
Minimum charge ($)	7.50	7.50
Fuel cost adjustment	0.00	0.00
Tax rate (%)	6.544	6.544
No. of energy blocks	1	3
Block 1 energy size (kWh)	0	400
Block 2 energy size (kWh)	0	400
Block 3 energy size (kWh)	0	800
Block 1 energy charge ($/kWh)	0.088	0.0874
Block 2 energy charge ($/kWh)	0.000	0.1209
Block 3 energy charge ($/kWh)	0.000	0.1406

Based on a comprehensive survey of existing rate structures offered by U.S. utilities (GRI, 1993), block pricing rate structures are commonly used. Indeed, almost 60% of the surveyed electric utilities offer descending or inverted block pricing structure for the energy charges. For the residential customers, a combination of descending, flat, and inverted rate structures is used throughout the United States. However, descending energy rate structures are used almost exclusively for commercial and industrial customers. Some electric utilities offer block pricing rates for demand as well as energy charges.

Example 9.10

Problem

Using utility structure A of Table 9.7, calculate the utility bill for a residence during a summer month when the energy use is 800 kWh.

Solution

Considering all the charges imposed by utility rate *A* (see Table 9.7), the monthly bill can be calculated as follows:

a. Customer charge = \$7.50
b. Energy charge = 400 kW × \$0.0874/kWh + 400 kWh × \$0.1209/kWh = \$83.32
c. Fuel cost adjustment = \$0.00
d. Taxes = 6.544% [(a) + (b) + (c)] = \$5.94

Total monthly charges = (a) + (b) + (c) + (d) = \$96.76

Thus, the average cost of electricity for the month is \$96.76/800 kWh = \$0.12095/kWh.

9.7.3 Seasonal Pricing Rates

Some electric utilities offer seasonal rate structures to reflect the monthly variations in their generation capacity and energy cost differences. Generally, the utilities that provide seasonal rate structures use different energy and/or demand charges during winter and summer months. The summer charges are typically higher than the winter charges for most electric utilities due to higher energy consumption attributed to the cooling of buildings. Table 9.8 shows an example of seasonal pricing rates while Example 9.11 illustrates the calculation procedure for monthly utility bills.

Based on a survey conducted by GRI (1993), more than 55% of U.S. electric utilities offer seasonal pricing rates for residential customers. Only 5% of electric utilities had residential rates where the winter rate is actually higher than the summer rate. These utilities are located in the Northeast and the West regions of the United States. The same survey reveals that over 42% of U.S. utilities use seasonal pricing rates for commercial and industrial customers. Only 7% of the utilities surveyed offer rates with higher winter prices for their commercial and industrial customers.

TABLE 9.8
An Electric Utility Rate with Demand Charges for Commercial General Service (Utility Rate B)

Billing Item	Winter (Oct–May)	Summer (Jun–Sep)
Customer charge	0.00	0.00
Minimum charge ($)	25.00	25.0
Fuel cost adjustment	0.01605	0.01377
Tax rate	0.00	0.00
No. of energy blocks	3	3
Block 1 energy size (kWh)	40,000	40,000
Block 2 energy size (kWh)	60,000	60,000
Block 3 energy size (kWh)	>100,000	>100,000
Block 1 energy charge ($/kWh)	0.059	0.065
Block 2 energy charge ($/kWh)	0.042	0.047
Block 3 energy charge ($/kWh)	0.039	0.042
No. of demand blocks	2	2
Block 1 demand size (kW)	50	50
Block 2 demand size (kW)	>50	>50
Block 1 demand charge ($/kW)	12.39	13.71
Block 2 demand charge ($/kW)	11.29	12.52
Reactive demand charge ($/kVAR)	0.20	0.20

Example 9.11

Problem

Using utility structure B of Table 9.8, calculate the utility bill for a commercial facility during a winter month when the energy use is 70,000 kWh, the billing demand is 400 kW, and the average reactive demand is 150 kVAR.

Solution

Accounting for all the charges considered by utility rate B (see Table 9.8), the electric energy bill for the winter month can be calculated as follows:

a. Customer charge = $0.00
b. Energy charge = 40,000 Kw × $0.059/kWh + 30,000 kWh × $0.042/kWh = $3,620
c. Demand charge = 50 kW × $12.39/kW + 350 kW × $11.29/kW = $4,571.00
d. Fuel cost adjustment = 70,000 kWh × $0.01605 = $1,123.50
e. Reactive demand charge = 150 kVAR × $0.20/kVAR = $30.00
f. Taxes = $0.00

Total monthly charges = (a) + (b) + (c) + (d) + (e) + (f) = $9,344.50

The average cost of electricity for the month is then $9,344.50/70,000 kWh or $0.1335/kWh.

9.7.4 Time-of-Use Rates

The time-of-use (TOU) rates are time-differentiated rates with the cost of electricity varying during specific times of the day and/or the year. The TOU rates, which appeared first during the 1940s, set "on-peak" and "off-peak" periods with different energy and/or demand charges. Generally, the on-peak periods occur during daytime hours and have a higher cost of energy and demand than the off-peak periods which occur during the nighttime. Table 9.9 presents a time-of-use rate structure for residential customers. Table 9.10 compares the TOU rates for commercial customers for various U.S. utilities. Example 9.12 illustrates the calculation approach of a utility bill for the TOU rate presented in Table 9.9.

Example 9.12

Problem

Using utility structure C of Table 9.9, calculate the utility bill for a residence during a summer month when the energy use is 2,000 kWh with 55% of the energy consumption occurring during the on-peak period.

Solution

With all the charges considered by utility rate *C* (see Table 9.9), the electric energy bill for the winter month can be calculated as follows:

a. Customer charge = \$9.85
b. Energy charge = 2,000 kW × [0.55 × \$0.1500/kWh + 0.45 × \$0.335/kWh] = \$195.15
c. Taxes = 3% × [(a) + (b)] = \$6.15

TABLE 9.9
An Electric Time-of-Use Rate for Residential Service (Utility Rate C)

Billing Item	Winter (Oct–May)	Summer (Jun–Sep)
Customer charge (\$)	9.85	9.85
Minimum charge (\$)	9.85	9.85
Tax rate (%)	3.00	3.00
Energy charge (\$/kWh)		
On-peak hours	0.1412	0.1500
Off-peak hours	0.0335	0.0335

Notes: On-peak hours are defined as 6:00 a.m.–1:00 p.m. and 4:00 p.m.–9:00 p.m. (Monday–Friday) during October–March. 10:00 a.m.–9:00 p.m. (Monday–Friday) during April–September. Off-peak hours are the remaining hours, including holidays.

TABLE 9.10
Comparison of TOU Rates Offered by Some U.S. Utilities for Commercial Customers

	Summer					
	Peak		Partial Peak		Off Peak	
Utility	Period	$/kWh	Period	$/kWh	Period	$/kWh
A	Jun–Sept	$0.04	Jun–Sept	$0.04	Jun–Sept	$0.04
B	May–Oct, M–F, 12–6 p.m.	$0.14	May–Oct, M–F, 8 a.m.–12 p.m. and 6–9 p.m.	$0.10	May–Oct, M–F, 9 p.m.–8 a.m. and all day, Sat and Sun	$0.07
C	Jun–Sept	$0.08	Jun–Sept	$0.08	Jun–Sept	$0.08
D	5 p.m.–9 p.m., M–F	$0.17	7 a.m.–5 p.m., M–F and 7 a.m.–9 p.m. Sat and Sun	$0.17	9 p.m.–7 a.m. daily	$0.16
E	Jun–Sept, M–F, 2 p.m.–7 p.m.	$0.10	Jun–Sept, M–F, 12 p.m.–2 p.m. and 7 p.m.–9 p.m.	$0.05	Jun–Sept, M–F, 9–12 p.m. and all day, Sat and Sun	$0.02

	Winter					
	Peak		Peak		Peak	
Utility	Period	Period	Period	Period	Period	Period
A	Oct–May	Oct–May	Oct–May	Oct–May	Oct–May	Oct–May
B	NA	NA	NA	NA	NA	NA
C	Oct–May	Oct–May	Oct–May	Oct–May	Oct–May	Oct–May
D	5–9 p.m., M–F	5–9 p.m., M–F	5–9 p.m., M–F	5–9 p.m., M–F	5–9 p.m., M–F	5–9 p.m., M–F
E	Oct–May	Oct–May	Oct–May	Oct–May	Oct–May	Oct–May

Total monthly charges = (a) + (b) + (c) = $211.15

The average cost of electricity for the month is then $211.15/2,000 kWh or $0.1056/kWh.

Utilities offer other rates to encourage customers to either reduce energy consumption or generate electricity at certain periods of a day or a season. In addition to real-time-pricing (RTP) rates that are discussed in detail in the following section, utilities have proposed the following rates.

9.7.4.1 The End-Use Rates

To encourage customers to install and operate specific energy-consuming equipment, some U.S. utilities offer end-use rates. With these rates, the utilities can impose operation periods and/or efficiency standards for selected and predefined equipment. For instance, the air conditioning rate allows electric utilities to interrupt service or cycle off the air conditioning equipment during specific times. Typically, the end-use rates require separate metering of the equipment.

9.7.4.2 Specialty Rates

The specialty rates are provided by utilities for specific purposes such as energy conservation and dispatchable customer generation. Energy conservation rates are offered by a limited number of U.S. utilities to foster the use of energy-efficient equipment and/or high standards of building materials. Dispatchable customer generation rates are provided to customers that have standby generators on their premises. In exchange for a reduced rate or a credit, the customers are requested to operate the generators whenever the utility needs additional generating capacity.

9.7.4.3 Financial Incentive Rates

Financial incentive rates encompass economic development rates, displacement rates, and surplus power rates. The economic development rates are typically offered to encourage new customers to locate or existing customers to expand in specific areas that need to be economically revitalized. The displacement rates are offered to customers that are capable of generating electricity to entice them to use utility-provided electricity. Finally, the surplus power rates are highly reduced energy rates that are offered to large commercial and industrial customers when the utility has an excess in electric capacity.

9.7.4.4 Non-Firm Rates

The non-firm rates include interruptible rates, standby rates, and load management rates. Interruptible rates are offered to customers that can reduce or even eliminate (interrupt) their electricity needs from the utility. The electricity pricing rates depend on several factors, such as the capacity that can be interrupted, the length of interruption, and the notification before interruption. Standby rates are intended for customers that require utility-provided electricity on an intermittent basis since they are capable of generating most of their electricity needs (by using, for instance, cogeneration systems, as discussed in Chapter 13). The standby rates can be offered using three options:

1. Maintenance rates when the customers need electricity from the utility during predefined downtime periods of the generating equipment to perform routine maintenance
2. Supplementary rates for customers that use regularly more electricity than they are capable of generating
3. Backup rates intended for customers that prefer to have backup power from the utility in case of unexpected events (i.e., outages in the customers' generators)

Load-management rates are offered by utilities to control the usage of specific equipment such as space conditioning systems during peak periods. Generally, the load management rates are offered in combination with TOU rates.

9.7.4.5 Energy Purchase Rates

The energy purchase rates, also known as buy-back rates, are offered by utilities that want to purchase specific levels of energy or generate capacity from customers. The customers are nonutility electricity generators that qualify under the requirements of Public Utility Regulatory Policies Act (PURPA) such as cogeneration facilities and independent power producers. More detailed discussion on the generation and cogeneration systems is provided in Chapter 13.

9.7.5 Real-Time Pricing Rates

Based on a survey performed by Mont and Turner (1999), most of the existing RTP rate structures in the United States remain restricted to a selected number of customers with large power demands (varying from 250 kW to 10 MW depending on the utility). The available RTP rate structures can be classified into four categories in the survey, as described briefly here.

9.7.5.1 Category 1: Base Bill and Incremental Energy Charge Rates

Under these rates, the customers receive firm hourly electricity prices for the next day before a warning time. Some utilities update their prices daily (day-ahead-pricing) while other utilities provide hourly prices only once a week (week-ahead-pricing). Typically, the category 1 RTP rates include the following charges:

- Base bill charge so that the utility recovers its revenue requirements. This charge depends on the customer baseline load (CBL) which defines the customer's typical energy use (kWh) for each hour of the year and demand values (kW) for each month.
- Incremental energy charge (or credit) reflects the cost of the energy used by the customer above (or below) its CBL profile.

In this category of RTP, an adequate estimation of the CBL profile is an important factor to determine the final energy bill. Therefore, the customer should negotiate for the most favorable CBL values. In general, the customer would benefit from smaller CBL values since the hourly prices of the RTP rate typically fall below those of the standard rate. Moreover, flexible customers can change their electrical load profile to avoid high costs and even obtain a credit for usage below their CBL.

9.7.5.2 Category 2: Total Energy Charge Rates

The hourly electricity prices for this category are applied to the total energy consumption. In addition, the utility imposes an additional charge to recover its revenue requirements. Some utilities set a monthly fixed charge based on the customer's average apparent demand (kVA) and actual demand (kW) determined using the last 12 months load profile. Other utilities charge for actual demand usually ratcheted for the last 12-month period.

The category 2 RTP rates are not suitable for inflexible customers that cannot change their load profile since they cannot protect themselves against higher electricity prices.

9.7.5.3 Category 3: Day-Type Rates

These rates are similar to the category 2 RTP rates but with several predefined day types (such as off-peak day, normal day, etc.) associated with different hourly firm electricity prices. These hourly prices are based on the total energy consumption with possibly a charge for actual demand. The utility informs the customer of the applicable day-type rate one day in advance. Again these rates are suitable only for flexible customers that can vary their energy use profile to accommodate the changes in the electricity price.

9.7.5.4 Category 4: Index-Type Rates

The electricity prices for these rates are determined based on the financial market electricity indexes (in Dow Jones) or the trading prices for electricity future contracts (in New York Mercantile Exchange). Generally, the customers have to forecast the electricity prices (using various tools and sources) since the utility does not provide the prices in advance.

9.7.5.5 Case Study of RTP Rates

In this section, a detailed description is presented to calculate a monthly utility bill using RTP rate structures. The RTP rates used in this section are based on the day-ahead-pricing (DAP) rates proposed by a U.S. utility. Typically, there are two DAP types of rates offered under either curtailment or noncurtailment services. Customers with curtailment service are allowed to buy electricity (referred to as "buy-through power" or "emergency power") above the contracted customer base load (CBL) profile during selected hours (or a curtailment period). The total cost of the buy-through power during a curtailment period is known as "buy-through period." Typically, the utility notifies in advance the availability and the hourly prices of emergency power. Table 9.11 shows the hourly prices for DAP rates offered by a U.S. utility for both noncurtailment and curtailment days (Krarti, 2010). The curtailment period is set between 3:00 p.m. and 8:00 p.m. with a contracted demand of 500 kW. To illustrate the calculation procedure of utility bills using the RTP rates compared to those based on the conventional rate structures, Example 9.13 is presented using DAP rates. Cost savings in the utility bills based on RTP rate structures can be achieved by shifting the load during the curtailment hours to noncurtailment hours. Example 9.14 illustrates the magnitude of savings achieved by a customer under DAP rates.

Example 9.13

Problem

Calculate the utility bills for a customer service under two rates:

i. Conventional rate with the following charges:
   a. Customer charge: $150.00
   b. Energy charge: $0.0264/kWh
   c. Demand charge: $13.1/kW
ii. RTP rate structure with the hourly prices given in Table 9.11. In addition, a customer charge (referred to as an administrative charge) of $300 is incurred monthly by the customer under the RTP rate.

TABLE 9.11
Day-Ahead-Price Rates for Typical Curtailment and Noncurtailment Days

Hours	Noncurtailment Rate Price ($/kWh)	Curtailment Rate Price ($/kWh)
0:00–1:00	0.01673	0.01673
1:00–2:00	0.01683	0.01683
2:00–3:00	0.01730	0.01730
3:00–4:00	0.01780	0.01780
4:00–5:00	0.01900	0.01900
5:00–6:00	0.02130	0.02130
6:00–7:00	0.02730	0.02730
7:00–8:00	0.02890	0.02890
8:00–9:00	0.02932	0.02932
9:00–10:00	0.03213	0.03213
10:00–11:00	0.03575	0.03575
11:00–12:00	0.03423	0.03423
12:00–13:00	0.03456	0.03456
13:00–14:00	0.03677	0.03677
14:00–15:00	0.03977	0.03977
15:00–16:00	0.04254	1.06350
16:00–17:00	0.04354	1.08950
17:00–18:00	0.03595	0.89875
18:00–19:00	0.02587	0.64675
19:00–20:00	0.01950	0.48750
20:00–21:00	0.01523	0.01523
21:00–22:00	0.01532	0.01532
22:00–23:00	0.01723	0.01723
23:00–24:00	0.01722	0.01722

The hourly energy use and the contracted CBL profiles (for the RTP rate) are summarized in Table 9.12 for typical week and weekend days.

Assume that the month has 21 weekdays (including two curtailment days) and nine weekends. In the utility bill calculations, do not account for any other charges such as taxes and credits.

Solution

i. First, the utility bill is determined using the conventional rate structure based on the actual energy use (see Table 9.12):
   a. Customer charge = $150.00
   b. Energy charge = $0.0264 kWh × [21 × 42,461/kWh + 9 × 10,510/kWh] = $26,037.55
   c. Demand charge = $131.1/kW × 3,015 kW = $39,496.50

   Total monthly charges = (a) + (b) + (c) = $65,684.05

   The average cost of electricity for the month based on the conventional rate structure is then $65,684.05/[21 × 42,461 kWh + 9 × 10,510 kWh] or $0.06660/kWh.

TABLE 9.12
Actual and CBL Hourly Energy Use Profiles for the Industrial Facility Assumed in Example 9.13

Hours	Typical Weekday Actual Energy Use (kWh)	Typical Weekend Day Actual Energy Use (kWh)	Noncurtailment Typical Day CBL Energy Use (kWh)	Curtailment Typical Day CBL Energy Use (kWh)
0:00–1:00	345	250	220	220
1:00–2:00	455	270	230	230
2:00–3:00	676	305	325	325
3:00–4:00	785	320	360	360
4:00–5:00	980	360	455	455
5:00–6:00	1,200	405	645	645
6:00–7:00	1,800	460	1,715	1,715
7:00–8:00	2,905	510	2,450	2,450
8:00–9:00	2,825	525	2,545	2,545
9:00–10:00	2,800	520	2,480	2,480
10:00–11:00	2,845	535	2,560	2,560
11:00–12:00	3,015	550	2,765	2,765
12:00–13:00	2,500	565	2,745	2,745
13:00–14:00	2,765	560	2,730	2,730
14:00–15:00	2,890	535	2,650	2,650
15:00–16:00	2,815	525	2,590	500
16:00–17:00	2,825	530	2,520	500
17:00–18:00	2,780	510	2,370	500
18:00–19:00	1,755	495	2,175	500
19:00–20:00	1,100	480	2,080	500
20:00–21:00	845	395	1,945	1,945
21:00–22:00	635	345	1,680	1,680
22:00–23:00	535	290	1,230	1,230
23:00–24:00	385	270	575	575
Total	42,461	10,510	42,040	32,805

ii. The utility bill based on the RTP rate is described here:
   a. Customer charges = \$300.00
   b. Energy charge for week days (non-curtailment days) (see Table 9.13) = \$19 × 55.99
   c. Energy charge for week days (curtailment days) (see Table 9.14) = \$2 × 8,196.79
   d. Standard charge based on the CBL profile (see note below)
      – Customer charge = \$150.00
      – Energy charge = \$0.0264/kWh × 958,960 (see note) = \$26,037.55
      – Demand charge = \$13.1/kW × 2,765 kW = \$36,221.50

Total monthly charge = i. + ii. + iii. = \$62,409.05
Total monthly charge = (a) + (b) + (c) + (d) = \$80,166.44

TABLE 9.13
RTP Hourly Energy Cost During a Typical Non-Curtailment Day for the Industrial Facility of Example 9.13

Hour	Typical Weekday Actual Energy Use (kWh)	Non-Curtailment Typical Day CBL Energy Use (kWh)	Non-Curtailment Typical Day Hourly Energy Price ($/kWh)	Non-Curtailment Typical Day Hourly Energy Cost ($)
0:00–1:00	345	220	0.01673	2.09
1:00–2:00	455	230	0.01683	3.79
2:00–3:00	676	325	0.01730	6.07
3:00–4:00	785	360	0.01780	7.57
4:00–5:00	980	455	0.01900	9.98
5:00–6:00	1,200	645	0.02130	11.82
6:00–7:00	1,800	1,715	0.02730	2.32
7:00–8:00	2,905	2,450	0.02890	13.15
8:00–9:00	2,825	2,545	0.02932	8.21
9:00–10:00	2,800	2,480	0.03213	10.28
10:00 –11:00	2,845	2,560	0.03575	10.19
11:00–12:00	3,015	2,765	0.03423	8.56
12:00–13:00	2,500	2,745	0.03456	-8.47
13:00–14:00	2,765	2,730	0.03677	1.29
14:00–15:00	2,890	2,650	0.03977	9.54
15:00–16:00	2,815	2,590	0.04254	9.57
16:00–17:00	2,825	2,520	0.04354	13.28
17:00–18:00	2,780	2,370	0.03595	14.74
18:00–19:00	1,755	2,175	0.02587	-10.87
19:00–20:00	1,100	2,080	0.01950	-19.11
20:00–21:00	845	1,945	0.01523	-16.75
21:00–22:00	635	1,680	0.01532	-16.01
22:00–23:00	535	1,230	0.01723	-11.97
23:00–24:00	385	575	0.01722	-3.27
Total	42,461	42,040		55.99

The average cost of electricity for the month based on the RTP rate structure is then $80,166.44/[21 × 42,461 kWh + 9 × 10,510 kWh] = $0.08128/kWh.

Therefore, the conventional rate structure is more cost-effective for the industrial facility considered in this example (at least during the month for which the utility bills were calculated).

NOTE: The total monthly energy use based on the CBL profiles is calculated as detailed here:

$$\text{kWh(CBL)} = 19 \text{ days } \times 42{,}040 \text{ kWh / day} + 2 \text{ days } \times 32{,}805 \text{ kWh / day}$$

$$+ 9 \text{ days } \times 10{,}510 \text{ kWh / day}$$

$$\text{kWh(CBL)} = 968{,}960 \text{ kWh}$$

TABLE 9.14
RTP Hourly Energy Cost during a Typical Curtailment Day for the Industrial Facility of Example 9.13

Hours	Typical Weekday Actual Energy Use (kWh)	Curtailment Typical Day CBL Energy Use (kWh)	Curtailment Typical Day Hourly Energy Price ($/kWh)	Curtailment Typical Day Hourly Energy Cost ($)
0:00–1:00	345	220	0.01673	2.09
1:00–2:00	455	230	0.01683	3.79
2:00–3:00	676	325	0.01730	6.07
3:00–4:00	785	360	0.01780	7.57
4:00–5:00	980	455	0.01900	9.98
5:00–6:00	1,200	645	0.02130	11.82
6:00–7:00	1,800	1,715	0.02730	2.32
7:00–8:00	2,905	2,450	0.02890	13.15
8:00–9:00	2,825	2,545	0.02932	8.21
9:00–10:00	2,800	2,480	0.03213	10.28
10:00–11:00	2,845	2,560	0.03575	10.19
11:00–12:00	3,015	2,765	0.03423	8.56
12:00–13:00	2,500	2,745	0.03456	–8.47
13:00–14:00	2,765	2,730	0.03677	1.29
14:00–15:00	2,890	2,650	0.03977	9.54
15:00–16:00	2,815	500	1.06350	2,462.00
16:00–17:00	2,825	500	1.08950	2,533.09
17:00–18:00	2,780	500	0.89875	2,049.15
18:00–19:00	1,755	500	0.64675	811.67
19:00–20:00	1,100	500	0.48750	292.50
20:00–21:00	845	1,945	0.01523	–16.75
21:00–22:00	635	1,680	0.01532	–16.01
22:00–23:00	535	1,230	0.01723	–11.97
23:00–24:00	385	575	0.01722	–3.27
Total	42,461	32,805		8,196.79

Example 9.14

Problem

Calculate the cost savings in electric utility bills for an industrial customer serviced by the RTP rate used in Example 9.13 if the excess electrical load experienced during the curtailment hours (between 3 p.m. and 8 p.m.) is shifted uniformly to the next 4 h (between 8 p.m. and midnight). With this load shifting measure, the demand during the curtailment hours would be 500 kW and the total daily energy use for typical days would remain the same, as shown in Table 9.12.

Solution

a. The excess load during the curtailment hours is first determined as follows:

$$\begin{aligned}\text{Excess load} &= 2{,}815\ \text{kWh} + 2{,}825\ \text{kWh} + 1{,}755\ \text{kWh} + 1{,}100\ \text{kWh} - 4 \times 500\ \text{kWh} \\ &= 6{,}495\ \text{kWh}\end{aligned}$$

b. For each hour during the next 4h (i.e., between 8 p.m. and midnight), the additional load will be 6,495 kWh/4 = 1,623.75 kWh. The additional energy cost during the noncurtailment hours associated to this shift is then

$$\begin{aligned}\text{Energy cost increase} &= 1{,}623.75\ \text{kWh}\big[(0.01523 + 0.01532 \\ &\quad + 0.01723 + 0.01722)\$/\text{kWh}\big] \\ &= \$105.54\end{aligned}$$

c. The shift of electrical load during the curtailment hours has the effect of eliminating any hourly energy cost during the curtailment hours from 3 p.m. to 7 p.m., as shown in Table 9.14.

That is, $2,533.09 + $2,049.15 + $811.67 + $292.50 = $5,686.41.

d. Since all the other charges remain unchanged, the reduction in cost savings is simply the energy charge avoided during the curtailment hours minus any energy cost increase during noncurtailment hours, or

$$\begin{aligned}\text{Utility cost savings} &= \left[\$5{,}686.41/\text{day} - 105.54/\text{day}\right] \times 2\ \text{days} \\ &= \$11{,}161.74\ \text{for the month}\end{aligned}$$

9.8 SUMMARY

This chapter describes a general but a systematic approach to perform the economic evaluation of various alternatives of energy projects including the integration of energy efficiency and renewable energy technologies in designing and retrofitting buildings. In particular, several analysis methods described in this chapter (such as the NPW, the benefit–cost ratio, the rate of return (*ROR*), and the discounted payback period) are suitable for evaluating single alternatives but may not be used to rank among the alternatives. The LCC analysis method is more suitable for evaluating several alternatives and should be used to select the most economical option among several possible alternatives for an energy efficiency or distributed generation project. Under certain market conditions, the simple payback method can lead to erroneous conclusions and thus should be used only to provide an indication of the cost-effectiveness of an energy project. For sound economic analysis, the simple payback analysis method should not be used.

In addition, the energy price rate structures proposed by U.S. utilities are outlined in this chapter with a special emphasis on electricity pricing features. Several calculation examples are presented to illustrate the effects of various components of rate structures on monthly utility bills.

9.9 PROBLEMS

1. A homeowner is considering adding a sunspace to his home. The project is completed during 4 years and requires $ 20,000 to be spent the first year, $15,000 the second year, $10,000 the third year, and $5,000 the fourth year. Determine how much money the homeowner needs to set aside to pay for the sunspace addition project:
   a. If the interest rate is 4% per year, compounded annually.
   b. If the interest rate is 8% per year, compounded annually.
2. You want to buy a new home at a cost of $250,000. Determine how much money you should pay every year for 20 years to pay for the house based on equal annual payments:
   a. If the interest rate is 5% per year, compounded annually.
   b. If the interest rate is 10% per year, compounded annually.
3. An energy audit indicates that a home consumption is 20,000 kWh/year. By using Energy Star appliances and CFLs, a savings of 20% can be achieved for the electrical energy use. The cost for the home retrofit is about $15,000. Assuming that the average energy charge is $0.09/kWh, is the expenditure justified based on a minimum rate of return of 18% before taxes. Assume a 20-year life cycle and use the present worth, annual cost, and rate of return methods.
4. Determine the cost-effectiveness using NPW method for investing $10,000 to add variable speed drives to 100-HP motor with an annual load factor of 70% and an efficiency of 0.90. For the analysis, assume that the electricity price is $0.15/kWh, the life cycle is 30 years and the discount rate is 5%.
5. Determine the breakeven incremental cost for replacing an existing 1,000-kVA transformer with an annual load factor of 85% and an energy efficiency of 0.98 with a premium transformer with an energy efficiency of 0.99. For the analysis, assume that the electricity price is $0.12/kWh, the life cycle if 40 years and the discount rate is 5%. Consider two options for replacing the existing transformer: (1) transformer is at the end of its useful life, and (2) transformer is still operational.
6. An electrical energy audit indicates that lighting consumes 5,000 kWh/year. By using more efficient lighting fixtures, electrical energy savings of 15% can be achieved. The additional cost for these energy-efficient lighting fixtures is about $900. Assume that the average energy charge is $0.08/kWh, is the expenditure justified based on a minimum rate of return (i.e., a discount rate) of 7.0%, and a 10-year life cycle. Use the present worth, annual cost, and rate of return methods.
7. An air conditioner (AC) consumes 7,500 kWh annually with an overall efficiency of $COP=3.0$. If this AC is replaced by a more energy-efficient AC ($COP=3.7$) at a cost of $3,500.
   a. Determine the SPP of the chiller replacement.

b. If the expected life of the old chiller is 15 years, is it cost-effective to replace the chiller in and economy with a discount rate of 7%?

8. Two air conditioner systems are proposed to cool a house. Each AC has a rated capacity of 5 tons and is expected to operate 500 full-load equivalent hours per year. Air conditioner A has a standard efficiency with a COP of 2.8 and costs $1,600 while AC B is more efficient with a COP of 3.5 at a cost of $2,750. The electricity cost is estimated to be $0.09/kWh. For a lifetime of 20 years and a discount rate of 5%,
   a. Estimate the payback time for using air conditioner B instead of A.
   b. Calculate the LCCs for both air conditioners. Conclude.
   c. Estimate the rate of return for using air conditioner B instead of A.
   d. Determine the highest cost for air conditioner B at which air conditioner A is more competitive.
   e. Determine the cut-off electricity price at which both air conditioners A and B have the same LCC.

   Assume that the average cost of electricity is $0.08/kWh.

9. For a generic facility connected to the grid,
   i. Calculate the utility bill for the facility during one month characterized by the following parameters:
      - Actual demand: 450 kW
      - Energy consumption: 85,000 kWh
      - Average power factor: 70%

   The facility is subject to the following rate structure:
      - Customer charge: $150/month
      - Billed demand charge: $10/kW
      - Energy charge: $0.025/kWh

      The rate structure includes a power factor clause that states that the billing demand is determined as the actual demand multiplied by the ratio of 85 and the actual average power factor expressed in percent.
   ii. Determine the cost savings achieved for the month if the facility power factor is improved to be at least 85%.

10. A Company consumed about 355,000 kWh of electrical energy during April. The billed demand is estimated to be 600 kW for April. If the Sales tax is 7%, calculate the electric utility bill for the company. Estimate the reduction in April's electric utility bill for the company following a peak-shaving measure that lowered April's billed peak demand to 450 kW. The company is subject to the following electric rate structure: *Rate-1*:
    a. Customer charge: $145.00/month
    b. Energy charge: $0.0325/kWh/month
    c. Demand charge applicable to the billing demand:
       i. *On-peak season*: $500 for the first 100 kW + $5.50/kW for any additional kW
       ii. *Off-peak season*: $300 for the first 100 kW + $3.50/kW for any additional kW

    The on-peak season extends from May to October and the off-peak season spans from November to April.

11. A manufacturing company consumes on average 250,000 kWh/month and has a billed demand of 1,200 kW during each month. The company can select between two rates: Rate-1 (described in Problem 9.8) and Rate-2 (described here). Determine the best rate suitable for the company. *Rate-2*:
    a. Customer charge: $250.00/month
    b. Energy charge: $0.0225/kWh/month
    c. Demand charge applicable to the billing demand:
        i. *On-peak season*: $400 for the first 100 kW + $6.00/kW for any additional kW
        ii. *Off-peak season*: $250 for the first 100 kW + $4.00/kW for any additional kW

    The on-peak season extends from June to September and the off-peak season spans from October to May.
12. The monthly energy consumption and peak demand history of a commercial building is shown here for 1 year.

Month	Energy Use (kWh)	Peak Demand (kW)
Jan	125,000	455
Feb	137,500	505
Mar	155,500	610
Apr	176,000	675
May	185,500	755
Jun	201,000	920
Jul	235,500	1,150
Aug	240,000	1,200
Sep	197,500	895
Oct	184,500	650
Nov	164,000	605
Dec	141,000	550

    The commercial building is on rate structure defined in Table 9.13 (utility rate B).
    a. Estimate the annual electric utility bill for the commercial building assuming that the average power factor is 70% during each month.
    b. Determine the reduction in the annual electric utility bill if the power factor is improved to 90%.
13. A utility offers the following rates to its commercial customers:
    Service level 1:
    a. *Customer charge*: $500/bill/month
    b. *Energy charge*: $0.025 applicable to all kWh/month
    c. Demand charge applicable to all kW/month of billing demand:
        i. *On-peak season*: $350.00 for first 75 kW or less, and $3.75/kW for all additional kW
        ii. *Off-peak season*: $210.00 for first 75 kW or less, and $2.50/kW for all additional kW

Service level 2:

a. *Customer charge*: \$200/bill/month
b. *Energy charge*: \$0.027 applicable to all kWh/month
c. Demand charge applicable to all kW/month of billing demand:
    i. *On-peak season*: \$400.00 for first 75 kW or less, and \$4.50/kW for all additional kW
    ii. *Off-peak season*: \$250.00 for first 75 kW or less, and \$2.90/kW for all additional kW

Service level 3:

a. *Customer charge*: \$110/bill/month
b. *Energy charge*: \$0.028 applicable to all kWh/month
c. Demand charge applicable to all kW/month of billing demand:
    iii. *On-peak season*: \$450.00 for first 75 kW or less, and \$4.90/kW for all additional kW
    iv. *Off-peak season*: \$290.00 for first 75 kW or less, and \$3.30/kW for all additional kW

The on-peak season includes the months of June through October. The off-season includes the remainder of the months of any year. The billing demand is defined as the maximum demand for each month corrected for the power factor (refer to the power factor clause defined in the following paragraph), provided no billing demand shall be considered as less than the percentage of the highest on-peak season maximum demand corrected for the power factor previously determined during the 12 months ending with the current month. The following power factor clause applies to all the utility rates structures: *Power factor clause*: The consumer shall at all times use power so that the power factor shall be as close to 100% as possible, but when the average power factor as determined by continuous measurement of lagging reactive kilovolt-ampere hours is less than 80%, the billing demand shall be determined by multiplying the maximum demand, shown by the demand meter for the billing period, by 80 and dividing the product thus obtained by the actual average power factor expressed in percent.

a. A large manufacturing company is on the rate schedule of level 3. Their peak demand for the last year is shown here. They found a way to reduce their demand in the off-peak season by 100 kW, but the peak season demand will be the same (i.e., the demand in November through May would be reduced by 100 kW). Assuming they are on the 65% ratchet clause specified in the rate schedule, what is their dollar savings? Assume the high month was July of the previous year at 1,150 kW. If the demand reduction of 100 kW occurred in the peak season, what would be the dollar savings (i.e., the demand in June through October would be reduced by 100 kW).

Month	Demand (kW)	Month	Demand (kW)
Jan	495	Jul	1,100
Feb	550	Aug	1,000
Mar	580	Sep	900
Apr	600	Oct	600
May	610	Nov	500
Jun	900	Dec	515

b. Determine among the three available rate structures, the rate level that minimizes annual utility bill for the manufacturing company before any demand reduction.

14. A company is charged based on the utility rate structure defined as level 1 in Problem 9.11. In auditing a company, you find they have averaged 65% power factor over the past year. They are on the rate schedule given in the class handout and have averaged 1,000 kW/month. Neglecting any ratchet clause and assuming their demand and power factor are constant each month, calculate the savings for correcting to 80% power factor. How much capacitance would be necessary to obtain this correction? (Assume level 1.)
15. A company has contacted you regarding their rate schedule. The company is on the rate schedule defined in Problem 9.11 as service level 3 (secondary service) but is located near existing utility transmission lines and so can accept service at a higher level (service level 1) if it buys its own transformer. Assuming the company consumes 300,000 kWh/month and is billed for 1,000 kW each month, how much would it save by owning its own transformer. Ignore any charges other than demand and energy.

10 Energy-Efficient Electrical Systems

This chapter describes energy efficiency measures specific to electrical equipment and loads common for residential and commercial buildings, including appliances, lighting, and motors. Moreover, proven measures and strategies to improve the energy efficiency of various components of building electrical distribution systems are outlined with specific examples. Throughout this chapter, simplified calculation methods are presented to evaluate the cost-effectiveness of the proposed energy efficiency measures.

10.1 INTRODUCTION

The United States building sector is responsible for 40% of the national energy demand and 40% of the national CO_2emissions (IEA, 2022). Residential buildings account for 21% of US energy consumption with over 60% attributed to space and water heating end-uses (IEA, 2022). The high energy demand, in tandem with the growing climate crisis, growth in population, and increase in global access to reliable power, is causing an energy supply crisis (IEA, 2022).

A wide range of efficient and renewable technologies have been developed over the last decades specifically to enhance the sustainability of the built environment. However, the deployment rates of these technologies are rather limited for both residential and commercial buildings in the US as summarized in Table 10.1 (Schwatz and Krarti, 2022). As indicated in Table 10.1 some technologies have already been widely adopted and are commonly utilized today. Specifically, energy-efficient envelopes can save energy by reducing heating and cooling thermal loads mostly due to the requirements set by building energy codes. Moreover, energy-efficient appliances use less energy and are adopted due to their brand recognition, affordability, savings incentives, and code requirements. For on-site renewable energy generation, solar panels provide an alternative option from the grid and can reduce significantly utility costs. On-site PV systems are becoming a social norm due to their energy cost savings and code requirements in some states.

However, several energy efficiency and renewable technologies have yet to be commonly adopted. For instance, innovations such as smart windows as noted in Table 10.1 have not made it to become standard for builders, despite their energy efficiency benefits because of lack of knowledge, high installation costs, and policy requirements. In addition, electric vehicles and battery energy storage systems have not become mainstream because of their high capital costs and limited confidence in their performance as their fuel-based counterparts. Heat pumps also have significant energy, carbon, and cost-saving benefits, yet are far less common than their fuel-based heating systems due mostly to the lack of knowledge, large initial cost, and code requirements. Other renewable energy technologies

DOI: 10.1201/9781003276999-10

TABLE 10.1
Adoption Rates for Energy Efficiency and Renewable Technologies in US Residential Buildings

Technology	Adoption Rate	Reasons for Adoption	Reasons Against Adoption
Envelopes/ insulation	~100%	IECC code, certification incentives, financial incentives, social norm	Slow code adoption updates
Efficient appliances	~50%	IECC code, certification incentives, financial incentives, social norm	Slow code adoption updates
Photovoltaics	12.9%	Code, certification incentives, financial incentives, social norm (based on location)	Financial cost, knowledge, accessibility
Smart windows	2.4%	Financial incentives	Lack of code, knowledge, financial cost
Electric vehicles	1.4%	Financial incentives, environmental morals, electrification	Lack of code, knowledge, financial cost, vehicle charging
Storage solutions	1.5%	Financial incentives, electrification, resiliency	Financial cost, lack of need
Heat pumps	8.7%	Financial incentives, environmental morals, electrification, health	Lack of code, knowledge, financial cost, no social norm, compatibility
Geothermal systems	4%	Financial incentives, environmental morals, electrification, health	Lack of code, knowledge, financial cost, no social norm, compatibility

such as geothermal are not as popular as PV solar panels due to their limited site availability and high capital cost requirements. However, there is great promise in the future for several energy efficiency and renewable technologies listed in Table 10.1 as well as others (Krarti, 2020) would be widely adopted with additional research and developed to lower their implementation costs, standardize their installation, and educate the public about their energy and cost benefits (Schwartz and Krarti, 2022).

In this chapter, a brief description of energy efficiency measures and technologies specific to building electrical loads such as motors, lighting fixtures, electrical vehicles and appliances is first presented. Then, a series of measures are discussed to reduce the energy consumption of the main components of electrical distribution systems, including transformers and wires.

10.2 ELECTRICAL MOTORS

10.2.1 Introduction

As outlined in Chapter 4, motors convert electrical energy to mechanical energy and are typically used to drive machines. The driven machines can serve a myriad of purposes in buildings, including moving air (supply and exhaust fans), moving liquids (pumps), and compressing gases (refrigerators). To select the type of motor to be used for a particular application, several factors have to be considered, which include the following:

1. The form of the electrical energy that can be delivered to the motor: direct current (DC) or alternating current (AC), single phase or three phases
2. The requirements of the driven machine, such as motor speed and load cycles
3. The environment in which the motor is to operate: normal (where a motor with an open-type ventilated enclosure can be used), hostile (where a totally enclosed motor must be used to prevent outdoor air from infiltrating inside the motor), or hazardous (where a motor with an explosion-proof enclosure must be used to prevent fires and explosions)

Two simple measures to reduce energy use and improve the energy efficiency of motors used in buildings are described in the following sections.

10.2.2 Energy-Efficient Motors

Based on their energy efficiency ratings, motors can be classified into two categories: (1) standard-efficiency motors and (2) high- or premium-efficiency (i.e., energy-efficient) motors. The energy-efficient motors are 2%–10% more efficient than standard-efficiency motors, depending on their size. Table 10.2 summarizes

TABLE 10.2
Typical Motor Efficiencies

Motor Mechanical Power Output kW (HP)	Average Nominal Efficiency for Standard-Efficiency Motors	Nominal Efficiency for Premium-Efficiency Motors
0.75 (1.0)	0.730	0.855
1.12 (1.5)	0.750	0.865
1.50 (2.0)	0.770	0.865
2.25 (3.0)	0.800	0.895
3.73 (5.0)	0.820	0.895
5.60 (7.5)	0.840	0.917
7.46 (10)	0.850	0.917
11.20 (15)	0.860	0.924
14.92 (20)	0.875	0.930
18.65 (25)	0.880	0.936
22.38 (30)	0.885	0.936
29.84 (40)	0.895	0.941
37.30 (50)	0.900	0.945
44.76 (60)	0.905	0.950
55.95 (75)	0.910	0.954
74.60 (100)	0.915	0.954
93.25 (125)	0.920	0.954
111.9 (150)	0.925	0.958
149.2 (200)	0.930	0.962

Source: Adapted from NEMA, Energy Management Guide for Selection and Use of Polyphase Motors, Standard MG-10-1994, National Electrical Manufacturers Association, Rosslyn, VA, 2006 for Premium Motors.

the average efficiencies for both standard-efficiency and energy-efficient motors for commercially available standard sizes. The improved efficiency for the high or premium motors is mainly due to better design and the use of better materials to reduce losses. However, this efficiency improvement comes with a higher price of typically 10%–30% more than standard-efficiency motors. These higher prices may be the main reason that not all motors sold in the United States are in the premium-efficiency category.

However, the installation of premium-efficiency motors is becoming a common method of improving the overall energy efficiency of buildings. The potential for energy savings from premium-efficiency motor retrofits is significant. In the United States alone, in 1991, there were about 125 million operating motors that consumed approximately 55% of the electrical energy generated in the United States (Andreas, 1992). It was estimated that replacing all these motors with premium-efficiency models would save approximately 60 ThW of energy per year (Nadel et al., 1991).

To determine the cost-effectiveness of motor retrofits, there are several tools available to perform the calculations for the potential energy use and cost savings, including the MotorMaster developed by the Washington State Energy Office (WSEO, 1992) and currently available at the Department of Energy (DOE) website (DOE, 2009). These tools have the advantage of providing large databases for cost and performance information for various motor types and sizes.

A simplified method to calculate the energy savings due to the utilization of energy-efficient instead of standard motors is provided by Equation 10.1. Inherent to this method, two assumptions are made: (1) the motor is fully loaded and (2) the change in motor speed is neglected.

$$\Delta P_R = P_M \left(\frac{1}{\eta_S} - \frac{1}{\eta_E} \right) \tag{10.1}$$

where

P_M is the mechanical power output of the motor
η_S is the design (i.e., full-load) efficiency of the standard motor
η_E is the design (i.e., full-load) efficiency of the energy-efficient motor

The electric energy savings incurred from the motor replacement is thus

$$\Delta \text{kWh} = \Delta P_R \times N_h \times LF_M \tag{10.2}$$

where

ΔP_R is the reduction in motor real power demand estimated using Equation 10.1
N_h is the number of hours per year during which the motor is operating
LF_M is the load factor of the motor's operation during 1 year

Example 10.1 indicates the potential energy use, cost savings, and payback period when a standard-efficiency 10 HP (7.46 kW) motor is replaced with a premium-efficiency motor.

Example 10.1

Problem

Determine the cost-effectiveness of replacing a 10 HP motor with an efficiency of 85% with a premium-efficiency motor with a rated full-load efficiency of 91.70%. Assume that

- The cost of electricity is $0.10/kWh
- The differential cost of premium versus standard motor is $300
- The average load factor of the motor is 0.80
- The average full-load operating hours of the motor are 5,000 h/year

Solution

To determine the cost-effectiveness of installing a premium-efficiency motor instead of a standard-efficiency motor, a simplified economic analysis is used to estimate the simple payback period. The savings in energy use in *kWh* for the premium-efficiency motor can be calculated using Equations 10.1 and 10.2:

$$\text{kWh}_{\text{saved}} = N_h \times HP \times 0.746 \times LF_M \left(\frac{1}{\eta_S} - \frac{1}{\eta_E} \right)$$

where

N_h is the total number of hours per year during which the motor is operating (5,000 h/year).

HP is the rated motor power output (10 HP)

LF_M is the annual average load factor of the motor ($LF_M = 0.80$)

η_S and η_E are the efficiencies of the standard and efficient motors, respectively (0.850 and 0.917)

Thus, the energy savings in *kWh* are calculated as follows:

$$\text{kWh}_{\text{saved}} = 5{,}000 \times 10 \times 0.746 \times 0.80 \left(\frac{1}{0.850} - \frac{1}{0.917} \right) = 2{,}565 \text{ kWh/year}$$

Therefore, the simple payback period, *SPP*, for using the premium-efficiency instead of the standard-efficiency motor is

$$SPP = \frac{\$300}{2{,}565 \text{ kWh} \times \$0.10 \text{ kWh}} = 1.2 \text{ years}$$

10.2.3 Motor Controls

Frequent turning on and off of HVAC systems (and in particular fan motors) may be actually detrimental and may not be cost-effective over the life cycle of the equipment due to added maintenance and repair costs. However, on/off cycling of motor

TABLE 10.3
Allowable Number of Starts Per Hour and Minimum Off Time for Motor Loads

	Two-Pole Motors		Four-Pole Motors		Six-Pole Motors	
Motor Size HP (kW)	**Max. Starts/h**	**Min. Off Time (s)**	**Max. Starts/h**	**Min. Off Time (s)**	**Max. Starts/h**	**Min. Off Time (s)**
2.0 (1.5)	11.5	77	23.0	39	26.1	35
5.0 (3.75)	8.1	83	16.3	42	18.4	37
7.5	7.0	88	13.9	44	15.8	39
15.0	5.4	100	10.7	50	12.1	44
20.0 (15.0)	4.8	110	9.6	55	10.9	48
25.0 (18.75)	4.4	115	8.8	58	10.0	51
30.0 (22.5)	4.1	120	8.2	60	9.3	53
40.0 (30.0)	3.7	130	7.4	65	8.4	57
50.0	3.4	145	6.8	72	7.7	64

loads can be performed safely without long-term damages when minimum on and off times are respected. The National Electrical Manufacturers Association (NEMA) provides a set of recommendations for minimum on/off times of duty cycling of motor loads. Some of these recommendations are summarized in Table 10.3. It is highly recommended, however, that the motor manufacturers be directly consulted to determine their suggested minimum off time and the allowable number of starts per hour.

To fully benefit from duty cycling energy cost savings (due mostly to reduction in demand charges), two methods can be used. In the first method, known as parallel duty cycling, all the motors are cycled on and off at the same time. This method can provide energy cost savings when the duty period is less than the demand period (typically 15 minutes in most utility rates). However, when the duty period exceeds the demand period, there is no reduction in demand changes if all the motors are cycled on and off at the same time. In this case, it is recommended to use the second method of duty cycling called staggered duty cycling, which alternates the on and off times of the motors. It should be noted that most motor manufacturers for building applications recommend extended duty cycling periods (higher than typical demand periods of 15 minutes) to reduce the wear and tear effects. Therefore, staggered duty cycling approach should be considered in most applications to ensure the safety of HVAC equipment while reducing operating costs. Example 10.2 illustrates the calculation procedure for energy cost savings incurred from staggered duty cycling control strategies.

Example 10.2

Problem

Determine the reduction in the annual energy costs due to staggered duty cycling of three identical fan motors (each rated at 30 kW [40 HP]). The motor manufacturer specifies that a cycle has 20 minutes on and 10 minutes off as the minimum duty cycle. The utility monthly demand charge is \$10/kW. First, determine the recommended NEMA duty period assuming 2-pole motors.

Solution

a. Using Table 10.3, the allowable number of starts per hour for 40 HP, two-pole motors is 3.7 starts/h. Thus, each start should last as in the following:

$$\text{Start period} = \frac{60 \text{ min/h}}{3.7 \text{ starts/h}} = 14.4 \text{ min/start}$$

The duty period is thus about 15 minutes. Since the duty period is the sum of the off time and the on time and since the minimum off time is about 2 minutes (130 seconds based on Table 10.2), the maximum on time allowed by NEMA standard is 13 minutes. Thus, the manufacturers' on time is longer than that recommended by NEMA.

b. The reduction in the electrical demand peak due to staggered duty cycling approach is 1/3 of the total demand of all the three motors (or $1/3 \times 90$ kW=30 kW). Indeed, at any given time, only two out of the three motors are operating. Thus, the annual savings in electrical demands charges are given as follows:

$$\Delta kW = 12 \times 30 \text{ kW} = 360 \text{ kW/year}$$

Therefore, the staggered duty cycling control provides an annual energy cost savings of \$3,600.

10.2.4 Variable Speed Drives

As explained in Section 4.5.1, the deployment of variable frequency drives (VFDs), also referred to as variable speed drives (VSDs), can be effective to improve the energy efficiency of motors. VFDs allow the motors to vary their speeds to match the load requirements. For buildings, VFDs are effective to reduce electricity consumption for heating, ventilating, and air conditioning (HVAC) equipment including fans and pumps. As discussed in this chapter, VFDs are solid-state power devices that can be integrated as part of overload protection for the motors. In addition to their energy efficiency benefits, VFDs can also reduce the initial inrush currents allowing a soft start for the induction motors. However, VFDs can degrade the power quality since they can generate and produce harmonics if they are not properly specified and installed. As outlined in Chapter 11 on power quality, motors with VSDs may need to be filtered out to reduce magnetic noise, reduce motor vibration and prevent harmonics from being fed back into the system. The energy savings that could be achieved by VFDs are discussed in Chapter 4 with a detailed calculation example.

10.3 LIGHTING SYSTEMS

10.3.1 Introduction

Lighting can account for a significant portion of the energy use in residential and commercial buildings. In addition, heat generated by lighting contributes to additional thermal loads that need to be removed by the cooling equipment. Typically, energy retrofits of lighting equipment are very cost-effective, with payback periods of less than 2 years in most buildings.

To better understand the options that need to be considered in order to improve the energy efficiency of lighting systems, a simple estimation of the total electrical energy use due to lighting is first considered:

$$\text{kWh}_{Lit} = \sum_{j=1}^{J} N_{Lum,j} \times WR_{Lum,j} \times N_{h,j} \tag{10.3}$$

where

$N_{Lum,j}$ is the number of lighting luminaires of type j in the building or part of the building. It should be noted that a luminaire can include several components, such as a ballast, electric wiring, housing, and lamps.

$WR_{Lum,j}$ is the wattage rating for each luminaire of type j. In this rating, the energy use due to both the lamps and ballast should be accounted for.

$N_{h,j}$ is the number of hours per year when the luminaires of type j are operating.

J is the number of luminaire types in the building.

It is clear from Equation 10.3 that there are three options to reduce the energy use attributed to lighting systems, as discussed briefly in the following:

1. Reduce the wattage rating for the luminaires, including both the lighting sources (e.g., lamps) and the power transforming devices (e.g., ballasts) (therefore, decrease the term $WR_{Lum,j}$ in Equation 10.3). In the last decade, technological advances such as light-emitting diodes (LEDs), compact fluorescent lamps (CFLs), and electronic ballasts have increased the energy efficiency of lighting systems.
2. Reduce the time of use of the lighting systems through lighting controls (therefore, decrease the term $N_{h,j}$ in Equation 10.3). Automatic controls have been developed to decrease the use of a lighting system, so illumination is provided only during times when it is actually needed. Energy-efficient lighting controls include occupancy sensing systems and light dimming controls through the use of daylighting.
3. Reduce the number of luminaires (therefore, decrease the term $N_{Lum,j}$ in Equation 10.3). This goal can be achieved only in cases where minimizing the lighting fixtures is possible.

In this section, only measures related to the general actions described in items (1) and (2) are discussed. To estimate the energy savings due to any energy efficiency measure for the lighting system, Equation 10.3 can be used. The energy use due to lighting has to be calculated for the baseline design option and for the energy efficiency alternative and the difference between the two estimated energy uses would represent the energy savings.

10.3.2 Energy-Efficient Lighting Systems

Improvements in the energy efficiency of lighting systems have provided several opportunities to reduce electrical energy use in buildings. In this section, energy-saving calculations for the following technologies are discussed:

- High-efficiency fluorescent lamps
- Compact fluorescent lamps
- Compact halogen lamps
- Electronic ballasts
- LED lighting

First, a brief description is provided of the factors that a designer should consider to achieve and maintain an acceptable quality and level of visual comfort from the lighting system. Second, the design and the operation concepts are summarized for each available lighting technology. Then, the energy savings that can be expected from energy-efficient lighting systems using any of the new technologies are estimated and discussed.

Typically, three factors determine the proper level of light for a particular space. These factors include the age of the occupants, speed and accuracy requirements, and background contrast (depending on the task being performed). It is a common misconception to consider that overlighting a space provides higher visual quality. Indeed, it has been shown that overlighting can actually reduce illuminance quality and visual comfort level within a space, in addition to wasting energy. Therefore, it is important, when selecting a lighting system, to determine the adequate illuminance level that needs to be maintained within various indoor spaces, as recommended by appropriate authorities. Table 10.4 summarizes the lighting levels recommended for various activities and applications in selected countries, including the United States, based on the most recent illuminance standards.

10.3.2.1 High-Efficiency Fluorescent Lamps

Fluorescent lamps have high efficacy (i.e., light output per watt consumed), diffuse light distribution, and long operating life. A fluorescent lamp consists generally of a glass tube with a pair of electrodes at each end. The tube is filled at very low pressure with a mixture of inert gases (primarily argon) and with liquid mercury. When the lamp is turned on, an electric arc is established between the electrodes. The mercury vaporizes and radiates in the ultraviolet spectrum. This ultraviolet radiation excites a phosphorous coating on the inner surface of the tube, which emits visible light. High-efficiency fluorescent lamps use a krypton–argon mixture, which increases the efficacy output by 10%–20%, from a typical efficacy of 70 lu/W to about 80 lu/W. Improvements in the phosphorous coating can further increase the efficacy to 100 lu/W.

It should be mentioned that the handling and disposal of fluorescent lamps are highly controversial because mercury inside the lamps can be toxic and hazardous to the environment. A new technology is being tested to replace mercury with sulfur to generate the radiation that excites the phosphorous coating of fluorescent lamps. The sulfur lamps are not hazardous and would present an environmental advantage to the mercury-containing fluorescent lamps.

TABLE 10.4
Recommended Lighting Levels for Various Applications in Selected Countries (in Lux Maintained on Horizontal Surfaces)

Application	France AEF	Germany DIN5035	Japan JIS	U.S./Canada IESNA
		Offices		
General	425	500	300–750	200–500
Reading tasks	425	500	300–750	200–500
Drafting (detailed)	850	750	750–1,500	1,000–2,000
		Classrooms		
General	325	300–500	200–750	200–500
Chalkboards	425	300–500	300–1,500	500–1,000
Retail stores				
General	100–1,000	300	150–750	200–500
Tasks/till areas	425	500	750–1,000	200–500
		Hospitals		
Common areas	100	100–300	—	—
Patient rooms	50–100	1,000	150–300	100–200
		Manufacturing		
Fine knitting	850	750	750–1,500	1,000–2,000
Electronics	625–1,750	100–1,500	1,500–3,000	1,000–2,000

CFLs are miniaturized fluorescent lamps with small diameter and shorter length. The compact lamps are less efficient than full size fluorescent lamps with only 35–55 lu/W. However, they are more energy efficient and have longer life than incandescent lamps. Indeed, CFLs have an average life between 6,000 and 15,000h, while incandescent lamps have a life of 750 or 1,000h. Currently, CFLs are being heavily promoted as energy savings alternatives to incandescent lamps, even though they may have some drawbacks. In addition to their high cost, CFLs are cooler and thus provide less pleasing contrast than incandescent lamps.

The most common retrofit measures in lighting systems for residential buildings over the last decade have been the upgrade of conventional incandescent lamps to more energy-efficient CFLs. Now LED lamps are utilized instead of CFLs for several lighting retrofit projects. It should be noted that for lighting retrofit applications, it is recommended that a series of tests be conducted first to determine the characteristics of the existing lighting system. For instance, it is important to determine the illuminance level at various locations within the space, especially in work areas such as benches and/or desks.

10.3.2.2 Compact Halogen Lamps

CFLs are adapted for use as direct replacements for standard incandescent lamps. Halogen lamps are more energy efficient, produce whiter light, and last longer than incandescent lamps. Indeed, incandescent lamps convert typically only 15% of their electrical energy input into visible light since 75% is emitted as infrared radiation and 10% is used by the filament as it burns off. In halogen lamps, the filament is encased inside a quartz tube that is contained in a glass bulb. A selective coating on the exterior surface of the quartz tube allows visible radiation to pass through but reflects the infrared radiation back to the filament. This recycled infrared radiation permits the filament to maintain its operating temperatures with 30% less electrical power input.

The halogen lamps can be dimmed and present no power quality or compatibility concerns as can be the case for CFLs.

10.3.2.3 Electronic Ballasts

Ballasts are integral parts of fluorescent luminaires since they provide the voltage level required to start the electric arc and regulate the intensity of the arc. Before the development of electronic ballasts in the early 1980s, only magnetic or "core and coil" ballasts were used to operate fluorescent lamps. While the frequency of the electrical current is kept at 60 Hz (in countries other than the United States, the frequency is set at 50 Hz) by the magnetic ballasts, electronic ballasts use solid-state technology to produce high-frequency (20–60 MHz) current. The use of high-frequency current increases the energy efficiency of the fluorescent luminaires since light is cycling more quickly and appears brighter. When used with high-efficiency lamps (T-8, for instance), electronic ballasts can achieve 95 lu/W as opposed to 70 lu/W for conventional magnetic ballasts. It should be mentioned, however, that efficient magnetic ballasts can achieve similar lumen/watt ratios as electronic ballasts.

Other advantages that electronic ballasts have relative to their magnetic counterparts include

- *Higher power factor*: The power factor of electronic ballasts is typically in the 0.90–0.98 range. Meanwhile, the conventional magnetic ballasts have low power factor (less than 0.80), unless a capacitor is added, as discussed in Chapter 2.
- *Less flicker problems*: Since the magnetic ballasts operate at 60 Hz current, they cycle the electric arc about 120 times/s. As a result, flicker may be perceptible, especially if the lamp is old during normal operation or when the lamp is dimmed to less than 50% capacity. However, electronic ballasts cycle the electric arc several thousands of times per second and flicker problems are avoided even when the lamps are dimmed to as low as 5% capacity.
- *Less noise problems*: The magnetic ballasts use electric coils and generate audible hum, which can increase with age. Such noise is eliminated by the solid-state components of the electronic ballasts.

10.3.2.4 Light-Emitting Diode Lamps

Light-emitting diodes, commonly called LEDs, are the newest lighting technologies available in the market. While they were first introduced as indicator lights, due to advances in their optics and their luminous output, LEDs are now used as lighting fixtures for a wide range of building applications. Some of the advantages of LEDs include their small size, reliability, flexibility, and especially energy efficiency. However, LEDs are still expensive and may not provide the desired color as lighting sources for some building spaces such exhibit rooms in museums. LEDs are semiconductors with light created by the flow of electrons in *p-n* junctions and releasing energy in the form of photons.

LEDs are typically divided into three different size categories including miniature LEDs, mid-range LEDs, and high-power LEDs. The miniature LEDs are very small, usually only a few millimeters in diameter, and are made up of a single semiconductor. Because these lamps are so small, they are not suitable for illuminating spaces. Typical current ratings for miniature LEDs range between 1 and 20 mA. The mid-range LEDs usually have a current rating of 100 mA and are used most often when little luminance levels are desired in specific spaces. High-intensity LEDs are currently suitable for lighting buildings and require typically currents in the range of 100 mA to 1.0 A. With these relatively high currents, overheating can be a concern for LEDs. Thus, they often need a heat sink.

LED lamps are quickly becoming popular because of their energy efficiency. Even though they are still expensive, LED lamps can be a cost-effective lighting alternative for several buildings. Moreover, these lamps offer significant flexibility due to their size and can be placed in a wide range of locations within buildings. Example 10.3 illustrates the cost-effectiveness of LED lamps compared to CFLs in residential buildings.

Example 10.3

Problem

Consider a home with a total of 10 incandescent lamps rated at 60 W. Determine, the energy-saving potential if these lamps are replaced with (1) 13 W CFLs and (2) 7 W LED lamps. The lifespan of the incandescent lamp, CFL, and LED lamp is, respectively, 1,300, 8,000, and 30,000 h. Perform the analysis using the lifespan of the LED lamps. Determine the best lamps to use, CFLs or LED lamps, if the costs of the incandescent lamp, CFL, and LED lamp are \$0.30, \$1.50, and \$10, respectively. Assume the cost of electricity is \$0.10/kWh and perform a simple payback analysis.

Solution

Incandescent lamps: Over the lifespan of 30,000 h, 230 incandescent lamps are needed. The energy cost for using incandescent lamps over 30,000 h would be

$$\text{kWh} = 230 \times 60 \times 1{,}300 \times \frac{1}{1{,}000} \times 0.10 = \$1{,}794$$

Thus, the total cost for using incandescent lamps is \$1,794 + \$0.30 × 230 = \$1,863.

CFLs: Over the lifespan of 30,000 h, about 38 CFLs are needed. The energy cost for using CFLs over 30,000 h would be

$$\text{kWh} = 38 \times 13 \times 8{,}000 \times \frac{1}{1{,}000} \times 0.10 = \$395$$

Thus, the total cost for using CFLs is $395 + $1.50 × 38 = $452.

LED lamps: Over the lifespan of 30,000 h, about 10 LEDs are needed. The energy cost for using LEDs over 30,000 h would be

$$\text{kWh} = 10 \times 7 \times 30,000 \times \frac{1}{1,000} \times 0.10 = \$210$$

Thus, the total cost for using LED lamps is $210 + $10 × 10 = $310.

Therefore, LEDs would be the best option to replace the incandescent lamps even though they are significantly more expensive than CFLs.

10.3.3 Lighting Controls

Significant energy use savings can be achieved by not operating the lighting system in cases when illumination becomes unnecessary. The control of the operation of a lighting system can be achieved by several means, including dimming switches, occupancy sensing systems, and daylighting.

While energy savings can be achieved by manual dimming, the results are typically unpredictable since they depend on the occupant's behavior. Scheduled or sensor-based lighting controls provide a more efficient approach to energy savings, but can also be affected by frequent adjustments from the occupants. Some of the dimming switches and automatic controls available for lighting systems are briefly discussed as follows.

- *Dimming switches*: One of the simplest and most cost-effective ways to reduce energy use associated with the lighting of residential and commercial buildings is the use of dimming switches. Indeed, these types of switches can save up to 50% of the electrical lighting energy used for a typical building compared to conventional toggle switches. Moreover, they can extend the lifespan of lighting lamps.
- *Occupancy or vacancy sensors*: These sensors can be installed in spaces with intermittent occupancy in buildings such as conference rooms or bathrooms. Occupancy sensors save energy by automatically turning off the lights in spaces that are not occupied. For several building applications, occupancy sensors using passive-infrared, or PIR devices can simply replace traditional wall switches and can detect motion for spaces of up to 60 m^2 (600 ft^2). Occupancy sensors can save up to 60% of energy use associated with lighting. It should be noted that these infrared sensors operate adequately only if they are in direct line-of-sight with the occupants. As alternatives to infrared sensors, ultrasound occupancy sensors can be used, especially for commercial buildings. Ultrasound sensors operate on a sonar principle, like submarines and airport radars. A device emits a high frequency sound (25–40 kHz), so it is beyond the hearing range of humans. This sound is reflected by the surfaces inside a space (including furniture and occupants) and is sensed by a receiver. When an occupant makes any noticeable movement within a space, the pattern of sound waves changes and is detected by ultrasound sensors. The lights remain on until no movement is detected for a preset period of time (after 5 minutes). Unlike

TABLE 10.5
Energy-Saving Potential with Occupancy Sensors

Space Application	Range of Energy Savings (%)
Offices (private)	25–50
Offices (open space)	20–25
Rest rooms	30–75
Conference rooms	45–65
Corridors	30–40
Storage areas	45–65
Warehouses	50–75

infrared radiation, sound waves are not easily blocked by obstacles such as wall partitions. However, the ultrasound sensors may not operate properly in large spaces, which tend to produce weak echoes.

Table 10.5 provides typical energy savings to be expected from occupancy sensors installed in various indoor spaces. As shown in Table 10.5, significant energy savings can be achieved in spaces where occupancy is intermittent such as conference rooms, restrooms, storage areas, and warehouses.

- *Photosensor-based controls*: Automatic controls of outdoor lighting can save significant electrical energy use for residential buildings. Timers and motion detection can be used for controlling outdoor lighting. However, photosensors offer an energy-efficient and practical means to control outdoor lighting, while ensuring safety and security around the buildings. Typically, photosensor-based controls can save 20% of energy use associated with outdoor lighting.

10.3.4 Daylighting Harvesting Systems

For several buildings, the use of special controls is required to take advantage of the natural light to illuminate indoor spaces through daylighting harvesting. The illuminance levels available inside building spaces are typically expressed as percentages of the illuminance outdoors. These percentages, referred to as demand factors (or DFs), are recommended to be 1.5%–2.5% for visual tasks, 2.5%–4.0% for moderately difficult tasks, and 4.0%–8.0% for difficult and prolonged tasks.

Spaces that require good quality lighting throughout the year, such as office spaces for commercial buildings as well as kitchens and bedrooms for homes, should be placed along the north–south perimeter of the buildings. Other rarely used spaces, such as conference rooms for commercial buildings and laundry rooms for homes, can be placed in the interior of the buildings. In addition to large windows, skylights, and roof monitors, daylighting can be better harvested for spaces with reflective surfaces, especially the ceiling. It is generally recommended that the ceiling surface has a reflectance of at least 0.90 by using high reflectance paints or tiles.

To estimate the potential annual energy use savings, f_d, associated with daylighting harvesting in reducing the need for electrical lighting in any perimeter space, the following expression can be used (Krarti et al., 2005):

$$f_d = b\left[1 - \exp\left(\frac{-a\tau_w A_w}{A_f}\right)\right] \tag{10.4}$$

where

A_w/A_f is the ratio of the window area over the spacious floor area (this parameter provides a good indicator of the window size relative to the daylight floor area).

a and b are the coefficients that depend only on the building location and are given in Table 10.6 for various sites throughout the world.

τ_w is the visible transmittance of the glazing.

Example 10.4 outlines the calculation procedures to estimate the annual energy savings associated with installing daylighting harvesting.

TABLE 10.6
Coefficients *a* and *b* of Equation 10.4 for Various Locations throughout the World

Location	*a*	*b*	Location	*a*	*b*
Atlanta	19.6	74.3	Casper	19.2	72.7
Chicago	18.4	71.7	Portland	17.8	70.9
Denver	19.4	72.9	Montreal	18.8	69.8
Phoenix	22.3	74.8	Quebec	19.1	70.6
New York City	18.7	67.0	Vancouver	16.9	68.7
Washington, DC	18.7	70.8	Regina	20.0	70.5
Boston	18.7	67.1	Toronto	19.3	70.5
Miami	25.1	74.8	Winnipeg	19.6	70.9
San Francisco	20.6	74.0	Shanghai	19.4	67.3
Seattle	16.6	69.2	K-Lumpur	20.2	72.4
Los Angeles	22.0	74.2	Singapore	23.3	73.7
Madison	18.8	70.0	Cairo	27.0	74.2
Houston	21.6	74.7	Alexandria	36.9	74.7
Fort Worth	19.7	72.9	Tunis	25.2	74.1
Bangor	17.9	70.7	Sao Paulo	29.4	71.2
Dodge City	18.8	72.6	Mexico	28.6	73.6
Nashville	20.0	70.4	Melbourne	20.0	67.7
Oklahoma City	20.2	74.4	Roma	16.0	72.4
Columbus	18.6	72.3	Frankfurt	15.2	69.7
Bismarck	17.9	71.5	Kuwait	22.0	65.3
Minneapolis	18.2	72.0	Riyadh	21.2	72.7
Omaha	18.9	72.30			

Example 10.4

Problem

Consider an open office space with 150 m² located in Denver, CO, with a total window area of 20 m², with double-pane glazing, which has a visible transmittance of 0.55. The office space has 10 fluorescent fixtures, each rated at 52 W. Determine the energy use and cost savings due to daylighting harvesting strategies. Assume that without daylighting harvesting, each fixture is operated 8 h/day when daylighting can be harvested, 5 days/week, 250 weeks/year. The cost of electricity is 0.10/kWh.

Solution

Based on Equation 10.4, using the coefficients $a=19.4$ and $b=72.9$ (refer to Table 10.6), the percent reduction in electrical lighting associated with daylighting harvesting is

$$f_d = 72.9\left[1-\exp\left(-19.4\times 0.55\times\frac{20}{150}\right)\right]=55.3\%$$

Therefore, the savings in annual electrical energy use for the office space are computed as follows:

$$\Delta \text{kWh} = 55.3\%\times 10\times 52\times\frac{1}{1{,}000}\times 8\times 5\times 250 = 2{,}876\ \text{kWh/year}$$

Thus, the annual energy use cost saving is $287.6/year which can justify the installation of one or two photosensors for implementing daylighting control in the open office space since one sensor costs about $250.

10.4 OTHER ELECTRICAL SYSTEMS

10.4.1 Electrical Appliances

Appliances account for a significant part of the energy consumption in buildings that used about 41% of electricity generated worldwide in 1990 (IPCC, 1996). Moreover, the operating cost of appliances during their lifetime (typically 10–15 years) far exceeds their initial purchase price. However, consumers—especially in developing countries where no labeling programs for appliances are enacted—do not generally consider energy efficiency and operating cost when making purchases since they are not well informed. Figure 10.1 indicates the appliances that have become more common in U.S. households over the last decade. In particular, microwave oven, clothes dryers, and dishwasher, are available in 83%, 71%, and 50% of U.S. households, respectively, in 1997. Almost all U.S. households have at least one refrigerator. Figure 10.2 shows the distribution of the capacity of the refrigerators used in U.S. households. It is clear that the size of the refrigerators used in U.S. homes has increased over the years.

Average daily energy use profiles as well as the runtime variations for selected electrical appliances are shown in Figures 10.3–10.6 for a refrigerator, an oven, a dishwasher, and a clothes dryer (Surles, 2011).

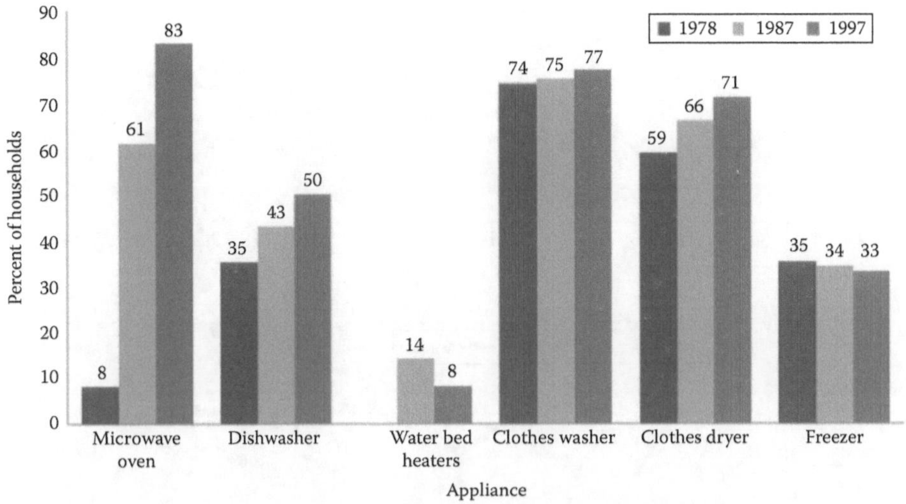

FIGURE 10.1 Penetration levels of selected appliances in U.S. households.

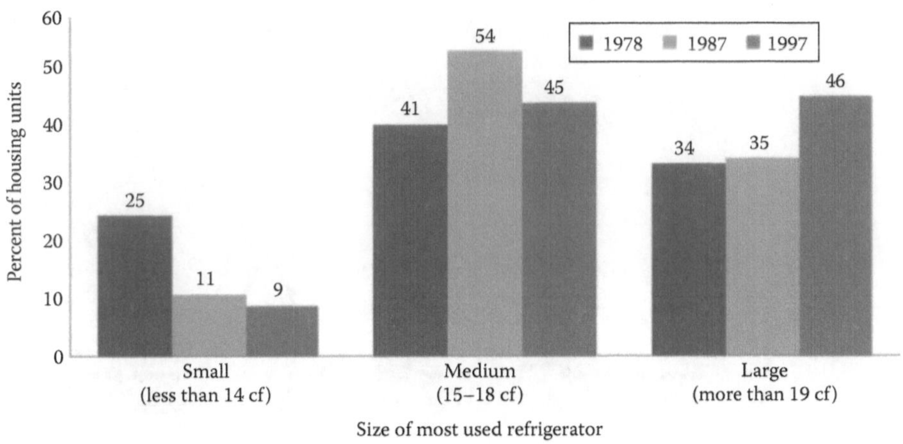

FIGURE 10.2 Size of refrigerators used in U.S. households.

As expected, the refrigerator consumes the most electrical energy during a typical day among the four appliances considered in Figures 10.3–10.6.

10.4.2 Electrical Vehicles

As part of the electrification of the transportation sector, electrical vehicles are becoming more popular in the US and throughout the world due to their environmental impacts as well as their operational cost benefits. According to the US DOE's Alternative Fuels Data Center, 879,320 electric vehicles (EVs) have been sold since

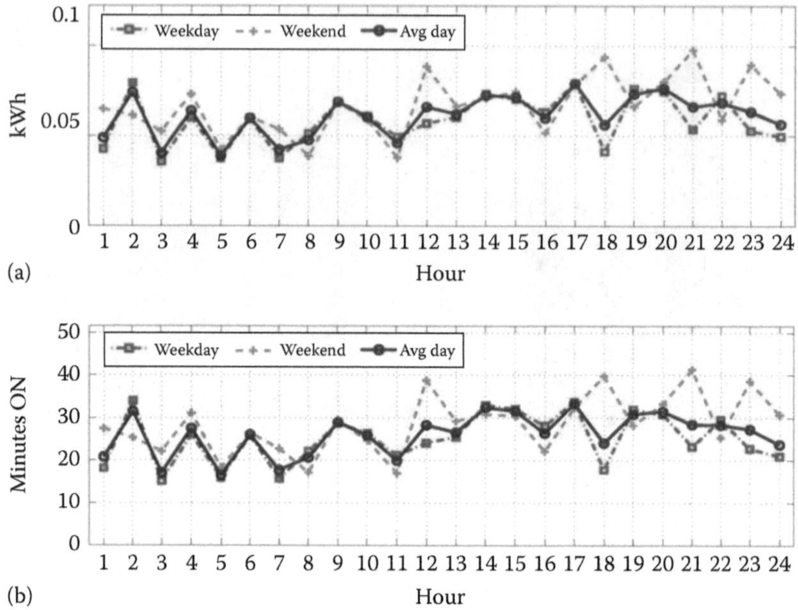

FIGURE 10.3 Average daily profiles for (a) energy use and (b) runtime for a refrigerator.

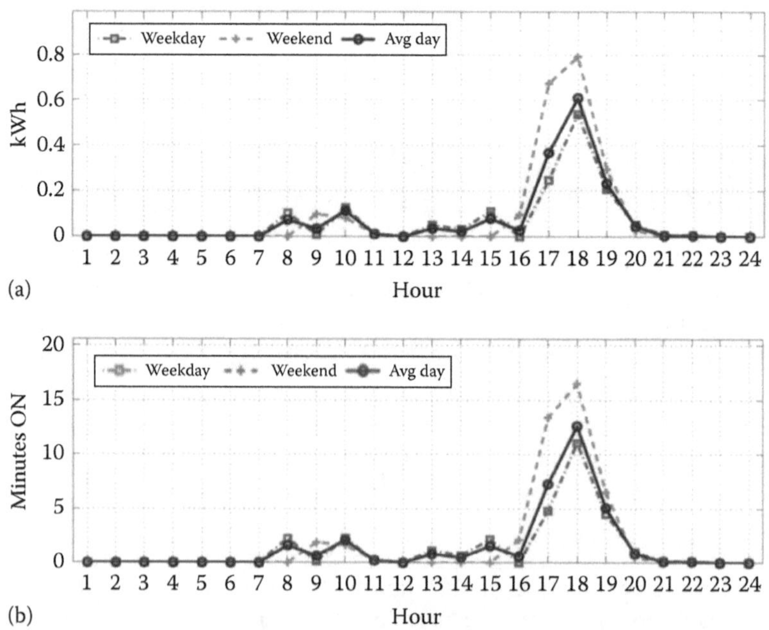

FIGURE 10.4 Average daily profiles for (a) energy use and (b) runtime for an oven.

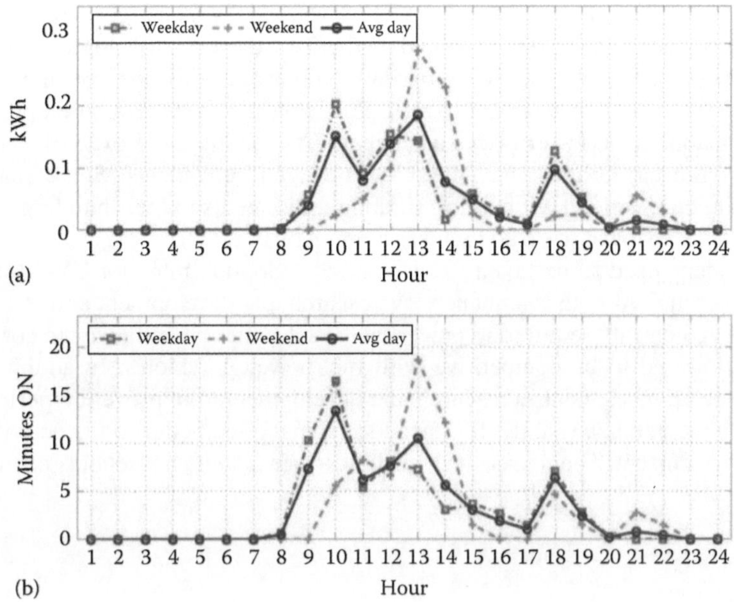

FIGURE 10.5 Average daily profiles for (a) energy use and (b) runtime for a clothes dryer.

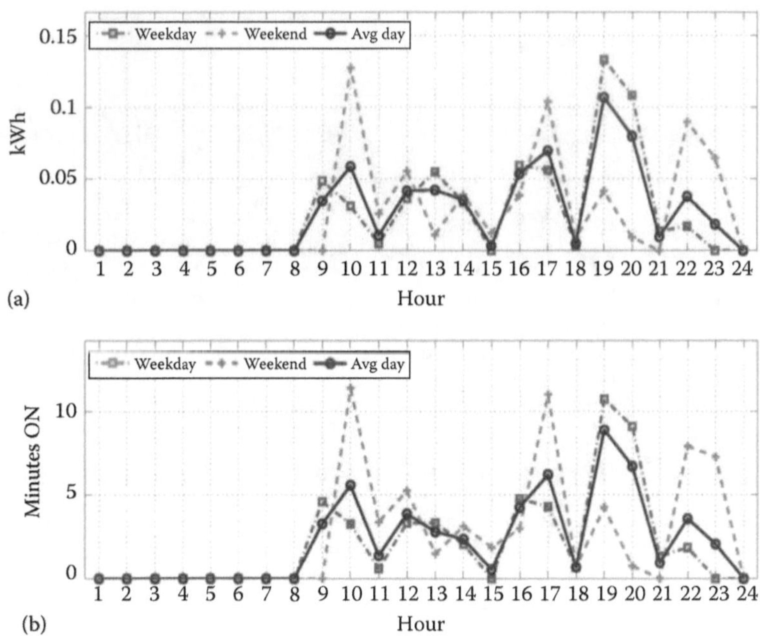

FIGURE 10.6 Average daily profiles for (a) energy use and (b) runtime for a dishwasher.

2011, with 241,912 made in 2019 alone. However, the EV stock represents only 1.4% of the total 16.9 million vehicles sold in 2019 in the US [80]. Figure 10.7 shows the number of EVs sold each year, by EV automakers. This does not include plug-in hybrid vehicles (Schwartz and Krarti, 2022). While there are more options with higher capabilities at lower prices each year, EVs remain more expensive and have fewer driving capabilities than their fuel-powered equivalents. Some states, like California, have added EV charging station requirements to their building code, but other states are slower to do so (Schwartz and Krarti, 2022).

More steps need to be taken to increase the adoption rates for EVs. Additional funding is required both to enhance EV research and development and to subsidize the cost of electricity required to power the EVs. Studies have shown the cost of EVs must be lowered to be competitive with fuel-powered vehicles. In addition, there needs to be an investment in research and development to improve EV performance and charging stations. Finally, EVs need to be marketed better to the demographics that are not currently being reached, such as lower income households or those that

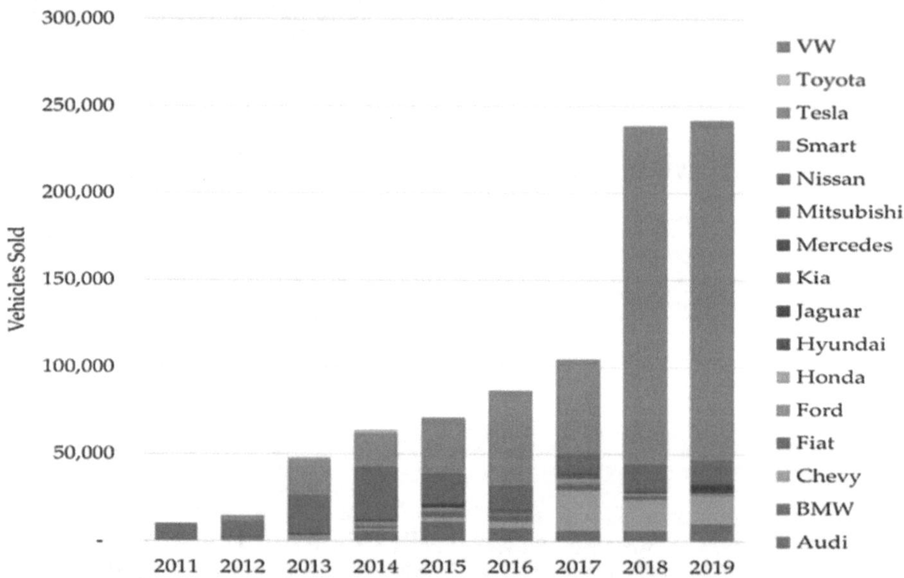

FIGURE 10.7 Electric vehicles sold by manufacturer and year (Schwartz and Krarti, 2022).

TABLE 10.7
Characteristics of Selected Electrical Vehicles Available in US market

Vehicle Manufacturer – Model	Driving Range [km (miles)]	Battery Type	Battery Size (kWh)	Charger Capacity (kW)
Ford – Focus	153	Lithium Ion	23	6.6
Nissan – Leaf	117	Lithium Ion	24	6.6
Mitsubishi I-Miev	100	Lithium Ion	16	3.6
Tesla S	426	Lithium Ion	85	20.0

believe they prefer specific fuel-type characteristics over EVs (Schwartz and Krarti, 2022). Several reported analyses have indicated that EVs have the capacity to grow and serve as a vital step towards complete electrification, lowering carbon footprints, and alleviating the discrepancy between grid energy production and demand.

The battery capacities are specific to EVs and the charging rates depending on the model and the manufacturers of the electrical vehicles. Table 10.7 lists some of the basic characteristics of commercially available EVs in the US market including their driving ranges as well as their battery type, size, and charging rate (Schwartz and Krarti, 2022).

10.5 ENERGY-EFFICIENT ELECTRICAL EQUIPMENT

10.5.1 Home Appliances

Recognizing the significance and the impact of appliances on national energy requirements, a number of countries have established energy efficiency programs. In particular, some of these programs target improvements in energy efficiency for residential appliances. Methods to achieve these improvements include energy efficiency standards and labeling programs.

Minimum efficiency performance standards for residential appliances have been implemented in some countries for a number of residential end-users. Energy savings associated with the implementation of these standards are found to be substantial. For instance, studies have indicated that in the United Kingdom, the energy consumption of average new refrigerators and freezers in 1993 was about 60% of the consumption in 1970. Similar improvements have been obtained in Germany (Waide et al., 1997). In the United States, savings due to the standards are estimated to be about 0.7 EJ/year during the period extending from 1990 to 2010 (1 EJ = 10^{18} J = 1 quadrillion of Btu = 10^{15} Btu).

Investing in energy-efficient appliances in the residential sector is generally highly cost-effective. In the United States, it is estimated that the average benefit–cost ratios for promoting energy-efficient appliances is about 3.5. In other terms, each U.S. dollar of federal expenditure on implementing the standards is expected to contribute $165 of net present-valued savings to the economy over the period of 1990–2010. In addition to energy and cost savings, minimum efficiency standards have environmental benefits with significant reduction in carbon emissions. From 2000 to 2010, it is estimated that energy efficiency standards will result in an annual carbon reduction of 4% (corresponding to 9 million metric tons of carbon/year) relative to the 1990 level.

Currently, energy efficiency standards are utilized with various degrees of comprehensiveness, enforcement, and adoption in a limited number of countries, as summarized in Table 10.8. However, several other countries are in the process of developing national standards or labeling programs to promote energy-efficient residential appliances.

Minimum efficiency performance standards have been in use in the United States and Canada for more than three decades and cover a wide range of products. In the last few years, standards programs have spread to several other countries like Brazil,

TABLE 10.8
Status of International Residential Appliance Energy Efficiency Standards

Country/Region	Compliance Status	Products
Australia	Mandatory	R, FR, WH
Brazil	Voluntary	R, FR
Canada	Mandatory	All
China	Mandatory	R, CW, RAC
European Union	Mandatory	R, FR
India	Voluntary	R, RAC, A/C
Japan	Voluntary	A/C
Korea	Mandatory	R, A/C
Mexico	Mandatory	R, FR, RAC
Philippines	Mandatory	A/C
United States	Mandatory	All

Notes: Products are: Refrigerators (R), Freezers (FR), Clothes Washers (CW), Cooking, Water Heaters (WH), Room Air Conditioners (RAC), and Central Air Conditioning (A/C). All include all the previous appliances including Dishwashers and Clothes Dryers.

China, Korea, Mexico, and the Philippines. Most of these countries have currently limited products subject to standards. Other countries are in the process of implementing or considering energy efficiency standards for household appliances. Among these countries are Colombia, Egypt, Indonesia, Malaysia, Pakistan, and Thailand.

Many of the currently existing standards are mandatory and prohibit the manufacturers from selling or delivering non-complying products. However, standards in some countries are voluntary and thus are not mandatory, such as in the case of product quality standards established in India.

As indicated in Table 10.6, most countries have established minimum efficiency performance standards for refrigerators and freezers since this product type has one of the highest growth rates, both in terms of sales value and volume. The existing international energy efficiency standards for refrigerators and freezers set a limit on energy use over a specific period of time (generally, 1 month or 1 year). This energy use limit may vary depending on the size and the configuration of the product. Table 10.9 shows maximum limits for the allowable annual energy use for U.S. refrigerators and freezers. Two standards are shown in Table 10.9: the standards that are currently effective in 2001 and the updated standards effective in April 2008. It should be noted that models with higher energy efficiencies than those listed in Table 10.7 do exist and are sold in the U.S. market. To keep up with the technological advances, U.S. standards are typically amended or changed periodically. However, any new or amended U.S. standard for energy efficiency has to be based on improvements that are technologically feasible and economically justified. Typically,

TABLE 10.9
Maximum Allowable Annual Energy Use (kWh/Year) for Refrigerators and Freezers Sold and Manufactured in the United States

Product Category	2001 Standard	2008 Standard
Manual defrost R/FR	248.4 + 8.82 AV	198.4 + 7.056 AV
Partial auto-defrost R/FR	248.4 + 8.82 AV	198.4 + 7.056 AV
Top-mount auto-defrost R/FR	276 + 9.8 AV	220.8 + 7.84 AV
Top-mount auto-defrost with through the door features R/FR	356 + 10.20 AV	284.8 + 8.16 AV
Side-mount auto-defrost R/FR	507.5 + 4.91 AV	406 + 3.928 AV
Side-mount auto-defrost with through the door features R/FR	406 + 10.10 AV	324.8 + 8.08 AV
Bottom-mount auto-defrost R/FR	459 + 4.6 AV	367.2 + 3.68 AV
Upright manual FR	258.3 + 7.55 AV	232.47 + 8.08 AV
Upright auto-defrost FR	326.1 + 12.43 AV	293.49 + 11.87 AV
Chest RF	143.7 + 9.88 AV	129.33 + 8.892 AV

Source: Energy Star, US Environmental Protection Agency, US Environmental Protection Agency, Washington, DC, http://www.energystar.gov, 2009.

Notes: R, refrigerators; FR, freezers; AV, adjusted volume = volume of R + 1.63 × volume of FR (for R/FR) = 1.73 × volume of FR (for FR only).

energy-efficient designs with payback periods of less than 3 years are periodically incorporated into new U.S. standards.

In addition to standards, labeling programs have been developed to inform consumers about the benefits of energy-efficient products. There is a wide range of labels used in various countries to promote energy efficiency for appliances. These labels can be grouped into three categories:

1. Efficiency-type labels are used to allow consumers to compare the performance of different models for a particular product type. For instance, a common label used for refrigerators indicates the energy use and operating cost over a specific period, such as 1 month or 1 year. Other labels such as labels for clothes washers show efficiency expressed in kWh of energy used per pound or kilogram of clothes washed. Another feature of efficiency labels is the ability to provide consumers with a comparative evaluation of the product models by showing the energy consumption or efficiency of a particular model on a scale of the lowest (or highest) energy use (or efficiency) models.
2. Ecolabels provide information on more than one aspect (i.e., energy efficiency) of the products. Other aspects include noise level, waste disposal, and emissions. Green Seal in the United States is an example of an ecolabel program that certifies that the products are designed and manufactured in an environmentally responsible manner (Green Seal, 1993). Certification

standards have been established for refrigerators, freezers, clothes washers, clothes dryers, dishwashers, and cooktops/ovens.

3. Efficiency seals of approval, such as the Energy Star program in the United States, are labels that indicate that a product has met a set of energy efficiency criteria but do not quantify the degree by which the criteria were met. The Energy Star label, established by the U.S. Environmental Protection Agency (EPA), indicates for instance that a computer monitor is capable of reducing its standby power level when not in use for some time period.

In recent years, labeling of appliances is becoming a popular approach around the world in order to inform consumers about the energy use and energy cost of purchasing different models of the same product. Presently, Australia, the United States, and Canada have the most comprehensive and extensive labeling programs. Other countries such as the European Union, Japan, Korea, Brazil, Philippines, and Thailand, have developed labels for select products.

In addition to energy efficiency, standards have been developed to improve the performance of some appliances in conserving water. For instance, water-efficient plumbing fixtures and equipment have been developed in the United States to promote water conservation. It should be noted that the reduction of water use by some household appliances can also increase their energy efficiency. Indeed, a large fraction of the electrical energy used by both clothes washers and dishwashers is attributed to heating the water (85% for clothes washers and 80% for dishwashers).

10.5.2 Office Equipment

Over the two decades, the energy used by office equipment has increased significantly and accounts currently for more than 10% of the total commercial sector electricity use. Recognizing this problem, the U.S. EPA, in cooperation with the U.S. DOE, has launched over a decade ago the Energy Star Office Program to increase the energy efficiency of commonly used office equipment such as computers, fax machines, printers, and scanners. To help consumers in identifying energy-efficient office equipment, Energy Star labels are provided to indicate the energy-saving features of the products.

It is estimated that Energy Star–labeled products can save as much as 75% of the total electricity use depending on the type and the usage pattern of the office equipment. Almost all office equipment manufacturers currently integrate power management features in their products. For instance, computers can enter a low-power "sleep" mode when idle for a specific period of time. Similarly, copiers can go into a low-power mode of only 15–45 W after 30–90 minutes of activity. Table 10.10 summarizes the Energy Star features of common office equipment.

A study has estimated that the Energy Star–labeled products have reduced significant carbon emissions through its voluntary labeling efforts (Sanchez et al., 2008). Specifically, and since its inception in 1992 through 2006, the program has saved 4.8 EJ of primary energy (1 EJ = 10^{18} J = 1 quadrillion of Btu = 10^{15} Btu). The same study indicates that the program will save 12.8 EJ over the period 2007–2010 (Sanchez et al., 2008).

TABLE 10.10
Summary of the Energy Star Specifications for Copiers, Scanners, and Fax Machines

Product	Typical Electrical Consumption in kWh/Week	Power Management Preset Default Times (min)
Color copiers, scanners, and fax machines (specifications depend on the images per minute [ipm])	0.1 ipm + 2.8 (ipm < 32)	5–30
	0.35 ipm − 5.2 (32 < ipm < 58)	5–60
	0.70 ipm − 26.0 (ipm > 58)	5–60
Monochrome printers, scanners, and fax machines (specifications depend on the pages per minute)	1.0 (ipm < 15)	5–15
	0.1 ipm − 0.5 (15 < ipm < 40)	5–60
	0.35 ipm − 10.3 (40 < ipm < 82)	5–60
	0.70 ipm − 39.0 (ipm > 82)	5–60

TABLE 10.11
Measured Average Standby Power and Energy Standard Requirements for Selected Electrical Devices

Electrical Device	Average Standby Power (W)	Energy Star Standby Power (W)
TV (CRT)	3	1
Set-top box, digital cable	17.8	1
Audio mini-system	8.3	1
VCR	4.7	1
DVD player	1.6	1
Desktop computer	2.8	2
Laptop computer	8.9	1
Printer	1.4	1
Fax machine	5.3	2
Copier	1.5	1
Microwave oven	2.9	2

10.5.3 Standby Power Losses

Some electrical devices have a built-in clock as well as battery units that consume energy even when the devices are turned off. Standby power loads, commonly called phantom loads, can be significant in residential homes. Typical values for standby power measured for selected electrical devices are outlined in Table 10.11 (LBNL, 2004). The Energy Star requirements for standby power are also provided in Table 10.11.

To estimate the energy used by phantom loads for any electrical device with a standby power, W_{stby}, the number of hours when the device is turned off, $N_{h,off}$, should be used:

$$\text{kWh}_{stby} = W_{stby,j} \times N_{h,off} \tag{10.5}$$

Example 10.5 shows the application of Equation 10.5 to estimate the electrical energy use due to phantom loads for TV sets.

Example 10.5

Problem

Estimate the annual phantom load for a TV with a set-top box if the TV is operated 2 h/day. Determine the annual cost of the phantom load energy if the electricity cost is 0.10/kWh.

Solution

Based on Table 10.6, the standby power for a TV is 3 W, while of the set-top box is 17.8 W; thus, $W_{stby}=20.8$ W. Since the TV and its set-top box are operated 2 h/day, it is turned off 22 h/day. Therefore, $N_{h,off}=22$ h/day * 365 days/year = 8,030 h/year. Using Equation 10.5, the annual energy used by the phantom load associated with the TV and the set-top box is

$$\text{kWh}_{Stby} = (20.8\text{ W})(8{,}030\text{ h/year}) = 167{,}024\text{ W/year} = 167\text{ kWh/year}$$

Thus, the annual cost of the phantom load is \$16.7/year, based on \$0.10/kWh electricity cost. The use of power strip would save this annual cost.

10.6 ELECTRICAL DISTRIBUTION SYSTEMS

10.6.1 Introduction

All electrical distribution systems such as transformers, wires, and protection devices have to be properly designed and installed in order to provide electrical energy to the utilization equipment as safely, reliably, and economically as possible. However, these systems offer some opportunities for increasing the energy efficiency of the electrical systems for buildings. In this section, some of the potential energy efficiency improvements are described for selecting transformers and conductors in typical electrical installations for new and existing buildings.

10.6.2 Transformers

As noted in Chapter 3, transformers are not ideal (i.e., 100% efficient) since they waste energy through copper and leakage losses. Table 10.10 compares full-load energy efficiency values as the function of rating for NEMA premium transformers, compared to the current DOE regulations and baseline efficiency levels for three-phase transformers (NEMA, 2010). It should be noted that energy efficiency varies with the load served by the transformer. Typically, most regulations provide the minimum allowable full-load energy efficiency specifications as shown in Table 10.12.

TABLE 10.12
Comparison of Energy Efficiency Values for Baseline, Department of Energy Regulations, and National Electrical Manufacturers Association Premium Classifications

Rating (kVA)	Baseline Efficiency (%)[a]	Efficiency DOE Regulations (%)[b]	NEMA Premium Efficiency (%)[c]
15	97.00	97.89	97.90
30	97.50	98.23	98.25
45	97.70	98.40	98.39
75	98.80	98.60	98.60
112.5	98.20	98.74	98.74
150	98.30	98.83	98.81
225	98.50	98.94	98.95
300	98.60	99.02	99.02
500	98.70	99.14	99.09
750	98.80	99.23	99.16
1,000	98.90	99.28	99.23

[a] These values are based on NEMA TP-1 2002.
[b] These values are based on DOE regulations of final rule of 10 CRR 431 effective January 2016.
[c] These values are based on NEMA EL-3, CSL-3.

The potential energy savings associated with installing energy efficient rather than standard-efficiency transformers can be estimated as follows:

$$kWh_{saved} = N_h \times kVA \times LF \times pf \times \left(\frac{1}{\eta_{std}} - \frac{1}{\eta_{eff}} \right) \quad (10.6)$$

where

N_h is the operating hours per year for the transformer
kVA is the rating of the transformer
η_{std} and η_{eff} are the energy efficiency values for the standard and the efficient transformers
LF is the annual average load factor for the transformer
pf is the annual average power factor for the electrical load served by the transformer

While more expensive than standard transformers, high-efficiency or premium-efficiency systems are generally cost-effective to install especially if they are old and are used continuously. Table 10.13 compares typical efficiency values and energy losses at full load for transformers made in 1973 and 2013 (Ferris and Johnson, 2014). Example 10.6 illustrates the cost analysis of replacing a standard transformer by a new one with better energy efficiency.

TABLE 10.13
Comparison of Typical Energy Efficiency Values and Losses for 1973 and 2013 Transformers

	1973 Transformers (Full Load)		2013 Transformers (Full Load)	
Rating (kVA)	Energy Efficiency (%)	Total Losses (W)	Energy Efficiency (%)	Total Losses (W)
15	95.3	740	96.8	551
30	96.0	1,250	97.2	904
45	96.6	1,620	97.9	1,027
75	96.8	2,570	97.7	1,782
112.5	97.6	1,400	97.9	2,521
150	97.6	1,830	98.3	2,760
225	98.0	4,775	98.3	4,047
300	98.3	5,400	98.9	5,338
500	98.4	8,300	98.9	5,858

Example 10.6

Problem

Determine the cost-effectiveness of using a unit with an efficiency of 98.50% rather than 98.95% (NEMA Premium) for a 225 kVA rated transformer. Assume that

- The cost of electricity is \$0.12/kWh
- The installed costs of 98.5% and 98.95% efficient transformers are \$5,950 and \$7,100, respectively
- The average power factor of the load 0.90
- For the analysis, consider that the transformer is used on average at 75% of its capacity for 8,760 h/year.

Solution

To determine the cost-effectiveness of installing an energy-efficient transformer, a simplified economic analysis based on the payback period is used. The savings in energy losses in kWh for the high-efficiency transformer can be calculated using Equation 10.6:

$$\text{kWh}_{\text{saved}} = 8,760 \times 225 \times 0.90 \times 0.75 \times \left(\frac{1}{0.9850} - \frac{1}{0.9895} \right) = 6,143 \text{ kWh/year}$$

Therefore, the payback period, *SPP*, for using the efficient transformer is

$$SPP = \frac{\$7,100 - \$5,950}{6,143 \text{ kWh} \times \$0.12 \text{ kWh}} = 1.56 \text{ years}$$

Note, for retrofit application, and if the existing one is operating normally (and thus does not need to be replaced), the payback would be estimated as follows:

$$SPP = \frac{\$7{,}100}{6{,}143\text{ kWh} \times \$0.12\text{ kWh}} = 9.63\text{ years}$$

It is clear that it can be cost-effective to consider investing in a more energy-efficient transformer for new constructions. However, it may not be cost-effective to replace an existing and functional transformer for retrofit applications. However, more rigorous economic analysis using life cycle cost approach is highly recommended to assess the cost-effectiveness of the transformer replacement even when high simple payback periods are estimated.

10.6.3 Electrical Wires

As discussed in Chapter 6, the number of electrical conductors that can be installed in any one conduit is restricted to avoid any damage to cables (especially when the cables are pulled through the conduit). The NEC restricts the percentage fill to 40% for three or more conductors. The percentage fill is defined as the fraction of the total cross-sectional area of the conductors—including the insulation—over the cross-sectional area of the inside of the conduit. When selecting the size of the conductors, the operating costs, in addition to the initial costs, should be considered. As illustrated in Example 10.7, the cost of energy encourages the installation of larger conductors than those required typically by the NEC, especially when smaller size conductors are involved (i.e., numbers 14, 12, 10, and 8). Unfortunately, most designers do not consider the operating costs in their design due to several reasons, including interest in lower first costs and uncertainty in electricity prices.

Example 10.7

Problem

Determine if it is economically feasible to install number 10 (AWG) copper conductor instead of number 12 (AWG) on a 100 ft branch circuit that feeds a load of 16 A. Assume that

- The load is used 10 h/day and 300 days/year
- The cost of electricity is $0.10/kWh
- The installed costs of No. 12 and No. 10 conductors are, respectively, $60.00 and $90.00/1,000 ft long cable

Solution

In addition to the electric energy used to meet the load, there is an energy loss in the form of heat generated by the flow of current, *I*, through the resistance of the conductor, *R*. The heat loss in *Watts* can be calculated as follows:

$$\text{Watts} = R \times I^2$$

Using Table 2.2 or NEC Tables, the resistance of both conductors, No. 12 and 10, can be determined to be, respectively, 0.193 and 0.121 Ω/100 ft. Thus, heat loss for the 400 ft branch circuit if No. 12 conductor is used can be estimated as follows:

$$\text{Watts}_{12} = 0.193 \times \frac{100}{100} \times (16)^2 = 49.4 \text{ W}$$

Similarly, the heat loss for the 400 ft branch circuit when No. 10 conductor is used is found to be

$$\text{Watts}_{10} = 0.121 \times \frac{100}{100} \times (16)^2 = 31.0 \text{ W}$$

The annual cost of copper losses for both cases can be easily calculated:

$$\text{Cost}_{12} = 49.4\text{W} \times 300 \text{ days/year} \times 10 \text{ h/day} \times \frac{1 \text{ kW}}{1{,}000 \text{ W}} \times \$0.10/\text{kWh} = \$14.8/\text{year}$$

$$\text{Cost}_{10} = 31.0\text{W} \times 300 \text{ days/year} \times 10 \text{ h/day} \times \frac{1 \text{ kW}}{1{,}000 \text{ W}} \times \$0.10/\text{kWh} = \$9.3/\text{year}$$

Therefore, if No. 10 is used instead of No. 12, the simple payback periods, *SPP*, for the higher initial cost for the branch circuit conductor is

$$SPP = \frac{\left(\frac{\$90}{1{,}000 \text{ ft}} - \frac{\$60}{1{,}000 \text{ ft}}\right) \times 100 \text{ ft}}{(\$14.8 - \$9.3)} = 0.54 \text{ year} = 6.5 \text{ months}$$

Savings in energy consumption through the use of larger conductors can thus be cost-effective. A life cycle cost analysis would provide more compelling arguments for the cost-effectiveness of increasing the conductor size. Moreover, it should be noted that the larger size conductors reduce the voltage drop across the branch circuit that permits the connected electrical utilization equipment to operate more efficiently. However, the applicable code has to be carefully consulted to determine if larger size conduit is required when larger size conductors are used.

10.7 SUMMARY

In this chapter, an overview is provided for the basic characteristics of electrical systems in residential and commercial buildings. In particular, an overview of various electrical end-use equipment including motors, lighting fixtures, and appliances is outlined. Moreover, several energy efficiency options specific to electrical loads are described to reduce not only the energy use cost but the peak demand associated with building electrical systems. The reduction in peak demand allows the designer to specify smaller capacities for the electrical distribution systems. Specific guidelines for selecting energy-efficient components for building electrical distribution systems are provided especially for transformers and wires. Moreover, illustrative examples are presented to evaluate the cost-effectiveness of selected energy efficiency measures specific to electrical systems for residential and commercial buildings.

10.8 PROBLEMS

1. Two motors operate 5,000 h/year at full load. One is 10 HP with an electrical efficiency of 0.75 and a power factor of 0.65 while the other is 7.5 HP with an electrical efficiency of 0.935 and a power factor of 0.85. Determine
   a. The overall power factor
   b. The simple payback period of replacing each motor by and energy-efficient motor

   Assume that the cost of electricity is \$0.12/kWh. The incremental costs for energy-efficient motors are \$150 for 10 HP motor and \$100 for 7.5 motor.
2. Determine the capacitor ratings (in kVAR) to add to a 10 HP motor with an efficiency of 0.85 to increase its power factor from 0.80 to 0.85, 0.90, and 0.95, respectively.
3. For an apartment building with a total of 20 incandescent lamps rated at 60 W. Determine the energy-saving potential if these lamps are replaced with 13 W CFLs. The lifespan of incandescent lamp and CFL is, respectively, 1,300 and 8,000 h. Perform the analysis using the lifespan of CFLs. Determine the cost-effectiveness of replacing the incandescent lamps with CFLs if the cost of the incandescent lamp and CFL is \$0.20 and \$1.40, respectively. Assume the cost of electricity is \$0.10/kWh. Perform a simple payback period analysis.
4. For a house with a total 20 of incandescent lamps rated at 60 W, determine the energy-saving potential if these lamps are replaced with (1) 13 W CFLs and (2) 7 W LED lamps. The lifespan of incandescent lamp, CFL, and LED lamp is 1,300, 8,000, and 30,000 h, respectively. Perform the analysis using the lifespan of the LED lamps. Determine the best lamps to use, CFLs or LED lamps, if the cost of the incandescent lamp, CFL, and LED lamp is \$0.25, \$1.45, and \$45, respectively. Assume the cost of electricity is \$0.12/kWh and perform a life cycle cost analysis using the proper life cycle period and a 5% discount rate.
5. Estimate the annual phantom load for a TV with a set-top box if the TV is operated 5 h/day. Determine the annual cost of the phantom load energy if the electricity cost is 0.12/kWh.
6. Estimate the annual phantom load for an audio mini-system operated 3 h/day. Determine the annual cost of the phantom load energy if the electricity cost is 0.10/kWh.
7. Consider a ranch house of 100 m^2 located in Phoenix, CO, with a total window area of 15 m^2 with double-pane glazing that has a visible transmittance of 0.35. The home has 20 incandescent lamps, each rated at 60 W.
   a. Estimate the energy use and cost savings when the incandescent lamps are replaced with 13 W CFLs
   b. Determine the energy use and cost savings due to daylighting harvesting strategies for both options with the existing incandescent lamps and with CFLs.

   Assume that without daylighting harvesting, each lamp is operated 4 h/day when daylighting can be harvested, 6 days/week, 50 weeks/year. The cost of electricity is 0.12/kWh.

8. Determine the energy and cost savings of daylighting harvesting to a thin two-story house building (each floor is 24 ft × 50 ft) with a window-to-floor area ratio of 30%. The building has clear windows (i.e., visible transmittance = 0.78) and is located in Denver, CO. The lighting system has a density of 0.5 W/ft^2 and operates at 3,000 h/year. Assume that the cost of electricity is $0.10/kWh.
9. Determine the breakeven costs for installing a 1,000-kVA transformer for a new building with an efficiency of 99.28% rather than 98.90%. If the annual load factor is 80%, the average building's power factor is 0.90, and the electricity cost is $0.12/kWh. Consider two options for the economic analysis:
   - Vary for the life cycle from 10 years to 60 years with the discount rate kept at 5%.
   - Vary the discount rate from –10% to 10% with the life cycle of the transformer is assumed to be 50 years.
10. Determine if it is economically feasible to install number 8 (AWG) copper conductor instead of number 12 (AWG) on 200 ft branch circuit that feeds a load of 12 A. Assume that:
    a. The load is used 14 h/day and 320 days/year.
    b. The cost of electricity is $0.12/kWh.
    c. The installed costs of No. 12 and No. 8 conductors are, respectively, $60.00 and $105.00/1,000 ft long cable.
11. Determine the breakeven incremental cost for using 10 (AWG) copper conductors instead of number 12 (AWG) on 5,000 ft branch circuit that feeds a load of 15 A. For the analysis, consider the following assumptions:
    a. The load is used 12 h/day and 300 days/year.
    b. The cost of electricity is $0.15/kWh.
    c. The life cycle for the conductors is 30 years and the discount rate is 5%.

11 Power Quality in Buildings

Disturbances in power supply to buildings can significantly affect the reliability, safety, and integrity of not only the electrical distribution system components but also of the connected loads. This chapter outlines common power quality disturbances and means to mitigate them in order to minimize significant damages to the building electrical systems and maintain reliable performance of electrical equipment. Voltage sags, voltage swells, interruptions, transients, voltage unbalances, harmonic distortions, and electrical noises are some of the power quality disturbances common in buildings and are presented in this chapter. In particular, a detailed description of the harmonic distortions is provided, with specific metrics to assess their magnitudes and mitigation methods to reduce their effects.

11.1 INTRODUCTION

As discussed in Chapter 2, both AC voltage and current follow sinusoidal waveforms when all the electrical loads are linear and are operating under normal conditions. However, disturbances may occur in the electrical distribution systems and can affect the time variations of voltage and/or current waveforms. Two types of disturbances are common in building power systems that include (1) slight voltage and/or current variations and (2) sudden and significant voltage and/or current deviations that are often called events (Bollen, 2000). Voltage unbalances, harmonic distortions, voltage fluctuations, and electrical noises are considered variations and may affect the normal operation of some electrical equipment if these disturbances are not corrected within short time periods. Voltage sags, voltage swells, interruptions, and transients are considered events that greatly affect the current and voltage waveforms and can result in significant damage to, especially, electronic devices connected to the electrical distribution systems.

To protect against power quality disturbances, the following are some preventive and corrective devices and strategies that can be considered:

- Pulsed rectifiers—utilized to improve the shape of voltage and current sinusoidal waveforms, especially when operating motors
- Power harmonic filters—commonly used to reduce the effects of harmonic distortions
- Dynamic voltage compensators—installed between power sources and loads to protect electrical systems against voltage sags
- Capacitor banks—specified to provide several benefits to the electrical distribution systems, including increase in the power factor, improvement of

DOI: 10.1201/9781003276999-11

voltage regulation, reduction in energy losses, and prevention of harmonic resonances
- Special transformers—utilized to control harmonic currents and mitigate voltage sags
- Uninterruptible power supply (UPS) systems—commonly specified as backup power sources for critical loads that require stable voltage levels.

It should be noted that by improving power quality, the reliability and efficiency of the electrical systems for buildings can be noticeably enhanced. Thus, it is important to consider potential power quality problems when designing electrical systems for buildings. In this chapter, common power quality disturbances are first described. Then, a detailed description of the harmonic distortions is provided, including common indicators and metrics used to evaluate the level of the disturbances and their potential impacts on the building electrical distribution systems and their components, including common loads.

11.2 ELECTRICAL DISTURBANCES

In this section, a summary of types and effects of common power disturbances is provided.

11.2.1 Voltage Sags

The voltage sag occurs when the root mean square (rms) voltage value decreases over a limited time period before it settles and returns to its average value. There are several types of sags including (Caramia et al., 2009)

- Instantaneous sags lasting typically from 0.5 to 30 cycles
- Momentary sags that can last 30 cycles to 3 seconds
- Temporary sags that last 3 seconds to 1 minute

If the sag lasts for more than a minute, it is then classified as an undervoltage. A voltage sag is generally caused by a short-circuit fault or by a sudden change in the characteristics of a load, such as the case during the starting of a motor. If the voltage does sag, some equipment may not be supplied with the amount of voltage they require and their performance may be affected (Kusko and Thompson, 2007). Figure 11.1 provides a typical time variation of voltage during a sag event.

11.2.2 Voltage Swells

A voltage swell refers to the case when the rms voltage increases during a time period before it comes back to its average value. Similar to the sag events, several swell types have been identified:

- Instantaneous swells lasting from 0.5 to 30 cycles
- Momentary swells that can last 30 cycles to 3 seconds
- Temporary swells that last 3 seconds to 1 minute

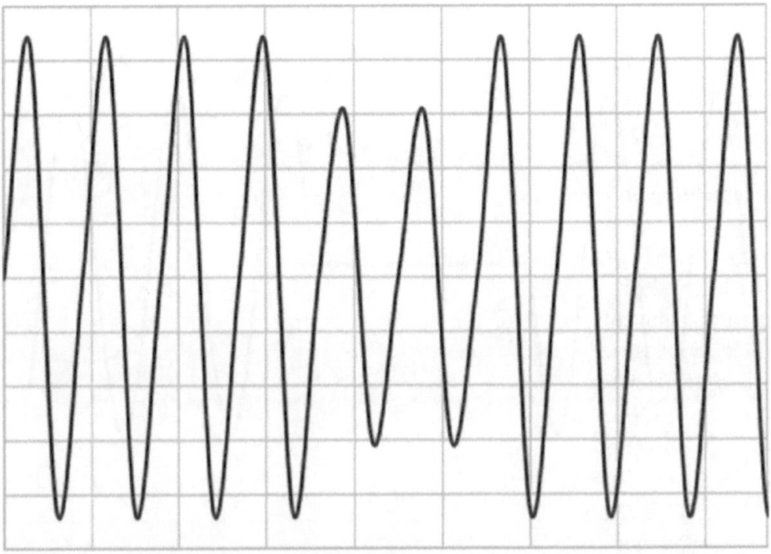

FIGURE 11.1 Variation of voltage during a sag event.

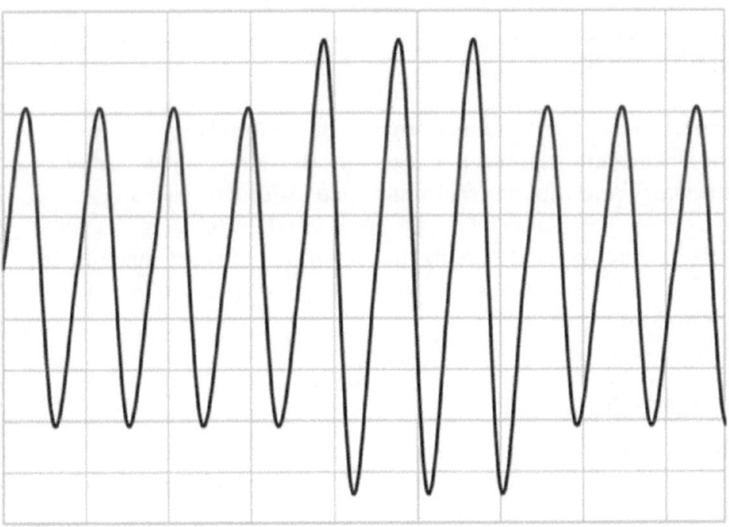

FIGURE 11.2 Variation of voltage during a swell event.

Again, if the swell lasts for more than a minute, it is then classified as an overvoltage event. Voltage swells occur mostly due to faults in the electrical distribution systems. They do not occur as often as voltage sags, but they can cause more damage to select devices that are sensitive to higher voltages than their rating levels (Kusko and Thompson, 2007). Figure 11.2 illustrates an example of time variation of voltage in case of a swell event.

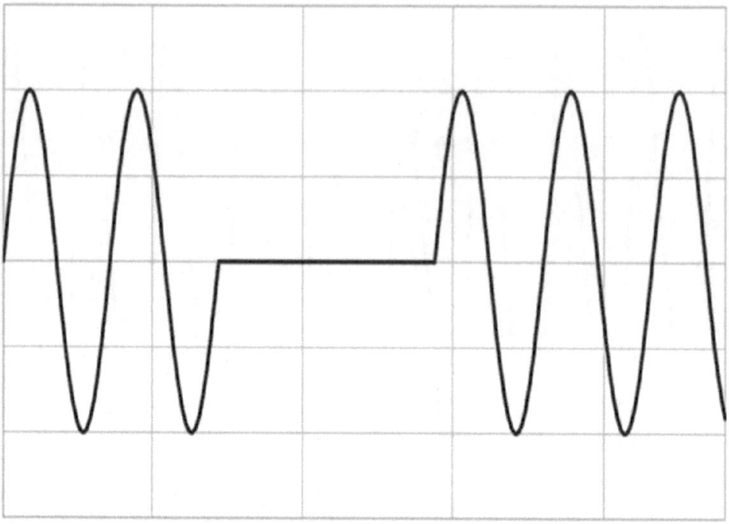

FIGURE 11.3 Variation of voltage during an interruption event.

11.2.2.1 Interruptions

An interruption is an event when the supply voltage decreases to less than 0.1 pu (i.e., 0.1% of its normal value) for a limited period of time. Several types of interruptions can occur:

- Instantaneous interruptions lasting from 0.5 to 30 cycles
- Momentary interruptions that typically last 30 cycles to 3 seconds
- Temporary interruptions that last 3 seconds to 1 minute

Like swells, interruptions are events that occur due to faults that may trip protection devices. Figure 11.3 shows an example of time variation of voltage in case of an interruption event.

11.2.3 Transients

Transients in electrical power systems can take two forms: impulsive transients and oscillatory transients. Impulsive transients refer to cases when there are nearly instantaneous changes in the steady-state conditions of voltages or currents. In particular, the voltages or currents suddenly spike and deviate by 30%–90% from their typical values. The most common cause of impulsive transients is a lightning strike. An example of an impulsive transient is illustrated in Figure 11.4.

When oscillatory transients occur, voltages and/or currents oscillate positively and negatively relative to their typical values. These oscillations are classified into low, medium, or high frequencies. Low-frequency transients are usually associated to switching from the utility power source to capacitor banks resulting in frequencies ranging from 300 to 900 Hz. Medium-frequency transients can be caused by

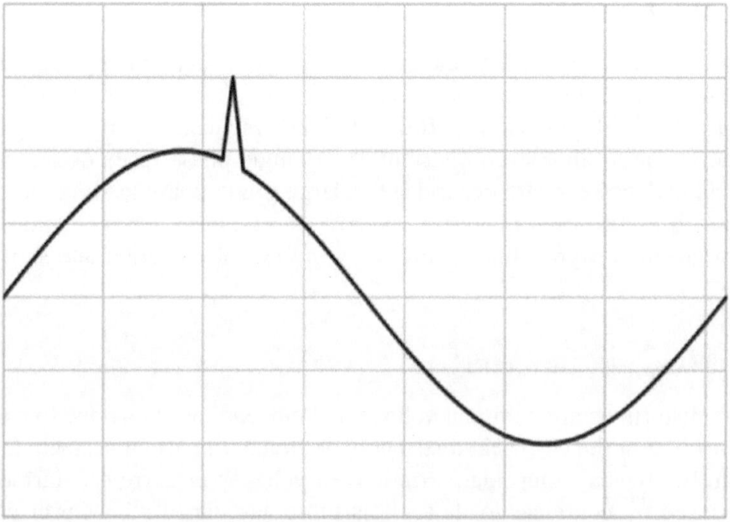

FIGURE 11.4 Variation of voltage during an impulsive transient.

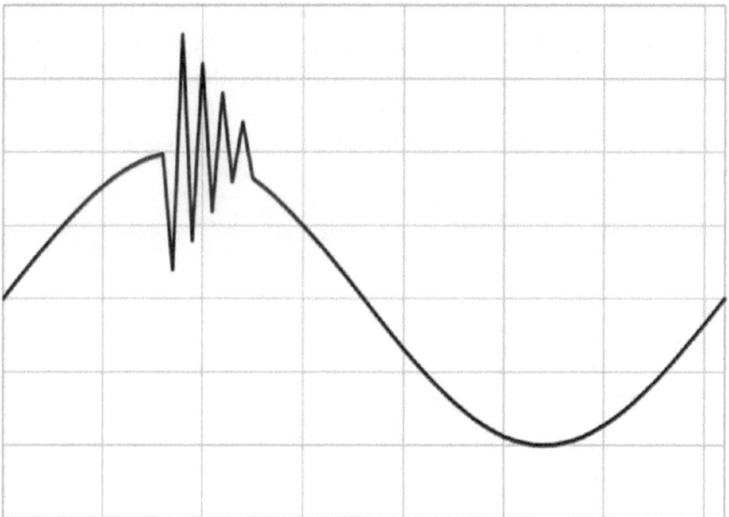

FIGURE 11.5 Variation of voltage during an oscillatory transient.

back-to-back capacitor switching and can result in frequencies of 5–500 kHz. High-frequency transients are typically caused by impulsive transients and result in frequencies of more than 500 kHz. The damages due to transients can be significant and can result in the immediate failure of some electrical equipment. An example of an oscillatory transient is shown in Figure 11.5.

11.2.4 Voltage Unbalances

Voltage unbalances are associated with variations in voltages and currents for three-phase systems when the rms voltage values or the phase angles become unequal between the three phases (Bollen, 2000). Voltage unbalances can occur when load imbalance is due to an uneven distribution of single-phase loads over three-phase power systems. They can also occur due to a large single-phase loading such as when one of the fuses becomes nonoperational for three-phase motors. Voltage unbalances can generate overheating within the windings of induction and synchronous machines.

11.2.5 Harmonic Distortions

Harmonic distortions are common voltage and current variations due to changes in frequencies within the electrical distribution systems. In particular, there are deviations from the typical sinusoidal variations in voltages or currents. Harmonics are mainly caused by nonlinear loads such as those associated with power electronic converters including variable frequency drives (VFDs) installed for fans and pumps serving building air conditioning systems (Vedam and Sarma, 2009). Specifically, these converters draw a nonsinusoidal current/voltage and introduce nonlinear loads throughout the electrical distribution systems. The problems due to harmonic distortions are nowadays common in buildings since computers and other power electronics are heavily utilized. In particular, harmonic distortions can cause several disturbances and damages, including wire overheating and power loss. More detailed description of harmonic distortions, including causes, assessments, and corrective measures, is provided in Section 11.3. Figure 11.6 shows typical voltage variation due to harmonic distortion.

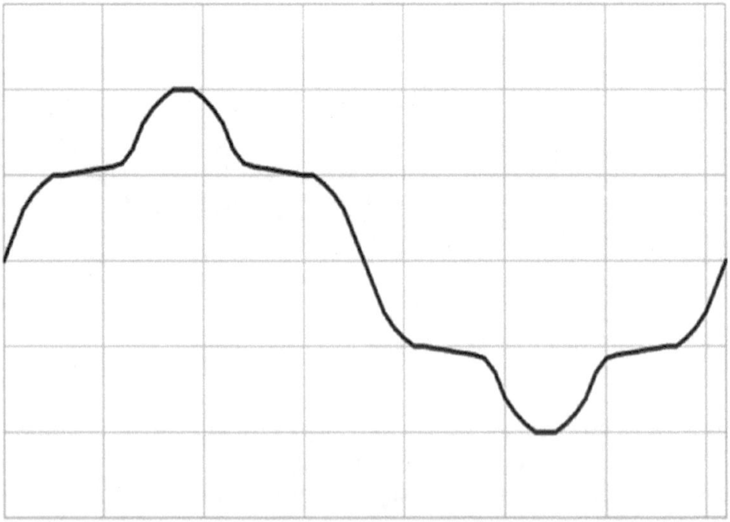

FIGURE 11.6 Variation of voltage during a harmonic distortion.

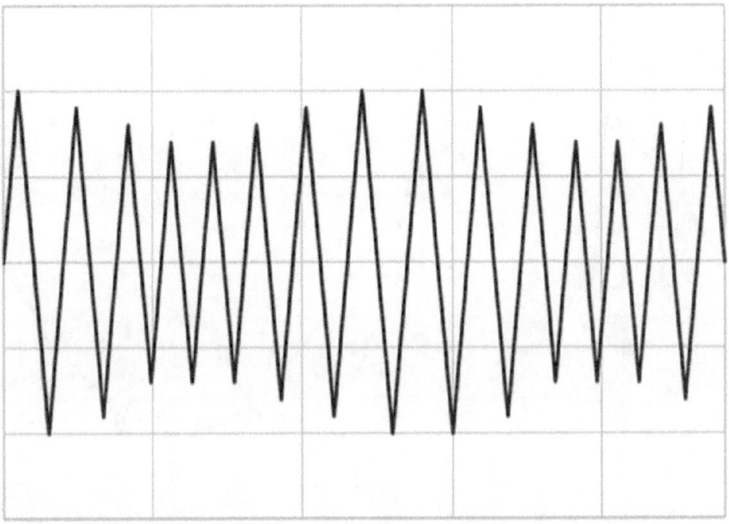

FIGURE 11.7 Variation of voltage during a voltage fluctuation condition.

11.2.6 Voltage Fluctuations

Voltage fluctuations present different types of voltage and current variations that can cause damage to the electrical distribution systems. These fluctuations, often called flickers, are the result of the application of large loads that can distort the voltages. Common loads that cause voltage fluctuations are arc furnaces used in steel industrial facilities. Flickers due to voltage fluctuations can be readily observed in some lighting fixtures. Figure 11.7 shows an example of voltage fluctuations.

11.2.7 Electrical Noises

Electrical noise is associated with an unwanted electrical signal with a frequency generally lower than 200kHz (Caramia et al., 2009). It should be noted that noise disturbances are different from harmonic distortions and transients since they do not act like waveforms. Sources of noise include power electronic devices, control circuits, arcing equipment and switching power supplies (Vedam and Sarma, 2009). It has been noted that the effects due to noises can be made worse when grounding is introduced to the electrical distribution system. An example of electrical noise is outlined in Figure 11.8.

11.3 MITIGATION OPTIONS

Power quality disturbances are becoming common occurrences in buildings, especially those with complex electrical distribution systems. Indeed, the wide range of electrical systems and devices present within modern buildings has increased unwanted and uncontrolled variations of electrical voltages and currents. To eliminate or reduce these disturbances, several strategies and techniques are currently available and can

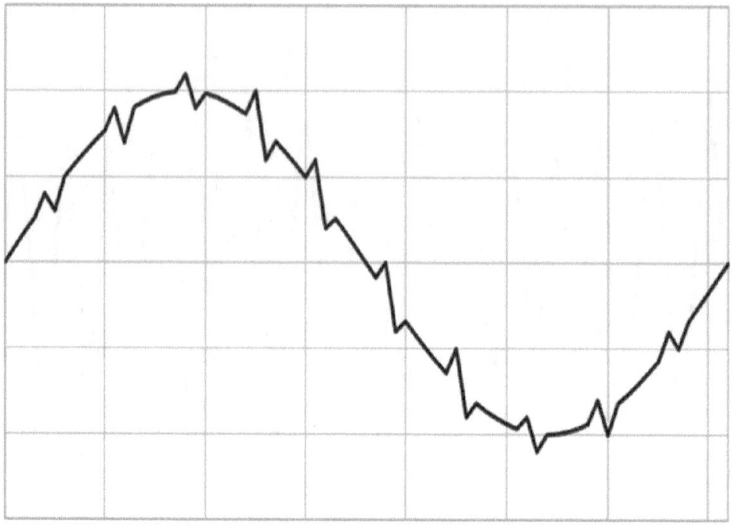

FIGURE 11.8 Variation of voltage during a noise event.

be employed to ensure acceptable and reliable power quality to vital electrical systems within buildings. First, it is important to properly design and select high-quality devices and equipment for the building electrical systems to ensure good power quality. When disturbances occur, methods of correcting power quality can be considered depending on the main causes. In particular, and to control voltages, power quality mitigation approaches include the addition of power harmonic filters and the use of voltage compensators. Moreover, capacitor banks can be installed to reduce the reactive power generated by some loads. Special transformers with phase-shifting connections can also diminish harmonic currents. Finally, the integration of UPS systems can minimize power quality issues within parts of the electrical distribution systems of the building.

In summary, two main criteria should be considered when designing and selecting equipment for building electrical systems. First, it is important to install high-quality electrical equipment with low possibilities of causing power quality issues. Second, it is crucial to use electrical systems with low sensitivity to voltage and current fluctuations. When harmonic distortions do occur, several mitigation systems can be considered, as briefly discussed in the following sections.

11.3.1 Multipulse Rectifiers

The use of VFDs, also known as variable speed drives (VSDs), is becoming common to operate motors used in heating, ventilating, and air conditioning (HVAC) systems. For three-phase motors, VFDs are typically made up of six-pulse rectifiers (i.e., sets of diodes) that allow ACs to be drawn in pulses and converted to DC currents. These pulses generate harmonic current distortions with frequencies h that are multiple of the fundamental frequency with

$$h = np \pm 1 \tag{11.1}$$

where

n is any integer number (i.e., 1–3)
p is the pulse number

Thus, for the conventional six-pulse rectifiers ($p=6$), the significant harmonics are 5, 7, 11, 13, 17, 19, etc. Typically, the magnitude of each harmonics is inversely proportional to the harmonic number so that 5th is more significant than 7th, which is more significant that 11th.

To reduce the generated harmonics, higher pulse rectifiers are utilized, including 12-pulse, 18-pulse, or 24-pulse rectifiers. For 12-pulse rectifiers ($p=12$), significant harmonics can be determined using Equation 11.1 and consist of 11th, 13th, 23rd, 25th, etc. Thus, and relative to the 6-pulse rectifiers, the 5th, 7th, 17th, 19th, and so on are cancelled by the use of 12-pulse rectifiers. While they cost more than the six-pulse rectifiers, the high-pulse rectifiers when connected to the VFDs represent an effective mitigation solution for harmonic distortions. Typically, multiple pulse rectifiers are coupled with transformers, which are referred to as multipulse transformers, in order to reduce the magnitude of harmonic distortions (Kusko and Thompson, 2007).

11.3.2 Harmonic Filters

Power harmonic filters are mainly used to either block harmonic currents from flowing through electrical distribution systems or locally isolate and cancel them (Fuchs and Masoum, 2008). Specifically, harmonic filters are designed to have low impedances to ensure that the harmonic currents are flowing between the loads and the filters, rather than reaching the power source and the other components of the electrical distribution systems. There are three main classifications of filters: passive, active, and hybrid filters, as briefly discussed in the following:

1. Passive filters are made up of reactive components like capacitors, inductors, and resistors. They are tuned to provide a low impedance path for specific currents with undesirable harmonic frequencies. Some potential issues can occur with passive systems if not properly selected, resulting in resonance circuits parallel with power source leading to worse power quality conditions.
2. Active filters rely on active power-conditioning electronic devices to compensate for unfavorable harmonic currents. Specifically, active filters correct for the distorted sinusoidal current waves to effectively eliminate harmonics. Active filters have the advantage to respond to dynamic changes in the electrical distribution systems (Fuchs and Masoum, 2008). Therefore, these can be considered for more complicated and critical systems because they have a better capability to act in response to drastic disturbances. They are, however, more expensive to install than passive filters.
3. Hybrid filters are combinations of both active and passive filters. They are used in specific applications to take advantage of the benefits of both filters.

11.3.3 Voltage Compensators

Voltage compensators can be used to improve power quality by placing voltage sources between power sources and loads. For instance, voltage sags can be eliminated by inserting voltage components into the voltage wave forms to adjust their profiles. Voltage compensators usually can store limited amounts of energy so that they can only provide specific voltage compensation range. This range is typically limited to 2 seconds for 50% dips in voltage and 12 cycles for 10% dips (Kusko and Thompson, 2007). Voltage compensators are relatively inexpensive and are easy to install since they are not bulky.

11.3.4 Capacitor Banks

The installation of capacitor banks is a common mitigation strategy to correct power quality problems. Indeed, capacitors, when optimally sized and placed to compensate heavily inductive building loads, can vastly reduce the reactive power drawn by the electrical system. This reduction in reactive power has several benefits, including increase of the overall power factor, improve voltage regulation, lower harmonic distortion, and reduce energy loss (Fuchs and Masoum, 2008). One of the challenges in utilizing capacitor banks for power quality improvements is determining the optimum location, size, and number of capacitors for a specific electrical distribution system. Indeed, several factors need to be taken into account to design and control the overall power quality throughout the system.

11.3.5 Harmonic Mitigation Transformers

Transformers are crucial elements of electrical distribution systems to transfer power from its source to the loads. In the United States, voltage from power plants ranges from 13.8 kV to 480 V, while buildings normally require 480 V three-phase and/or 277 or 120 V single-phase system. To achieve the necessary drop in voltage, multiple step-down transformers are used. Specially designed transformers can also reduce some harmonic current distortions, including the triplens (i.e., third, sixth, ninth). These transformers are set up to fit delta–delta connections to wye–delta connections so that secondary voltages have a 30° phase shift (Kusko and Thompson, 2007). Other types of transformers can also help in mitigating voltage sags. For instance, constant voltage transformers use secondary transformers and capacitors to achieve a constant output voltage.

11.3.6 Uninterruptible Power Supply

For critical building loads, the most common effective protection option is a UPS system. For several buildings, UPS systems consist typically of battery-powered devices that are charged from the utility power source using inverters and rectifiers. When the source voltage level is reduced due to any disturbances, a bypass switch is activated and power is supplied from the battery. The battery continues to provide steady power to the critical loads until the main power source returns to its normal operating condition, unless the battery is totally discharged. Battery-powered UPS systems come in a variety of sizes from 100 W to 500kW. It should be noted that these standard sizes of batteries can only maintain the necessary voltage level for short periods ranging from 3 to 10 minutes. For large critical building loads, such as data centers or operating rooms in hospitals, UPS systems include, instead of batteries, standby engine generators. The generators are connected to the critical load through transfer switches and can provide power for extended periods of time. Chapter 13 presents an overview of different engine types used for generators and cogeneration systems.

11.3.7 Codes and Standards

In the United States, there are several standards related to power quality of electrical systems. These standards are mostly developed by Institute of Electrical and Electronic Engineering (IEEE), American National Standards Institute (ANSI), and National Electrical Manufacturing Association (NEMA). It should be noted that the National Electrical Code (NEC), developed by the National Fire Protection Agency (NFPA), does not generally provide any specific guidelines on methods and strategies to reduce or eliminate power quality issues. Indeed, the NEC emphasis is to define minimum safety requirements, rather than to ensure reliable performance of electrical systems. Although several power quality problems are disruptive to normal operation of electrical distribution systems, they do not necessarily lead to safety issues. For instance, one of the common power quality issues in buildings is an excessive voltage drop that can affect significantly the operation of electrical equipment such as lighting systems and motors, but is not considered a safety issue. NEC provides some specifications to systems that can assist in eliminating power quality programs such as surge arrestors and grounding systems. Otherwise, NEC has no special rules about power conditioners, filters, special-purpose transformers, or other mitigation devices.

As one of the main guidelines to improve power quality, the IEEE Standard 519 provides specifications for controlling harmonic currents. Specifically, the standard sets requirements for power quality from utilities and limits the harmonic currents caused by the end users. Table 11.1 summarizes voltage distortion and total harmonic distortion (THD) limits as outlined by the IEEE Standard 519. Table 11.2 provides the allowable harmonic current limits for the end users. These limits are based on the ratio of the maximum available short-circuit current to the maximum demand load current at the point of common coupling (PCC) between the building and the utility. As this ratio increases, the allowable total harmonic distortion for current increases.

Some other standards are available and can be considered for power quality specifications when designing power electrical systems for buildings:

- *IEEE Standard 1159*: Monitoring and definition of electric power quality
- *IEEE Standard 1250*: Guide on equipment sensitive to momentary voltage disturbances
- *IEEE Standard 1346*: Guide on compatibility for adjustable speed drives (ASDs) and process controllers

TABLE 11.1
Limits for Voltage Distortion Set by IEEE Standard 519

Bus Voltage at PCC	Individual Voltage Distortion (%)	Voltage *THD*(%)
69 kV and below	3.0	5.0
69.001–161 kV	1.5	2.5
161.001 kV or above	1.0	1.5

TABLE 11.2
Limits for Harmonic Currents for 120 V–69 kV Distribution Systems Set by IEEE Standard 519

I_{sc}/I_L	$h<11$	$11\leq h<17$	$17\leq h<23$	$23\leq h<35$	$35\leq h$	THD
<20	4.0	2.0	1.5	0.6	0.3	5.0
20–50	7.0	3.5	2.5	1.0	0.5	8.0
50–100	10.0	4.5	4.0	1.5	0.7	12.0
100–1,000	12.0	5.5	5.0	2.0	1.0	15.0
>1,000	15.0	7.0	6.0	2.5	1.4	20.0

Notes: I_{sc}, maximum short-circuit current; I_L, maximum fundamental frequency demand load at the point of common coupling or PCC; h, individual harmonic number; *THD*, total harmonic distortion for currents.

- *ANSI Standard C62*: Guides and standards on surge protection
- *ANSI Standard C84.1*: Voltage ratings for power systems and equipment
- *ANSI Standard C 57.100*: Transformer derating for supplying nonlinear loads
- *NEMA-UPS*: Uninterruptible power supply specifications

11.4 HARMONIC DISTORTIONS

11.4.1 Overview

Harmonic distortions are disturbances associated with variations of voltage or current from their sine waveforms due to the operation of some electrical systems. The voltage and current for an electrical system with ideal power quality vary as a simple sine function of time, often referred to as the fundamental harmonic, and are expressed as discussed in Chapter 2 by Equations 11.2 and 11.3, respectively:

$$e(t) = \sqrt{2}E\cos(\omega t + \theta) \tag{11.2}$$

$$i(t) = \sqrt{2}I\cos(\omega t + \phi) \tag{11.3}$$

where

- E and I are the rms values for the voltage and current, respectively
- $\omega = 2\pi f$ is the angular velocity of the voltage and current periodic variations, with f being the frequency ($f = 60$ Hz for the United States and $f = 50$ Hz for most other countries)
- θ and ϕ are the phase angles for, respectively, the voltage and current waveforms

When the power is distorted due, for instance, to electronic ballasts (which change the frequency of the electricity supplied to the lighting systems) or VFDs, several harmonics need to be considered in addition to the fundamental harmonic to represent the voltage or current time variation as shown in Equations 11.4 and 11.5:

$$e(t) = \sum_{k=1}^{N_V} \sqrt{2} E_k \sin(k\omega t + \theta_k) \tag{11.4}$$

$$i(t) = \sum_{k=1}^{N_I} \sqrt{2} I_k \sin(k\omega t + \phi_k) \tag{11.5}$$

Highly distorted waveforms contain numerous harmonics associated with frequencies that are multiple of the fundamental frequency $f = 2\pi/\omega$. As noted earlier, in the United States, $f = 60$ Hz and in the rest of the world, $f = 50$ Hz. Thus, the third harmonics correspond to a frequency of 180 Hz in the United States and 150 Hz in other countries. It should be noted that the even harmonics (i.e., second, fourth) tend to cancel each other's effects, while the odd harmonics (i.e., third, fifth) have their peaks coincide and significantly increase the distortion effects. To better understand the impact of various harmonic frequencies, current time variations are considered for three-phase systems.

As outlined in Chapter 2, the time variations for the fundamental harmonic current in the three phases A, B, and C under balanced conditions are out of phase by 120°C (i.e., $2\pi/3$) of each other:

$$\begin{aligned} i_{A,1}(t) &= I_1 \times \sin \omega t \\ i_{B,1}(t) &= I_1 \times \sin(\omega t - 120°) \\ i_{C,1}(t) &= I_1 \times \sin(\omega t + 120°) \end{aligned} \tag{11.6}$$

where I_1 is the peak current common to all three phases. The sum of the three currents $i_{A,1}(t)$, $i_{B,1}(t)$, and $i_{C,1}(t)$ is equal to zero for a balanced system (i.e., the impedances connected to all three phases are the same):

$$i_{A,1}(t) + i_{B,1}(t) + i_{C,1}(t) = 0 \tag{11.7}$$

Thus, the neutral wire does not carry any current. Moreover, the phase sequence between A through C is positive since phase C angle is higher than that of phase A, and current phase C trails current phase A.

For the third harmonics, the current variations for the three phases under balanced conditions can be expressed as follows:

$$\begin{aligned} i_{A,3}(t) &= I_3 \times \sin 3\omega t \\ i_{B,3}(t) &= I_3 \times \sin(3\omega t - 360°) \\ i_{C,3}(t) &= I_3 \times \sin(3\omega t + 360°) \end{aligned} \tag{11.8}$$

The sum of the three currents $i_{A,3}(t)$, $i_{B,3}(t)$, and $i_{C,3}(t)$ becomes

$$i_{A,3}(t) + i_{B,3}(t) + i_{C,3}(t) = 3 \times I_3 \sin 3\omega t \tag{11.9}$$

Thus, the neutral wire carries three times the phase current. Moreover, the phase sequence is zero since all the phase angles are the same.

Now considering the fifth harmonics, the current variations for the three phases under balanced conditions can be expressed as follows:

$$\begin{aligned} i_{A,5}(t) &= I_5 \times \sin 5\omega t \\ i_{B,5}(t) &= I_5 \times \sin(5\omega t - 600°) = I_5 \times \sin(5\omega + 120°) \\ i_{C,5}(t) &= I_5 \times \sin(5\omega t + 600°) = I_5 \times \sin(5\omega - 120°) \end{aligned} \tag{11.10}$$

The sum of the three currents $i_{A,5}(t)$, $i_{B,5}(t)$, and $i_{C,5}(t)$ is equal to zero:

$$i_{A,5}(t) + i_{B,5}(t) + i_{C,5}(t) = 0 \tag{11.11}$$

Thus, the neutral wire does not carry current. Moreover, the phase sequence is negative since phase C current leads phase A current.

Finally, and for the seventh harmonics, current variations for the three phases under balanced conditions can be expressed as follows:

$$\begin{aligned} i_{A,7}(t) &= I_7 \times \sin 7\omega t \\ i_{B,7}(t) &= I_7 \times \sin(7\omega t - 840°) = I_7 \times \sin(7\omega - 120°) \\ i_{C,7}(t) &= I_7 \times \sin(5\omega t + 840°) = I_7 \times \sin(7\omega + 120°) \end{aligned} \tag{11.12}$$

Like the case for the fundamental and the fifth harmonics, the sum of the three currents $i_{A,7}(t)$, $i_{B,7}(t)$, and $i_{C,7}(t)$ is equal to zero:

$$i_{A,7}(t) + i_{B,7}(t) + i_{C,7}(t) = 0 \tag{11.13}$$

Thus, the neutral wire does not carry current associated with the seventh harmonics. Moreover, the phase sequence is positive as was the case for the fundamental harmonic.

TABLE 11.3
Phase Sequences and Neutral Currents for Various Harmonics

Harmonic Number	Phase Sequence	Ratio of Neutral to Phase Currents
$3n+1$ (i.e., 1, 4, 7, 10, 13)	Positive	0
$3n+2$ (i.e., 2, 5, 8, 11, 14)	Negative	0
$3n+3$ (i.e., 3, 6, 9, 12, 15)	Zero	3

Table 11.3 summarizes various harmonics of the neutral wire current and the phase sequence.

11.4.2 Total Harmonic Distortion

To quantify the level of distortion for both voltage and current, a dimensionless number referred to as total harmonic distortion (THD) is determined through a Fourier series analysis of the voltage and current waveforms, as depicted by Equations 11.4 and 11.5, respectively. The THD for voltage and current are defined, respectively, as follows:

$$\mathrm{THD}_V = \sqrt{\frac{\sum_{k=2}^{N_V} E_k^2}{E_1^2}} \tag{11.14}$$

$$\mathrm{THD}_I = \sqrt{\frac{\sum_{k=2}^{N_I} I_k^2}{I_1^2}} \tag{11.15}$$

Note that in the United States, the reference for THD is the fundamental harmonic rms values as defined by the IEEE Standard 519. However, other reference values can be utilized. For instance, the European International Electrotechnical Commission (IEC) considers the total rms values as the references defined for voltage and current, respectively:

$$E_{tot} = \sqrt{\sum_{k=1}^{N_V} E_k^2} \tag{11.16}$$

$$I_{tot} = \sqrt{\sum_{k=1}^{N_I} I_k^2} \tag{11.17}$$

The individual harmonic distortion (IHD) values can also be defined and estimated for both voltage and current. Specifically, each IHD value represents the contribution of each harmonic frequency associated with the order n to the distorted variations for voltage and current, respectively:

$$\mathrm{IHD}_{V,n} = \frac{E_n}{E_1} \tag{11.18}$$

$$\mathrm{IHD}_{I,n} = \frac{I_n}{I_1} \tag{11.19}$$

Example 11.1 illustrates the calculations for both THD and IHD values using both U.S. IEEE and European IEC definitions for voltage variations associated with a building power panel.

Example 11.1

Problem

Based on the analysis of voltage measured for a power panel, it was found that the rms values for the fundamental, the third, the fifth, the seventh, and the ninth harmonics that make up the voltage waveform are the following:

$$V_1 = 115\text{ V},\ V_3 = 5\text{ V},\ V_5 = 2\text{ V},\ V_7 = 1\text{ V},\ V_9 = 0.5\text{ V}$$

Determine the IHD and THD values using the IEEC and IEC definitions.

Solution

Using the IEEE definition for the IHD and THD, the following values can be estimated:

$$\mathrm{IHD}_3 = \frac{5}{115} = 4.35\%$$

$$\mathrm{IHD}_5 = \frac{2}{115} = 1.74\%$$

$$\mathrm{IHD}_7 = \frac{1}{115} = 0.87\%$$

$$\mathrm{IHD}_9 = \frac{0.5}{115} = 0.43\%$$

$$\mathrm{THD}_V = 4.78\%$$

Using the European IEC definition, first, the reference E_{tot} is estimated using Equation 11.16:

$$E_{tot} = 115.13\text{ V}$$

Then, the IHD and THD values can be estimated using Equation 11.18:

$$\text{IHD}_3 = \frac{5}{115.13} = 4.34\%$$

$$\text{IHD}_5 = \frac{2}{115.13} = 1.74\%$$

$$\text{IHD}_7 = \frac{1}{115.13} = 0.87\%$$

$$\text{IHD}_9 = \frac{0.5}{115.13} = 0.43\%$$

Then, the THD value can be estimated using Equation 11.14, using E_{tot} instead of E_1 as a reference:

$$\text{THD}_V = 4.77\%$$

For this example, there is very little difference between the calculations using IEEE and IEC definitions of IHD and THD.

Table 11.4 provides current THD for selected but specific lighting and office equipment loads (Krarti, 2012). Generally, it is found that devices with high current THD values contribute to voltage THD in proportion to their share to the total building electrical load. Therefore, the electrical engineer should consider low-wattage devices before higher wattage devices to reduce the voltage THD for the entire building or facility. Example 11.2 shows a simple calculation procedure that can be used to assess the impact of specific electrical equipment on the THD current. Thus, one can determine which devices need to be corrected first to improve the power quality of the overall electric system. Typically, harmonic filters are added to electrical devices to reduce the current THD values.

Example 11.2

Problem

Assess the impact on the current THD of a building of two devices: 13 W compact fluorescent lamp (CFL) with electronic ballasts and a laser printer while printing. Use the data provided in Table 11.4.

Solution

Both devices have an rms voltage of 120 V (i.e., $E_{rms} = 120\text{ V}$); their rms current can be determined using the real power used and the power factor given in Table 11.4:

$$I_{rms} = \frac{P_R}{E_{rms} \times pf}$$

This equation gives an rms current of 0.22 A for the CFL and 6.79 A for the printer. These values correspond actually to the rms of each device's fundamental current

TABLE 11.4
Typical Power Quality Characteristics (Power Factor and Current Total Harmonic Distortion) for Selected Electrical Loads

Electrical Load	Real Power Used (W)	Power Factor	Current *THD* (%)
Incandescent lighting systems			
100 W incandescent lamp	101	1.0	1
Compact fluorescent lighting systems			
13 W lamp w/magnetic ballast	16	0.54	13
13 W lamp w/electronic ballast	13	0.50	153
Full-size fluorescent lighting systems (two lamps per ballast)			
T12 40 W lamp w/magnetic ballast	87	0.98	17
T12 40 W lamp w/electronic ballast	72	0.99	5
T10 40 W lamp w/magnetic ballast	93	0.98	22
T10 40 W lamp w/electronic ballast	75	0.99	5
T8 32 W lamp w/electronic ballast	63	0.98	6
T5 28 W lamp w/electronic ballast	62	0.95	15
High-intensity discharge lighting systems			
400 W high-pressure sodium lamp w/magnetic ballast	425	0.99	14
400 W metal halide lamp w/magnetic ballast	450	0.94	19
Office equipment			
Desktop computer w/o monitor	33	0.56	139
Color monitor for desktop computer	49	0.56	138
Laser printer (in standby mode)	29	0.40	224
Laser printer (printing)	799	0.98	15
External fax/modem	5	0.73	47

Sources: Adapted from NLPIP, *Newsletter*, 2(2), 5, 1995; NLPIP, *Newsletter*, 7(5), 17, 2005.

waveform and can be used in the THD equation, Equation 11.15, to estimate the total harmonic current of each device:

$$I_{tot} = I_{rms} \times \text{THD}_I$$

The resultant values of 0.33 A for the CFL and 1.02 A for the printer show that although the printer has relatively low current THD (15%), the actual distortion current produced by the printer is more than three times that of the CFL because the printer uses more power.

The IEEE Standard 519 recommends a maximum allowable voltage THD of 5% at the building service entrance (i.e., point where the utility distribution system is connected to the building electrical system). Based on a study by Verderber et al. (1993), the voltage THD reaches the 5% limit when about 50% of the building electrical load has a current THD of 55% or when 25% of the building electrical load has a current THD of 115%. It should be noted that when the electrical device has a power factor of unity (i.e., $pf=1$), there is little or no current THD (i.e., $THD_I=0\%$) since the device has only a resistive load and effectively converts input current and voltage into useful electrical power. As shown in Table 11.4, the power factor and the current THD are interrelated and both define the characteristics of the power quality. In particular, Table 11.4 indicates that lighting systems with electronic ballasts have typically high power factor and low current THD. This good power quality is achieved using capacitors to reduce the phase lag between the current and voltage and filters to reduce harmonics.

Possible problems that have been reported due to poor power quality include the following:

1. Overload of neutral conductors in a three-phase system with four wires. In a system with no THD, the neutral wire carries no current if the system is well balanced. However, when the current THD becomes significant, the currents due to the odd harmonics do not cancel each other and rather add up on the neutral wire that can overheat and cause a fire hazard.
2. Reduction in the life of transformers and capacitors. This effect is mostly caused by distortion in voltage.
3. Interference with communication systems. Electrical devices that operate with high frequencies such as electronic ballasts (that operate at frequencies ranging from 20 to 40 kHz) can interfere and disturb the normal operation of communication systems such as radios, phones, and energy management systems (EMS).

A more detailed discussion of the impact of harmonic distortions on building electrical systems is provided in the following section.

11.5 IMPACT HARMONIC DISTORTIONS

For several cases, harmonic distortions are detected only when their effects occur. It is therefore important to have insights on the potential impact of harmonics on various electrical systems for buildings. In this section, the effects of harmonics are presented on specific components of electrical distribution systems including transformers, wires, and protection devices. Moreover, a brief discussion of harmonic distortions on motors and capacitors is also provided.

11.5.1 Impact on Transformers

Voltage harmonics can increase core losses for electrical transformers by forming additional hysteresis loops from higher frequencies. Moreover, current harmonics increase winding wire losses (which are proportional to RI^2) and ultimately can cause overheating and reduce the rating of transformers. Thus, transformers supplying nonlinear loads should be derated in order to handle safely the effects of harmonic distortions. To determine the suitability of electrical transformers to handle

TABLE 11.5
Recommended Transformer *k*-Factors for Selected Loads

Load Type	Transformer *k*-Factor
Lighting with electric discharge	*k*-4
UPS with input filtering	*k*-4
Induction heating units	*k*-4
Telecommunication equipment	*k*-13
UPS without filtering	*k*-13
Receptacles for health care facilities	*k*-13
Computer servers	*k*-20
VFDs	*k*-20

harmonics, the concept of *k*-factor was introduced by ANSI/IEEE C57.100 standard. The *k*-factor is defined as follows:

$$k = \frac{1}{I_{tot}^2} \sum_{h=1}^{N_I} h^2 I_h^2 \tag{11.20}$$

where I_{tot} is the total rms value for all the measured harmonics defined by Equation 11.17. From Equation 11.20, it is clear that when there is no harmonics, $k=1$. As more harmonics are present, k increases. Typical *k*-factors for transformers are 4, 9, 13, 20, 30, 40, and 50. Table 11.5 lists for typical loads, the recommended *k*-factors for the transformers supplying these loads. Example 11.3 outlines the estimation of the transformer *k*-factor supplying a load with significant third, fifth, and seventh harmonics.

Example 11.3

Problem

Determine the *k*-factor required for a transformer supplying a load made up of 400 A for fundamental harmonic, 195 A for the third harmonics, 125 A for the fifth harmonics, and 95 A for the seventh harmonics.

Solution

First, the total rms value for all the available harmonics, I_{tot}, is estimated using Equation 11.17:

$$I_{tot} = \sqrt{400^2 + 195^2 + 125^2 + 95^2} = 471.9 \text{ A}$$

Then, the *k*-factor can be estimated using Equation 11.20:

$$k = \frac{1}{471.9^2}\left[(1)^2 \times (400)^2 + (3)^2 \times (195)^2 + (5)^2 \times (125)^2 + (7)^2 \times (95)^2\right]$$

Thus,

$$k = 6.0$$

A transformer with k-factor of 9 should be selected to handle the effects of harmonics for this load.

11.5.2 Impact on Electrical Wires

Current harmonics increases RI^2 losses for branch circuits and feeders. Indeed, harmonics increases the total rms current value. Moreover, the presence of high-frequency harmonics increases the effective resistance of wires due to the skin effect (Sankaran, 2002). Specifically, ACs flow mostly along the outer surface (i.e., skin) of wires. The penetration depth of the AC, d, is inversely proportional to the square root of the current frequency, f:

$$d = \frac{S}{\sqrt{f}} \tag{11.21}$$

where S is a constant that depends on the wire characteristics including its conductor type and its magnetic permeability. Thus, the higher the frequency, the smaller is the penetration depth d and the closer it is to the skin of the wire where the AC flows. As a result, the AC resistance for the wire increases with the harmonic frequency. The AC resistance can be estimated from the DC resistance for most conductors used in buildings as follows (Sankaran, 2002):

$$R_{AC} = K \times R_{DC} \tag{11.22}$$

The constant K is the skin effect factor and is typically provided by the cable manufacturers. Table 11.6 shows cable skin effect factors using a parameter X, which is a function of the frequency f, DC resistance R_{DC}, and the magnetic permeability of the conductor μ. When the unit of R_{DC} is Ω per mile, X can be obtained as follows:

TABLE 11.6
Typical Cable Skin Effect Factors

X	K
0	1.00000
0.5	1.00032
1.0	1.00519
1.5	1.02558
2.0	1.07816
2.5	1.17538
3.0	1.31809
3.5	1.49202
4.0	1.67804

$$X = 0.0636\sqrt{\frac{\mu \times f}{R_{DC}}} \tag{11.23}$$

When R_{DC} is expressed in Ω/km, X is determined using the following equation:

$$X = 0.05013\sqrt{\frac{\mu \times f}{R_{DC}}} \tag{11.24}$$

Note that for copper conductors in nonmagnetic conduits, $\mu = 1.0$. Example 11.4 indicates the impact of the harmonic frequency on the AC resistance for a feeder copper conductor.

Example 11.4

Problem

Determine the AC resistance values for a 250 MCM copper conductor serving a panel. Consider the fundamental, the third, the fifth and the seventh harmonics.

Solution

First, the DC resistance for 250 MCM copper conductor is obtained (refer to Table 2.2), $R_{DC} = 0.0431$ Ω/1,000 ft = 0.2276 Ω/mile = 0.1414 Ω/km. Using Equation 11.23, the values for X are estimated for the various harmonics for the copper conductor:

$$X_1 = 0.0636 \times \sqrt{\frac{(1)\times(60)}{0.2276}} = 0.05013 \times \sqrt{\frac{(1)\times(60)}{0.1414}} = 1.03$$

$$X_3 = 0.0636 \times \sqrt{\frac{(1)\times(180)}{0.2276}} = 0.05013 \times \sqrt{\frac{(1)\times(180)}{0.1414}} = 1.79$$

$$X_5 = 0.0636 \times \sqrt{\frac{(1)\times(300)}{0.2276}} = 0.05013 \times \sqrt{\frac{(1)\times(300)}{0.1414}} = 2.31$$

$$X_7 = 0.0636 \times \sqrt{\frac{(1)\times(420)}{0.2276}} = 0.05013 \times \sqrt{\frac{(1)\times(420)}{0.1414}} = 2.73$$

From Table 11.6, the cable skin effect factors can be obtained through interpolation:

$$K_1 = 1.006$$

$$K_3 = 1.052$$

$$K_5 = 1.131$$

$$K_7 = 1.232$$

The AC resistances for various harmonics are determined using Equation 11.22:

$$R_{AC,1} = K_1 \times R_{DC} = 1.006 \times 0.2276\ \Omega/\text{mile} = 0.2290\ \Omega/\text{mile} = 0.1423\ \Omega/\text{km}$$

$$R_{AC,3} = K_3 \times R_{DC} = 1.052 \times 0.2276\ \Omega/\text{mile} = 0.2394\ \Omega/\text{mile} = 0.1488\ \Omega/\text{km}$$

$$R_{AC,5} = K_5 \times R_{DC} = 1.131 \times 0.2276\ \Omega/\text{mile} = 0.2574\ \Omega/\text{mile} = 0.1599\ \Omega/\text{km}$$

$$R_{AC,7} = K_7 \times R_{DC} = 1.232 \times 0.2276\ \Omega/\text{mile} = 0.2804\ \Omega/\text{mile} = 0.1742\ \Omega/\text{km}$$

Thus, the AC resistance of 250 MCM copper conductor increases by 23% for the seventh harmonics when compared to the fundamental harmonic.

To account for the effect of harmonics on wires, a current rating factor, q, is introduced to estimate an equivalent fundamental frequency load current. Specifically, the q current rating factor is defined as follows (Sankaran, 2002):

$$q = \frac{1}{R_{AC,1} \times I_1^2} \sum_{k=1}^{N_I} R_{AC,k} I_k^2 \qquad (11.25)$$

Note that the value of q can be also obtained using Equation 11.25 by replacing the R_{AC} values by the cable skin effect factors K estimated for all the harmonics. Example 11.5 indicates how the value of the current rating factor q is calculated for the same conductor considered in Example 11.4.

Example 11.5

Problem

For the 250 MCM copper conductor considered in Example 11.4, estimate the current rating factor, q, if the rms current values for the various harmonics are found to be: $I_1 = 225$ A (fundamental harmonic), $I_3 = 50$ A (third harmonics), $I_5 = 75$ A (fifth harmonics), and $I_7 = 40$ A (seventh harmonics).

Solution

Using the values of AC resistances found in Example 11.4, the current rating factor, q, can be estimated using Equation 11.25 and the cable skin effect factors K:

$$q = (1)\left(\frac{225}{225}\right)^2 + \left(\frac{1.052}{1.006}\right)\left(\frac{50}{225}\right)^2 + \left(\frac{1.131}{1.006}\right)\left(\frac{75}{225}\right)^2 + \left(\frac{1.232}{1.006}\right)\left(\frac{40}{225}\right)^2$$

Thus,

$$q = 1.215$$

The fundamental harmonic equivalent load that the feeder has to handle is 1.215×225 A $= 273$ A.

11.5.3 Impact on Protection Devices

Current harmonics can affect protection devices including circuit breakers, fuses, and overload relays for motors. Indeed, the variations of current flowing through the building distribution system due to harmonics can interrupt the normal operation of the protection devices. The response of the protection devices can be either faster or slower due to higher-frequency harmonics, compared to their operation under their rated settings for the fundamental harmonic. Therefore, protection devices have to be properly designed and selected when harmonic distortions are expected to be present in any section of the electrical distribution system.

11.5.4 Impact on Motors

The voltage variations due to harmonics result in additional magnetic core losses for induction motors. Moreover, current harmonics increase the copper (i.e., RI^2) losses in the motor windings. The most significant impact, however, of harmonic distortions for motors is the increase in oscillations and vibrations. Failures and damages can occur due to excessive vibration and noise if no action is taken. As noted earlier, VFDs commonly coupled with motors for air conditioning systems are significant sources of harmonics. As discussed earlier, filters should be installed when VFDs are specified to prevent any damage to motors due to harmonic distortions.

11.5.5 Impact on Capacitor Banks

The capacitor banks are typically used to correct poor power factor that represents another common power quality issue, especially in commercial buildings. Chapter 2 discusses the application of capacitors or a bank of capacitors to improve the power factor for buildings. In particular, it was shown in Chapter 2 that the reactance of a capacitor is inversely proportional to the frequency:

$$X_C = \frac{1}{2\pi fC} \tag{11.26}$$

Thus, the reactance of a capacitor bank varies and decreases with higher-frequency harmonics. Since the electrical loads for buildings are dominated by inductive loads (associated with induction motors and discharge-type lighting fixtures), the higher frequencies may lead to the conditions of harmonic resonance when the reactance of the capacitor bank, X_C, becomes equal to the building inductive reactance, X_L, as expressed by the following equation, as outlined in Chapter 2:

$$X_C = \frac{1}{2\pi fC} = 2\pi fL = X_L \tag{11.27}$$

If one of the harmonic frequencies coincides with the resonance frequency, f_R, as expressed by Equation 11.28, resonance conditions occur and significant damages may result not only to the capacitor bank but to the entire electrical distribution system, if no actions are taken to prevent this occurrence:

$$f_R = \frac{1}{2\pi}\sqrt{LC} \tag{11.28}$$

Properly sizing the bank of capacitors (expressed typically in kVAR) is required to avoid the harmonic resonance conditions. Example 11.6 shows the calculation of resonance frequency for two ratings of capacitor banks.

Example 11.6

Problem

A 1,500 kVA transformer rated at 13.8 kVA–480Y/277 V for a commercial building has a per unit reactance of 5%. Determine the resonance frequency for two capacitor banks: (1) 450 kVAR and (2) 600 kVAR. Identify the best option for the capacitor bank to be connected to the building electrical distribution system.

Solution

First, the inductance X_L for the building is estimated based on the per unit reactance of the transformer using the reference resistance at the secondary windings with a line voltage of 480 V (refer to Chapters 3 and 8):

$$R_{ref} = \frac{E_{L-N,S}}{I_{L,S}}$$

With the line-to-neutral voltage at the secondary side of the transformer: $E_{L\text{-}N,\,S}$ = 277 V and the line current,

$$I_{L,S} = \frac{P_T}{\sqrt{3}E_{L-L,S}} = \frac{1{,}500}{1.732\times(0.48)} = 1{,}804 \text{ A}$$

Thus,

$$R_{ref} = \frac{277}{1{,}804} = 0.154\ \Omega$$

Therefore, the reactance of the transformer is

$$X_L = Z_L(\%)\times R_{ref} = (0.05)\times(0.154) = 0.0077\ \Omega$$

The equivalent inductance L for the transformer is then estimated:

$$L = \frac{X_L}{2\pi f} = \frac{0.0077}{2\pi(60)} = 20.36\times10^{-6}\text{H}$$

For the capacitor bank, the reactance can be determined as follows:

$$X_C = \frac{E_{L-N,S}}{I_C}$$

With the current flowing through the bank estimated using its *kVAR* rating,

$$I_C = \frac{kVAR}{\sqrt{3}E_{L-L,S}}$$

The capacitance of the bank is calculated from the reactance (refer to Chapter 2):

$$C = \frac{1}{2\pi f X_C}$$

Finally, the resonance frequency, f_R, can be obtained using Equation 11.28.

The results for estimating X_C, I_C, C, and f_R are provided for two sizes of the capacitor bank:

For *kVAR*=450,

$$I_C = 581 \text{ A},\ X_C = 0.511\ \Omega,\ C = 0.00518 \text{ F, and } f_R = 490 \text{ Hz}$$

For *kVAR*=600,

$$I_C = 722 \text{ A},\ X_C = 0.384\ \Omega,\ C = 0.00691 \text{ F, and } f_R = 424 \text{ Hz}$$

In the first option, the resonance frequency is close to that of the eighth harmonics (8×60=480 Hz), while in the second option, it coincides with the frequency of the seventh harmonics (7×60=420 Hz).

Since the seventh harmonics are more common in building electrical systems than the eighth harmonics, selecting a 450 kVA capacitor bank would be safer to prevent any significant impacts associated with harmonic distortions.

11.6 MEASUREMENTS OF HARMONIC DISTORTIONS

Using a power analyzer, the characteristics and electrical power use for a wide range of electrical systems can be measured in buildings. For instance, Figure 11.9 shows the current drawn by home appliances including a washing machine, a dryer, a dishwasher, a refrigerator, a garbage disposal, and a microwave. The measurements using power analyzers provide also the THD, real power, and total power for the appliances. Table 11.7 summarizes some of the electrical measurements carried out for the common appliances used in a residential building. As noted in Table 11.7, some of the appliances have high current THD values, including garbage disposal, microwave, and washing machine. Figures 11.10–11.14 show the time variation of THD values for both voltage and currents from select home appliances considered in Figure 11.9.

Motors in buildings are often equipped with VFDs to change their speed. Indeed, fans and pumps used for HVAC systems are set to vary fluid flows to meet variable thermal loads. As noted earlier, VFDs are one of the main sources of harmonics in buildings. Figure 11.15 illustrates the real power, total power, and the current THD measured for an air conditioning system used in a residential building during a period of 3 days. As shown in Figure 11.15, the air conditioning system has both low power factor and poor power quality. Indeed, the real power averages to 3.9 kW with a power factor of 58% and THD-neutral current of 41% during the testing period.

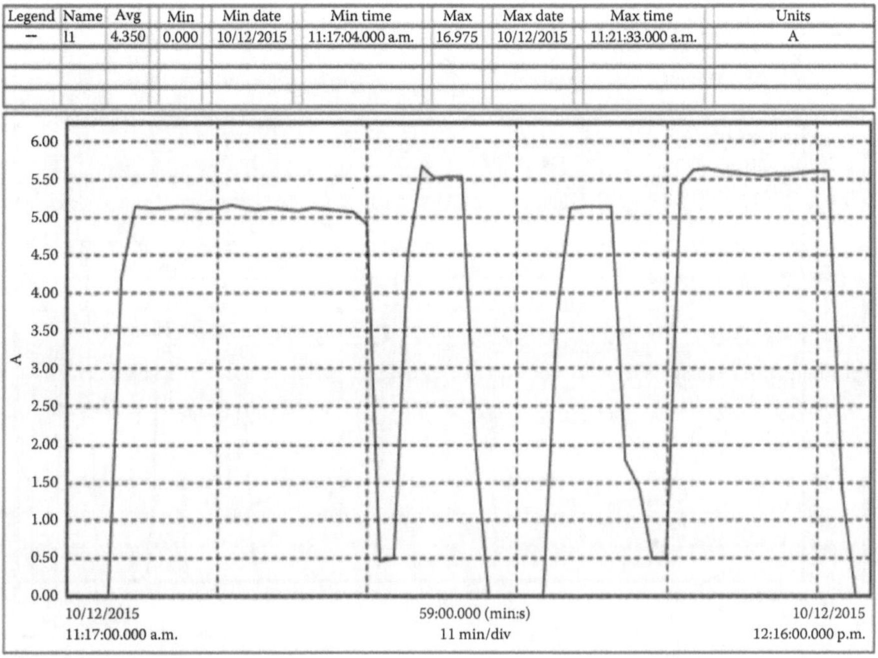

(a)

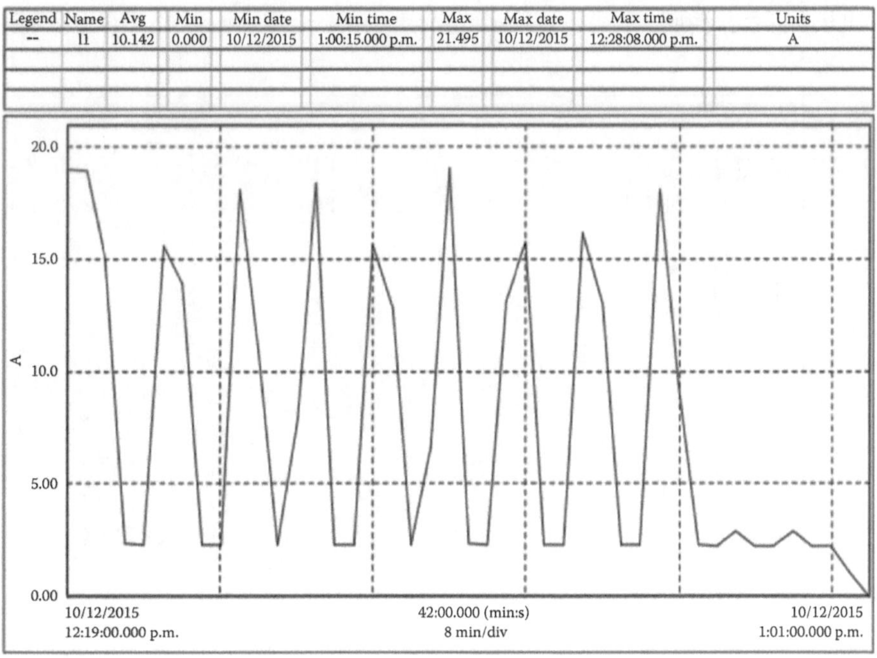

(b)

FIGURE 11.9 Time variation of current drawn under normal operation of home appliances. (a) Washing machine, (b) dryer. (c) dishwasher, (d) refrigerator. (e) disposal, and (f) microwave.

(Continued)

Legend	Name	Avg	Min	Min date	Min time	Max	Max date	Max time	Units
--	I1	0.171	0.000	10/12/2015	2:40:27.000 p.m.	0.708	10/11/2015	4:05:58.000 p.m.	A

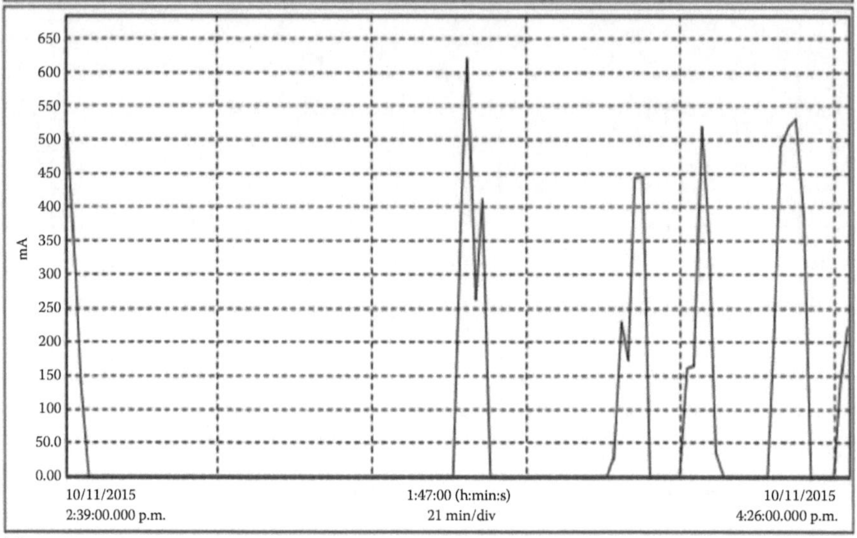

(c)

Legend	Name	Avg	Min	Min date	Min time	Max	Max date	Max time	Units
--	I1	0.795	0.000	10/12/2015	4:57:52.000 p.m.	6.344	10/14/2015	9:46:52.000 p.m.	A

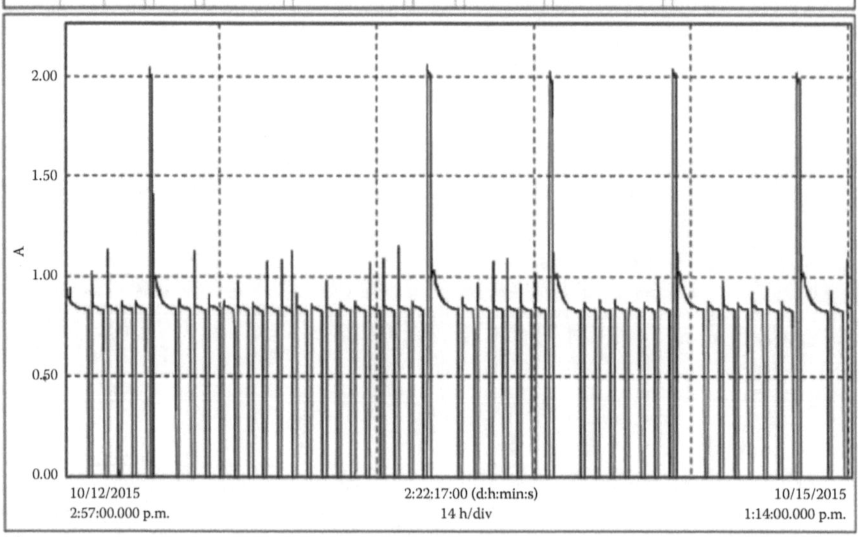

(d)

FIGURE 11.9 (*Continued*) Time variation of current drawn under normal operation of home appliances. (a) Washing machine, (b) dryer. (c) dishwasher, (d) refrigerator. (e) disposal, and (f) microwave.

(Continued)

Legend	Name	Avg	Min	Min date	Min time	Max	Max date	Max time	Units
--	I1	0.371	0.321	10/15/2015	2:13:54.000 p.m.	0.452	10/15/2015	1:54:42.000 p.m.	A

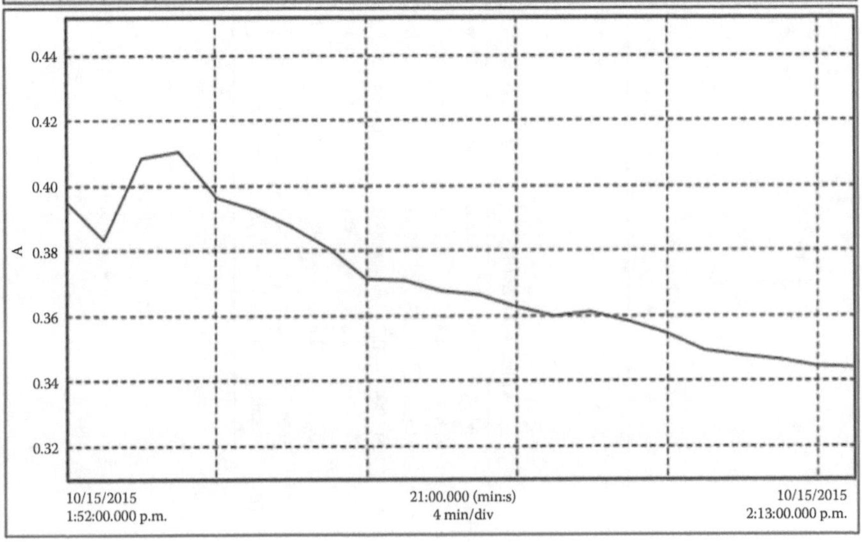

(e)

Legend	Name	Avg	Min	Min date	Min time	Max	Max date	Max time	Units
--	I1	5.890	0.000	10/15/2015	1:17:04.000 p.m.	9.105	10/15/2015	1:24:16.000 p.m.	A

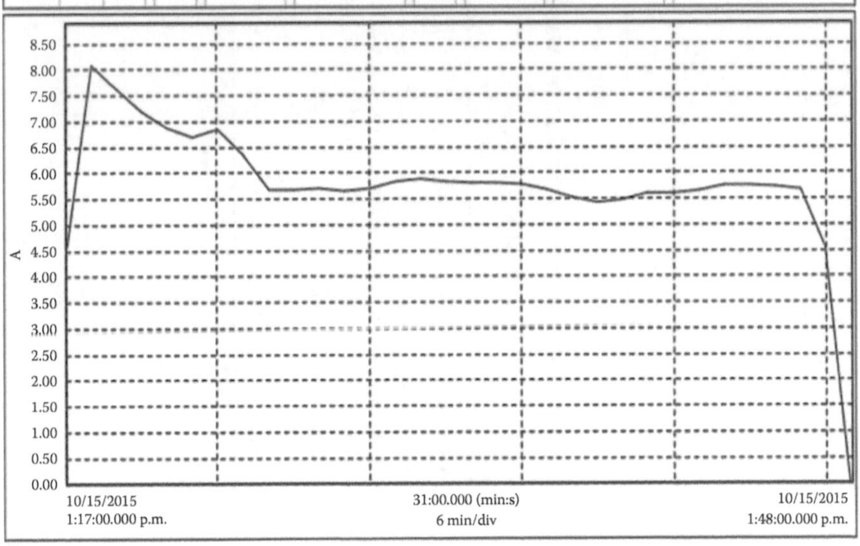

(f)

FIGURE 11.9 (*Continued*) Time variation of current drawn under normal operation of home appliances. (a) Washing machine, (b) dryer. (c) dishwasher, (d) refrigerator. (e) disposal, and (f) microwave.

TABLE 11.7
Summary of Testing Results for Common Appliances Used in Homes

		Real Power (kW)			Total Power (kVA)			Voltage *THD* (%)			Current *THD* (%)		
Equipment	**Test Duration**	**Min.**	**Avg.**	**Max.**	**Min.**	**Avg.**	**Max.**	**Min.**	**Avg.**	**Max.**	**Min.**	**Avg.**	**Max.**
Cooker	41 minutes	0.057	6.836	9.202	0.058	6.845	9.203	0.840	0.862	0.980	1.440	2.072	5.800
Dishwasher	107 minutes	0.000	0.012	0.123	0.000	0.014	0.149	0.650	0.783	0.920	0.000	1.093	8.790
Disposal	21 minutes	0.062	0.068	0.075	0.083	0.089	0.099	0.600	0.781	0.870	74.00	77.52	80.18
Washing Machine	59 minutes	0.000	0.241	0.378	0.000	0.868	1.374	0.740	0.892	1.040	0.000	3.163	16.52
Dryer	42 minutes	0.000	1.385	4.603	0.000	1.643	4.626	0.730	0.876	0.980	0.540	2.516	6.330
Microwave	31 minutes	0.000	0.453	1.632	0.000	1.354	1.931	0.560	0.696	0.820	0.000	47.17	55.82
Refrigerator	71 hours	0.000	0.107	0.501	0.000	0.164	0.501	0.540	0.771	1.170	0.000	7.130	10.22

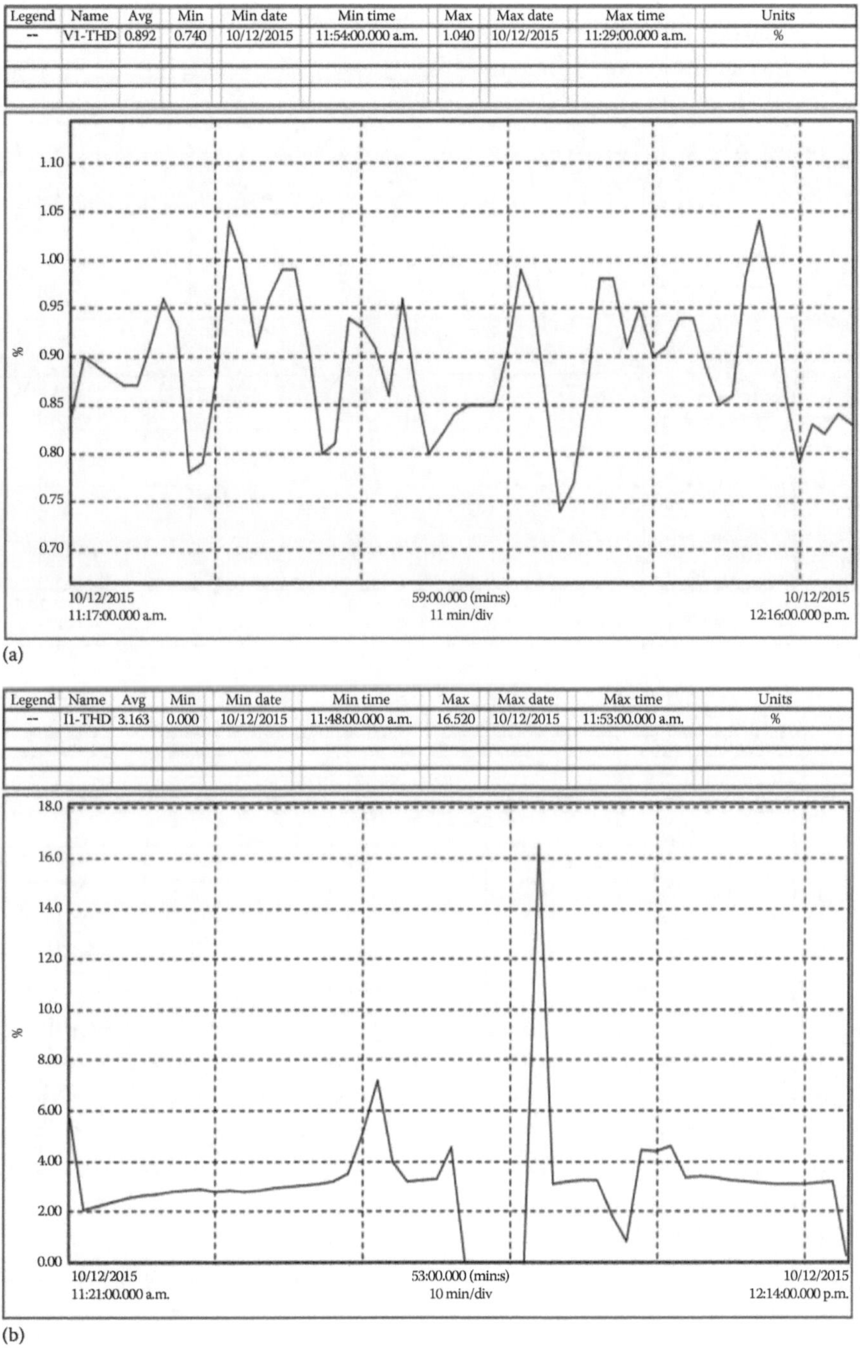

FIGURE 11.10 Time variation of (a) voltage THD and (b) current THD for a washing machine in a home.

Legend	Name	Avg	Min	Min date	Min time	Max	Max date	Max time	Units
--	V1-THD	0.876	0.730	10/12/2015	12:41:00.000 p.m.	0.980	10/12/2015	12:33:00.000 p.m.	%

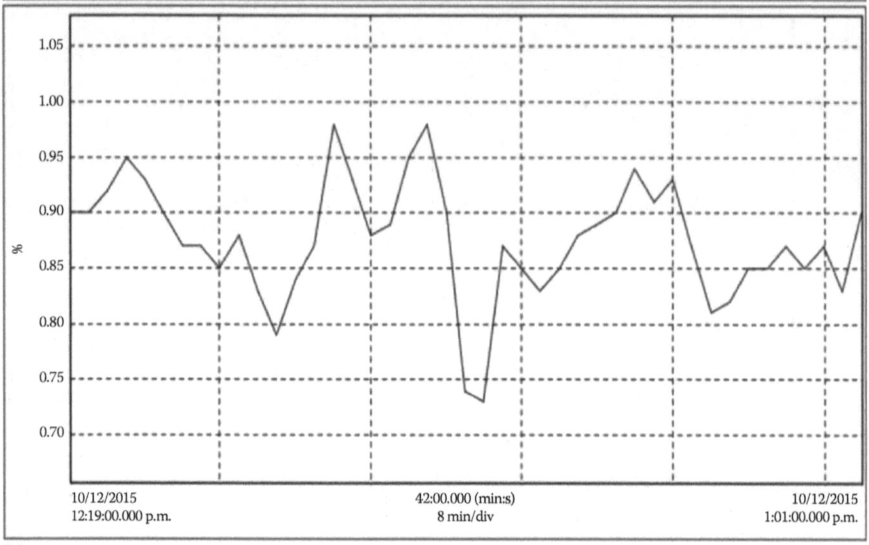

(a)

Legend	Name	Avg	Min	Min date	Min time	Max	Max date	Max time	Units
--	I1-THD	2.516	0.540	10/12/2015	1:00:00.000 p.m.	6.330	10/12/2015	12:54:00.000 p.m.	%

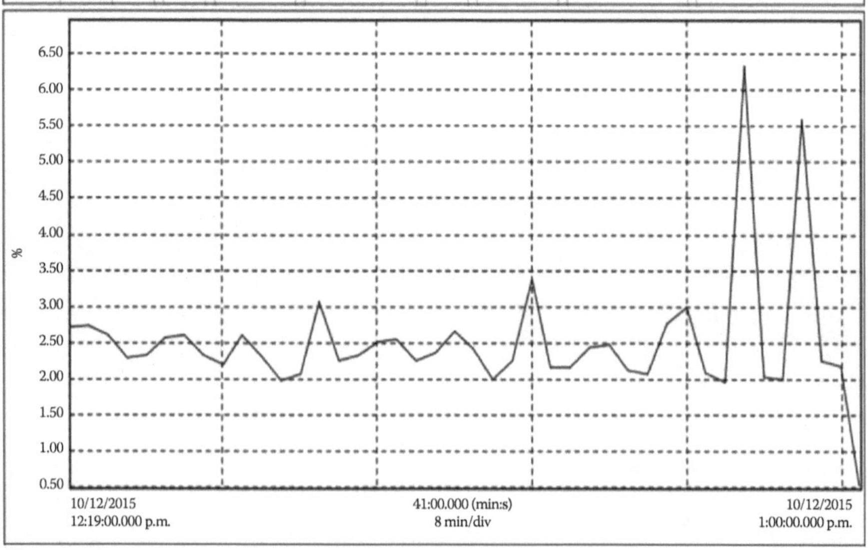

(b)

FIGURE 11.11 Time variation of (a) voltage THD and (b) current THD for clothes dryer in a home.

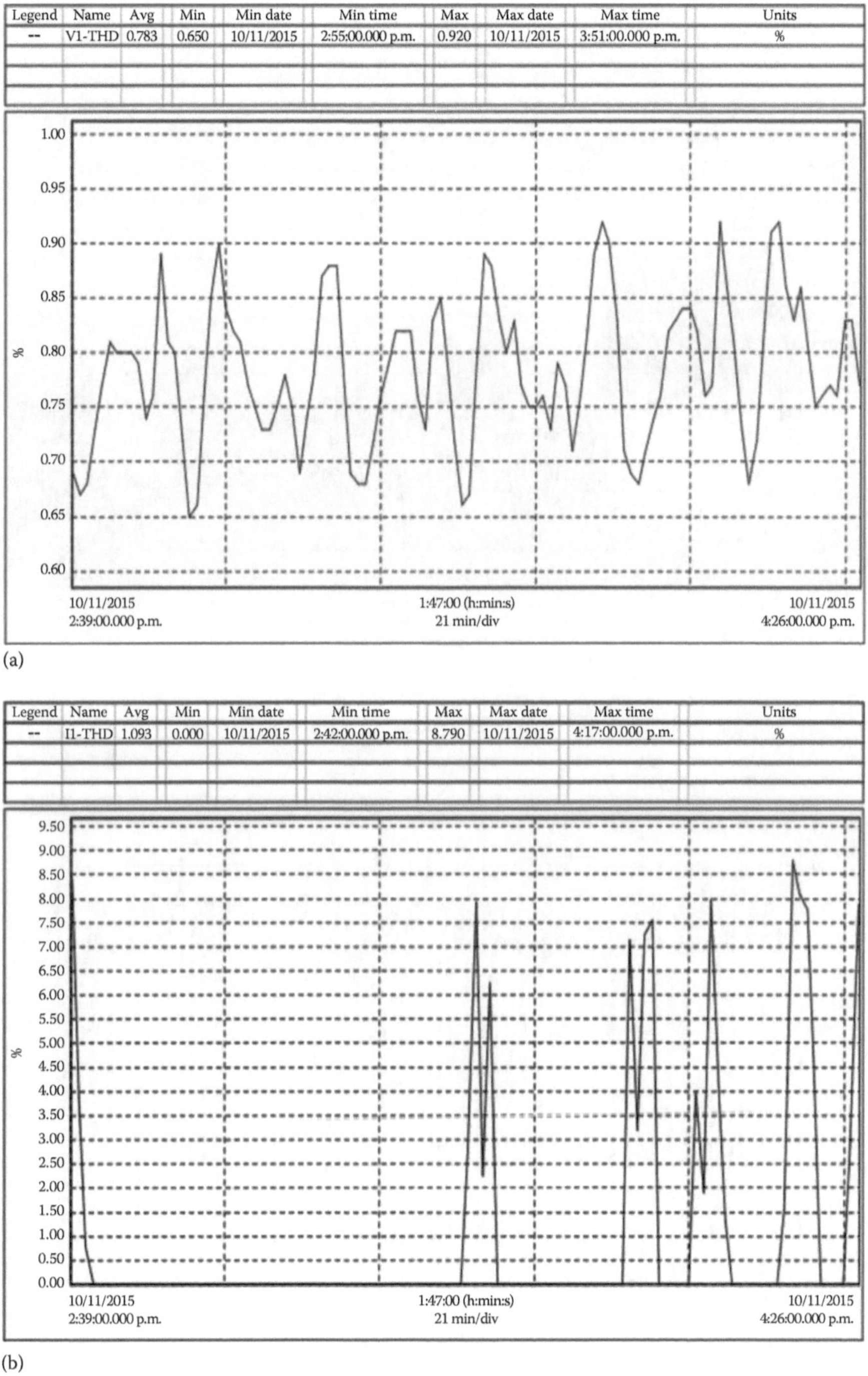

FIGURE 11.12 Time variation of (a) voltage THD and (b) current THD for a dishwasher in a home.

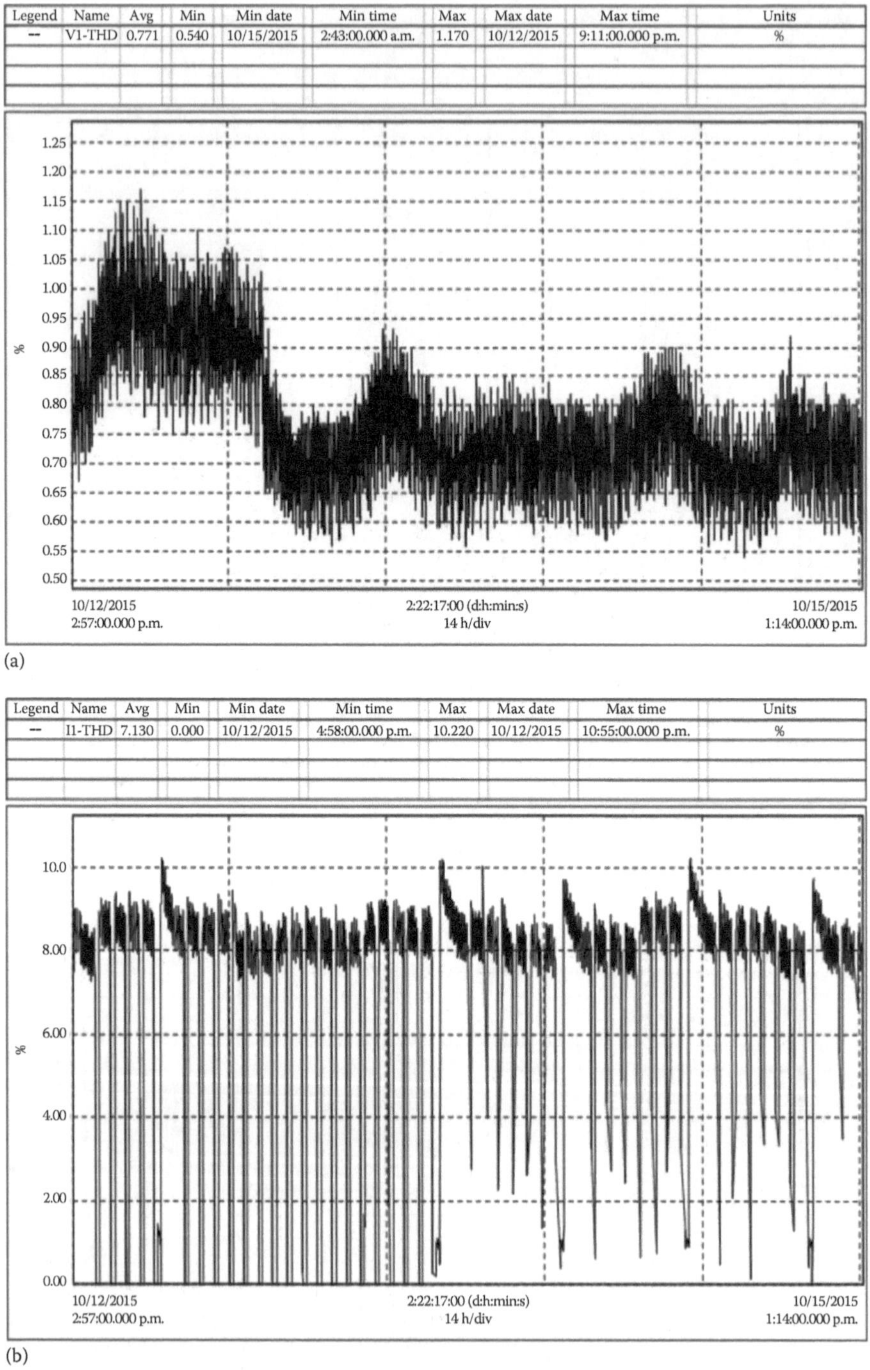

FIGURE 11.13 Time variation of (a) voltage THD and (b) current THD for a refrigerator in a home.

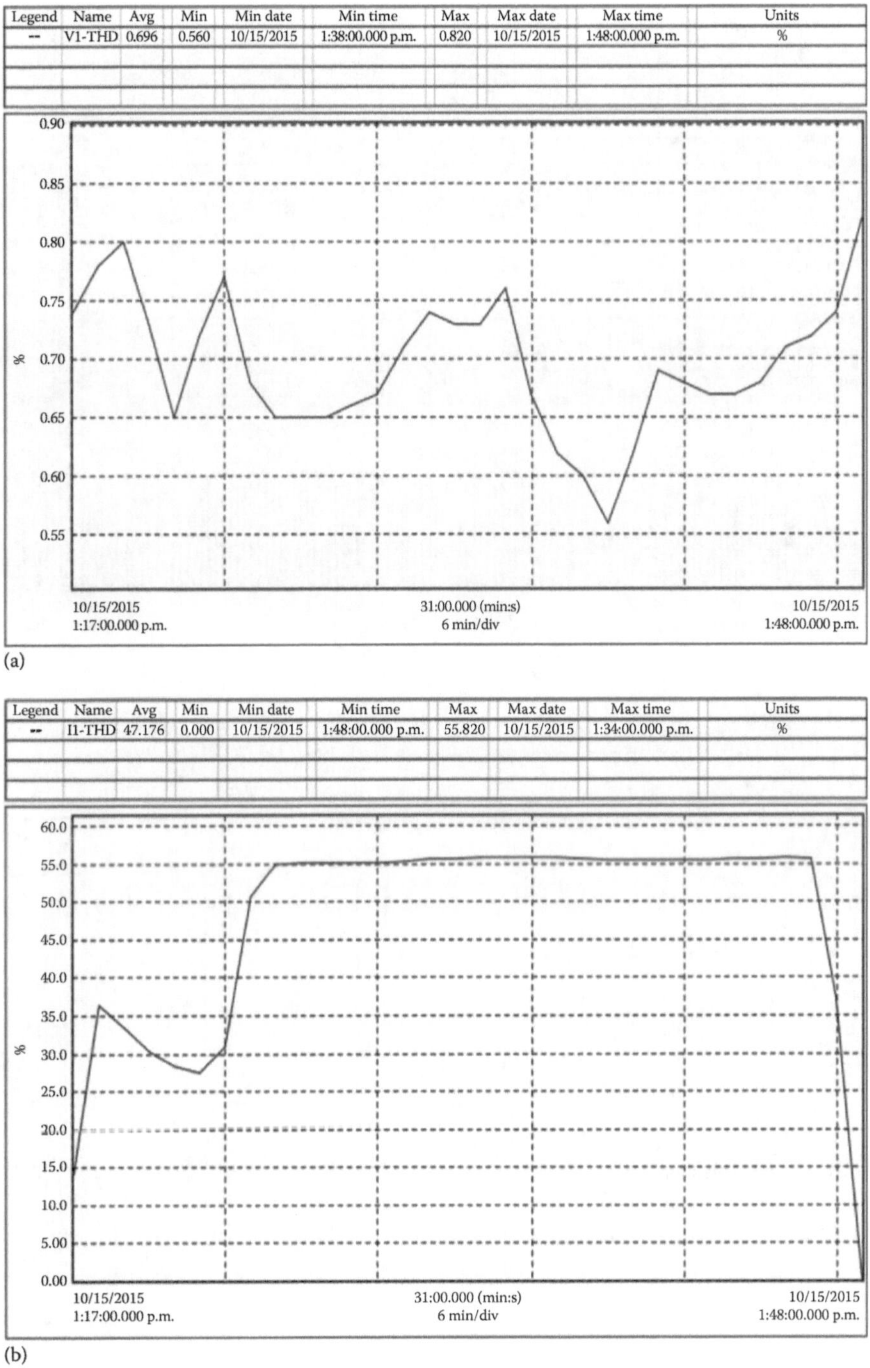

FIGURE 11.14 Time variation of (a) voltage THD and (b) current THD for a microwave in a home.

Legend	Name	Avg	Min	Min date	Min time	Max	Max date	Max time	Units
--	P (W)	3886.507	1005.000	10/21/2015	1:35:00.000 p.m.	8489.000	10/20/2015	1:40:00.000 p.m.	kW
--	P1 (W)	1177.676	287.000	10/21/2015	5:20:00.000 p.m.	2621.000	10/20/2015	1:40:00.000 p.m.	kW
--	P2 (W)	1463.578	353.000	10/20/2015	9:35:00.000 a.m.	3152.000	10/20/2015	1:40:00.000 p.m.	kW
--	P3 (W)	1245.281	332.000	10/21/2015	4:45:00.000 a.m.	2716.000	10/20/2015	1:40:00.000 p.m.	kW

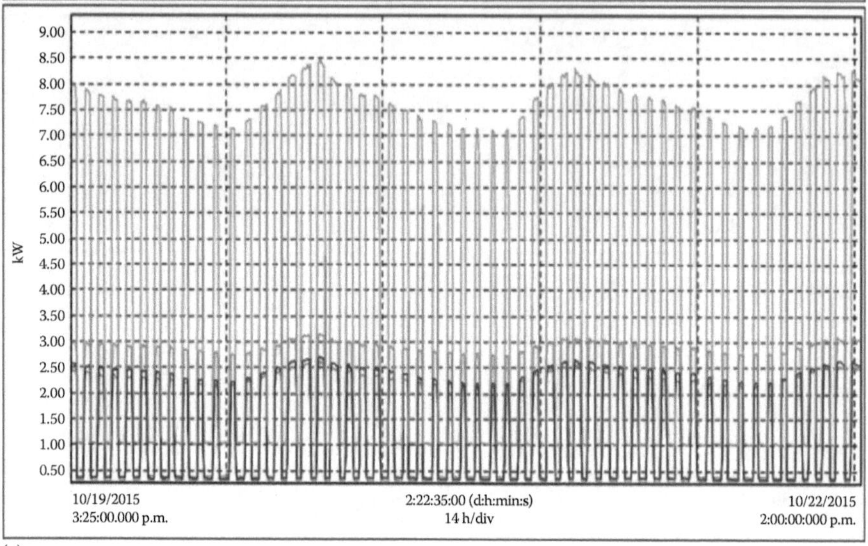

(a)

Legend	Name	Avg	Min	Min date	Min time	Max	Max date	Max time	Units
--	S (VA)	6683.678	2393.000	10/21/2015	11:00:00.000 a.m.	12686.000	10/20/2015	12:50:00.000 p.m.	kVA
--	S1 (VA)	2175.121	808.000	10/21/2015	11:00:00.000 a.m.	4096.000	10/20/2015	12:40:00.000 p.m.	kVA
--	S2 (VA)	2297.619	744.000	10/20/2015	2:35:00.000 p.m.	4469.000	10/20/2015	12:20:00.000 p.m.	kVA
--	S3 (VA)	2210.933	831.000	10/21/2015	11:00:00.000 a.m.	4136.000	10/20/2015	12:50:00.000 p.m.	kVA

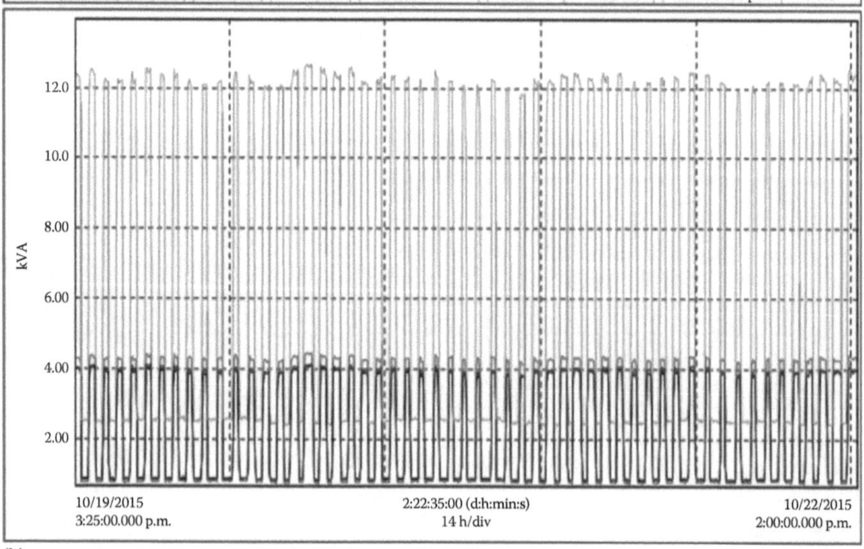

(b)

FIGURE 11.15 Time variation of (a) real power, (b) total power. (c) current THD for the air conditioning system of a residence.

(Continued)

Legend	Name	Avg	Min	Min date	Min time	Max	Max date	Max time	Units
—	I1-THD	3.087	2.390	10/20/2015	9:35:00.000 a.m.	3.920	10/22/2015	8:50:00.000 a.m.	%
—	I2-THD	3.052	2.330	10/22/2015	5:15:00.000 a.m.	4.130	10/22/2015	12:40:00.000 p.m.	%
—	I3-THD	2.855	1.870	10/21/2015	3:15:00.000 a.m.	4.200	10/21/2015	8:00:00.000 a.m.	%
—	IN-THD	41.332	26.020	10/21/2015	6:55:00.000 a.m.	75.740	10/22/2015	1:20:00.000 p.m.	%

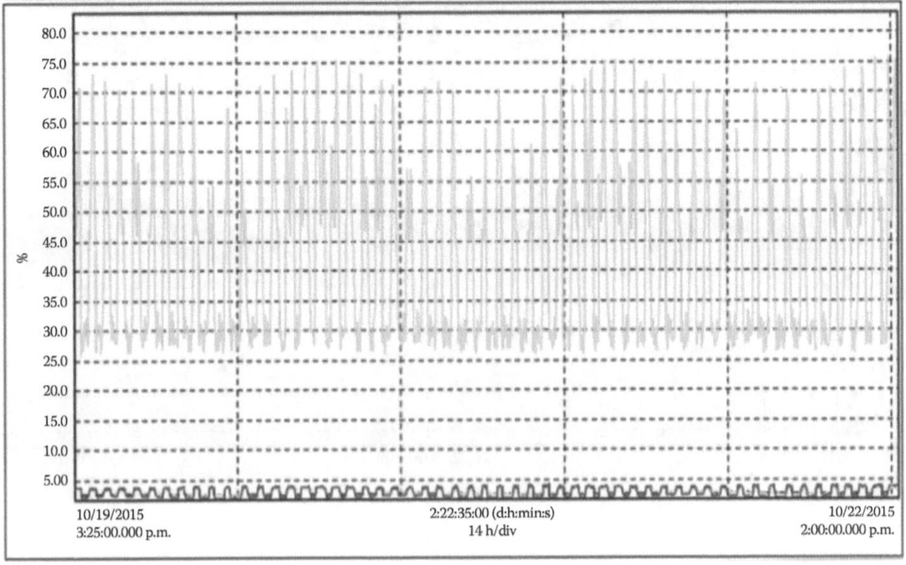

(c)

FIGURE 11.15 (*Continued*) Time variation of (a) real power, (b) total power. (c) current THD for the air conditioning system of a residence.

Monitoring devices are now integral parts of building energy management system (EMS) to measure power consumption, reactive power, and power quality indicators for the main electrical services to buildings. The monitored data are utilized to improve building energy efficiency and thermal comfort as well as to detect any faults related to electrical and mechanical systems distributed throughout the building. Figure 11.16 illustrates the voltages, currents, real power, reactive power, voltage THD, and the current THD measured for a residential building. As indicated in Figure 11.16, while the average THD values for voltage are less than 1%, the average THD value for neutral current is high and is estimated to be 50.8% during the testing period.

11.7 SUMMARY

In this chapter, some of the causes for power quality problems in building electrical distribution systems are discussed and described. While power quality issues affect the reliability of power delivery, they are not considered typically as safety problems and are mostly ignored by the NEC. However, it is important to consider the power quality effects for various loads when designing electrical systems for buildings. This chapter provided some mitigation options to eliminate or at least reduce the power quality problems. Throughout the chapter, a special emphasis is given to the causes and mitigation solutions for current and voltage harmonic distortions. Indeed,

Legend	Name	Avg	Min	Min date	Min time	Max	Max date	Max time	Units
—	V1	242.520	238.090	10/17/2015	9:38:36.000 a.m.	247.260	10/17/2015	11:26:59.000 p.m.	V
—	V2	241.293	236.010	10/17/2015	9:38:35.000 a.m.	246.320	10/19/2015	5:36:09.000 a.m.	V
—	V3	245.171	239.870	10/18/2015	10:10:57.000 a.m.	250.040	10/17/2015	1:07:17.000 a.m.	V

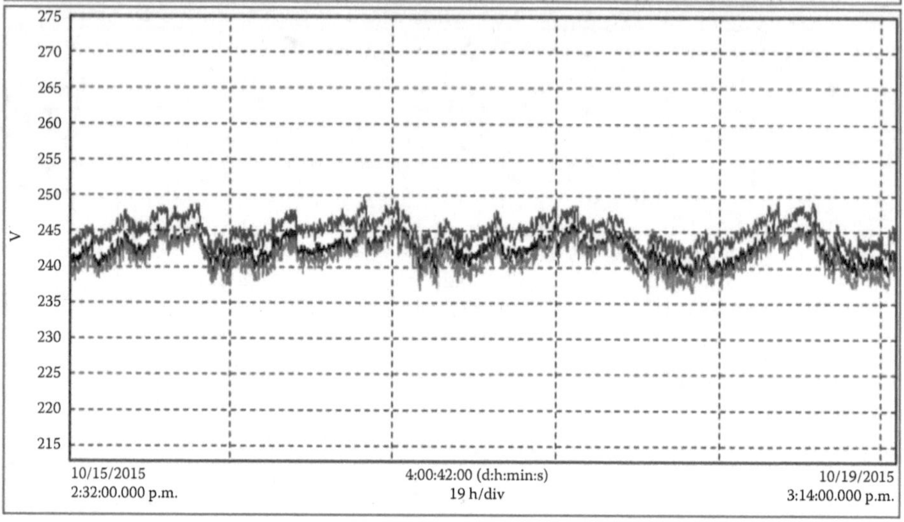

(a)

Legend	Name	Avg	Min	Min date	Min time	Max	Max date	Max time	Units
--	I1	12.014	3.943	10/16/2015	7:27:15.000 a.m.	20.896	10/19/2015	12:56:35.000 a.m.	A
	I2	13.453	3.521	10/19/2015	10:42:14.000 a.m.	23.467	10/16/2015	10:42:08.000 a.m.	A
--	I3	13.574	3.520	10/18/2015	1:08:34.000 p.m.	32.343	10/18/2015	2:06:11.000 p.m.	A
--	IN	5.550	0.904	10/16/2015	7:27:15.000 a.m.	18.883	10/15/2015	9:35:44.000 p.m.	A

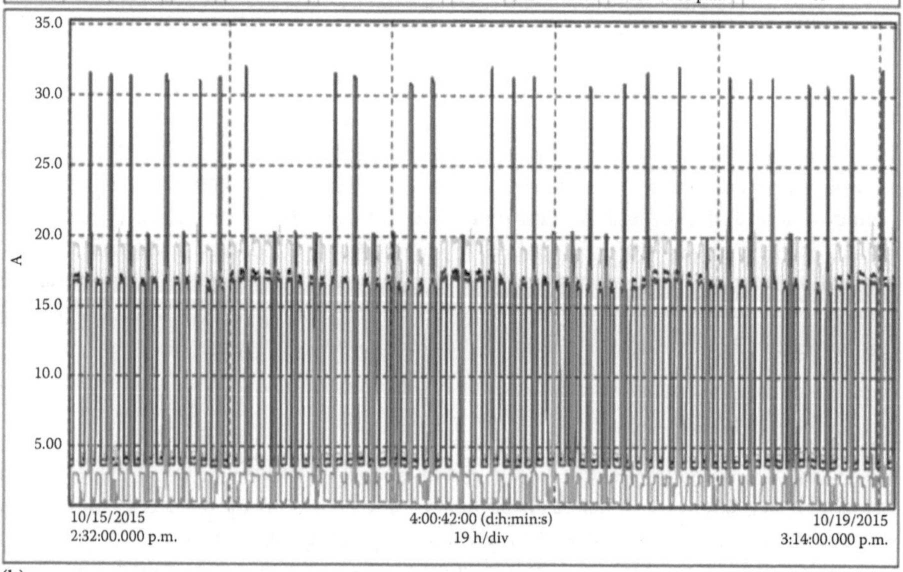

(b)

FIGURE 11.16 Time variation of (a) voltage, (b) currents. (c) real power, (d) reactive power. (e) voltage THD, and (f) current THD for service entrance for an apartment building.

(Continued)

Legend	Name	Avg	Min	Min date	Min time	Max	Max date	Max time	Units
--	P (W)	5009.166	1309.000	10/18/2015	2:50:00.000 p.m.	13428.000	10/18/2015	2:02:00.000 p.m.	kW
--	P1 (W)	1439.217	453.000	10/16/2015	10:25:00.000 p.m.	2871.000	10/18/2015	2:02:00.000 p.m.	kW
--	P2 (W)	1795.169	456.000	10/18/2015	4:01:00.000 p.m.	3621.000	10/19/2015	12:46:00.000 p.m.	kW
--	P3 (W)	1774.780	370.000	10/19/2015	5:55:00.000 a.m.	7181.000	10/17/2015	4:16:00.000 p.m.	kW

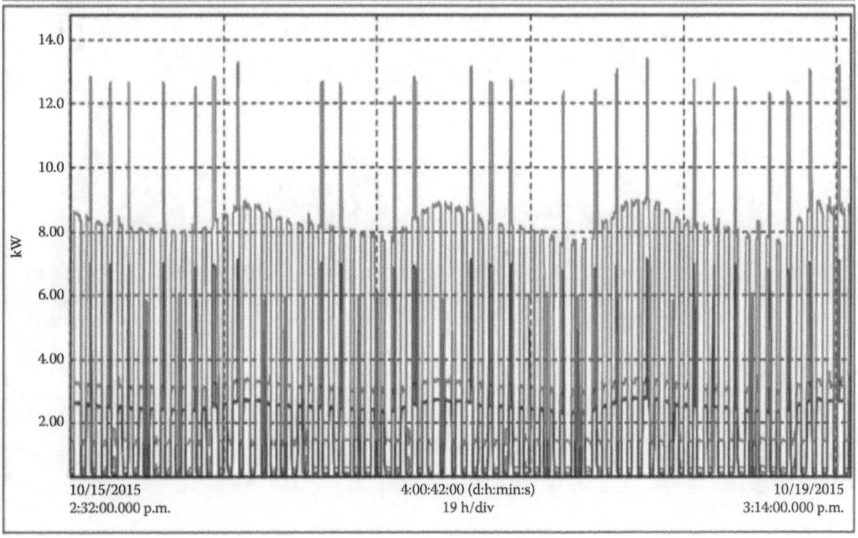

(c)

Legend	Name	Avg	Min	Min date	Min time	Max	Max date	Max time	Units
--	Q (var)	5900.086	2269.000	10/18/2015	5:19:00.000 p.m.	10519.000	10/16/2015	5:35:00.000 a.m.	kVAR
--	Q1 (var)	1978.238	824.000	10/16/2015	7:27:00.000 a.m.	3465.000	10/17/2015	1:21:00.000 a.m.	kVAR
--	Q2 (var)	2014.369	703.000	10/18/2015	5:19:00.000 p.m.	3784.000	10/16/2015	5:35:00.000 a.m.	kVAR
--	Q3 (var)	1907.441	737.000	10/18/2015	5:19:00.000 p.m.	3432.000	10/19/2015	3:43:00.000 a.m.	kVAR

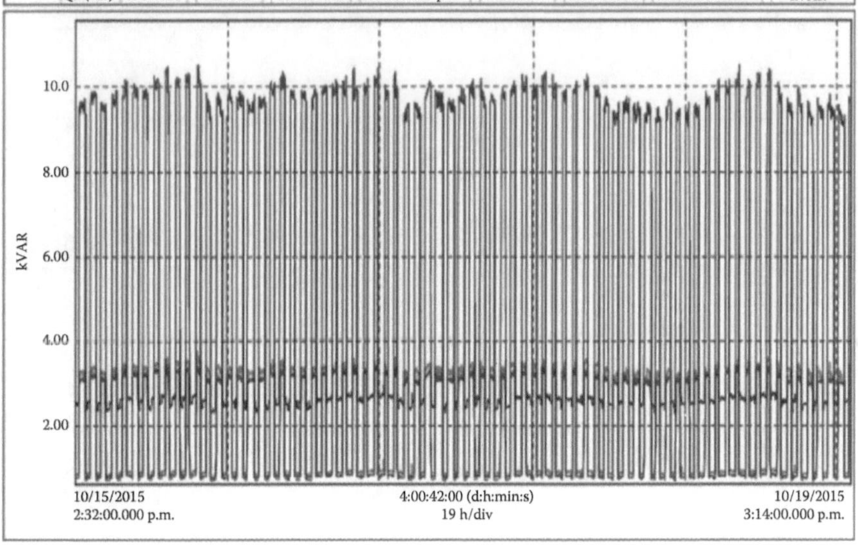

(d)

FIGURE 11.16 (*Continued*) Time variation of (a) voltage, (b) currents. (c) real power, (d) reactive power. (e) voltage THD, and (f) current THD for service entrance for an apartment building.

(Continued)

Legend	Name	Avg	Min	Min date	Min time	Max	Max date	Max time	Units
--	V12-THD	0.692	0.570	10/15/2015	2:38:00.000 p.m.	0.840	10/18/2015	9:20:00.000 p.m.	%
--	V23-THD	0.685	0.550	10/15/2015	2:43:00.000 p.m.	0.820	10/18/2015	9:21:00.000 p.m.	%
--	V31-THD	0.702	0.540	10/15/2015	2:32:00.000 p.m.	0.880	10/19/2015	3:42:00.000 a.m.	%

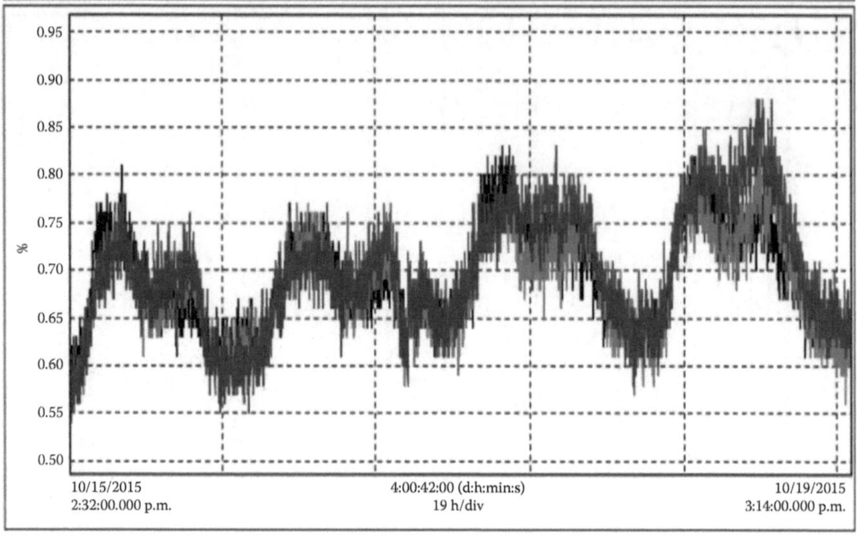

(e)

Legend	Name	Avg	Min	Min date	Min time	Max	Max date	Max time	Units
--	I1-THD	8.455	3.290	10/15/2015	7:22:00.000 p.m.	21.710	10/17/2015	4:43:00.000 a.m.	%
--	I2-THD	3.307	1.870	10/15/2015	8:10:00.000 p.m.	15.100	10/18/2015	1:41:00.000 a.m.	%
--	I3-THD	2.921	1.740	10/16/2015	1:55:00.000 a.m.	7.890	10/17/2015	5:34:00.000 a.m.	%
--	IN-THD	50.797	7.050	10/19/2015	5:13:00.000 a.m.	93.240	10/18/2015	1:36:00.000 p.m.	%

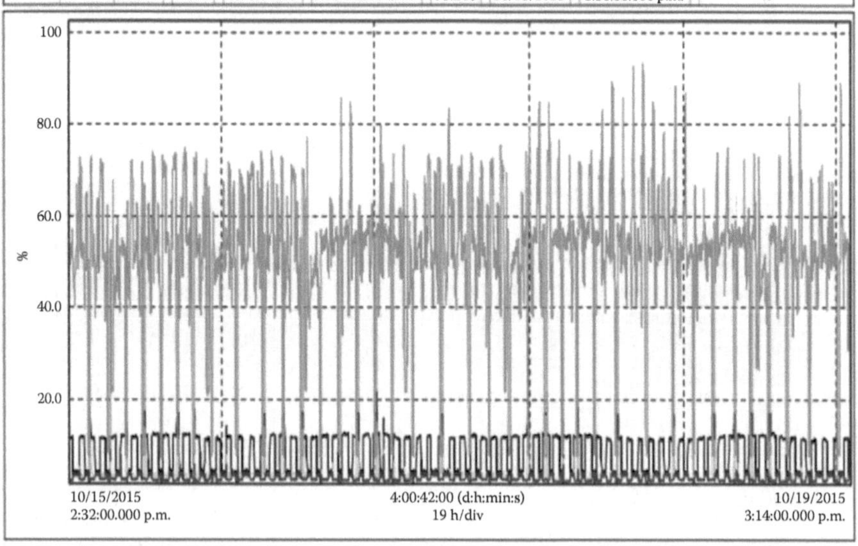

(f)

FIGURE 11.16 (*Continued*) Time variation of (a) voltage, (b) currents. (c) real power, (d) reactive power. (e) voltage THD, and (f) current THD for service entrance for an apartment building.

several loads commonly utilized in modern buildings can generate noticeable harmonic distortions and may cause, if not adequately mitigated, significant damages to various components of the distribution systems.

11.8 PROBLEMS

1. Based on the analysis of voltage measured for a lighting panel, it was found that the rms values for the fundamental, the third, the fifth, the seventh, and the ninth harmonics that make up the voltage waveform are the following:

$$V_1 = 475 \text{ V},\ V_3 = 15 \text{ V},\ V_5 = 7 \text{ V},\ V_7 = 3 \text{ V},\ V_9 = 1 \text{ V}$$

Determine the IHD and THD values using the IEEC and IEC definitions.
2. Based on the analysis of current measured for a power panel, the rms values for the fundamental, the third, the fifth, and the seventh harmonics that make up the voltage waveform are estimated to be the following:

$$I_1 = 175 \text{ A},\ I_3 = 85 \text{ A},\ I_5 = 50 \text{ A},\ I_7 = 25 \text{ A}$$

Determine the IHD and THD values using the IEEC and IEC definitions.
3. Estimate the current THD of a desktop computer with its monitor. Use the data provided in Table 11.2.
4. Determine the k-factor required for a transformer supplying a load made up of 120 A for fundamental harmonic, 60 A for the third harmonics, and 20 A for the fifth harmonics.
5. Determine the *k*-factor required for a transformer supplying a load made up of 245 A for fundamental harmonic, 115 A for the third harmonics, 75 A for the fifth harmonics, 50 A for the seventh harmonics, and 30 A for the ninth harmonics.
6. Determine AC resistance values for a 1/0 AWG copper conductor serving a. Consider the fundamental as well as the third, the fifth, the seventh, and the ninth harmonics.
7. Determine AC resistance values for a 400 MCM copper conductor serving a house panel. Consider the fundamental as well as the third, the fifth, and the seventh harmonics.
8. For the 400 MCM copper conductor, estimate the current rating factor q if the rms current values for the various harmonics are $I_1 = 310$ A (fundamental harmonic), $I_3 = 120$ A (third harmonics), $I_5 = 165$ A (fifth harmonics), $I_7 = 70$ A (seventh harmonics), and $I_9 = 40$ A (ninth harmonics).
9. For 1,000 kVA transformer rated at 13.2 kVA–208Y/120 V with a per unit impedance of 3%. Determine the resonance frequency for three capacitor banks (1) 300 kVAR, (2) 400 kVAR, and (3) 500 kVAR.
10. A 2,500 kVA transformer rated at 13.8 kVA–480Y/277 V with an impedance of 5%. Determine the resonance frequency as a function of the size in kVAR capacitor banks ranging from 100 to 1,000 kVAR. Determine the best option for the capacitor bank.

12 Design of Photovoltaic Systems

In this chapter, an overview of various components of photovoltaic (PV) systems is provided. In particular, the basic types and operations of PV modules, batteries, inverters, and controllers are described. In addition, typical stand-alone and grid-connected PV system configurations suitable for buildings are discussed. Finally, the design approach of various components of PV systems is outlined. Several calculation examples are presented to illustrate the design guidelines for PV systems integrated as part of the electrical power distribution for buildings.

12.1 INTRODUCTION

According to the US Department of Energy, as of 2020, there are 97.2 gigawatts (GW) of solar photovoltaic (PV) capacity, or enough to power 18 million (12.9%) of the 139 million homes in the US. This solar capacity includes non-residential PV arrays. Nevertheless, the PV capacity has grown nearly 286 times larger since 2008 when it was only 0.34 GW (DOE, 2022). By 2030, it is predicted that one in seven US homes will have a rooftop PV system (DOE, 2022).

Currently, the adoption rate for PV systems in US households is estimated at 12.9% (Schwartz and Krarti, 2022). The adoption of PV both for early adopters (i.e., before 2007) and late adopters is a prime example of how motivations and reasons to adopt renewable energy technologies have changed over time (Schwartz and Krarti, 2022). Early adopters were mostly interested in the environmental impacts and love for new technology, whereas late adopters were primarily incentivized by the cost benefits of PV systems. Apart from environmental beliefs and educational levels, there are no clear correlations between psychological or demographic characteristics and the adoption of PV systems in buildings.

Building location characteristics and prevalent grid electricity prices impact the adoption of PV systems since their cost-benefit increases and their impact on reducing utility costs is greater. Location can be a good indicator for the adoption level of PV systems among US households. Indeed, energy costs and incentives for installing PV panels vary greatly from state to state. While the median cost for PV systems (i.e., PV, inverter, and controls) has decreased from about \$12.00/Watt in 2000 to \$3.80 in 2019 and the median size of the PV arrays has increased from 2.5kW in 2000 to 6.5kW in 2019, not every state has the same optimal solar radiation conditions (Schwartz and Krarti, 2022). Nevada and Arizona have the lowest cost of energy, including system installation and generation price, at \$2.30/Watt and Colorado has the highest at \$3.15/Watt. Median size also varies by state, with Washington DC having the smallest PV systems at 7.8 kW total and Florida having the largest systems at 13.4 kW total. It should also be noted that larger PV systems correlate to higher

DOI: 10.1201/9781003276999-12

monthly electricity consumption. However, the overall cost across all US states is steadily decreasing and is on average \$2.75/Watt as of 2020, a 27% drop since 2014 (Schwartz and Krarti, 2022)

The PV technology is accessible but is not yet affordable to all US households. Indeed, there is only one PV array for every 65 residents in California based on a survey conducted in 2019 (Lukanov and Krieger, 2019). In areas where PV is common like cities, residents only spend 3.5% of their income on utilities, but in rural areas where it is uncommon to install rooftop PV systems and households are poorer, they are spending as much as 14% of their income on utilities. In a way, the lack of PV not only contributes to poverty but also causes poorer health conditions since the alternate energy sources often include fuel-based power systems that emit toxic gasses. The initial capital cost to install a PV system can be high, even with rebates. While there are subsidies for installing PV systems, the cost of these systems is still not affordable for all US households (Lukanov and Krieger, 2019).

The photovoltaic power industry has originated from the efforts of a few determined individuals with a commitment to clean and renewable energy sources. Unfortunately, most of these pioneers had little electrical systems expertise. Even until recently, several PV dealers and installers lacked adequate knowledge and training in building electrical systems. Contrary to a common popular belief, PV systems are at least as complicated to be properly designed and installed as the electrical distribution systems for buildings. Not only must the National Electrical Code (NEC) requirements for building electrical systems be followed, but additional requirements for PV systems and outdoor, sunlight-exposed locations must be met. Another complication is the inherent presence of direct current (DC) electricity within PV systems. DC systems are much less familiar to most electricians than alternating current (AC) circuits. Moreover, despite their lower voltage, DC systems can be as dangerous as AC circuits, as noted in Chapter 5. It should be noted that NEC is not as nearly as comprehensive on DC as it is on AC systems. Components rated for DC are not as widely available as those rated for AC and are often expensive due to both low production levels and the inherent difficulty of handling DC currents. Finally, some PV equipment manufacturers do not have their equipment properly listed to specify their electrical performance and requirements.

As noted in the previous chapters, for building electrical distribution systems, following all NEC requirements is absolutely necessary to achieve safe and reliable PV installations. These requirements are not always straightforward, and the longevity of PV power systems makes durability a special concern. This chapter first reviews typical PV system components and their functions. Then, a detailed discussion is provided to highlight NEC requirements and the best practices for PV modules, conductors, overcurrent devices, disconnects, grounding, lightning and surge protection, terminals and connections, wiring and load centers, and system labeling. Finally, some design guidelines are presented to ensure safe, durable, code-compliant PV systems.

12.2 PHOTOVOLTAIC SYSTEM COMPONENTS

PV systems are made up of several components that are not common in most residential and commercial electrical installations. In this section, a brief overview is provided of the major components for building integrated PV systems.

12.2.1 PV Modules

PV modules are made up of a set of PV cells that are connected in series and/or parallel and are encapsulated in environmentally protective laminates and are essentially the building blocks of all types of PV systems. PV panels represented a set of modules assembled in prewired units that can be easily field installed. Finally, PV arrays constitute the entire set of PV panels that are installed to generate power as part of building-integrated electrical systems or electrical distribution systems.

Typical PV cells are made of either crystalline silicon or amorphous (thin-film) silicon. Crystalline silicon is typically grown in either monocrystalline (single crystal) or polycrystalline (multicrystal) arrangements. Amorphous silicon is applied to surfaces in layers often through vapor deposition processes. Silicon is doped with small impurities of boron or phosphorus to form p–n junctions. These p–n junctions create a flow of electrons when energized by photons. The resultant solar cells produce DC electricity when sunlight strikes their surface. The silicon cells are typically protected by tempered glass on the surface and a polymer backer. Aluminum framing adds strength and allows for a simple and secure mounting.

The PV modules are typically mounted using additional aluminum or steel members. There are five common mounting methods: roof, pole, ground, bracket, and building integrated. Most residential and several commercial PV systems are roof-mounted. The panels are attached directly to the roof surface with brackets. In ground mounting, the same method is used but on a footing, instead of a roof deck. Pole mounts are fixed on the top of a pole at least 3 m (10 ft) above the ground and are frequently used to gain solar access on obstructed sites. Pole-mounted PV modules are sometimes fitted with one- or two-axis trackers to increase total generated electricity output. Building-integrated photovoltaics (BIPV) are installed as an integral part of buildings. Transparent modules, roofing products, shading devices, and many other designs are being used to both produce electricity and serve as an element of the building envelope. Crystalline silicon is used in some BIPVs, but amorphous silicon is more flexible and is more suitable for BIPV applications. Moreover, amorphous silicon is cheaper to manufacture than crystalline forms. However, it is also less efficient and is not necessarily less expensive on a cost per watt basis. Monocrystalline cells are somewhat more efficient than polycrystalline, but are also more expensive to manufacture. Given their popularity, it is the properties of crystalline silicon cells that drive the requirements in the NEC for electrical PV system requirements (NEC, 2014).

PV modules are typically rated under standard test conditions (STC). These tests are carried out in a laboratory setting with an irradiance of 1,000 W/m^2 and a cell temperature of 25°C. The spectral properties of the radiation are standardized and designed to approximate sunlight. It should be noted that these parameters do not represent the extremes of the environmental conditions likely to be experienced by the PV modules in operation. Therefore, when designing PV system components, module ratings must be adjusted to the extreme climatic conditions to ensure the safe and reliable operation of all system components. Figure 12.1 shows typical PV cell performance expressed in I–V curves. The output current varies almost directly with solar radiation, while the voltage increases as the cell temperature decreases. When batteries are present, the operating voltage is typically regulated by the battery voltage, except when a maximum power point tracker (MPPT) system is used.

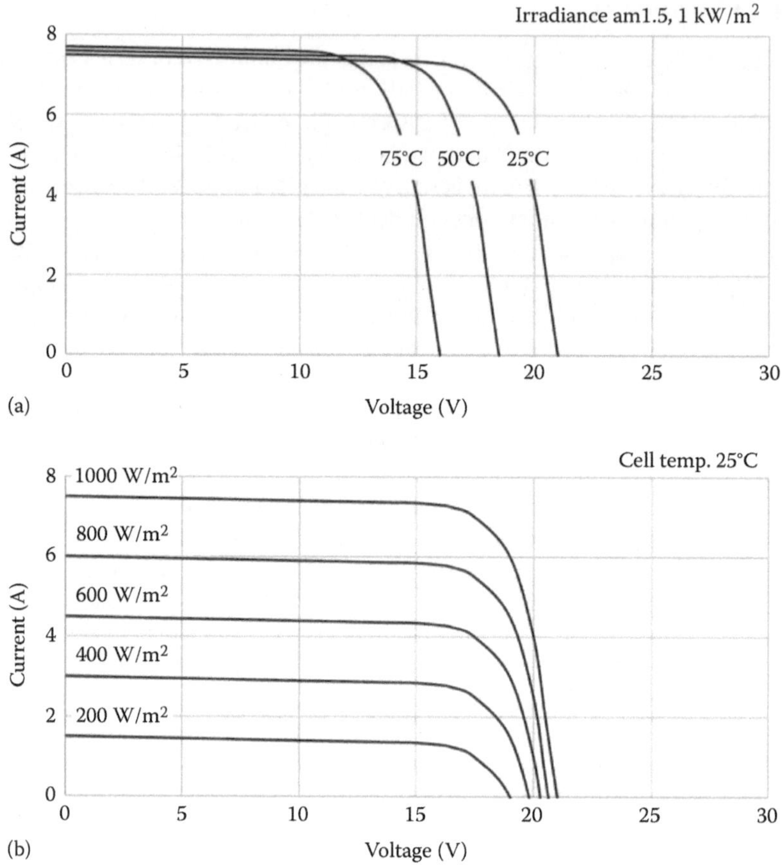

FIGURE 12.1 Variation of voltage and current for various (a) PV cell temperature and (b) solar radiation.

Figure 12.2 shows five key values that are required to be specified on PV modules by the Underwriters Laboratories (UL) as part of the listing process. These include open circuit voltage (V_{OC}), short-circuit current (I_{SC}), maximum power voltage (V_{MP}), maximum power current (I_{MP}), and maximum power (P_{MAX}). Open circuit voltage and short-circuit current represent the maximum values achieved under the STC. Under normal operating conditions, the array can experience higher values of both voltage and current.

Table 12.1 presents typical PV module characteristics for common cell types, including rated power, short-circuit current, open circuit voltage, and efficiency.

As noted in Table 12.1, the conversion efficiency of commercially available PV modules is in the 10%–13% range under rated conditions. It should be noted that thin-film PV modules are typically used for building-integrated PV applications (such as shingle-PV systems) and their efficiency is improving. A simple analysis to assess the

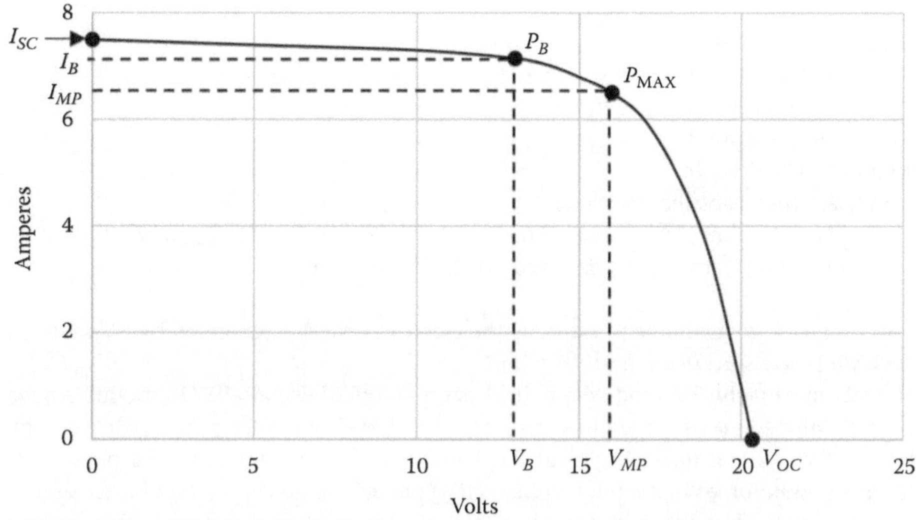

FIGURE 12.2 Typical current–voltage curve for a photovoltaic module.

TABLE 12.1
Typical PV Module Characteristics

PV Cell Type	Rated Power (W)	Short-Circuit Current, I_{SC} (A)	Open Circuit Voltage, V_{OC} (V)	Efficiency (%)
Single crystal	175	5.55	44.4	13.5
Multicrystal	75	4.75	21.8	11.6
Thin film	40	2.68	23.3	9.4

performance of PV systems in generating electricity is to combine the efficiencies for all of its components. In particular, for a typical PV system with a PV panel, an inverter, and MPP tracker, the overall system efficiency can be estimated as follows:

$$\eta_{PV,\text{system}} = \eta_{PV,\text{Module}} \times \eta_{INV} \times \eta_{MPPT} \times \eta_{VR} \tag{12.1}$$

where

$\eta_{PV,\text{Module}}$ is the PV panel efficiency (refer to Table 12.1 for typical efficiency values)
η_{INV} is the inverter efficiency (typically over 95%)
η_{MPPT} is the maximum power point tracker efficiency (typically over 95%)
η_{VR} is the voltage regulator efficiency (typically over 95%)

Then the electrical energy generated, E_{gen}, by the PV system can be estimated on an hourly basis, using the following expression:

$$E_{gen} = \eta_{PV,\text{system}} \times A_{PV} \times G_{net} \tag{12.2}$$

where

A_{PV} is the area of the PV panel

G_{net} is the net incident radiation hitting the PV panel and depending on the tilt angle and tracking system used for the PV system

Several tools are available to estimate the electrical energy generated by PV systems such PVWatts, TRNSYS, and PV-Chart.

Instead of only PV modules, hybrid photovoltaic/thermal (PV/T) modules have been developed and installed in several buildings. PVT modules convert solar energy into electricity and heat. A typical PV/T collector is analogous to a flat plate solar thermal collector, except a photovoltaic (PV) panel is attached on top of the metallic absorber plate. The PV cells are composed of semiconductor materials that convert high-energy photons of incident solar radiation into electricity. The lower energy photons are absorbed by the PV panel and generate heat within the cells. This heat is rejected to the ambient conditions while also being removed from behind the PV cells through convective heat transfer. If extracted, heat from the PV panels can offer multiple benefits. The heat can be used to meet various thermal loads while improving the PV electrical efficiency due to its lower operating temperature. Both air and fluids are used to recover heat from the PV systems. The amount of thermal energy and its quality are important factors when considering the value of the output heat. The performance of a PV/T system is characterized by both the amount of electrical energy and useful heat produced. Currently, there are only very limited PV/T systems commercially available.

12.2.2 Controllers

Several types of controllers are used in PV systems to regulate the power flow of the battery system. There are three basic controller types:

1. *Charge controllers* are defined as equipment devices that control DC voltage or DC current, or both, used to charge a battery (NEC, 2014). There are four common types of charge controllers: shunt, single stage, multistage, and pulse. Shunt controllers divert excess current through a resistor and release it as heat. These are generally specified for small systems. Single-stage controllers simply shut off current flow to the batteries at a specific voltage set point. Both shunt and single-stage controllers are simple and inexpensive, but they may lead to less flexibility, lower state of charge, and shorter battery life. Multistage controllers have several voltage ranges with different current settings resulting in higher battery state of charge and longer battery life. Finally, pulse controllers provide pulsing charges that decrease as the voltage set point is reached.

As batteries are charged and discharged, charge controllers generally adjust the current to achieve optimum battery conditions. Charge controllers also provide protection from overcharging and perform equalization, vital functions for the protection of battery life. Equalization restores all of the individual cells in the battery bank to approximately the same condition by delivering a high-voltage charge. A number of additional features are often incorporated with the controller, including monitoring features to provide indications on system performance, temperature compensation to ensure proper battery charging, LED charge indicators (for a low battery alarm, for instance), and automatic equalization. Another key feature of charge controllers is sometimes to adjust the PV array voltage to reach the maximum power point (MPP). In this case, the controllers are called maximum power point tracking or MPPT controllers. In particular, Figure 12.2 shows a common battery voltage of 13 V (VB) and the corresponding array current (IB) and power from the array (PB). The module maximum power point PMAX is 110 W under STC. An MPPT controller can cause the array to operate at or near this point, maximizing the useful energy output. If the array were allowed to operate at the battery voltage in Figure 12.2, the total power would be approximately equal to 13 V × 7.1 A, or 92 W. This is about 16% lower than the power produced by the MPPT controller. MPPT controllers can be substantially more expensive than other controllers, but can pay for themselves by optimizing the PV-generated electrical energy output.

2. *Load controllers* are typically charge controllers operating in load control mode. They protect the batteries from overdischarging by disconnecting the loads when the battery voltage gets too low. They can also be used to monitor system performance. It should be noted that some charge controllers can simultaneously operate as load controllers. Depending on a system's level of monitoring, a separate load controller is often desirable.
3. *Diversion controllers* are actually specific types of charge controllers and are defined as equipment devices that regulate the charging process of a battery by diverting power from energy storage to DC or AC loads or to an interconnected utility service (NEC, 2014). A diversion controller may be connected to DC loads or, when included in an inverter, to AC circuits. NEC requires that a diversion charge controller have a backup charge controller to ensure that the batteries are not overcharged. Indeed, when protection devices in the diversion circuits are manually or automatically opened, the diversion controllers would become inoperable.

12.2.3 Batteries

Batteries are used to provide electricity power when solar radiation is not available during nights and on cloudy days when the PV array is not producing enough energy to serve the connected loads. In a utility-interactive system, they are used to provide power during blackouts and brownouts. Batteries used in solar systems are typically deep cycle. Unlike starting, lighting, and ignition (SLI) batteries, deep-cycle batteries can be drawn down to below 50% of capacity repeatedly and still maintain a reasonable operating life. SLI batteries contain a large number of thin plates in each cell that can provide large currents over short periods of time. These battery types are rarely used in PV systems. They can have a useful lifetime of about 2 years if the depth of discharge is kept small (Dunlop, 1997). The three most common battery types include flooded lead–acid, captive electrolyte, or valve-regulated lead–acid, and nickel–cadmium (Ni–Cad).

1. *Flooded lead–acid batteries*: There are two types of flooded lead–acid batteries commonly used in PV systems. These include lead–antimony and lead–antimony/lead–calcium hybrid. Lead–antimony batteries perform well when subjected to deep discharge and high discharge rates. They also have good mechanical strength and maintain their lifetime under high-temperature conditions. The main disadvantage of lead–antimony batteries is their high self-discharge rate, which leads to significant water loss. Water loss can be reduced through the use of catalytic recombination caps that convert some of the escaping gases back into water. Lead–antimony/lead–calcium hybrid batteries have similar durability and deep-cycle performance as those of lead–antimony batteries, but with lower water loss. They often employ lead–calcium positive plates and lead–antimony negative plates and are used primarily in large PV systems. Lead–antimony/lead–calcium hybrids are however prone to sulfation and stratification, which can be addressed through frequent equalization charges.
2. *Valve-regulated lead–acid (VRLA) batteries*: VRLA batteries contain immobilized electrolyte and are sealed (i.e., no electrolyte can be added). Pressure-regulating valves provide for safe discharge of gases under abnormal conditions. These batteries are suitable for remote applications where maintenance is a concern since they are spill-proof and do not require regular water addition. However, since the electrolyte cannot be replenished, charge controller set points must be selected carefully and should be temperature-compensated. Overcharging can lead to electrolyte losses and shorten the battery's lifetime. Moreover, VRLA batteries have the additional benefit to be less susceptible to freezing. There are two common types of VRLA batteries: gelled cells and absorbed glass mat (AGM). Both are designed to recombine gases internally into water to preserve the electrolyte. Gelled cells typically use lead–calcium plates, although some use a lead–antimony/lead–calcium hybrid. If lead–calcium alone is used, the batteries may be susceptible to overcharging and be less amenable to very deep cycling. AGM batteries use glass mats to suspend a liquid electrolyte. The glass mats are placed between metallic plates with hydrogen and oxygen recombined internally. Similar to gelled cells, AGMs are susceptible to overcharging and high operating temperatures.
3. *Nickel–Cadmium (Ni–Cad) batteries*: Ni–Cad batteries have some advantages over lead–acid batteries. In particular, they have longer lifetimes and do not have outgassing problems of electrolyte. They can also be deeply discharged without serious detriment. Moreover, Ni–Cads retain their rated capacity well at low temperatures, have a low freezing point, and do not require careful voltage regulation. Their main disadvantages include high cost and limited availability.

12.2.4 Inverters

Inverters convert the DC output of the PV array, batteries and other DC power sources to AC to serve building loads, the utility grid, or both. The two major types of inverters are stand-alone and utility-interactive. Common inverter features include automatic low-voltage shutoff, load transfer switching, back-feed protection, and stack-ability (i.e., the ability to operate multiple units in series or parallel). Some inverters are also capable of battery charging, taking AC power from the grid or an AC generator and converting it to DC electricity at the battery voltage.

1. *Stand-alone inverters*: These inverters are designed to provide AC power to building loads with no connection to the utility. The two commonly available types are sine wave and modified sine wave. Sine wave inverters produce AC electricity with the same or higher quality as that generated by the utility. Despite this high power quality, stand-alone inverters are not specified to be connected to the grid. Indeed, these inverters lack the ability to synchronize with the grid. Typically, an automatic disconnect is needed to prevent feeding power onto a de-energized grid. Modified sine wave inverters create a stepped approximation of a true sine wave. Figure 12.3 shows a comparison of the modified sine and sine waveforms. While they are less expensive, modified sine wave inverters often cannot be used to operate sensitive electrical loads such as electronics, cordless tool chargers, or certain motors due to the possibility of damaging the end-use equipment. They can

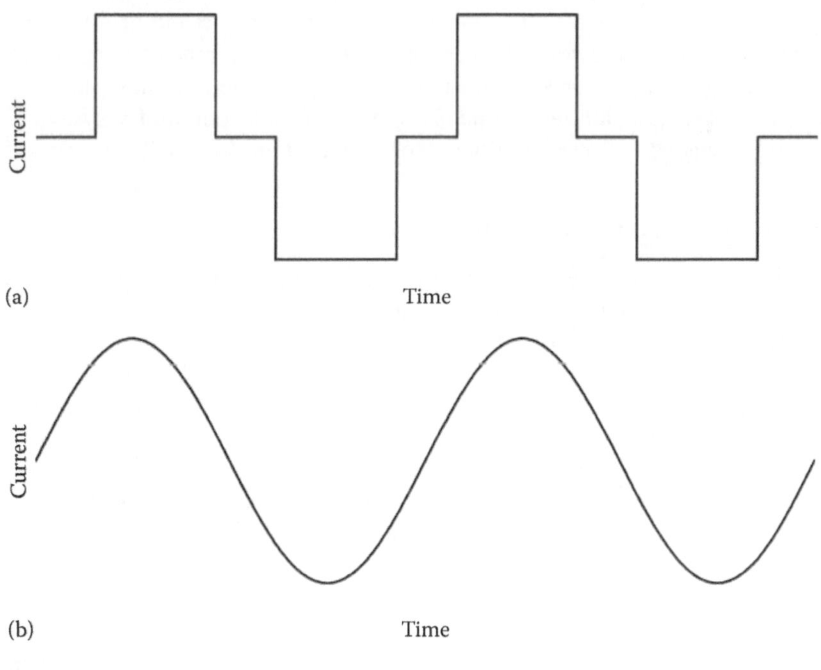

FIGURE 12.3 Comparison of a modified sine wave (a) and a sine wave (b) inverter outputs.

also generate noise when supplying electronic devices such as stereos and digital timers.

2. *Utility-interactive (U-I) inverters*: These inverters are typically connected to the utility power system. Power from the PV array is converted into AC electricity and synchronized with the voltage, frequency, and period of the power already present on the grid. In the case of a utility undervoltage, the inverter shuts down, preventing accidental energizing of the grid but making the rest of the PV system inoperable. Some utility-interactive inverters can be connected to multiple series strings of PV modules without the use of a combiner box. These are known as string inverters and typically contain the required series fuse to protect the PV modules.

The features of utility-interactive and stand-alone inverters are sometimes combined in a single unit. These combined units can provide power to the building loads and the utility system as well as charging and discharging a battery bank. Under normal conditions, these inverters behave as U-I inverters, taking PV array output and sending it to the building loads and the utility grid in AC form. In addition, they maintain the desired battery charge levels. When the grid experiences an undervoltage, the inverter automatically disconnects the grid and supplies the building loads from the PV array and the battery bank.

12.3 PV SYSTEM CONFIGURATIONS

PV systems can be specified for a wide range of applications, each requiring a unique design. The most common design categories are stand-alone and grid-tied PV systems. However, there are several variations of the two main design categories. For instance, several PV systems are designed to supply specific loads such as water pumps. In this section, a brief outline is provided of the two main PV system design categories, intended to power significant loads in buildings.

12.3.1 Stand-Alone PV Systems

In stand-alone systems, a PV array is typically used to charge a battery bank. This system is sometimes augmented by a backup generator to ensure that power is always available.

1. *PV-only systems*: The simplest form of stand-alone PV systems is powered exclusively by photovoltaic modules. While loads could be powered directly from the array without storage, batteries are included nearly universally to ensure continuous power supply. Since most loads and stand-alone inverters require a constant voltage, batteries are also required to stabilize the supply voltage delivered to the connected devices. A typical layout of a PV-only system is shown in Figure 12.4.
2. *Hybrid stand-alone systems*: The intermittent nature of solar resources makes meeting all building loads with a PV system difficult and expensive. One cost-effective solution is to meet a portion of the loads with an

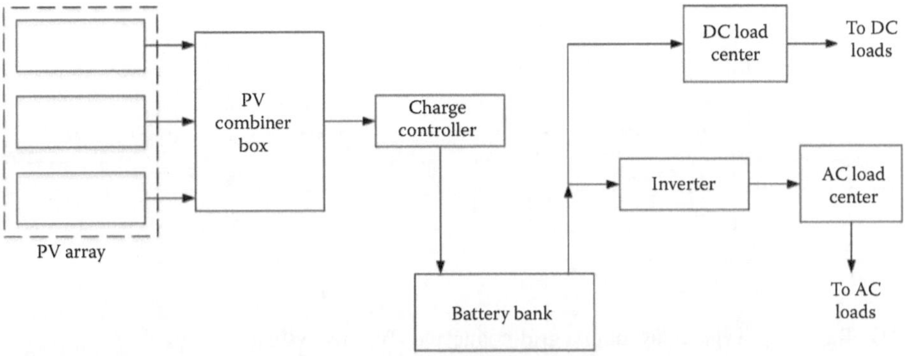

FIGURE 12.4 Typical layout of a stand-alone PV-only system.

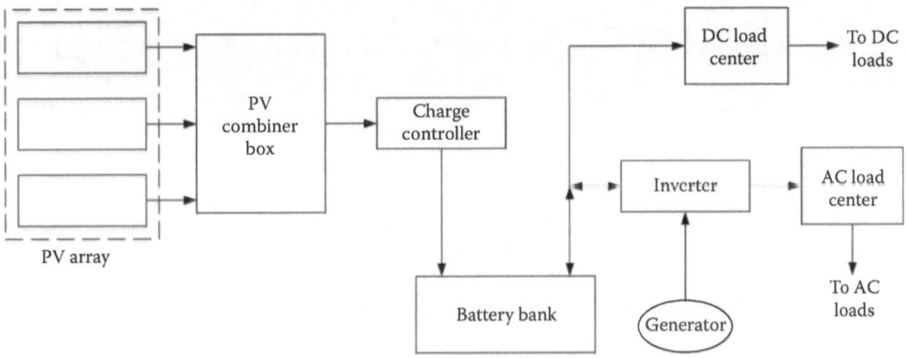

FIGURE 12.5 Typical layout of a hybrid stand-alone PV system.

auxiliary generator. Figure 12.5 shows a system layout of a hybrid stand-alone PV system. The generator typically provides AC power using fossil fuels such as gasoline, diesel, or propane. When the battery bank state of charge becomes low, the generator can be operated to recharge the bank to a high level. When the loads are too high, the generator can be operated to meet those loads and maintain a desired battery state of charge. To have this level of functionality, the inverter used must have battery charging capability. Automatic transfer switching can be used to allow hands-off system operation.

12.3.2 Grid-Connected PV Systems

Grid-connected PV systems contain a PV array that connects directly to an inverter. The inverter has to be of the utility-interactive type and include an automatic disconnect to prevent accidental energizing of the grid. Some grid-connected systems

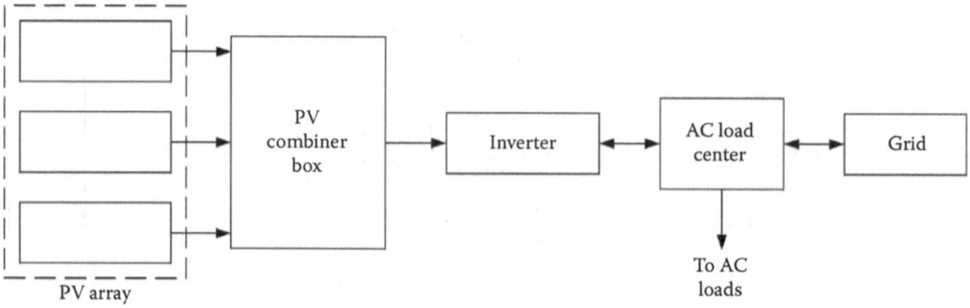

FIGURE 12.6 Typical layout of a grid-connected PV-only system.

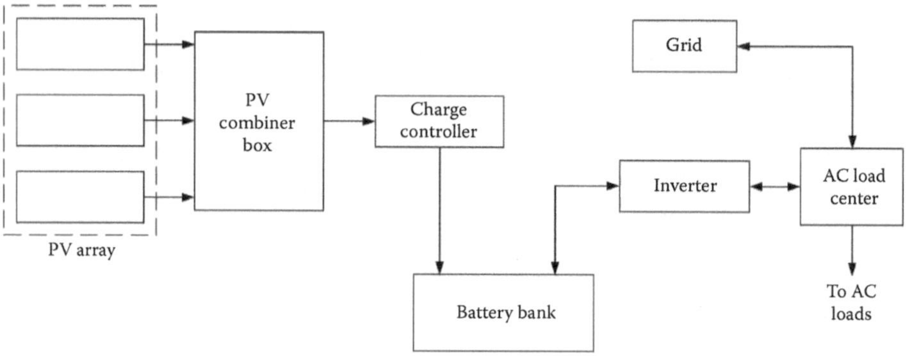

FIGURE 12.7 Typical layout of a grid-connected PV system with battery backup.

include a battery bank to provide backup power in the event of a utility power failure. Two main categories of grid-connected PV systems are discussed here.

1. *PV-only systems*: These systems are the simplest grid-connected PV systems and consist of a PV array connected directly to a utility-interactive inverter and then to the grid through an AC load center. No battery backup is included in this type of grid-connected PV system. When power from the grid is lost, the inverter shuts down and no power is available from the system. Figure 12.6 shows a typical layout of a PV-only grid-connected system. These directly grid-connected systems require only the PV modules and an inverter, making them very cost-effective, especially where net metering is available. If a string inverter is used, there would be no PV combiner box and each string of modules would be wired directly to the inverter.
2. *PV systems with backup*: When battery backup is included, as outlined in Figure 12.7, the grid-connected PV system becomes more complex to design and expensive to install. Indeed, a charge controller is needed to take DC power directly from the PV array and charge the batteries. Typically, the size of the battery bank is smaller than it would be for a stand-alone system. The inverter has to be of the utility-interactive type with a battery charger.

This arrangement is more expensive but provides an uninterrupted power supply for the AC load center. With this arrangement, the critical loads continue to be served even when grid power is not available. An alternative to batteries is to connect the system to an auxiliary generator. Batteries provide a more compact solution due to fuel storage requirements, but a generator may provide some initial cost savings.

12.4 DESIGN OF PV POWER SYSTEMS

12.4.1 Introduction

When its components are selected, the design of a safe, durable, and code-compliant PV power system needs to be considered before its installation and operation. In the United States, and as discussed in Chapters 5–8, the NEC provides the primary guidance for safe electrical installations including PV systems (NEC, 2014). Specifically, Article 690 covering solar photovoltaic systems was added to the NEC in 1984. It has been updated every 3 years since then and represents the current consensus on PV safety. While Article 690 is the only section of the NEC that specifically discusses PV installations, most of the code requirements also apply. Article 300 on wiring materials and methods, and Sections 210–250 on building electrical systems and grounding are particularly important. It should be noted that when Article 690 conflicts with other sections of the NEC, Article 690 takes precedence.

12.4.2 Listing of Electrical Components

Typically, PV modules are warranted for up to 25 years and can produce power for 50 years or more. However, the main focus of NEC is to ensure not the longevity but the safety of the PV installations. Some parts of PV systems experience enormous temperature and moisture swings, exposure to ultraviolet light, and variation of local weather, including wind, rain, snow, and ice. The durability of each component in a PV system installation is critical. To reliably verify the performance of an equipment device, it is important to ensure that it has been listed by an accredited testing laboratory. In the United States, Underwriter's Laboratories (UL) is the accepted listing entity. Components that are listed by UL have generally met a rigorous set of standards for safety and reliability applied to all products listed under that standard. UL also tests components that meet manufacturer-specified standards. All devices used in PV systems in the United States should be UL listed and identified, specifically for the application in which it is used. Just as importantly, the manufacturer's instructions must be followed carefully. These instructions are often mandated by the listing organization and are essential to ensure that the component would perform as expected in the field.

12.4.3 Common Design Issues

PV systems often operate at voltages up to 600 V and can be as dangerous and may cause damage as any building electrical systems. Systems containing batteries or those that are grid-connected can be subjected to high short-circuit currents in the tens of thousands of amperes. Lightning-induced surges can also pose appreciable danger to safety and equipment. The following list summarizes common issues in PV installations, as found by Sandia National Laboratory (Wiles, 2005):

- Improper ampacity of conductors
- Improper types of conductors
- Improper or unsafe wiring methods
- Lack of or improper specification of overcurrent protection devices on conductors
- Inadequate number and placement of disconnects
- Improper application of listed equipment
- No, or underrated, short-circuit or overcurrent protection devices for batteries
- Use of nonlisted components when listed components are available
- Improper system grounding
- Lack of, or improper, equipment grounding
- Use of underrated hardware or components
- Use of AC components (fuses or switches) in DC applications

The remainder of this section addresses the appropriate design approach of PV systems, including specifying conductors, locating and specifying overcurrent protection devices, selecting short-circuit protection, estimating voltage drop, and defining grounding locations. It should be noted that the basic design guidelines for PV systems are the same as those for typical electrical systems outlined for residential buildings (i.e., Chapter 7) and commercial buildings (i.e., Chapter 8). Moreover, most of the design calculations noted for AC systems are applicable for DC systems, including for protection devices and grounding (i.e., Chapter 5), as well as for conductors (i.e., Chapter 6). In this section, the design guidelines specific to PV systems are highlighted.

12.4.4 Design of Conductors

As noted in Chapter 6, the common materials for conductors used in building electrical distribution systems are copper, aluminum, and copper-clad aluminum. Copper conductors are widely used for building electrical wiring, including PV systems, mostly due to their lower electrical resistance and higher resistance to corrosion, even though they can be more expensive than the other types. Several components of PV systems are listed for use with copper only. Indeed, aluminum conductors are more sensitive to faulty installation due to their thermal properties and have been known to cause fires in dwellings.

For PV power systems, a variety of different installation environments are encountered. Indeed, PV modules are typically located outdoors and can be exposed to extreme weather conditions, including freezing, rain, high temperatures, and solar radiation. Each conductor selected must have the appropriate characteristics to withstand these operating conditions. Table 12.2 summarizes some key properties of

TABLE 12.2
Properties of Different Types of Conductors and Their Applications

Wire Type[a]	Temperature Rating (°C)	Dry Locations	Damp Locations	Wet Locations	Conduit Required	Flame Retardant	Sunlight Resistant
THHN	90	Yes	Yes	No	Yes	Yes	No
THWN	75	Yes	Yes	Yes	Yes	Yes	No
THWN-2	90	Yes	Yes	Yes	Yes	Yes	No
THW	75	Yes	Yes	Yes	Yes	Yes	No
THW-2	90	Yes	Yes	Yes	Yes	Yes	No
RHH	90	Yes	Yes	No	Yes	Yes	No
RHW	75	Yes	Yes	Yes	Yes	Yes	No
RHW-2	90	Yes	Yes	Yes	Yes	Yes	No
SE	75	Yes	Yes	Yes	No	No	Yes
USE	75	Yes	Yes	Yes	No	No	Yes
USE-2	90	Yes	Yes	Yes	No	No	Yes
UF	60	Yes	Yes	Yes	No	No	Marked
XHHW	90/75	Yes	Yes	Yes	Yes	Yes	No
XHHW-2	90	Yes	Yes	Yes	Yes	Yes	No
NM	60	Yes	No	No	No	Yes	No

[a] Refer to Chapter 6 for the designation of conductors and cables. The designation "–2" is of particular note for PV installers since it denotes the capability of operating at 90°C even in outdoor locations.

various types of conductors. The temperature rating refers to the maximum operating temperature of the conductor. Where two temperatures are given, the first is for dry applications and the second for wet. The next three columns indicate the suitability of each cable for use in dry, damp, and wet conditions, respectively. A conduit is typically required for all cables not rated for service entrances. Single conductors almost always require conduits. Flame retardants are required for all cables used in buildings whether or not they are included in a conduit. Sunlight resistance is required for exposed cables and conductors used outdoors.

Table 6.1 can be used to determine the ampacity of common conductor types and sizes. For PV installations, there are three common ampacity adjustments that need to be made: (1) ambient temperature, (2) number of conductors, and (3) terminal temperature rating. The first two adjustments are discussed in Chapter 6 and require the use of derating factors provided in Table 6.1 for ambient temperature and Table 6.2 for the number of conductors installed in the same conduit. A brief discussion of the adjustment related to terminal temperature rating is outlined here. Most equipment and overcurrent protection device terminals are rated for 60°C or 75°C conductors. If a higher temperature conductor is used, the ampacity at the lower terminal ratings temperature must be considered to ensure that the terminal would not overheat. Example 12.1 illustrates the ampacity adjustment for conductors serving PV modules.

Example 12.1

Problem

For a PV module wiring, #12 AWG copper USE-2 cables have been selected. The terminal for the protection device is rated at 75°C. The PV system is roof-mounted and can operate at 45°C during hot summer days. Determine if the terminal can operate safely at these conditions if the short-circuit current of the PV module is $I_{SC}=18.9$ A.

Solution

The total short-circuit current from the modules, $1.25\times I_{SC}$, is 23.6 A.

USE-2 cables are rated at 90°C even for wet conditions. However, the terminal is rated at 75°C; thus, the conductor's operating temperature has to be limited to 75°C. At this temperature, the ampacity of USE-2 cable would be 35 A (refer to Table 6.1).

If the terminal is in a roof-mounted enclosure, a temperature derating should be considered to account for high ambient air temperatures during the hot summer days.

Referring again to Table 6.1, the derating factor for 75°C conductors at 45°C ambient temperature is 0.82. So the ampacity of the terminal to remain operating at or below 75°C would be 35×0.82, or 28.7 A. This value is larger than the expected current of 23.6 A, so the terminals should operate at a safe temperature.

12.4.5 Voltage Drop Calculations

Due to the low voltage present in several PV systems, voltage drop is a major design and operation consideration. Voltage drop can also lead to losses in PV-generated power. For DC systems, the voltage drop, ΔE_{DC}, for any wire can be calculated using the basic Ohm's law:

$$\Delta E_{DC} = R\times I \tag{12.3}$$

where

I is current flowing through the wire

R is resistance of the wire taken from Table 8.1

The values listed in Table 8.1 are expressed in Ohms (Ω) per 1,000 m or 1,000 ft and thus should be adjusted for the actual circuit length, which is generally twice the wiring distance. Moreover, it should be noted that these resistances listed in Table 8.1 are specific to conductors rated at 75°C. Fine-stranded conductors have lower resistance since they can operate at temperatures higher than 75°C. In this case, the resistance can be adjusted based on the temperature using the following equation for copper conductors:

$$R = R_{ref}\left[1-0.00323(T-75)\right] \tag{12.4}$$

A typical good rule of thumb is to limit total voltage drop to 2% for 12 and 24 V systems and 3% for 48 V systems (Wiles, 2001). It is important to remember that each terminal, battery, and piece of equipment in a system adds resistance and thus increases voltage drop through the circuit.

Example 12.2 illustrates the voltage drop calculations to size wiring for PV systems.

Example 12.2

Problem

A PV array is mounted 100 ft (30.5 m) from a residence. It operates at 48 V nominal and has battery storage for the stand-alone system. The short-circuit current I_{SC} available from the array is 7.3 A from each of four series strings, yielding a total I_{SC} of 29.2 A. What wire size is needed to keep the voltage drop at or below 2%?

Solution

First, the system operates at 48 V, so the allowable voltage drop is 48 × 0.02, or 0.96 V. Since the system is operating at the battery voltage, we use the total short-circuit current of 29.2 A.

Using a simple calculation method for voltage drop using Equation 12.3, the allowable resistance, then, is

$$R = \frac{\Delta E_{DC}}{I} = \frac{0.96}{29.2} = 0.0329\ \Omega$$

The total wire run is 100 ft, so the circuit length is 200 ft. The resistances are provided per 1,000 ft, so we multiply our allowable resistance by 1,000/200, yielding 0.1644 Ω/1,000 ft.

From Table 8.1, #1 AWG copper wire is required.

12.4.6 Overcurrent Protection Devices

As noted in Chapter 5, overcurrent protection devices (OCPDs) are used to protect conductors from overloads and short-circuits that could cause premature failure and fire hazards. OCPDs are required on all ungrounded conductors in PV power systems. In grounded conductors, OCPDs are allowed only under specific conditions (NEC, 2014). Standard sizes for both circuit breakers and fuses are indicated in Chapter 5. To be used in PV systems, OCPDs must be listed for AC or DC as needed and rated for the appropriate voltage, current, and short-circuit interrupt (NEC, 2014).

For systems with batteries, short-circuit currents can be significant. Indeed, battery banks typically have 1,000–10,000 of amps of short-circuit current available. One approach to reduce the magnitude of these currents is to install a current-limiting RK-1, RK-5, or Class T fuse to limit the short-circuit current levels to less than 3,000–5,000 A. This allows the use of OCPDs in other parts of the system with low interrupt capabilities. Circuit breakers cannot operate quickly enough to be used as current-limiting devices. A possible alternate option is to use OCPDs with interrupt ratings of 20,000–25,000 A throughout the system. If a fuse is used, a disconnect will be required, but it is possible that less expensive OCPDs could be used elsewhere in the system. Commercially available circuit breakers can have a low interrupting capability and are inexpensive for 12–48 V DC systems.

12.4.7 Disconnects

Disconnects in PV systems serve two basic functions. First, they disconnect sources of power in case of an emergency. Second, they provide a way to deactivate power from part or all of the system for troubleshooting and maintenance.

The requirements for disconnects are similar to those for OCPDs. All devices must be rated for the proper type of current (AC or DC), voltage, and power. Disconnects must be provided in all current-carrying conductors sufficient to disconnect them from the rest of the building electrical distribution system. However, switches and OCPDs may not be placed in grounded conductors unless they are part of the ground-fault detection device. The required disconnect on grounded conductors may be accomplished using a bolted or terminal disconnecting means.

The main PV disconnect isolates the PV array from all other components in the system. This disconnect can be located outside or near the location where the PV conductors first penetrate the building. For utility-interactive systems, the main AC disconnect may be subjected to additional local and utility restrictions. In addition to the main disconnects, each piece of equipment must be provided with disconnects sufficient to isolate it from all power sources. This requirement allows for safe maintenance and operation of each system component. Since PV systems often contain multiple sources of power, fuses may be energized at both terminals. These fuses must be provided with disconnects to de-energize the terminals to allow safe maintenance.

All disconnects in PV systems and other electrical installations must satisfy four requirements: They should (1) be readily accessible, (2) be operated without exposing live contacts, (3) have the open and close positions clearly identified, and (4) have sufficient voltage and current interrupting ratings. Accessibility precludes spaces such as bathrooms, locked rooms, and any locations requiring the use of a ladder. For a PV array, the disconnect could be located in the attic if there are permanent (not pull down) stairs, in the garage, or in the basement. Some interior rooms such as living rooms and kitchens may also be acceptable. The most convenient and cost-effective location, particularly for a utility-interactive system, is near the existing AC supply and disconnects.

Disconnects must be grouped together in the same enclosure or a group of enclosures (NEC, 2014). A maximum of six motions of the hand are allowed for the removal of all power from the system. Each disconnect must be clearly marked regarding the load they control. This is particularly important for the main PV disconnect and the utility-interactive disconnect (if present). The PV disconnect must be marked as the main disconnect. The utility-interactive system AC disconnect must be marked with the maximum AC output current and AC operating voltage. When a building has multiple sources of power, a permanent plaque or markings must be placed at each source location denoting all of the power sources on the premises. If a disconnect can have all of its terminals energized even when the device is open, a warning sign must be posted (NEC, 2014).

12.4.8 Grounding

As noted in Chapter 5, grounding plays an important role in protection and safe operation of electrical systems. For PV systems, grounding is needed

- To limit the voltages caused by lightning or accidental contact of the supply conductors with conductors of high voltage.
- To stabilize the voltage under normal operating conditions, which maintains the voltage at one level relative to ground so that any equipment connected to the system is subjected to that same potential difference.
- To facilitate the operation of overcurrent devices, such as fuses, circuit breakers, or relays, when ground-fault conditions occur.

Two grounding types are required for PV systems:

1. *Equipment grounding*: Typically, the equipment grounding is recommended for the PV source and output circuits and they should be sized generally to carry at least 1.25 times the PV array short-circuit current.
2. *System grounding*: It is important to ensure that there is only one point of connection between the grounding system and the grounded conductor.

The same design requirements outlined in Chapter 5 are applicable to the size of both equipment grounding and system (or service) grounding conductors for PV installations.

12.5 PV MODULES AND THE BALANCE OF A SYSTEM

This section reviews the particular design considerations using NEC requirements associated with the components of PV systems. These include PV modules, inverters, batteries, generators, panelboards and enclosures, interior wiring, and system labeling.

12.5.1 PV Modules and Wiring

The characteristics of the PV modules in a system affect the design and size of several other components. In particular, the system voltage and its maximum currents are dictated by the module ratings. Additional requirements for the grounding of PV arrays are also pivotal to the overall system design.

When PV modules are listed by UL, they are required to have a minimum set of specifications marked on the module. These specifications include

- Maximum power output (P_{MAX})
- Open circuit voltage (V_{OC})
- Short-circuit current (I_{SC})
- Maximum power voltage (V_{MP})
- Maximum power current (I_{MP})

- Maximum system voltage
- Maximum series fuse rating

All of these values are obtained at standard test conditions (STC). Some manufacturers provide additional specifications, such as the required copper conductor size. The values of I_{SC}, V_{OC}, I_{MP}, and the maximum fuse rating are particularly important since they are used directly in the design calculations of electrical components of PV systems. The maximum system voltage is needed only for relatively large PV arrays with several panels connected in series.

To determine the size of the conductors of the PV modules, the array open circuit voltage needs to be estimated. According to the NEC, the maximum PV system voltage can be estimated as the sum of all the rated open circuit voltages of the series-connected PV modules corrected for the lowest expected temperatures (NEC, 2014). ASHRAE heating design temperatures data can be used to estimate the lowest expected temperatures at the site when the PV system would be installed (ASHRAE, 2013). Table 12.3 can be used to obtain the appropriate correction factors of the rated open circuit voltages, depending on the lowest temperatures.

There are several conductors specific to PV systems that should be sized, including

- *PV source circuits*: Wiring for the series-connected sections of the array
- *PV output circuits*: Wiring required to connect the series string connected in parallel to the charge controller or inverter
- *Inverter input circuits*: Wiring between the inverters and either batteries (for stand-alone PV systems) or PV output circuits (for interactive PV systems)

TABLE 12.3
Correction Factors for the PV Rated Open Circuit Voltage Rating as a Function of Ambient Temperatures

Ambient Temperature Range in °C (°F)	Correction Factors
24–20 (76–68)	1.02
19–15 (67–59)	1.04
14–10 (58–50)	1.06
9–5 (49–41)	1.08
4–0 (40–32)	1.10
−1 to −5 (31–23)	1.12
−6 to −10 (22–14)	1.14
−11 to −15 (13–5)	1.16
−16 to −20 (4 to −4)	1.18
−21 to −25 (−5 to −13)	1.20
−26 to −30 (−14 to −22)	1.21
−31 to −35 (−23 to −31)	1.23
−36 to −40 (−32 to −40)	1.25

- *Inverter output circuits*: Wiring between the inverters and either AC panelboards (stand-alone PV systems) or service entrance conductors (for interactive PV systems)

The maximum current in a PV source circuit can be estimated by first multiplying the short-circuit current of a module by 1.25. Then, the obtained value is multiplied by an additional factor of 1.25 because the currents within the PV systems are considered continuous:

$$I_{\max,PV-source} = 1.25 \times 1.25 \times I_{SC} \tag{12.5}$$

The maximum current in a PV output circuit is the sum of the paralleled source circuits connected to it with an additional factor of 1.25 for continuous loads:

$$I_{\max,PV-output} = 1.25 \sum_{\text{parellel source Circuits}} I_{\max,PV-source} \tag{12.6}$$

In a PV system operating at less than 50 V, a minimum conductor size of #12 AWG may be required, even though the NEC allows conductors as small as #18 AWG. The same current used to size conductors can also be utilized to size overcurrent devices in the PV source and output circuits. In particular, OCPDs should be rated at no less than $1.25 \times 1.25 \times I_{SC}$ for PV source circuits labeling. If OCPDs are rated for use at 100% of their rating, one of the 1.25 factors can be removed from the OCPD rating calculation. For PV output circuits, OCPDs should be rated based on the sum of the current calculations in the source circuits.

Example 12.3 illustrates the design of OCPDs for a PV system with several panels.

Example 12.3

Problem:

Consider an 880 W array composed of several panels. These panels have a 100 W peak power rating, V_{OC} = 20.7 V, I_{SC} = 7.5 A, and a maximum series fuse of 15 A. The panels are arranged into four source circuits, each operating at 24 V nominal. The ASHRAE 99% heating design temperature is below −36°C. Thus, the maximum operating voltage is 2 × 20.7 × 1.25 or 51.8 V. Standard 600 V conductors can be used, and all OCPDs must be rated for at least 51.8 V DC.

Solution:

Each series circuit has I_{SC} = 7.5 A, so the maximum current is 1.25 × 7.5 = 9.4 A.

If the OCPDs are listed for use at 100% of its rating, it can be rated as low as 10 A.

Otherwise, the OCPDs have to be rated at 9.4 × 1.25 or 11.7 A. Thus, a 15 A fuse or a 15 A circuit breaker can be used to satisfy both the current and maximum fuse requirements. For this system, a 15 A circuit breaker is chosen for convenience of operation.

It should be noted that if the OCPDs are operated at a higher ambient temperature than 40°C, they must be derated according to the manufacturer's instructions.

USE-2 cables are selected for the module interconnections to carry 11.7 A. These cables have to be derated to 70°C ambient temperature using a correction factor of 0.58 (refer to Table 6.1). Thus, the ampacity of the cables has to be at least 11.7 ÷ 0.58, or 20.2 A.

Referring again to Table 6.1, 18–12 AWG copper USE-2 conductors can carry 16, 24, 35, and 40 A, respectively. While 16 AWG is allowed by NEC, 14 AWG conductors are selected since this size is commonly used in residential buildings. Note that 14 AWG conductors have a derated ampacity of 35 × 0.58 or 20.3 A. This is higher than the OCPD rating, and the 15 A OCPD complies with the maximum rating for #14 AWG conductors required by NEC (NEC, 2014).

There are four paralleled source circuits, so the maximum current in the output circuit is 9.4 × 4, or 37.6 A. This is continuous current, so an additional factor of 1.25 is required, giving a total current of 47 A. These conductors can be selected to be THW copper to run in a metal conduit through the attic to the rest of the system. If the ambient temperature in the attic reaches 50°C, the ampacity of the conductor should be derated. For 50°C ambient air temperature, Table 6.1 indicates a derating factor of 0.75, so the required ampacity is 47 ÷ 0.75, or 62.7 A. Since their rated ampacity is 65 A, 6 AWG copper conductors can be selected.

Moreover, the OCPD for the output circuit must be rated at least at 47 A, so a 50 A circuit breaker can be used.

12.5.2 Grounding Conductors

As required for all exposed conductive components in an electrical distribution system, PV module frames must be grounded. Moreover, mounting racks may also have to be grounded. The equipment grounding conductor must be sized to carry at least 1.25 times the PV array short-circuit current, unless the system has a ground-fault protection device. Since the PV module frames are usually aluminum, care must be taken when connecting them to copper conductors. Grounding connections should be made at the designated grounding point, following the manufacturer's instructions.

If the grounding hardware provided is inadequate, Sandia National Laboratory has developed an acceptable method to achieve durable grounding connections (Wiles, 2005). A thread-forming 10–32 stainless steel screw can be attached to the aluminum frame, creating an oxygen-free connection to prevent oxidation.

Moreover, the NEC allows the use of supplementary grounding electrodes connected to the equipment grounding conductors (NEC, 2014). These supplementary electrodes need not be bonded to the rest of the grounding electrode system. PV array frames can be connected to supplementary grounding electrodes to provide lightning protection and prevent surges from flowing to the rest of the electrical system.

12.5.3 Ground-Fault Protection Devices

DC PV systems that are installed on the roofs of buildings require ground-fault protection devices (GFPDs) to reduce fire hazards. These GFPDs must perform four functions: (1) detect the ground fault, (2) interrupt ground fault, (3) disconnect all ungrounded conductors in the faulted circuit, and (4) indicate that a ground fault has occurred. To interrupt the fault current, a GFPD can disconnect the grounded conductor as long as all of the ungrounded conductors in the circuit are also disconnected. If a GFPD can disconnect the grounded conductor, a warning label must be placed nearby indicating this possibility. GFPDs are available both as internal functions of inverters or as independent devices.

12.5.4 Inverters

The inverter is the interface between the AC and DC sides of a PV power system. Thus, inverters have a DC input circuit and an AC output circuit. There are also special considerations for grid-tied inverters.

1. *DC input circuit*: The DC input circuit must be sized to carry 125% of the current required to supply the continuous output rating of the inverter. The input current is calculated as the maximum continuous output rating of the inverter divided by the inverter efficiency and the minimum input voltage. If the inverter is expected to operate near its maximum output rating, special consideration should be given to the sizing of the input conductors. The DC input current actually has a sinusoidal shape similar to AC power in response to the sinusoidal output of the inverter. This current can have an RMS value of 150% or more of the input current calculated using the maximum output rating, inverter efficiency, and minimum input voltage. The inverter manufacturer should provide the actual RMS value for the input current and the ampacity of the conductors and the OCPD adjusted accordingly. Additionally, if high surge loads are present, voltage drop should be checked at the surge wattage to ensure that the inverter does not shut down due to low voltage when attempting to start the load.
2. *AC output circuit*: The design of conductors of an inverter output circuit is carried out using the same approach used for any other AC wiring (refer to Chapter 6). In particular, the output current can be calculated as the rated continuous output divided by the line voltage. If the manufacturer gives a maximum output current, the larger the maximum current and calculated current should be used. An OCPD is required at the output of the inverter unless it is located very close to the AC load center, in which case the back-fed breaker in the load center may suffice. When connecting an inverter to a load center, any objectionable current should be avoided in the equipment grounding conductors. Moreover, the manufacturer's instructions should be checked in advance to be sure that a given inverter can be connected to an existing load center without creating parallel paths. This is especially true of inverters with plug connections, which often have the neutral bonded to the chassis, and therefore to the equipment grounding conductor. The inverter must have a hard wiring option. Receptacles cannot be used to connect the inverter to the load center. If an inverter with an internal AC ground bond must be used, the bond in the AC load center must be disconnected. This can be accomplished by using an isolated neutral bar with a separate equipment grounding bar. When a 120 V single-phase inverter in

a stand-alone system is used to feed both busbars of an AC load center, no 240 V or multiwire branch circuits can be connected.

3. *Utility interconnection*: Inverters used for utility connection should be listed to meet UL Standard 1741 specifications in order to ensure proper operation and safe interconnection with the local authorities and utility. A utility-interactive (U-I) inverter must synchronize the frequency and phase angle of its output to that of the grid. It also must automatically disconnect from the grid when low voltage is detected. Some utilities require a lockable disconnect under their control to facilitate safe line maintenance. The point of interconnection, usually the main disconnect, must be labeled with the AC operating voltage and current. The inverter may be connected to either the supply or the load side of the service disconnect. When the inverter is connected to the load side, four primary conditions must be met:
   a. Each connection must be made with a dedicated disconnect/OCPD.
   b. The sum of the OCPDs supplying the load center busbar must not exceed the busbar rating, except in dwelling units where the sum can be 120% of the busbar rating.
   c. All equipment containing OCPDs that are fed from multiple sources must be marked indicating those sources.
   d. Circuit breakers used for backfeeding must be identified for that purpose.

Circuit breakers can be identified for backfeeding if they are not marked "line" and "load". If a load center has insufficient free busbar capacity, there are three possible remedies. First, if there is available load capacity in the load center, the main breaker could be replaced with a breaker having a lower rating. If this is not possible, the load center could be replaced with one with a higher busbar rating with the same main breaker rating. Depending on the costs, a third option would be to install a second service entrance for the inverter output. This service would be connected on the supply side of the load center's main breaker, preventing overloading of the busbar. Where a 120 V inverter is connected to only one phase of a 120/240 or three-phase, four-wire service, the neutral must not be overloaded.

Example 12.4 illustrates the design procedure for the conductors and protection devices for inverters.

Example 12.4

Problem:

Determine the size of a 2,500 W inverter with 90% efficiency and a minimum input voltage of 22 V.

Solution:

The DC input current of the inverter is 2,500 ÷ 0.9 ÷ 22, or 126 A.

The ampacity of the cable should be 125% × 126, or 158 A.

From Table 6.1, 3/0 copper conductors protected by 175 A fuses can be selected.

12.5.5 Batteries and Generators

Batteries and generators provide backup power to supplement PV systems. They can provide large currents when needed and power at night or on cloudy days. Generators can also help maintain the battery state of charge when charge from the PV array is not available.

12.5.5.1 Batteries

NEC addresses storage batteries in Article 480, but the requirements in Article 690 can be more pertinent for PV system installations (NEC, 2014). A battery bank is considered grounded if one of the current-carrying conductors connecting the PV power source is grounded. In dwelling units, the battery bank voltage is limited to 48 V nominal unless all live parts are guarded during routine maintenance. Regardless of voltage, the live parts in a battery bank of any voltage must be guarded if the bank is located in a dwelling unit. As discussed earlier, current-limiting OCPDs must be installed in battery output circuits if the available short-circuit exceeds the interrupt capability of any of the equipment in the circuit.

If a battery bank operates over 48 V nominal, there are several special requirements. Conductive cases must not be used unless required for battery operation, disconnects must be provided to separate the bank into 48 V sections for maintenance, and a disconnect must be provided for the grounded conductor. For battery interconnections, building-type conductors must be used. Fine-stranded cables are typically used but flexible cables are also allowed.

Batteries should be enclosed in a sealed box that is large enough to provide adequate ventilation and strong enough to support the significant weight of the battery bank. The enclosure must also be separate from all controls and sources of spark and should be able to contain acid spills. Batteries should always be installed in well-ventilated areas such as garages, but should not be located in living spaces. Passive or active venting should be used to prevent hydrogen gas buildup especially when batteries are held at high states of charge. Recombination caps can reduce electrolyte loss and the release of hydrogen gas by up to 50% (Dunlop, 1997). Minimizing temperature swings in the battery enclosure can lead to longer life, better performance, and lower maintenance. Depending on where the enclosure is located, insulation and intentional passive cooling can help moderate battery operating temperatures. Monitoring the batteries with a charge controller, inverter, or dedicated battery monitoring system can prevent unforeseen problems with the battery bank.

12.5.5.2 Generators

Backup generators in PV systems are typically AC equipment and are used to supplement the power supplied by the inverter. When the inverter is equipped with a battery charger, the generator can be used to maintain the battery bank. NEC Article 445 covers generators and provides some additional requirements for their design and operation. AC generators are required to have certain information on the nameplate, including manufacturer, frequency, power factor, number of phases, power rating, operating voltage and current, and rotation speed in rpm. As with all installations, an OCPD must be provided to protect the conductors from overcurrent. A disconnect is required to isolate the generator from the rest of the system. The ampacity of the conductor connecting the generator to the rest of the system must be at least 115% of the nameplate current. If continuous operation is anticipated, the ampacity should be further increased by a factor of 125%.

12.6 CASE STUDIES

In this section, two case studies are presented to illustrate the design guidelines for the electrical components of PV systems suitable for buildings. The first case study is specific to a stand-alone PV system while the second case study provides an example for a grid-connected PV system.

12.6.1 Case Study 1: Stand-Alone PV System for a Residential Building

For this case study, the PV modules are mounted on the roof of a house. The maximum temperature experienced by the PV system is 68°C. The PV system includes

- A PV array made up of 10, 12 V, 51 W modules (each module has the following characteristics: $I_{SC}=3.25$ A and $V_{OC}=20.7$ V)
- A set of batteries rated at 800 Ah at 12 V
- An inverter rated at 500 W with 90% efficiency with a minimum voltage output of 10.75 V
- A charge controller
- Loads are 500 W AC and a 5 A DC load

The conductors for the modules are free air (USE-2) while UF cables are considered for the panels and the array. Temperature for the array cables is 40°C. THHN cables are used to connect batteries to inverter and the charge controller. Only fuses are used for protection devices.

The layout for the electrical system is shown in Figure 12.8. The design calculations for various components are highlighted in the following steps:

1. *Conductor and protection for the PV modules*: Each module has a short-circuit current of $I_{SC}=3.25$ A. Based on Equation 12.5, the maximum current from the PV source is

$$I_{\max,PV-source} = 1.25 \times 1.25 \times 3.25 = 5.08 \text{ A}$$

For USE-2 free-air conductors, the temperature derating factor at 68°C temperature is 0.58 (refer to Table 6.1). The ampacity and fuse size for various conductor sizes are summarized in Table 12.4.

Thus, 14 AWG USE-2 copper conductors and 15 A fuses can be used for all the modules.

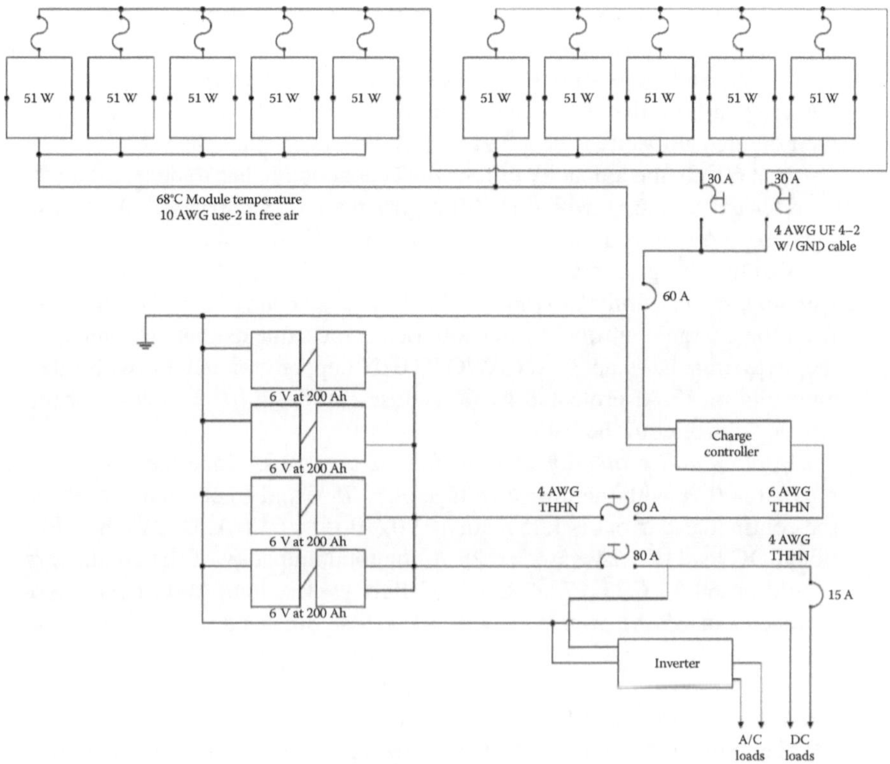

FIGURE 12.8 Electrical layout for the stand-alone PV system of Case Study 1.

TABLE 12.4
Rated and Derated Ampacity for Select Free-Air Conductor Sizes and Associated Fuse Ratings

Conductor Size (USE-2 in AWG)	Rated Ampacity (A)	Derated Ampacity at 68°C (A)	Fuse Size (A)
14	35	20.3	15
12	40	23.2	20
10	55	31.9	30
8	80	46.4	40

2. *Conductors and protection devices for the PV panels*: The PV array has 25-module panels. Each panel has a load of $5 \times 1.25 \times 3.25 = 20.3$ A. Using USE 75°C insulated conductors, the ampacity for #10 cables is 35 A (refer to Table 6.1), which is higher than $1.25 \times 20.3 = 25.4$ A. As noted in Table 12.4, fuse rating of 30 A is needed for #10 conductors. Thus, 10 AWG USE-2 conductors protected by two 30 A fuses can be used for the two panels on the roof.

3. *Conductors and protection devices for the PV array*: For the entire array, the ampacity requirement is 10×4.06=40.6 A. Using USE-2 cables at 40°C, (derating factor of 0.91 as noted in Table 6.1), four AWG copper conductors can be utilized to connect the array to the charge controller since their derated ampacity is 95×0.91=86 A. However, since the cables are UF (rated at 60°C), the ampacity of four AWG conductors has to be less than 70 A (refer to Table 6.1), which is still higher than 1.25×40.6=51 A. A fuse rated at 60 A can be used as a protection device for the cable connecting the array to the charge controller.
4. *Conductors and protection devices for the charge controller*: Since the load from the charge controller to the batteries is the same as that estimated for the array (refer to step 3), six AWG THHN copper conductors (with rated ampacity of 75 A) protected by 60 A fuse can be utilized to connect the charge controller to the batteries.
5. *Conductors and protection devices for the batteries*: Since the inverter is rated at 500 W with the lowest voltage of 10.75 V and an efficiency of 90%, the continuous current is 1.25×500/(0.90)(10.75)=64.6 A. Due to the additional DC load of 1.25×5 A=6.25 A, the total ampacity of the conductors should be 64.6+6.25=70.9 A. For THHN cables, four AWG conductors (ampacity of 85 A) protected with 80 A fuse are needed to connect the batteries to the inverter and the DC loads. A 15 A fuse or circuit breaker can be used to protect the conductors serving the DC loads.
6. *Grounding conductors*: For the array and the charge controller circuits, 10 AWG equipment grounding copper conductors can be used since the protection device rating is 60 A (refer to Chapter 5).

12.6.2 Case Study 2: Grid-Connected PV System for a Residential Building

This case study is specific to a rooftop utility-interactive PV system with the following features:

- A PV array made up of 24, 50 V, 240 W modules. For each module, $I_{SC}=5.6$ A and $V_{OC}=62$ V. The array consists of six parallel-connected panels with four modules each.
- An inverter rated at 5,000 W, 200 V AC output with 95% efficiency.

All conductors are set to be RHW-2 in conduit. The maximum temperature experienced by the PV system is 63°C. Figure 12.9 shows the design specifications of the grid-connected PV electrical system. The design calculations are summarized briefly in the following steps.

1. *Conductor and protection for the PV modules*: Each module has a short-circuit current of 5.6 A. Based on Equation 12.5, the maximum current from the PV source is

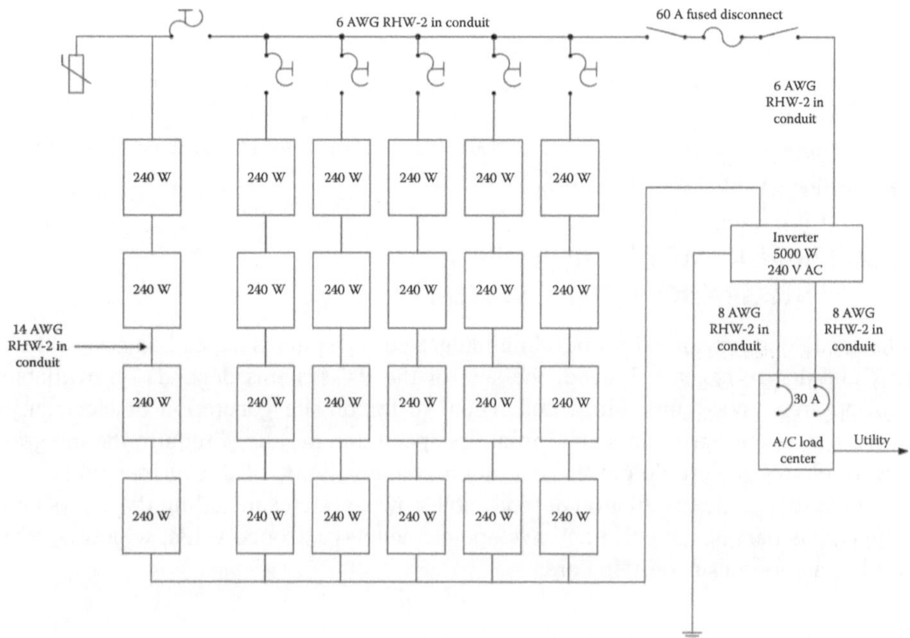

FIGURE 12.9 Electrical layout for the utility-interactive PV system of Case Study 2.

$$I_{\max,PV-source} = 1.25 \times 1.25 \times 5.6 = 8.75 \text{ A}$$

The modules in each string are connected in series and are operated at 63°C. Thus, the derating factor for RHW conductors is 0.58 (refer to Table 6.1). The required ampacity for the RHW-2 conductors serving the PV modules is then 8.75 A/0.58 = 15 A. Since the rated ampacity is (1) 25 A for 90°C insulated 14 AWG conductors, (2) 20 A for 75°C insulated 14 AWG conductors, and (3) 20 A for 60°C insulated 14 AWG conductors (refer to Table 6.1) and thus higher than 8.75 A, there is no issue connecting 14 AWG RHW-2 copper conductors to 9 A (or 15 A) fuse rated at 75°C or even 60°C.

2. *Conductors and protection devices for the PV array*: For the array made up of six parallel strings of modules, the ampacity requirement is 6 × 8.75 = 52.5 A. Using RHW-2 cables at 40°C, (derating factor of 0.91, as noted in Table 6.1), the required ampacity becomes 52.5/0.91 = 57.7 A. The rated ampacity for 6 AWG RHW-2 copper conductors is (1) 75 A when they are rated at 90°C insulation and (2) 65 A when they are rated at 75°C insulation. Therefore, 14 AWG RHW-2 copper conductors located in 2-in conduit protected by a 60 A fused disconnect can be used to connect the PV array to the inverter.
3. *Conductors and protection devices for the inverter*: The continuous inverter output current is 1.25 × 5,000 W/240 V = 26 A. The cable from the inverter to the load center operates at 30°C. A two-pole 30 A circuit breaker can

be used at the output of the inverter. A cable size of 8 AWG RHW-2 with an ampacity of 50 A when rated at 75°C insulation (refer to Table 6.1) is adequate for the conductors connecting the inverter to the AC load center.

4. *Grounding conductors*: Since the protection device rating is 60 A, the equipment grounding conductors for the system should be at least 10 AWG copper conductors (refer to Chapter 5).

12.7 PROGRAMS FOR BUILDING INTEGRATION OF PV SYSTEMS

The capacities considered for building integrated PV systems depend on several factors and design criteria. Indeed, the size of the PV systems depends on available rooftop areas, cost constraints, and incentive for on-site generation of electricity. Moreover, several programs and initiatives specific to buildings require the integration of on-site renewable power generation systems. Some of the current initiatives for encouraging the installation of renewable power systems including PV arrays and batteries as part of the building envelope elements (i.e., roofs, walls, windows, and shades) and/or its surrounding are:

- Grid-interactive Efficient Buildings (GEBs)
- Grid-Connected Communities (Microgrids)
- Electrification of Buildings and Cities
- Decarbonization of Buildings
- Resilient Buildings and Communities

Some of these programs are discussed in Chapter 14 with specific analysis and design approaches complemented with case studies.

12.8 SUMMARY

In this chapter, a general design approach of PV systems suitable for integration with buildings is provided using specific requirements and guidelines outlined in NEC. Moreover, the components of typical stand-alone and grid-connected PV systems are described. Sizing of wiring and protection devices of PV systems is illustrated with some calculation examples and case studies. The approach outlined in this chapter should assist the readers to complete the preliminary design phase of electrical components for roof-mounted PV systems suitable for integration with residential and commercial buildings.

12.9 PROBLEMS

1. Using your location, estimate the total electricity generate from a PV system per unit of roof area using a single crystal technology with a rated efficiency of 13.5%. You may use PV Watts or any other analysis tool to estimate the electricity generation for your PV rooftop PV system.

2. Compare the electricity generated by a rooftop PV system for various tilt angle for your specific house location. You can use PV Watts or a similar tool. Conclude on the optimal tilt angle.
3. Determine the wire size needed to maintain the voltage drop below 2% for a 48-V PV array located 200 ft away from the electrical panel. The short-circuit current for the array is estimated at 36.5 A.
4. Design the specifications and the layout for the following standalone PV system made up of six, 12 V, 51 W module. Each module has the following characteristics: I_{SC} = 2.85 A and V_{OC} = 18.2 V.
   - A set of batteries rated at 600 Ah at 12 V
   - An inverter rated at 400 W with 90% efficiency with a minimum voltage output of 8.75 V
   - A charge controller
   - Loads are 400 W AC and a 4 A DC load

   You can use THHN cables to connect batteries to inverter and the charge controller. Assume fuses are used for protection devices.
5. Design the specifications and the layout for a grid-connected PV system made up of 12, 50 V, 200 W modules. Each module has a short-circuit current of 4.5 A and open circuit voltage of 58 V. The array is configured to have 4 parallel connected panels with three modules each. An inverter rated at 3,000 W, 200 V AV output with an efficiency of 95% is used. Assume that all the conductors are RHW-2 and that the PV system is subject to a maximum temperature of 70°C.

13 Power Generation and Cogeneration Systems

This chapter provides a general description of the basic concepts of electrical generation systems as well as combined heat and power generation systems. First, an overview of the capital as well as operation and maintenance costs for various power generation technologies is presented. Then, a historical background of power generation and cogeneration is outlined. Then, a review of the fundamental theory of power cycles and cogeneration systems is outlined, followed by a description of commercially available cogeneration systems and their components. In addition, some guidelines and examples are presented to carry out feasibility analysis of cogeneration systems. Finally, a cogeneration case study is presented.

13.1 INTRODUCTION

Several technologies are available to generate electricity in large scale (utilities) or small scale (buildings). These technologies include the conventional fuel-based generators, nuclear reactors, and renewable energy technologies. Table 13.1 compares the capital and operating costs of various power plant capacities and technologies based on EIA estimates (US Energy Information Agency [EIA], 2013). As indicated in Table 13.1, while the capital and operation and maintenance (O&M) costs for hydroelectric and wind are competitive compared to those of fuel gas–based generating technologies, other generating technologies remain expensive to install and operate. Table 13.1 also indicates that most fuel-based generating technologies can be utilized for combined heat and power (CHP) production including fuel cells whose capital costs are still high. CHP systems offer energy-efficient alternative options to serve both the electricity and thermal needs of buildings and should be considered especially for large facilities such as hospitals and university campuses.

As noted in Chapter 9 and in order to fairly compare the cost-effectiveness of various generating technologies, detailed economic analysis should be conducted. A recent EIA evaluation has compared the levelized cost of energy or LCOE (refer to Chapter 9) for a wide range of generating technologies in the United States. Table 13.2 presents a summary of the total and specific components of LCOE values as well as capacity factors for the U.S. utility scale power plants listed in Table 13.1. Note that the total LCOE values listed in Table 13.2 include the transmission costs ranging from \$5.8/MWh for offshore wind farms to \$1.2/MWh for conventional fuel-based power plants. As noted in Table 13.2, renewable energy power plants have typically low capacity factors and are considered as nondispatchable systems and thus cannot be relied on as emergency systems for buildings unless combined with storage systems (i.e., electrical batteries).

DOI: 10.1201/9781003276999-13

TABLE 13.1
Capital and O&M Cost Estimates for U.S. Generating Power Plants

Fuel/Power Plant Type	Electrical Capacity (MW)	Heat Output Rate (Btu/kWh)	Capital Costs ($/kW)	Fixed O&M Cost ($/kW-Year)	Variable O&M ($/kW-Year)
Coal					
Pulverized coal (single unit)	650	8,800	$3,246	$37.80	$4.47
Integrated gasification combine cycle with carbon capture	520	10,700	$6,599	$72.83	$8.45
Natural gas					
Conventional combined cycle	620	7,050	$917	$13.17	$3.60
Advanced combustion turbine	210	9,750	$676	$7.04	$10.37
Fuel cells	10	9,500	$7,108	$0.00	$43.00
Uranium					
Dual unit nuclear reactor	2,234	N/A	$5,530	$93.28	$2.14
Biomass					
Combined cycle	20	12,350	$8,180	$356.07	$17.49
Wind					
Onshore turbine	100	N/A	$2,213	$39.55	$0.00
Offshore turbine	400	N/A	$6,230	$74.00	$0.00
Solar					
Solar thermal (CSP)	100	N/A	$5,067	$67.26	$0.00
Photovoltaic (PV)	150	N/A	$3,873	$24.69	$0.00
Geothermal					
Dual flash	50	N/A	$8,312	$132.00	$0.00
Hydroelectric					
Conventional	500	N/A	$2,939	$14.13	$0.00
Pumped storage	200	N/A	$5,288	$18.00	$0.00

Source: EIA, Updated capital cost estimates for utility scale electricity generating plants, Energy Information Agency, Department of Energy, Washington, DC, http://www.eia.gov/forecasts/capitalcost/pdf/updated_capcost.pdf, 2013.

In this chapter, only dispatchable generation technologies are presented with a focus on CHP systems to provide both electricity and thermal energy to buildings. First, the basic advantages of cogeneration systems are outlined with some brief outline of their history. Then, a detailed description is provided for technologies that can be used to generate power for building applications. An analysis including an estimation of the optimal capacity of cogeneration systems is presented with specific case studies. The integration of renewable energy to design net zero-energy buildings and carbon-neutral communities is discussed in Chapter 14.

TABLE 13.2
Estimates for Average LCOE Values and Capacity Factors for U.S. Generating Power Plants

Fuel/Power Plant Type	LCOE Capital Costs ($/MWh)	LCOEFixed O&M Costs ($/MWh)	LCOE Variable O&M Costs ($/MWh)	Total System LCOE ($/MWh)	Capacity Factor (%)
Coal					
Pulverized coal (single unit)	60.4	4.2	29.4	95.1	85
Integrated gasification combine cycle with carbon capture	97.3	9.8	36.1	144.4	85
Natural gas					
Conventional combined cycle	14.4	1.7	57.8	75.2	87
Advanced combustion turbine	27.8	2.7	79.6	113.5	30
Uranium					
Dual unit nuclear reactor	70.1	11.8	12.2	95.2	90
Biomass					
Combined cycle	47.1	14.5	37.6	100.2	83
Wind					
Onshore turbine	57.7	12.8	0.0	73.6	36
Offshore turbine	168.6	22.5	0.0	196.9	38
Solar					
Solar thermal (CSP)	191.6	42.1	0.0	239.7	20
Photovoltaic (PV)	109.8	11.4	0.0	125.3	25
Geothermal					
Dual flash	34.1	12.3	0.0	47.8	92
Hydroelectric					
With seasonal storage	70.7	3.9	7.0	83.5	54

Source: EIA, Updated capital cost estimates for utility scale electricity generating plants, Energy Information Agency, Department of Energy, Washington, DC, http://www.eia.gov/forecasts/capitalcost/pdf/updated_capcost.pdf, 2013.

13.2 BENEFITS OF COGENERATION

Cogeneration and combined heat and power (CHP) are terms used interchangeably to denote the simultaneous generation of power (electricity) and usable thermal energy (heat) in a single, integrated system. A CHP plant derives its efficiency, and hence lower costs, by recovering and utilizing the heat produced as a by-product of the electricity generation process that would normally be wasted in the environment. The overall fuel efficiency of typical CHP installations can be in the range of 70%–90%, compared with 35%–50% for conventional electricity generation. Overall, CHP achieves a 35% reduction in primary energy usage compared with remote power stations and heat-only boilers. Moreover, CHP also avoids transmission and distribution losses since it supplies electricity generally close to the generation site.

While the CHP concept is not new, it has been recently applied to a wide range of commercial buildings. Indeed, until the 1980s, cogeneration systems were used only in large industrial or institutional facilities with a high electricity demand (typically over 1,000 kW). After the energy crisis of 1973, during which fuel and electricity prices increased significantly (by a factor of 5), the U.S. government passed in 1978 the National Energy Act (NEA), which includes the Public Utility Regulatory Policies Act (PURPA). The PURPA regulations forced utilities to purchase electricity and to provide supplementary or backup power to any qualified cogeneration facility. The Energy Policy Act (EPACT) of 1992 increased even more the appeal of cogeneration systems by opening up transmission line access and retail wheeling.

In addition to the favorable regulations, the development of energy-efficient cogeneration systems and small pre-engineered packaged cogeneration units has provided the needed incentives to encourage the implementation of systems capable of generating electricity and heat for commercial, institutional, and even residential applications. Currently, cogeneration systems are available over a wide range of sizes, from less than 50 kW (microsystems) to over 100 MW. Moreover, advances in controls have provided better procedures to operate and integrate the various components of a cogeneration system (including prime movers, electrical generators, and heat recovery systems). More recently, energy service companies (ESCOs) have increased awareness and interest of the public in on-site generation technologies.

In addition to their better overall energy efficiency compared to conventional utility plants, cogeneration systems offer the following benefits:

- Cleaner power generation sources with reduced NO_x and carbon emissions
- Increased reliability and quality of electrical power since customers are less vulnerable to blackouts from utility power lines
- Added national energy security with diversified sources and locations of power generation
- Avoided transmission and distribution costs since no new lines would be needed if distributed cogeneration plants are available

To evaluate the feasibility of a cogeneration system, several technical and economical aspects as well as regulatory issues should be considered. While this chapter provides some discussion of the main regulatory considerations and financial options for cogeneration in the United States, the main focus of the chapter is to provide technical information and engineering principles that are necessary to understand the design and operation of cogeneration systems. First, an overview of the past, present, and future status of the electricity generation industry in the United States is presented.

13.3 HISTORY OF COGENERATION

Systems for CHP generation have been in existence since the 1880s in the United States and Europe. Indeed, several industrial facilities generated their own electricity and steam using coal-fired boilers and steam turbine generators. It is estimated that CHP systems produced up to 58% of the total electricity generated in the United States.

However, in the mid-1900s, large central electrical power plants were built with reliable utility grids. The electricity from these plants was produced at relatively low costs. As a consequence, industrial facilities started to purchase electricity from the power plants and thus reduced gradually their reliance on on-site generation. The contribution of CHP systems in the total U.S. electric power generation represented 15% in 1950 and only 4% in 1974 (EIA, 2000). Large electric utility holding companies have been formed and expanded since the early 1900s. In 1920, the majority of the U.S. power industry was controlled by a few privately owned electric power holding companies. These holding companies abused their power and charged consumers higher prices for electricity.

To reduce the monopoly of the privately owned electric utility holding companies, the Federal government intervened by passing the Public Utility Holding Company Act (PUHCA) in 1935. Under the provisions of the PUHCA, the electric utility holding companies came to be regulated by the Securities and Exchange Commission. In a further effort to reduce electricity prices, government-owned hydroelectric power facilities were built, including the Hoover dam in 1936. The Bonneville Project Act of 1937 provided the Federal government with the means to oversee transmission and marketing of power produced from hydroelectric facilities. By 1941, power produced from publicly owned facilities represented 12% of the total utility generation.

Until the early 1970s, utilities were able to meet the ever-increasing electrical power demands at decreasing prices due to the economies of scale, technological advances, and declining fuel costs. However, a series of events that occurred during the 1970s significantly impacted the U.S. electric power industry. These events included the oil embargo in 1973–1974, the passage of the Clean Air Act in 1970, as well as the Energy Supply and Environmental Coordination Act in 1974. As a result of these events, prices of electrical power increased dramatically in the 1980s. Between 1973 and 1985, prices of fuels and electricity increased by a factor of 5.

As a reaction to the oil embargo, the Federal government passed in 1978 the NEA in order to reduce U.S. dependence on foreign oil, develop alternative energy sources, and encourage energy conservation. The NEA comprises five different statutes:

- Public Utility Regulatory Policy Act (PURPA)
- Energy Tax Act
- National Energy Conservation Policy Act
- Powerplant and Industrial Fuel Use Act
- Natural Gas Policy Act

PURPA has had the most significant impact on the electric power industry and the development of cogeneration. Indeed, PURPA allowed nonutility facilities that met certain ownership and efficiency criteria to sell electric power to utility companies.

Specifically, PURPA set the following legal obligations for the electric utility companies toward cogenerators defined as Qualified Facilities (QFs):

- Utility companies have to purchase cogenerated energy and capacity from QFs.
- Utility companies have to sell energy and capacity to QFs.
- Utility companies have to provide to QFs supplementary power, backup power, maintenance power, and interruptible power.
- Utility companies have to provide access to transmission grid to "wheel" to other electric utility companies.

The term "wheeling" refers to the process by which utilities can buy or sell electricity to or from other utilities in order to meet peak demands or shed excess generation. The wheeling process enables utilities to spend less on peaking power plants, thereby lowering capital expenditures required to meet high periods of electric power demands.

Additional legislations passed in the 1990s, including the Clean Air Act Amendments (CAAA) of 1990 and the Energy Policy Act (EPACT) of 1992, have provided new opportunities for cogeneration. In particular, EPACT provides incentives for nonutility generators to enter the wholesale market for electrical power by exempting them from the PUHCA constraints. The law creates a new category of power producers known as Exempt Wholesale Generators (EWGs). It should be noted that EWGs are different from PURPA QFs since they are not required to meet PURPA's cogeneration criteria. In addition, utilities are not required to purchase electric power from EWGs.

Table 13.3 summarizes the major federal legislations that marked the U.S. power industry since the early 1930s. Because of the passage of PURPA, multiple megawatt CHP projects have been developed and built, especially at large industrial facilities, including pulp and paper, steel, chemical, and refining plants. Recent advances in reciprocating engines and micro-combustion turbines make CHP more cost-effective for small applications, such as fast-food restaurants, as well as commercial buildings.

It is estimated that in 2000, CHP accounted for about 7.5% of the electricity generation capacity and almost 9% of the electricity generated in the United States (EIA, 2009). Both DOE and EPA have set a goal to double U.S. CHP capacity between 1999 and 2010, from 46 to 92 GW. Even with this goal achieved, CHP represents only 14% of the U.S. electric generating capacity. Unlike the case in some European countries, the share of CHP in the total U.S. electricity generation remains rather small, as indicated in Table 13.4.

13.4 TYPES OF FUEL-BASED GENERATION SYSTEMS

There are several types of power generation systems that are commercially available and can be utilized for CHP for building applications. In general, three categories of cogeneration systems can be considered:

1. *Conventional cogeneration systems*: These systems consist of large cogeneration units (more than 1,000 kW) and require a thorough design process to select the size of all equipment and components (i.e., prime movers, electrical generators, and heat recovery systems).

TABLE 13.3
Major Federal Legislation Affecting U.S. Electric Power Industry

Legislation Name	Year	Main Scope
Public utility holding company act (PUHCA)	1935	To reduce utility industry abuses by giving the Securities and Exchange Commission the authority to oversee holding companies.
Bonneville project act	1937	To create the Bonneville power administration (BPA) to oversee the transmission and marketing of power produced in Northwest dams.
Clean air act	1970	To reduce sulfur dioxide and nitrogen dioxide emissions.
Energy supply and environmental coordination act (ESECA)	1974	To allow the Federal Government to prohibit electric utilities from burning natural gas or petroleum products.
Public utility regulatory policies act (PURPA)	1978	To promote conservation of electric energy by allowing non-utility generators and qualified cogenerators to sell power to utilities.
Energy tax act (ETA)	1978	To allow tax credits for investment in cogeneration equipment and renewable technologies. These incentives were curtailed in the mid-1980s.
National energy conservation policy act	1978	To require utilities to develop residential energy conservation plans in order to reduce growth in electricity demand.
Powerplant and industrial fuel use act	1978	To replace ESECA of 1974 and extend Federal Government prohibition on the use of natural gas and petroleum in new power plants.
Pacific northwest electric power planning and conservation (PNEPPC) act	1980	To create PNEPPC Council in order to coordinate the conservation and resource acquisition of the BPA.
Electric consumers protection act (ECPA)	1986	To set new environmental criteria in licensing hydroelectric power plants; to reduce significantly PURPA benefits for new hydroelectric projects; and to increase the enforcing powers of FERC.
Clean air act amendments (CAAA)	1990	To establish a new emissions-reduction program. Generators of electricity are made responsible for a large portion of the sulfur dioxide and nitrogen oxide reductions.
Energy policy act (EPACT)	1992	To create a new category of electricity producers and to authorize FERC to open up the national electricity transmission system to wholesale suppliers.

Source: EIA, The changing structure of the Electric Power Industry, Energy Information Administration, Washington, DC, http://www.doe.eia.gov, 2000.

TABLE 13.4
Share of CHP in Total National Power Generation for Selected Countries

Country	Denmark	Netherlands	Germany	Italy	US	UK	France
Contribution of CHP (%)	40	30	14	12	7	5	3

Source: EIA, The changing structure of the Electric Power Industry, Energy Information Administration, Washington, DC, http://www.doe.eia.gov, 2000.

2. *Packaged cogeneration systems*: These systems are small (below 1,000 kW) and are easy to design and install since they are pre-engineered and preassembled units.
3. *Distributed generation technologies*: Some cogeneration systems can use a number of technologies that have been recently developed to produce both electricity and heat including fuel cells.

13.4.1 Conventional Cogeneration Systems

A typical cogeneration plant consists of several pieces of equipment to produce electricity and heat (in the form of either steam or hot water). The number and the type of equipment in a cogeneration plant depend on the size of the system and the procedure used to generate electricity and heat. Generally, a conventional cogeneration system includes the following components:

1. *A prime mover*: This is the most important equipment in a cogeneration system. It is typically a turbine that generates mechanical power using a primary source of fuel. There are three turbine types that are commonly used in cogeneration plants: turbines operated by steam generated from boilers, gas turbines fueled by natural gas or light petroleum products, and internal combustion engines fueled by natural gas or distillate fuel oils.
2. *A generator*: A device that converts mechanical power to electrical energy.
3. *A heat recovery system*: A set of heat exchangers that can recover heat from exhaust or engine cooling and convert it into a useful form—typically hot water.

To operate a cogeneration plant, a robust control system is needed to ensure that all individual pieces of equipment provide the expected performance. Two basic operation cycles are used to generate electricity and heat: either a bottoming cycle or a topping cycle.

1. *Bottoming cycle*: In this cycle, the generation of heat is given the priority to supply process heat to the facility. Thermal energy is produced directly from fuel combustion (in the prime mover). Heat is then recovered and fed to the generator to produce electricity, as illustrated in Figure 13.1a. Industrial plants characterized by high-temperature heat requirements (such as steel,

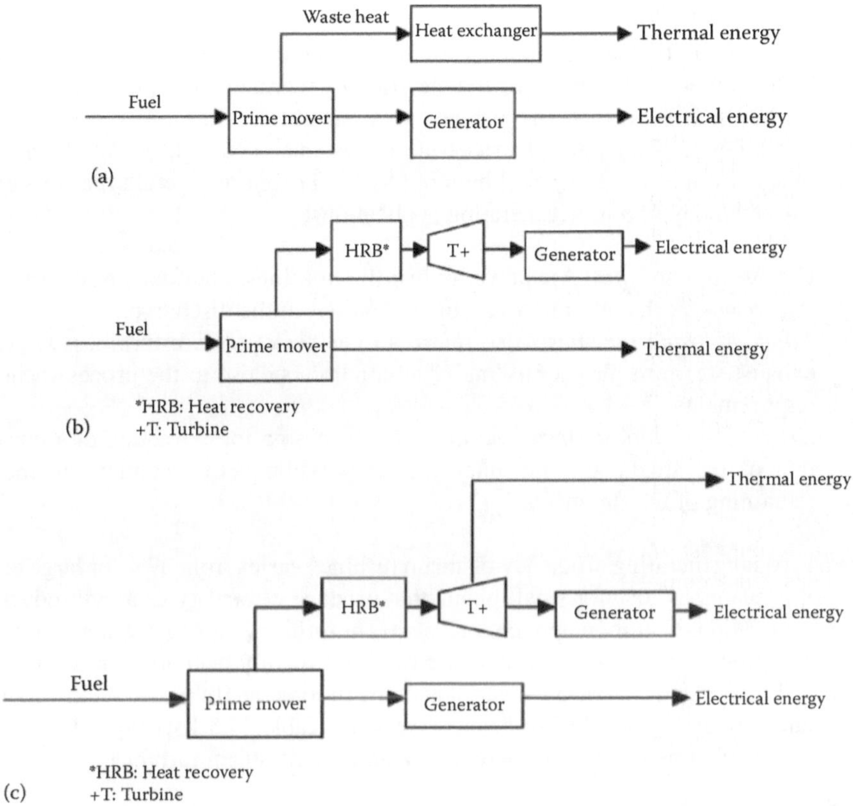

FIGURE 13.1 (a) Topping cycle cogeneration system. (b) Bottoming cycle cogeneration system. (c) Combined cycle cogeneration system.

aluminum, glass, and paper industries) typically use bottoming cycle cogeneration systems.

2. *Topping cycle*: Unlike the bottoming cycle, the generation of electricity in a topping cycle takes precedence over the production of heat, as indicated in Figure 13.1b. The waste heat is then recovered and converted to either steam or hot water. Most existing cogeneration systems are based on topping cycles. A hybrid of a topping cycle commonly used by several industrial facilities and even by electrical utilities is the combined cycle, as depicted in Figure 13.1c. In this cycle, a gas turbine is typically used to produce electricity. The exhaust gas is then fed to a heat recovery steam generator to generate more electricity, using a steam turbine. For a cogeneration plant, a small portion of steam can be converted into a useful form of thermal energy.

Three types of prime movers are generally considered for large and medium conventional cogenerations systems: steam turbines, reciprocating engines, and gas turbines. A brief overview of each prime mover type is provided here.

13.4.1.1 Steam Turbines

Steam turbines are the oldest and most versatile prime movers used in power generation. In the United States, most of the electricity generated from power plants use steam turbines. However, steam turbines are also utilized in CHP systems, combined cycle power plants, and district heating systems. The capacity of steam turbines ranges from 50 kW to several hundred MWs. There are several types of steam turbines used today in power generation applications:

- *Condensing turbines*: Are power-only utility turbines. They exhaust directly to condensers that maintain vacuum conditions at the discharge.
- *Noncondensing turbines*: Also referred to as back-pressure turbines. They exhaust steam to the facility mains at conditions close to the process heat requirements.
- *Extraction turbines*: Have openings in their casing for extraction of a portion of the steam at some intermediate pressure before condensing the remaining of the steam.

The electrical generating efficiency of steam turbines varies from 37% for large electric utility plants to 10% for small plants that produce electricity as a by-product of steam generation. Common applications of steam turbines for CHP systems involve industrial processes where solid or waste fuels are readily available. In the United States, it is estimated over 580 industrial and institutional facilities use steam engines to produce about 19,000 MW of electric capacity. Table 13.5 lists typical cost and performance parameters for cogeneration systems using steam turbines.

13.4.1.2 Reciprocating Engines

Reciprocating internal combustion engines are a widely used technology to generate power for several applications, including institutional and industrial applications, and CHP systems. The capacity of reciprocating engines varies from a few kilowatts

TABLE 13.5
Cost and Performance Characteristics for Selected Cogeneration Systems Using Steam Turbines

	System Capacity (kW)		
Cost and Performance Parameters	**500**	**3,000**	**15,000**
Total installed cost ($/kW)	$918	$385	$349
Fuel input (MMBtu/h)	26.7	147.4	549.0
Electric efficiency (%)	6.4	6.9	9.3
Steam to process (MMBtu/h)	19.6	107.0	386.6
Overall efficiency (%)	79.6	79.5	79.7

Source: Energy Nexus Group, Technology characterization: Steam turbines, A report prepared for EPA, Climate Protection Partnership Division, Washington, DC, 2002a.

to more than 5 MW. Two basic types of reciprocating engines are utilized for power generation applications:

- Spark ignition (SI) engines generally use natural gas. SI engines are now commonly used for duty-cycle stationary power applications.
- Compression ignition (CI) or diesel engines have been historically popular for both small and large power generation applications. However, diesel engines are now restricted to emergency and standby applications due to emission concerns.

Electric efficiency of reciprocating engines ranges from 28% for small engines (less than 100 kW) to 40% for large engines (above 3 MW). Waste heat can be recovered from four sources in reciprocating engines: exhaust gas, engine jacket cooling water, lube oil cooling water, and turbocharger cooling. High- and low-pressure steam can be produced from reciprocating engines for CHP applications. Overall efficiency for a CHP system using a reciprocating engine fueled by natural gas can exceed 70%.

Reciprocating engine technology has improved significantly over the last decades with increased fuel efficiency reduced emissions, improved reliability, and low first cost. The use of reciprocating engines for CHP generation applications is expected to continue to grow in the next decade. Currently, it is estimated that 1,055 CHP systems operating in the United States use reciprocating engines with an overall power capacity of 800 MW. Table 13.6 provides typical cost and performance parameters for commercially available reciprocating engines suitable for cogeneration applications.

13.4.1.3 Gas Turbines

Gas turbines provide one of the cleanest means for electric power generation with very low emissions of carbon dioxide (CO_2) and oxides of nitrogen (NO_x). The available capacity of gas turbines ranges from 500 kW to 250 MW. Gas turbines are well suited for CHP applications because high-pressure steam (as high as 1,200 psig) can

TABLE 13.6
Cost and Performance Characteristics for Selected Cogeneration Systems Using Reciprocating Engines

	System Capacity in kW				
Cost and Performance Parameters	**100**	**300**	**800**	**3,000**	**5,000**
Total installed cost (2,001 \$/kW)	\$1,515	\$1,200	\$1,000	\$920	\$920
Fuel input (MMBtu/h)	1.11	3.29	8.20	28.48	43.79
Electrical efficiency (%)	30.6	31.1	33.3	36.0	39.0
Total heat recovered (MMBtu/h)	0.57	1.51	3.50	11.12	15.28
Overall efficiency (%)	81.0	77.0	76.0	75.0	74.0

Source: Energy Nexus Group, Technology characterization: Reciprocating engines, A report prepared for EPA, Climate Protection Partnership Division, Washington, DC, 2002b.

TABLE 13.7
Cost and Performance Characteristics for Selected Cogeneration Systems Using Gas Turbines

Cost and Performance Parameters	System Capacity (kW)				
	1,000	**5,000**	**10,000**	**25,000**	**40,000**
Total installed cost (2,000 $/kW)	$1,780	$1,010	$970	$860	$785
Fuel input (MMBtu/h)	15.6	62.9	117.7	248.6	368.8
Electrical efficiency (%)	21.9	27.1	29.0	34.3	37.0
Steam output (MMBtu/h)	7.1	26.6	49.6	89.8	128.5
Overall efficiency (%)	68.0	69.0	71.0	73.0	74.0

Source: Energy Nexus Group, Technology characterization: Gas turbines, A report prepared for EPA, Climate Protection Partnership Division, Washington, DC, 2002c.

be generated from their high-temperature exhaust using heat recovery steam generators (HRSGs). It is estimated that more than 575 industrial and institutional facilities in the United States use gas turbines to generate power and heat with a total capacity of 40,000 MW (Energy Nexus Group, 2002c). Table 13.7 lists the performance characteristics of selected commercially available gas turbine CHP systems. As indicated in Table 13.7, both electrical efficiency and overall CHP efficiency increase with the size of the gas turbine.

13.4.2 Packaged Cogeneration Systems

For cogeneration facilities requiring small systems ranging from less than 50 kW to about 1 MW, pre-engineered and factory-assembled cogeneration units are currently available with reduced construction, installation, and operation costs. In addition, small packaged systems with capacities ranging from 4 to 25 kW have been developed and can be installed in short periods of time with little interruption of service. Almost all packaged cogeneration systems are equipped with advanced controls to improve the reliability and energy efficiency of the units.

Packaged cogeneration units are often sold as turn key installations. In particular, manufacturers are generally responsible for the testing and installation of the complete cogeneration systems. The development of packaged systems has enlarged the appeal of cogeneration to a wide range of facility types, including office buildings, restaurants, homes, and multifamily complexes. In addition, packaged cogeneration systems are now typically more cost-effective than conventional systems for small and medium hospitals, schools, and hotels. However, each facility has to be thoroughly evaluated to determine the economical feasibility of any packaged cogeneration system.

Packaged cogeneration systems use typically reciprocating engines. Recently, microturbines have been developed and have been commercially available since 2000. Microturbines are small electricity generators that operate at very high speeds

TABLE 13.8
Cost and Performance Characteristics for Selected Cogeneration Systems Using MicroTurbines

	System Capacity (kW)			
Cost and Performance Parameters	**30**	**70**	**100**	**350**
Total installed cost (2,000 $/kW)	$2,516	$2,031	$1,561	$1,339
Fuel input (MMBtu/h)	0.437	0.948	1.264	4.118
Electrical efficiency (%)	23.4	25.2	27.0	29.0
Heat output (MMBtu/h)	0.218	0.369	0.555	1.987
Overall efficiency (%)	73.0	64.0	71.0	77.0

Source: Energy Nexus Group, Technology characterization: Micro-turbines, A report prepared for EPA, Climate Protection Partnership Division, Washington, DC, 2002d.

(over 60,000 rpm). They are available in sizes ranging from 30 to 350 kW. They have been promoted as ideal generators for distributed generation applications, including cogeneration systems, due to their connection flexibility (they can be stacked in parallel), their reliability, and their low emissions. Table 13.8 summarizes performance characteristics of selected commercially available microturbines suitable for CHP applications.

13.4.3 Distributed Generation Technologies

Distributed generation is a relatively recent approach proposed to produce electricity using small and modular generators. The small generators with capacities in the range of 1 kW to 10 MW can be assembled and relocated in strategic locations (typically near customer sites) to improve power quality, reliability, and flexibility in order to meet a wide range of customer and distribution system needs. Some technologies have emerged in the last decade that allow the generation of electricity with reduced waste, cost, and environmental impact, which may make the future of distribution generation promising, especially in a competitive, deregulated market. Among these technologies are renewable energy sources (wind and solar), fuel cells, as well as microturbines, combustion turbines, gas engines, and diesel engines. Table 13.9 summarizes some distribution generation technologies with their average efficiencies, capacities, and applications (American Gas, 1999). Perhaps recent developments in fuel cells represent the best opportunity for distributed generation and CHP applications. It is expected that fuel cells will play a significant part in the twenty-first century electricity market.

The principle of fuel cell was first demonstrated over 150 years ago. In its simplest form, a fuel cell is constructed similar to a battery with two electrodes in an electrolyte medium, which serves to carry electrons released at one electrode (anode) to the other electrode (cathode). Typical fuel cells use hydrogen (derived from hydrocarbons) and oxygen (from air) to produce electrical power with other by-products (such

TABLE 13.9
Characteristics of Current Distributed Generation Technologies

Type	Efficiency (%)	Size	Applications
Combustion turbine	24–40	500 kW–30 MW	Cogeneration (commercial/industrial), transmission and distribution support
Diesel engine	36–42	50 kW–6 MW	Standby, remote and peak shaving power
Gas engine	28–38	5 kW–2 MW	Cogeneration (commercial/industrial), peak shaving and primary power
Microturbine	21–40	25 kW–300 kW	Cogeneration (commercial/light industrial), primary power
Fuel cell	40–65	1 kW–3 MW	Cogeneration (residential/commercial), primary power

TABLE 13.10
Basic Characteristics of Selected Types of Fuel Cells

Fuel Cell Name	Electrolyte	Fuel	Oxidant	Operating Temperatures (°C)
PAFC	Phosphoric acid	Pure hydrogen	Clear air (without CO_2)	200
AFC	Alkaline	Pure hydrogen	Pure oxygen and water	60–120
SPFC	Solid polymer	Pure hydrogen	Pure oxygen	60–100
MCFC	Molten carbonate	Hydrocarbons	Air and oxygen	650
SOFC	Solid oxide	Any fuel	Air	900–1,000

as water, carbon dioxide, and heat). High efficiencies (up to 73%) can be achieved using fuel cells.

Table 13.10 summarizes various types of fuel cells that are commercially available or under development. Each fuel cell type is characterized by its electrolyte, fuel (source of hydrogen), oxidant (source of oxygen), and operating temperature range. CHP systems using fuel cells are available in sizes ranging from few kW to 1,000 of kW. Table 13.11 lists some performance characteristics of selected commercially available CHP systems using fuel cell technology.

13.5 EVALUATION OF COGENERATION SYSTEMS

To evaluate the technical and the economical feasibility of a cogeneration system, it is important to collect accurate data about the facility and its energy consumption. In particular, current and projected future energy consumption and costs need to be available. For a detailed evaluation analysis, hourly electrical and thermal energy data are required. However, monthly and even yearly energy data can be sufficient for a preliminary feasibility analysis of cogeneration systems.

In this section, basic considerations are discussed to evaluate the feasibility of cogeneration systems.

TABLE 13.11
Cost and Performance Characteristics of Selected CHP Using Fuel Cells

Cost and Performance Parameters	System Capacity (kW)			
	10	**100**	**200**	**2,000**
Fuel cell type	PEM	SOFC	PAFC	MCFC
Total installed cost (2,002 \$/kW)	\$5,500	\$3,500	\$4,500	\$2,800
Fuel input (MMBtu/h)	0.10	0.80	1.90	14.80
Electrical efficiency (%)	30.0	45.0	36.0	46.0
Heat output (MMBtu/h)	0.04	0.19	0.74	3.56
Overall efficiency (%)	68.0	70.0	75.0	70.0

Source: Energy Nexus Group, 2002e.

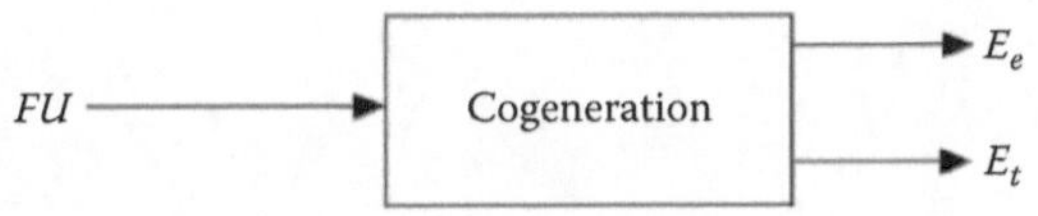

FIGURE 13.2 Energy input/output for a typical cogeneration system.

13.5.1 Efficiency of Cogeneration Systems

To account for the fact that a typical cogeneration system produces both electrical power, E_e, and thermal energy, E_t, from fuel energy, FU, as illustrated in Figure 13.2, the overall thermodynamic efficiency, η_{overall}, of the cogeneration system is defined as follows:

$$\eta_{\text{overall}} = \frac{E_e + E_t}{FU} \tag{13.1}$$

It should be noted that for a cogeneration facility to meet the criteria specified by the PURPA, it has to comply with certain efficiency standards. These standards use a "PURPA efficiency," η_{PURPA}, which is defined by Equation 13.2:

$$\eta_{PURPA} = \frac{E_e + \dfrac{E_t}{2}}{FU} \tag{13.2}$$

PURPA states that to be a Qualified Facility (QF) for cogeneration, the efficiency, η_{PURPA}, has to be at least

1. 145% for the cogeneration facilities with a useful thermal energy fraction that is larger than 5%
2. 242.5% for the cogeneration facilities with a useful thermal fraction that is larger than 15%

The main purpose of PURPA efficiency is to insure that a sufficient amount of thermal energy is produced so that the cogeneration facility is more efficient than the electric utility. A number of regulations within PURPA provide incentives to develop cogeneration facilities. Among these incentives are the legal obligations of the electric utilities toward the cogeneration facilities set by section 210 of PURPA (FERC, 1978). For instance, the electric utility has to

- Purchase cogenerated electrical energy from the Qualified Facilities (QFs)
- Sell electrical energy to the QFs
- Provide access for the QFs to transmission grid

Example 13.1 illustrates how to estimate overall thermodynamic and PURPA efficiencies for conventional cogeneration systems.

Example 13.1

Problem

Consider a 20 MW cogeneration power plant in a campus complex. An energy balance analysis indicates the following energy fluxes for the power plant:

- Electricity generation 33%
- Condenser losses 30%
- Stack losses 30%
- Radiation losses 7%

It is estimated that all the condenser losses and 12% of the stack losses can be recovered. Determine both the overall thermodynamic efficiency as well as the PURPA efficiency of the power plant.

Solution

First, the recovered thermal energy is determined:

$$E_t = \text{Condenser losses} + \text{Part of the stack losses}$$

This thermal energy output can be expressed in terms of the fuel use (FU) of the power plant:

$$E_t = 30\% \times FU + 12\% \times [30\% \times FU] = 0.34 \times FU$$

The energy flow for the campus power plant is summarized in this diagram:

Thus, the overall thermodynamic efficiency of the power plant can be easily determined using Equation 13.1:

$$\eta_{\text{overall}} = \frac{E_e + E_t}{FU} = \frac{0.33 \times FU + 0.34 \times FU}{FU} = 0.67$$

The PURPA efficiency can be calculated using Equation 13.2,

$$\eta_{PURPA} = \frac{E_e + \dfrac{E_t}{2}}{FU} = \frac{0.33 \times FU + 0.34 \times \dfrac{FU}{2}}{FU} = 0.50$$

Therefore, the campus power plant meets the PURPA criteria ($\eta_{PURPA} > 45\%$) and is thus a qualified cogeneration facility.

13.5.2 Analysis of Cogeneration Systems

To determine the capacity of a cogeneration system that is cost-effective to meet the loads for a building, simplified analysis procedures can be first used. A further evaluation with more detailed energy analysis tools may be warranted to determine the optimal design specifications of the cogeneration system. Example 13.2 illustrates one simplified calculation procedure that can be used to determine the cost-effectiveness of installing a cogeneration system for a hospital building.

Example 13.2

Problem

Consider a 60 kW cogeneration system that produces electricity and hot water with the following efficiencies: (1) 26% for the electricity generation and (2) 83% for the combined heat and electricity generation. Determine the annual savings of operating the cogeneration system compared to a conventional system that consists of purchasing electricity at a rate of \$0.08/kWh and producing heat from a boiler with 70% efficiency. The cost of fuel is \$5/MMBtu. The maintenance cost of the cogeneration system is estimated at \$1.00 per hour of operation. Assume that all the generated thermal energy and electricity are utilized during 6,500 h/year.

Determine the payback period of the cogeneration system if the installation cost is \$2,500/kW.

Solution

First, the cost of operating the cogeneration system is compared to that of the conventional system on an hourly basis:

Cogeneration system: For each hour, 60 kW of electricity is generated (at an efficiency of 26%) with fuel requirements of 0.787 MMBtu [= 60 kW × 0.003413 MMBtu/kW/0.26]. At the same time, a thermal energy of 0.449 MMBtu [= 0.787 MMBtu (0.83 − 0.26)] is obtained. The hourly flow of energy for the cogeneration system is summarized in this diagram:

Thus, the cost of operating the cogeneration on an hourly basis can be estimated as follows:

Fuel cost: 0.787 MMBtu/h × \$5/MMBtu = \$3.93/h

Maintenance cost: \$1.00/h

Total cost: \$4.93/h

Conventional system: For this system, the 60 kW electricity is directly purchased from the utility, while the 0.449 MMBtu of hot water is generated using a boiler with an efficiency of 0.65. Thus, the costs associated with utilizing a conventional system are as follows:

Electricity cost: 60 kWh/h × 0.08 = $4.80/h

Fuel cost (boiler): (0.449 MMBtu/h)/0.65 × $ 5/MMBtu $3.45/h

Total cost: $8.25/h

Therefore, the annual savings associated with using the cogeneration system are

$$\Delta\text{Cost} = (\$8.25\text{ h} - \$4.93\text{ h})6{,}500\text{ h/year} = \$21{,}580\text{ year}$$

Thus, the simple payback period for the cogeneration system:

$$SPP = \frac{\$2{,}500/\text{kW} \times 60\text{ kW}}{\$21{,}580\text{ year}} = 7.0\text{ years}$$

A Life Cycle Cost analysis may be required to determine if the investment on the cogeneration system is really warranted.

For a more detailed evaluation, it is important to determine the main goal of the cogeneration system. Ideally, the cogeneration system can be sized to match exactly both the electrical and thermal loads. Unfortunately, there is almost never an exact match. Therefore, the cogeneration system has to be designed to meet specific load requirements such as the base-load thermal demand, base-load electrical demand, peak thermal demand, or peak electrical demand. The main features of each design scenario are briefly described here:

- *Base-load cogeneration systems*: Produce only a portion of the facility's electrical and thermal requirements. Thus, production of supplemental thermal energy (using a boiler for instance) and purchase of additional electrical energy are generally required. Base-load cogeneration systems are suitable for facilities characterized by variable thermal and electrical loads but not willing or able to sell electrical power.
- *Thermal-tracking cogeneration systems*: Are those systems that produce all the thermal energy required by a facility. In case the generated electrical energy exceeds the electrical demand, the facility has to sell power to the utility. In case the generated electrical energy is lower than the electrical demand, additional power has to be purchased from the utility. Thermal-tracking cogeneration systems are increasingly becoming attractive to small buildings that have to pay higher utility rates than large industrial and commercial facilities.
- *Electricity-tracking cogeneration systems*: Are designed to match electrical loads. Any supplemental energy requirement is produced through boilers. These systems are typically suitable for large industrial facilities with fairly high and constant electrical loads and lower but variable thermal loads.
- *Peak-shaving cogeneration systems*: In the case where the cost associated with peak electrical demand is high, it may be cost-effective to design cogeneration systems, specifically for peak shaving, even though these systems may operate only a few hours (less than 1,000 h/year).

Example 13.3 illustrates a monthly analysis to determine the optimum size of a base-load cogeneration system that is designed without selling any generated electrical power. The following equations have been used to carry out the analysis summarized in Table 13.4.

The monthly electrical energy, kWh_{cogen}, and the monthly thermal energy, TE_{cogen}, produced by a cogeneration system of capacity kW_{cogen}, are estimated using Equations 13.3 and 13.4, respectively:

$$kWh_{cogen} = \min\left(24 \times N_d \times kW_{cogen}; kWh_{actual}\right) \tag{13.3}$$

$$TE_{cogen} = \min\left(24 \times N_d \times kW_{cogen}; TE_{actual}\right) \tag{13.4}$$

where

- N_d is the number of days in the month
- kW_{cogen} is the electrical power capacity of the cogeneration system
- kWh_{actual} is the actual electrical energy used by the facility during the month
- TE_{actual} is the actual thermal energy used by the facility during the month

Example 13.3

Problem

Provide a simple payback period analysis for implementing a cogeneration system to be installed in a hospital. Use the following characteristics for the cogeneration system:

Fuel input rate: 10,000 Btu/kWh
Heat recovery rate: 5,500 Btu/kWh
Maintenance cost: $0.02/kWh
Maximum electrical output: 200 or 300 kW
Installed equipment cost: $1,000/kW

Table 13.12 summarizes the energy usage and cost of the hospital. Assume that the boiler(s) efficiency is 70%. For this analysis, assume also that the cogeneration system requires diesel fuel only (the other option is dual fuel). Assume the heating value of diesel fuel is 140,000 Btu/gal.

Solution

For each month, the energy cost incurred with a cogeneration system is calculated using a step-by-step procedure based on Equations 13.3 and 13.4. Table 13.13 summarizes the results of the step-by-step analysis performed for the month of January. Table 13.14 provides the results for all the months with the payback period for each cogeneration size. In this example, the smaller cogeneration system (200 kW) is more cost-effective since there is no option to sell excess generated power to the utility. However, a detailed economic analysis should be carried out to optimize the size of the cogeneration system.

13.5.3 Financial Options

To finance a cogeneration system, several financial options are generally available. Selecting the most favorable financial arrangement is critical to the success

TABLE 13.12
Monthly Utility Data for the Hospital Used in Example 13.3

	Utility Summary				
	Electricity			Fuel Oil	
Month	**(kWh)**	**(kW)**	**($)**	**(Gal)**	**($)**
January	226,400	546	27,020	20,659	14,911
February	273,600	572	28,949	20,555	12,639
March	280,800	564	31,048	16,713	9,670
April	228,000	526	25,251	10,235	4,742
May	246,000	692	28,755	12,193	5,347
June	301,200	884	36,604	12,352	9,001
July	346,800	1,040	45,031	20,604	3,122
August	403,200	944	46,374	17,276	5,711
September	303,600	860	36,541	10,457	3,762
October	276,000	872	33,559	10,890	3,726
November	272,400	662	28,042	13,478	5,255
December	276,000	524	25,041	17,661	7,808
Total	3,434,000		393,215	183,073	85,694

TABLE 13.13
Details of Step-by-Step Analysis Performed for the Month of January in Example 13.3

	Cogeneration System	
Energy/cost requirements	200 kW	300 kW
Electrical energy requirements (kWh)	226,400	226,400
Thermal energy requirements (MMBtu)	2,024	2,024
Cogenerated electrical energy, kWh_{cogen} (kWh)	148,808	223,400
Cogenerated thermal energy, TE_{cogen} (MMBtu)	808	1,228
Electrical energy to be purchased from utility (kWh)	77,600	3,000
Thermal energy to be directly generated (MMBtu)	1,206	796
Fuel use for cogeneration (Gal)	10,628	15,957
Fuel energy for direct generation of thermal energy (Gal)	8,614	5,686
Total fuel use requirements (Gal)	19,242	21,643
Cost of utility electrical energy ($)	9,258	557
Cost of fuel ($)	13,893	15,627
Cost of maintenance ($)	2,976	4,468
Total cost ($)	26,127	20,450

TABLE 13.14
Cogeneration Cost Savings and Payback Periods

	Output (kW)	200	300
S. No.	**Conventional E Cost ($/Month)**	**Cogen E Cost ($/Month)**	**Cogen Cost, Cogen ($/Month)**
1	41,931	26,127	20,450
2	41,588	28,920	27,687
3	40,718	27,108	21,753
4	29,993	14,527	14,813
5	34,102	20,178	14,121
6	45,605	31,658	26,034
7	49,153	32,167	24,142
8	52,085	37,815	31,536
9	40,303	26,388	20,414
10	37,285	22,687	16,339
11	33,297	21,581	16,512
12	32,849	22,097	17,892
Total ($)	478,909	311,251	251,692
Equipment cost ($)	0	200,000	300,000
Payback period (year)	—	1.19	1.32

of a cogeneration project. A number of factors affect the selection of the best financial arrangement for a given cogeneration project. These factors include ownership arrangements, risk tolerance, tax laws, credit markets, and cogeneration regulations. In the United States, the most common financial approaches for cogeneration facilities are the following:

1. *Conventional ownership and operation*: In this financing structure, the owner of the cogeneration facility funds either totally or partially the project from internal sources. In the case of partial funding, the owner can borrow the remainder funds from a conventional lending institution. Operation and maintenance of the cogeneration system can be performed by an external contractor.
2. *Joint venture partnership*: This structure is an alternative to the conventional ownership and operation with shared financing and ownership with a second partner, such as an electric utility. Indeed, PURPA regulations provide the option for an electric utility to own up to 50% of a cogeneration facility. The joint venture financing structure reduces the risks for both partners but may increase the complexity of the various contracts between all the involved parties, including the owner and its partner, gas provider, electric utility, lending institution, and possible operation and maintenance contractor.
3. *Leasing*: In this financing option, a company builds the cogeneration facility with a leasing agreement from the owner to use part or all the thermal and electrical energy output of the cogeneration plant. The construction of

the cogeneration system by the lessor (i.e., the builder of the facility) can be financed through funds from lenders and/or investors. The owner is generally heavily involved in the construction phase of the cogeneration facility.

4. *Third-party ownership*: This financing structure is similar to that described for the leasing case. However, in the third-party ownership, the owner is not involved in both the financing and construction of the cogeneration facility. Instead, a third party or a lessee develops the project and arranges for gas/fuel supply, electrical power and heat sales, and operation and maintenance agreements. The finances can be arranged by a lessor through funds from investors and/or lenders.
5. *Guaranteed savings contracts*: In this financing option, a developer first builds and maintains the cogeneration facility. Then, the developer enters in a guaranteed savings contract with the energy consumer (the owner). This contract is typically made for a period ranging from 5 to 10 years with a guaranteed fixed savings per year. This type of financial structure is common for small cogeneration systems (i.e., packaged units) since it shifts all the financing and operation risks from the owner to the facility developer (i.e., the guaranteed savings contractor).

13.6 CASE STUDY 1: EVALUATION OF AN EXISTING COGENERATION SYSTEMS

A power plant, serving all the buildings of a university campus, utilizes natural gas with fuel oil No. 6 as a backup fuel. Due to increasing demands in electricity and steam use, the university retrofitted the power plant into a cogeneration plant. The specifications of the components for the cogeneration systems are summarized in Table 13.15.

The primary systems of the cogeneration plant include two combustion gas turbines, two heat recovery steam generators (HRSGs), high-pressure steam boilers, and low-pressure absorption chillers. The two gas turbines are each capable of producing 16 MW of power while providing sufficient hot exhaust gas to generate 80,000 lb/h of 300 psig steam in each heat recovery steam generator (HRSG). The steam turbine generator consists of dual topping turbines driving a common generator. The turbine reduces the incoming steam pressure by expanding it through the turbines, which in turn drive the generator. The exhaust steam is then exported to either the 130 or the 10 psig steam header. The 130 psig steam is delivered to various buildings in the campus for heating purposes. The 10 psig steam is utilized to operate three absorption chillers that deliver chiller water to the campus. The high-pressure steam boilers deliver 300 psig steam to the gas turbines for NO_x control and steam injection for power augmentation.

The cogeneration plant was designed to deliver over 30 MW of electrical power. Absorption chillers with over 3,000 tons capacity were added to increase the thermal load on the plant, especially during the summer period. A large portion of the cogenerated electricity is sold to the local utility. As stipulated by PURPA, the plant can purchase electricity from the same utility in case of maintenance or emergency periods. A typical monthly performance of the cogeneration plant is outlined in Table 13.16.

TABLE 13.15
Equipment Used for the Cogeneration Plant

Utility	Quantity	Equipment	Description
Electric power	2	Mitsubishi industrial gas turbine sets	16 MW each includes a dual fuel Mitsubishi Heavy Industries MF-111AB gas turbine driving a RENK single reduction gearbox coupled to a Brush two pole synchronous generator. Operates with a shaft speed of 9,645 rpm
Electric power	1	Dresser rand steam turbine set	One MW induction generator double-ended low- and high-pressure steam
Steam	2	Zurn (HRSG) heat recovery steam generators	Supplemental fire capability—Davis duct burners 80,000 lb/h maximum steam output 300 psig
Steam	1	Erie city 1,966 boiler	Front fired 150,000 lb/h maximum steam output 130 psig
Steam	1	Combustion engineering 1,957 boiler	Tangentially fired 115,000 lb/h maximum steam output 130 psig
Chilled water	3	Steam absorption chillers	(1) Trane—1,470 tons cooling; (2) York—900 tons cooling each utilizes 10 psig steam lithium bromide

TABLE 13.16
Monthly Fuel Used and Electricity/Steam Produced at the University Cogeneration

		Cogeneration System Performance			
		Electricity			
Month	Nat. Gas (MMBtu)	Used (kWh)	Sold (kWh)	Total (kWh)	130 psig Steam (1,000 lb)
Jan	195,300	9,803,800	5,952,000	15,755,800	84,932
Feb	161,000	9,332,820	5,376,000	14,708,820	79,452
Mar	141,600	10,011,670	5,952,000	15,963,670	84,932
Apr	141,600	10,055,890	5,760,000	15,815,890	82,192
May	150,350	9,776,410	5,952,000	15,728,410	84,932
Jun	141,000	9,899,130	5,760,000	15,659,130	82,192
Jul	181,757	10,752,740	5,952,000	16,704,740	84,932
Aug	201,802	10,604,980	5,952,000	16,556,980	84,932
Sep	180,480	10,492,980	5,760,000	16,252,980	82,192
Oct	150,474	10,389,270	5,952,000	16,341,270	84,932
Nov	166,230	10,553,990	5,760,000	16,313,990	82,192
Dec	154,039	9,441,120	5,952,000	15,393,120	84,932
Year	1,965,632	121,114,800	70,080,000	191,194,800	1,000,000

TABLE 13.17
Comparison of Annual Energy Use and Carbon Emissions in 1990 and 2000

	Year 1990		Year 2000	
	Energy Purchased or Sold	**Carbon Emissions (Tons)**	**Energy Purchased or Sold**	**Carbon Emissions (Tons)**
Natural gas purchases				
Central plant/cogeneration	634,159 MMBtu	10,115	1,936,341 MMBtu	30,885
Individual buildings	116,500 MMBtu	1,849	632,419 MMBtu	10,080
Electricity purchases				
Central plant/cogeneration	66,024,000 kWh	18,476	39,937,364 kWh	10,637
Individual buildings	20,782,510 kWh	5,816	9,970,008 kWh	2,656
Electricity sales				
Cogeneration plant	0 kWh	—	74,893,631 kWk	19,947
Total natural gas purchased	750,659 MMBtu	11,994	2,568,760 MMBtu	40,965
Net electricity Purchased/sold	86,806,510 kWh	24,292	24,986,259 kWh	6,655
Net carbons emissions	—	36,296	—	34,310

As indicated in Table 13.16, the university campus used about 63% of the electricity produced by the cogeneration plant, while the remaining 37% was sold to a local utility. Due to the ever increasing electricity demands at the university (due to new buildings and additional research laboratories), the cogeneration plant may not be able to meet the university load in the near future. The university is currently investigating plans to add a new cogeneration plant.

From the data provided in Table 13.17, the electrical and overall efficiency of the cogeneration plant during 2000 is estimated at about 33% and 80%, respectively. The PURPA efficiency of the plant is over 56%, well above the requirements set by PURPA.

Over one decade, electricity and steam consumption have increased substantially at the university with a growth rate of about 5%. Table 13.17 illustrates the progression of energy use and carbon emissions at the university between 1990 (before the construction of the cogeneration plant) and 2000 (after the construction of the cogeneration plant). As indicated in Table 13.17, the cogeneration plant has substantially decreased the carbon output per unit of energy consumed. Even though natural gas use has tripled between 1990 and 2000, the net carbon emissions were actually reduced by 5%.

13.7 CASE STUDY 2: DESIGN OF OPTIMAL HYBRID SYSTEMS

This case study outlines the design procedure for optimal hybrid distributed generation (DG) systems for three remote communities in Alaska. All three of these towns have isolated electrical grids and currently rely on diesel fuel for 100% of their electricity and heating requirements. The assessment considered for this case study includes an analysis of each community's electrical and thermal load, a wind and solar resource evaluation, modelling, and optimization of various DG systems using LCC analysis of these systems. For all three of the communities, hybrid wind–diesel systems have the potential to provide reductions in the cost of energy (COE). However, the economic feasibility of these systems is extremely site specific. In addition to providing possible decreases in the COE, these hybrid systems can also provide significant environmental benefits, such as reductions in diesel fuel use and CO_2 emissions.

13.7.1 Introduction

In the state of Alaska, large transmission systems exist in the most populated areas such as the Railbelt region, which runs from Fairbanks through Anchorage to the Kenai Peninsula. Transmission systems are also present in some areas in Southeast Alaska; however, over 150 rural Alaskan communities have their own isolated electrical grids. About 70% of the electricity requirements for the Railbelt region are met with natural gas generators whereas in the majority of the remote areas, most of the electricity and heating requirements are met with diesel generators (AEA, 2011a). In 2010, residents in these rural areas had electricity rates between 40¢ and 60¢ per kilowatt hour (kWh), with some paying over $1.00 per kWh (AEA, 2011b). Conversely, communities in the Railbelt region that get electricity from hydroelectric facilities or natural gas generators have electricity rates between 10¢ and 15¢ per kWh (Fay et al., 2011).

Due to rising electricity rates and diesel prices, many rural Alaskan communities are considering the addition of renewable energy sources and other DG technologies to form hybrid DG systems. Alaska's significant wind resource has led many coastal communities to invest in wind energy projects (AEA, 2011a). As of spring 2009, 10 hybrid wind–diesel systems were operating in rural communities, with 6 more systems in development (Baring-Gould and Dabo, 2009). Preliminary results are promising; hybrid DG systems have the potential to lower the cost of electricity, improve the reliability of electricity and heating systems, and lower the emissions from diesel generators. Based on their study in rural Alaska, Clark and Isherwood (2004) believe that hybrid DG systems can compete with conventional fossil fuel generation for communities with high electricity costs, available renewable resources, and no interconnection with a large grid. Indeed, the results indicated that a wind–diesel hybrid system could result in diesel fuel savings of over 50% and cost savings of over 30% compared to the base case (diesel only).

This section summarizes the results of an optimization analysis of hybrid DG systems for three rural communities in Alaska—Mountain Village, Deering, and Ambler—conducted by Willam and Krarti (2013). The study includes an analysis of each community's electrical and thermal load, a wind and solar resource evaluation, modelling, and optimization of hybrid DG systems, an economic analysis of these systems, and an evaluation of the environmental benefits that they can provide. In this section, a brief description of the analysis approach and design recommendations is provided mostly for the Mountain Village site. More details for the other two sites can be found in the published study by Willam and Krarti (2013).

13.7.2 Optimization Design Approach

The analysis to determine the best hybrid DG system to meet the energy requirements for a residential community considers the following steps:

1. Determine the hourly electrical and thermal loads for the community. The estimation can be based on measured data or simulation analysis.
2. Identify the main energy resources available within or close to the site where the community is located. These resources typically include solar, wind, geothermal, hydroelectric, fossil fuel, and biomass.
3. Collect specific analysis data including economic parameters such as fuel costs and electricity prices from the grid as well as the capital and operation and maintenance (O&M) costs for various power-generating technologies.
4. Carry out optimal analysis for various technically feasible options for power generation technologies, including fuel based as well as renewable energy technologies. Simulation tools are typically used in the analysis.
5. Rank and select the most cost-effective hybrid system, including the optimal capacities. The economic analysis using LCOE or LCC is typically considered to determine the cost-effectiveness of the various hybrid distributed systems considered in the evaluation. Chapter 9 discusses the various economic analysis methods suitable to assess the cost-effectiveness of energy projects.

A description of the steps and the results for the analysis is presented with a focus on one rural community in Alaska located in Mountain Village (Willam and Krarti, 2013). Note that all three communities have a stand-alone electrical grid and currently rely on diesel fuel for all of their electricity and heating requirements.

13.7.3 Electrical and Thermal Loads

The annual electricity generation, diesel fuel use for electricity generation, and CO_2 emissions from electricity generation for three Alaskan rural remote communities are shown in Table 13.18.

The total generating capacity for the community of Mountain Village is 2,212 kW (AVEC, 2011). However, the type and capacity of the diesel generators are unknown. As a result, Mountain Village was modeled as having four 455 kW Caterpillar generators and a 410 kW Cummins generator for a total capacity of 2,230 kW. Deering has four generators with a total capacity of 578 kW. These units include 100 and 138 kW John Deere generators and two 170 kW Cummins generators (Vaught, 2007). Ambler has three generators with a total capacity of 982 kW (AVEC, 2011), but the type and capacity of the generators are unknown. As a result, Ambler was modeled as having three Cummins generators with the following capacities: 270, 315, and 410 kW for a total generating capacity of 995 kW. Since hourly electricity and heating demand data were not available for any of the communities, a whole building simulation program was used to model the hourly electrical and thermal load for each village (Willam and Krarti, 2013). The results of the simulation including the peak and average electric and thermal loads are shown in Table 13.19.

TABLE 13.18
Electricity Generation, Diesel Fuel, and Carbon Emissions for Three Alaskan Communities

Community	Electricity Generation (kWh/Year)	Diesel Fuel Use (gal/Year)	CO_2 Emissions (Tons/Year)
Mountain village	2,799,595	189,184	1,920
Deering	711,319	55,145	560
Ambler	1,249,161	89,892	912

TABLE 13.19
Average and Peak Electrical and Thermal Loads for Three Alaskan Communities

Community	Average Electric Load (kW)	Average Thermal Load (kW)	Peak Electric Load (kW)	Peak Thermal Load (kW)
Mountain Village	303	2,277	512	7,846
Deering	74	775	132	2,232
Ambler	139	1,212	285	4,439

TABLE 13.20
Annual Average Wind Speed, Wind Power Density, and Solar Radiation for Three Alaskan Communities

Community	Annual Avg. Wind Speed (m/s)	Annual Avg. Wind Power Density (W/m^2)	Annual Avg. Solar Insolation (kWh/m^2/Day)
Mountain village	7.28	523	2.80
Deering	5.22	322	2.61
Ambler	3.60	88	2.58

13.7.4 Energy Resource Assessment

Using Alaska Energy Authority's (AEA) extensive study on renewable energy resources (AEA, 2011a), wind and solar resource assessment is performed for each of the villages. Although Alaska's solar resource is minimal during the winter, a solar resource assessment was also completed. Hourly wind and solar resource data were obtained from the AEA and the National Solar Radiation Database. The annual average wind speed, wind power density, and solar radiation values for each community are shown in Table 13.20.

13.7.5 Electricity and Fuel Prices

Power Cost Equalization (PCE) program statistics for fiscal year 2010 were obtained for each community (AEA, 2011b). This program provides subsidies for residential customers in rural areas in order to decrease the cost of electricity for the first 500 kWh used by a customer each month. Community facilities are also eligible for a subsidized rate, but state and federal customers and commercial facilities are not. PCE subsidies are based on both nonfuel and fuel costs (AEA, 2007). The PCE program statistics provide electricity price information for the electricity that is eligible for a subsidy and that which is not eligible. Current diesel fuel costs for each community are shown in Table 13.21 and PCE statistics are shown in Table 13.22.

The results from a study performed by the Institute of Social and Economic Research (ISER) at the University of Alaska Anchorage indicate that electricity rates and PCE subsidies decrease as wind penetration increases (Fay et al., 2010). As wind power supplies more electricity to a community, diesel fuel use decreases, which lowers the electricity rate. However, a reduction in the electricity rate also results in a lower subsidy from the PCE program. It should be noted that this has been viewed as a disincentive for communities to invest in wind–diesel systems. As diesel fuel prices increase, the PCE subsidy also increases. However, communities with wind–diesel systems are not protected by the PCE program to the same extent that those with diesel systems are. This effect has the most significant impact on communities whose wind–diesel systems operate at low wind penetration levels. For systems with high penetration levels, the effect is not as great since an increase in wind penetration protects the community from high fuel prices. Despite the issues surrounding the

TABLE 13.21
Average Diesel Fuel Prices Available for Three Alaskan Communities

Community	Diesel Fuel Price ($/gal)
Mountain village	3.16
Deering	4.69
Ambler	3.75

TABLE 13.22
Electricity Sales and Rates Based on PCE Program

Community	Mountain Village	Deering	Ambler
Total electricity sales (kWh)	2,650,142	647,749	1,216,727
PCE eligible sales (kWh)	1,094,841	Unknown	540,632
Non-PCE eligible sales (kWh)	1,555,301	Unknown	676,095
Electricity rate (¢/kWh)	50.62	77.37	53.80
PCE subsidy (¢/kWh)	29.20	30.74	32.22
PCE electricity rate (¢/kWh)	21.42	46.63	21.58

PCE subsidy, wind systems continue to be an attractive option for remote communities with an abundant wind resource due to their potential to lower electricity rates, reduce CO_2 emissions, and protect these areas from diesel price fluctuations.

13.7.6 Optimization Analysis

DG systems were modeled and optimized using HOMER software (Homer, 2013). Each town was modeled with its electrical and thermal load, a boiler, diesel generators with heat recovery, wind turbines, solar panels, batteries, and converters. An example of the structure of the system, as modeled in HOMER, is shown in Figure 13.3. For the sensitivity analyses, the diesel price was varied from \$1/L to \$3/L (about \$3.80/gal to \$11.40/gal). These values are consistent with the fuel price projections from ISER. Based on other wind–diesel system studies, a project lifetime of 25 years and a discount rate of 3% were chosen for the analysis.

The capital and O&M costs that were used for the simulations are shown in Table 13.23. The capital cost for the solar panels was obtained through the

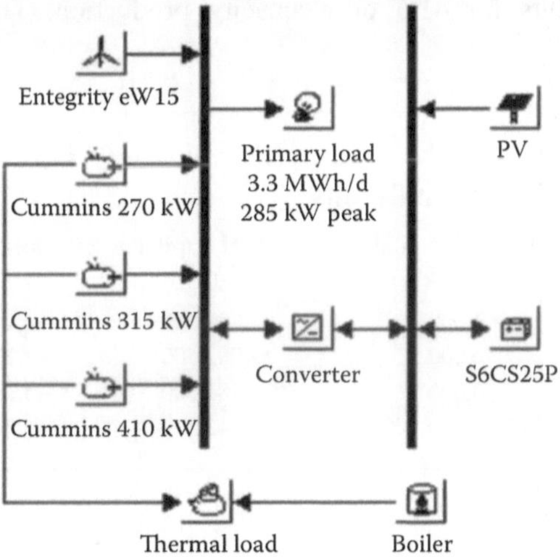

FIGURE 13.3 Hybrid system structure.

TABLE 13.23
Economic Parameters for HOMER Simulations

System Component	Capital Cost (\$/kW)	O&M (\$/kWh)	Reference
Diesel generator	0	0.02	Vaught (2007)
Wind turbine	4,000–15,000	0.04	Fay et al. (2010)
Solar panels	11,000	110 (\$/year)	Arvizu et al. (2013)
Battery (Surrette 6CS25P)	\$1,375 (total cost)	15 (\$/year)	
Converter	1,100	50 (\$/year)	

examination of grant applications for the Alaska Renewable Energy Fund. Battery system costs were reported by Susitna Energy and Sandia National Laboratories and converter costs were obtained from ABS Alaskan. Based on the types of wind turbines currently installed in Alaska, the Entegrity 50 kW turbine was considered for each hybrid system.

13.7.7 Selected Results

The analysis results for the Mountain Village community's current diesel fuel cost of \$3.16/gal are shown in Table 13.24. This table displays the capital cost, the net present cost (NPC), and the COE for the optimal system for each level of wind turbine capital costs. The first wind–diesel–battery (W-D-B) hybrid system listed in the table includes eight wind turbines, a 455 kW John Deere generator, a 410 kW Cummins generator, 100 batteries, and a 200 kW converter system.

The monthly average electric production profile for the hybrid system with eight wind turbines is shown in Figure 13.4. For this case, the annual electricity production from wind is 1,227,169 kWh and that from the diesel generators is 1,748,822 kWh. Wind accounts for 41% of electricity production. The diesel generator

TABLE 13.24
Mountain Village Simulation Results

Optimal System	Turbine Capital Cost (\$/kW)	Initial Capital (\$)	Total NPC (\$)	LCOE (\$/kWh)
W-D-B	4,000	1,887,500	44,168,904	0.210
W-D-B	6,000	2,082,500	44,776,100	0.223
W-D-B	8,000	1,787,500	45,186,308	0.232
D-B	10,000–15,000	247,500	45,315,064	0.235

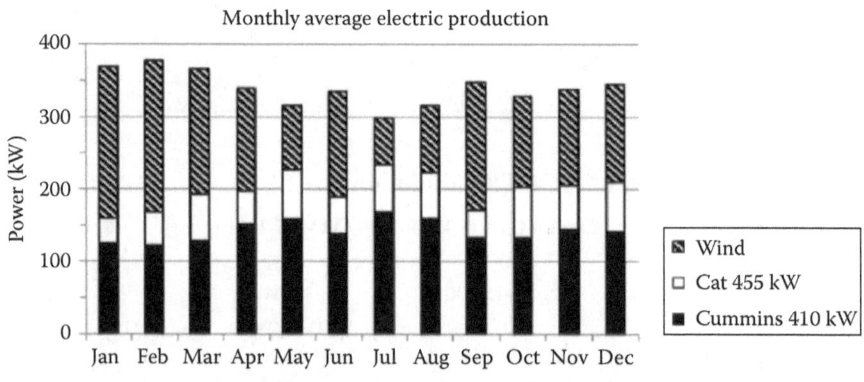

FIGURE 13.4 Electric production profile for the hybrid system.

TABLE 13.25
Comparison between Baseline System and the Hybrid DG System for Mountain Village community

System	Electricity Production from Diesel (kWh/Year)	Diesel Fuel Use (gal/Year)	CO_2 Emissions (Tons/Year)
Baseline	2,799,595	189,184	1,920
Hybrid	1,748,822	118,164	1,199

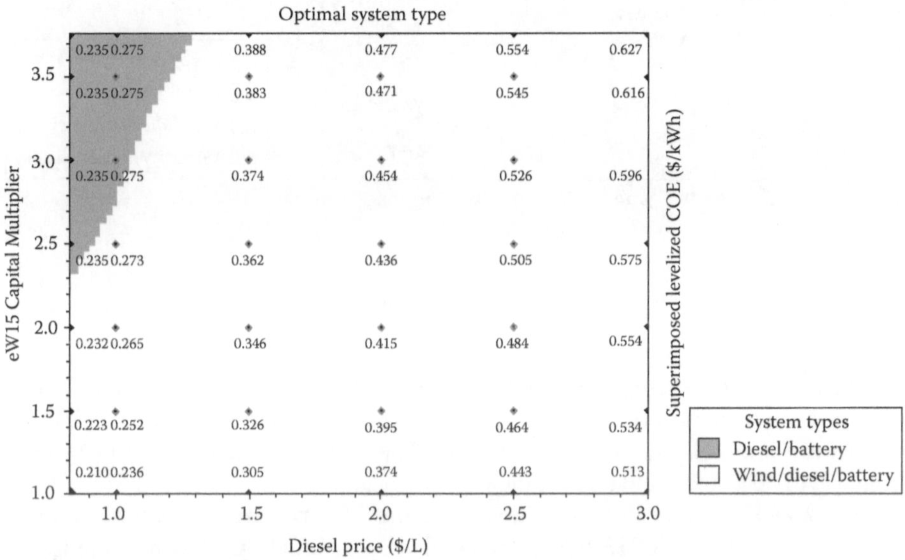

FIGURE 13.5 Mountain Village wind turbine sensitivity analysis.

electricity production, diesel fuel use for electricity generation, and CO_2 emissions from electricity generation for this hybrid system and the baseline system are shown in Table 13.25. The W-D-B hybrid system results in a reduction of 1,050,773 kWh of diesel generator electricity production. This equates to a reduction in diesel fuel use of about 71,020 gallons and a CO_2 emissions reduction of 721 metric tons per year.

The results from the sensitivity analysis are shown in Figure 13.5. The sensitivity variables are the diesel price and the capital cost of the wind turbines, denoted by eW15 Capital Multiplier. The capital cost ranges from $4,000 to $15,000 per kW of installed capacity. The LCOE for each system is also shown in the figure. A W-D-B hybrid system is the optimal system, except for low diesel prices and high capital cost values.

Even though PV is not included in the optimal system, it could still be economically feasible. According to the simulation results, the COE for solar hybrid systems

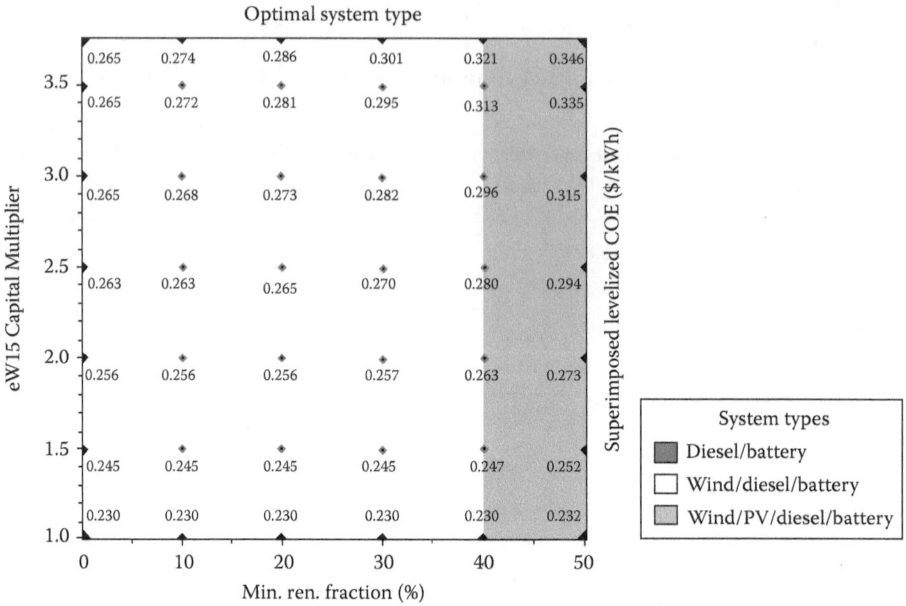

FIGURE 13.6 Mountain Village REF sensitivity analysis.

at a diesel price of \$3.16/gal ranges from \$0.214/kWh to \$0.265/kWh. Solar hybrid systems become part of the optimal system if renewable energy fraction (REF) constraints are imposed on the electric profile. The optimal system for REFs ranging from 0% to 50% and wind turbine capital costs ranging from \$4,000 to \$15,000 per kW are shown in Figure 13.6. Specifically, Figure 13.6 indicates the optimal systems for a fixed diesel price of \$3.16/gal, the current cost of fuel. As shown in the figure, a W-D-B system is the predominant optimal system; however, wind–PV–diesel–battery systems are the optimal system for high REF constraints. For all of these systems, the COE falls below Mountain Village's standard electricity rate of \$0.51/kWh.

The LCOE as a function of the percent reduction in CO_2 emissions from electricity generation is shown in Figure 13.7. As indicated in the chart, the optimal system in terms of the COE has an REF of 50%, a COE of \$0.23/kWh, and a reduction in CO_2 emissions of 44.3%. These values are achieved with a W-D-B system with 10 turbines at a capital cost of \$4,000 per kW, two generators, 200 batteries, and a 300 kW converter system. With a capital cost of \$10,000 per kW, the COE for this system increases to \$0.292/kWh. The system with the highest COE includes 10 turbines at a capital cost of \$15,000 per kW, a 20 kW PV system, 2 generators, and a 400 kW converter system.

The LCOE as a function of the percent reduction in CO_2 emissions from electricity generation obtained for the rural community in Deering is shown in Figure 13.8. As indicated in the chart, the optimal system in terms of the COE has an REF of 34%, a COE of \$0.317/kWh, and a reduction in CO_2 emissions of 35.4%. These values are achieved with a W-D-B system with two turbines at a capital cost of \$4,000 per

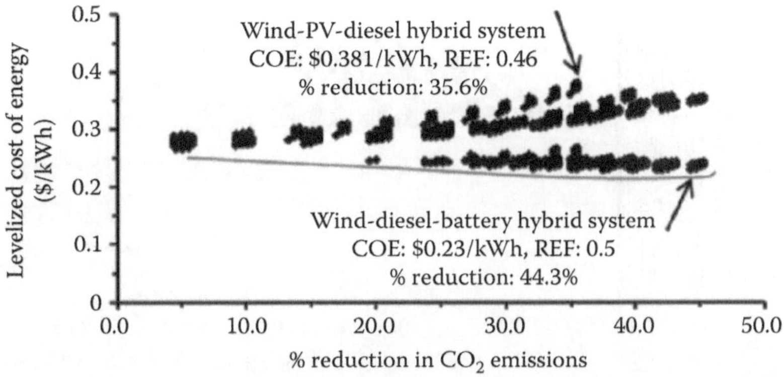

FIGURE 13.7 COE as a function of reduction in CO_2 emissions for Mountain Village community.

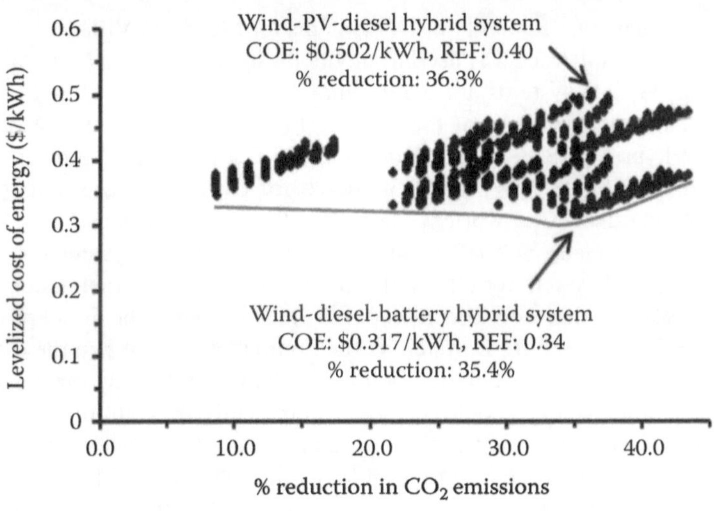

FIGURE 13.8 COE as a function of reduction in CO_2 emissions for Deering Community.

kW, three generators, 50 batteries, and a 75 kW converter system. With a capital cost of $10,000 per kW, the COE for this system increases to $0.369/kWh. The system with the highest COE has two turbines at a capital cost of $15,000 per kW, a 70 kW PV system, three generators, and a 100 kW converter system.

For the Ambler community, the LCOE obtained as a function of the percent reduction in CO_2 emissions from electricity generation is outlined in Figure 13.9. Unlike the other two communities, the COE increases with the percent reduction in CO_2 emissions for all systems. The optimal system in terms of the COE is a system with an REF of zero and therefore no reduction in CO_2 emissions.

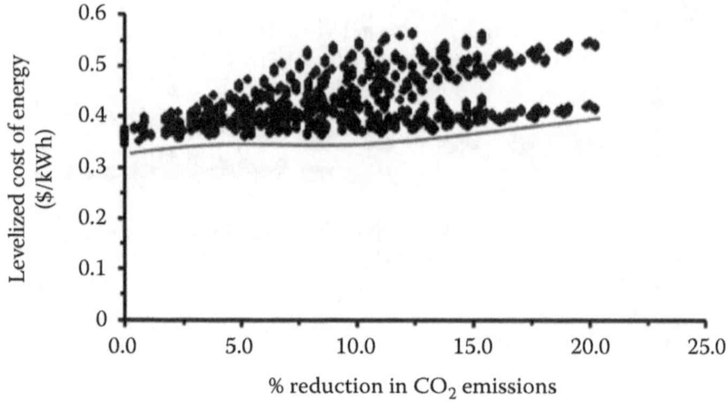

FIGURE 13.9 COE as a function of reduction in CO_2 emissions for Amber community.

13.7.8 Conclusions and Recommendations

In order to evaluate the feasibility of DG systems for Mountain Village, Deering, and Ambler, HOMER software was used to model these systems. For Mountain Village and Deering, W-D-B systems are the optimal system for current diesel fuel prices, depending on the capital cost of the wind turbines. The levelized cost of electricity for these hybrid systems ranges from \$0.210/kWh to \$0.232/kWh for Mountain Village and from \$0.288/kWh to \$0.305/kWh for Deering. These ranges are well below the standard electricity prices for each of these communities.

The sensitivity analysis results indicate that the optimal system for Mountain Village is a W-D-B system for almost the entire range of diesel fuel prices and wind turbine capital costs that were considered. For Deering, when the diesel price is high (above about \$2.3/L), a W-D-B system is the optimal system. When the diesel price is low, this hybrid system is only the optimal system for relatively low wind turbine capital costs. Solar systems could also be economically feasible for both Mountain Village and Deering, but they are not part of the optimal system unless an REF constraint is imposed on the electric profile. Unlike Mountain Village and Deering, W-D-B system is not the optimal system for Ambler. For this community, a diesel system with heat recovery is the optimal system for almost the entire range of diesel fuel prices and wind turbine capital costs that were considered in the sensitivity analysis. However, it should be noted that although a wind–diesel system is not the optimal system for Ambler, this type of system could still be economically feasible for this community. This is also true for systems with PVs.

The range of COE results supports the findings from previous studies that have shown that the economic feasibility of wind–diesel systems is extremely site-specific. The hybrid systems for Mountain Village have a lower COE than those for Deering. It is the wind resource that has the largest effect on this result, since Mountain Village has a significantly better wind resource than Deering. The wind resource also has an effect on the optimal system for Ambler. This community has the worst wind resource; as a result, a wind–diesel system is not the optimal system.

Additionally, solar hybrid systems may be more economically feasible than wind systems for Ambler, especially when the cost of wind turbines is high.

Overall, the results from this assessment indicate that hybrid wind–diesel systems may be an economically beneficial option for rural Alaskan communities with an abundant wind resource. However, it should be noted that these results are based on simulated electricity and heating demand data since real hourly demand data were not available for any of the communities. The electricity and heating demand have a significant impact on the optimization, and any changes in the demand may alter the HOMER results for the optimal system type and the LCOE for each system. As a result, a more detailed analysis with real hourly electricity and heating demand data should be completed in order to more accurately evaluate the economic feasibility of these hybrid systems. Additionally, more research is needed to determine the impact that the PCE program may have on communities that develop these systems. Despite the issues surrounding the PCE subsidy, wind systems continue to be an attractive option for remote communities due to their potential to lower electricity rates and provide significant environmental benefits, such as reductions in diesel fuel use and CO_2 emissions. In some cases, these benefits may be achieved without an increase in the COE.

13.8 SUMMARY

In this chapter, existing types and designs for power generation as well as cogeneration systems are discussed. Moreover, simplified feasibility analyses of cogeneration systems are described with some illustrative examples. In the future, cogeneration is expected to become more attractive for small buildings, especially with new developments in fuel cell technologies and microprocessor-based control systems. These developments will make small cogeneration systems cost-effective, reliable, and efficient for even non traditional cogeneration applications, such as residential complexes and office buildings.

13.9 PROBLEMS

1. Provide the actual efficiency and the PURPA efficiency for 10 MW cogeneration system with an electrical efficiency of 30% and thermal efficiency of 40%.
2. A small cogeneration system rated at 100 kW rated at 80% overall efficiency and 30% electrical efficiency. Estimate the potential cost-effectiveness of deploying this system for a building that purchased electricity at $0.15/kWh and generate heat from 85% efficient boiler using natural gas at $15/therm. The cost of the system is estimated at $200,000. For the analysis, consider a maintenance cost of $0.85/hour with an operation period at full load of 6,000 hours per year and a lifetime of 40 years with a discount rate of 5%.
3. Provide the optimal size for a cogeneration system for a hospital. The cogeneration system has the following characteristics:
    - Fuel input rate: 8,000 Btu/kWh
    - Heat recovery rate: 4,800 Btu/kWh

- Maintenance cost: \$0.02/kWh
- Maximum electrical output: 200, 300, 400, 500, 600, 700, or 800 kW
- Installed equipment cost: \$700, \$1,000, or \$1,500/kW

Table 13.10 summarizes the energy usage and cost of the hospital. Assume that the boiler(s) efficiency is 75%. The heating value of diesel fuel is 140,000 Btu/gal. Present the results in one graph: the payback period versus the equipment cost for various equipment sizes.

4. Same as Problem 13.1. Assuming that the electric cost varies from \$0.05 to \$0.15/kWh (including demand charge), determine the variation of the cogeneration system payback period versus the electricity cost. For this question, assume that the capacity and the cost of the cogeneration system are 300 kW and \$2,000/kW, respectively.
5. Same as Problem 13.3. If the life of the cogeneration system is 40 years, provide the optimal size of the cogeneration system for the hospital when the average interest rate is 8% and the inflation rate is 4%.
6. A 400 kW cogeneration system has an overall efficiency of 86%. The cogeneration system produces hot water at a thermal efficiency of 24%. Determine the cost-effectiveness of installing the cogeneration system in a facility at a cost of \$60,000. The cost of electricity, if purchased directly from the utility, is \$0.09/kWh and the cost of natural gas, used to operate the cogeneration system, is \$1.50/therm. Without the cogeneration system, a gas-fired boiler with an efficiency of 85% is utilized to produce hot water needed by the facility. Assume that:
    - The cogeneration system, as well as the facility, is operated 65,000 h/year
    - The maintenance cost of the cogeneration system is \$0.015/kWh

    Perform a life cycle cost analysis of the cogeneration system using a life cycle of 20 years and a discount rate of 5%.

14 Optimal Designs of Energy Efficient and Resilient Power Systems

In this chapter, design approaches of energy systems are outlined to develop energy efficient and resilient building. Specifically, energy efficiency measures (EEMs) are combined with renewable energy technologies to design and retrofit buildings or communities to reach low-energy, zero-energy, and resiliency targets. In particular, overviews of grid interactive efficient buildings (GEBs) as well as electrification and net zero energy designs are provided. In addition, optimization methodologies are introduced to determine the best combination of energy efficiency strategies and renewable energy systems to reduce the reliance on fossil fuel resources. Both source energy use and life cycle cost are accounted for in the optimization procedure to design energy efficient and net zero-energy buildings and communities. In addition to the optimization techniques, specific case studies are described to design and retrofit buildings and communities to enhance their energy efficiency and resiliency.

14.1 INTRODUCTION

In the United States, the electricity grid continues to experience rapid changes both on the supply and demand sides. On the supply side, its generation fuel mix is shifting away from coal and nuclear sources towards natural gas and renewable energy sources. At the same time, grid operators must balance an increasing supply of non-dispatchable and intermittently available renewable sources such as wind and solar, which are projected to be the fastest growing electricity resource through 2050 (EIA, 2020). On the demand side, an increasing share of demand is now met by distributed energy resources (DER) such as solar photovoltaics (PV), which are also supplying energy to the grid through net metering. These changes are driven mostly by retiring less efficient fossil fuel-based generation sources, low natural gas prices, policies supporting renewable energy, and continued decline in renewable energy costs. At the same time, concerns about climate change have also begun to shift both supply-side and demand-side fuel sources away from fossil fuel-based sources to renewable energy sources. Utilities are committing to 100% renewable energy goals as municipalities adopt electrification policies for buildings in an effort to reduce carbon emissions.

Fundamentally, grid operators must balance electricity supply with demand. In order for the operators to have enough generation capacity in reserve to meet periods of high peak demands, utilities need to build new generation capacity requiring costly and long-term investments. To defer the construction of new generation

DOI: 10.1201/9781003276999-14

capacity, utilities often implement programs that reduce, shed, or shift load through demand-side management (DSM), energy efficiency, or demand response (DR) programs. Buildings, which collectively consumed 63% of delivered electricity in the U.S. in 2019 (15% residential, 12% commercial, 35% industrial), have the potential to offer grid services through implementation of DSM strategies that enhance electrical load flexibility.

Buildings have to meet increasingly stringent energy-efficiency regulations, and society's expectations for improved sustainability in communities and cities. Indeed, future buildings and cities are expected not only to be highly energy efficient, adaptive, and resilient to meet occupant needs, but also be robust and resilient in case of extreme and catastrophic events. In a highly competitive industry, energy-efficient buildings are quickly becoming a market differentiator for residential and commercial properties. Indeed, investments in building energy efficiency and renewable energy technologies have been significant in the last decade and have created millions of jobs throughout the US. It is estimated that the US economic impact of sustainable and intelligent buildings can reach over $200 billion mostly related to avoiding health problems and the associated lost productivity. Several initiatives and programs have been implemented worldwide to design net zero energy, adaptive, and resilient buildings. In the US, grid-interactive efficient buildings (GEBs) are promoted to use existing and new technologies to provide demand flexibility have recently emerged to balance the grid's supply and demand and a source of value through avoided electricity system costs (DOE, 2019a). Several countries are requiring new buildings to be designed as net zero energy (NZE) and even positive energy through the integration of on-site renewable power systems with energy efficiency strategies.

This section summarizes various energy efficiency and renewable energy technologies and operation strategies suitable for GEB and NZE applications. In addition to energy efficiency benefits, the overviewed technologies can enhance demand management capabilities and grid services for buildings (Steen and Krarti, 2020). Moreover, this section provides case studies are presented to illustrate design approaches for on-site renewable energy systems to achieve NZE buildings as well as electrified and resilient communities.

14.2 GRID INTERACTIVE EFFICIENT BUILDINGS

14.2.1 Overview of GEBs

The U.S. Department of Energy (DOE) defines a GEB as an energy efficient building that uses smart technologies and on-site DERs to provide demand flexibility while co-optimizing for energy cost, grid services, and occupant needs and preferences, in a continuous and integrated manner. Specifically, the DOE characterizes GEBs further as efficient, connected, smart, and flexible buildings (DOE, 2019a). First, GEBs can reduce demand on the grid by using less energy through efficiency. Second, they are well connected to allow two-way communication with the grid by sending and receiving signals in order to respond to time-dependent needs. Third, GEBs are smart by integrating sensors and controls that allow for the co-optimization of cost functions from the perspective of the building owner, occupants, and grid through the use of analytics. Finally, they are flexible so their energy loads can be shaped dynamically to meet any grid services and/or signals. These characteristics are facilitated by capabilities at the component level and system level where individual components can monitor and send information about their status and receive control commands to shed, shift, or modulate loads from the building's control system.

Two features are generally associated with GEBs including demand-side management and grid services as detailed in the following sections.

14.2.1.1 Demand-Side Management

GEBs incorporate DSM strategies that provide services to the grid. For the purposes of GEBs, the DOE identifies load shed, load shift, and modulation as demand flexibility strategies as illustrated in Figure 14.1 (DOE, 2019a). Energy efficiency is a reduction in energy use while providing the same or improved level of energy service(s). Graphically, energy efficiency is represented by a shift downward in demand. Load shedding is the ability of a building to reduce electricity use during a short period, which typically occurs during times of peak demand or emergencies and on short notice. Graphically, load shedding is represented by flattening of peak areas. Load shifting is the ability to change the timing of electricity use, which is graphically represented by decreasing demand during one period while increasing demand during another. Modulation is the ability to balance real power supply with demand or the ability to balance reactive power draw with supply in response to a grid signal. In contrast to the previous strategies, modulation occurs at a smaller time scale within seconds or sub-seconds. Generation is the ability to produce electricity for consumption on-site or to dispatch electricity to the grid in response to a signal from the grid operator.

14.2.1.2 Grid Services

Grid services provide value through avoided electricity system costs by supporting the generation, transmission, or distribution of electricity. The potential value of a DER can be quantified by estimating the avoided cost of acquiring the next least expensive energy resource that provides comparable grid services. Table 14.1 summarizes the services GEBs can provide to the grid and maps these services to DSM strategies. Examples of current mature demand-side resources that provide grid services are energy efficiency and DR.

The benefits for GEBs involve primarily both energy efficiency and demand response:

1. *Energy efficiency*: Energy efficiency improves grid reliability by decreasing peak demand and reducing strain on the transmission and distribution (T&D) system. By reducing load, energy efficiency improves reliability on the supply side by increasing the system's reserve margin by offsetting generation that would otherwise be needed. Energy Efficiency also improves T&D reliability by increasing available capacity in both low-voltage and high-voltage systems. To defer costly T&D upgrades, some utilities have used geographically targeted energy efficiency programs to reduce strain (NEEP, 2017).

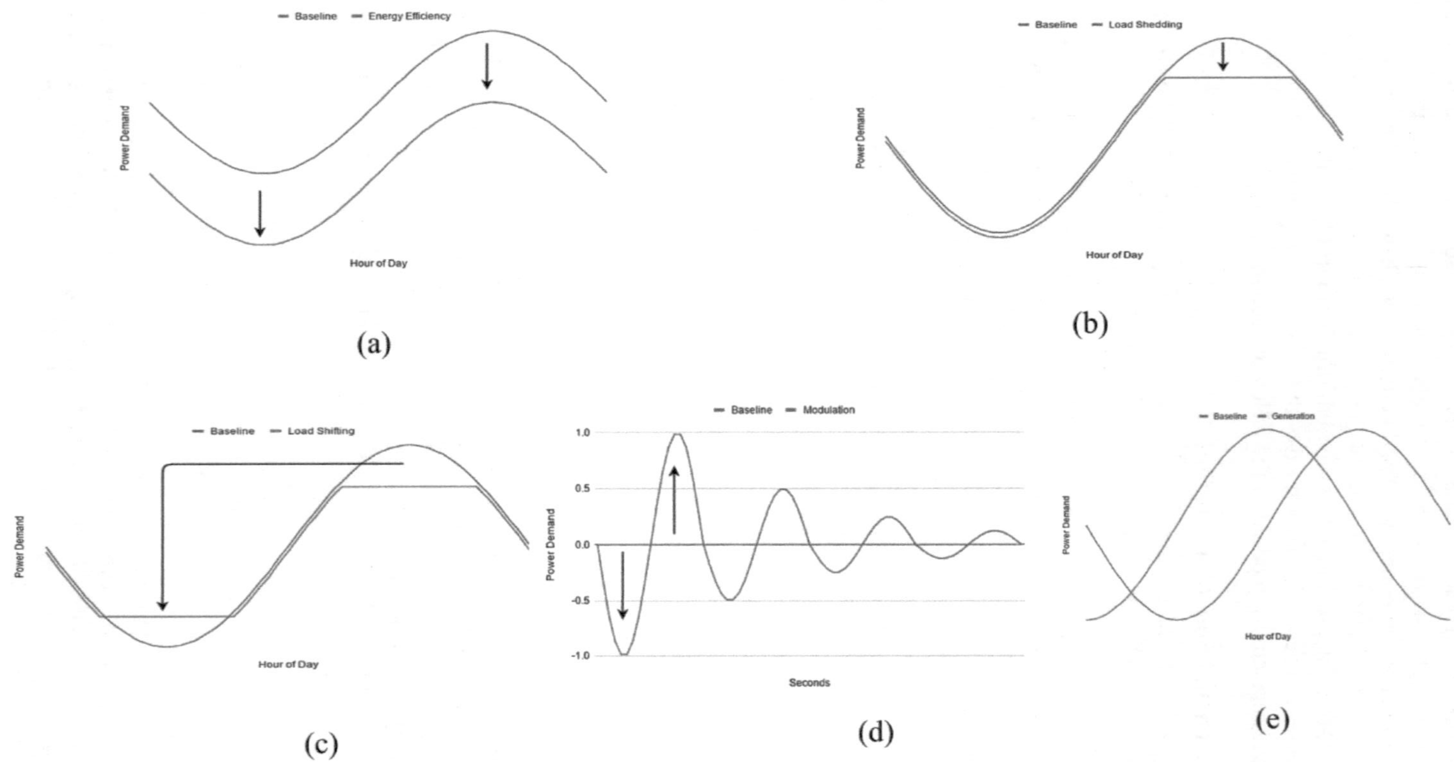

FIGURE 14.1. Changes in building energy demand profiles for (a) energy efficiency, (b) load shifting, (c) load shedding, (d) modulation, and (e) generation.

TABLE 14.1
Summary of Grid Services Associated with GEBs

Grid Service Category	Grid Service	Energy Efficiency	Load Shedding	Load Shifting	Modulation
Generation	Energy	●	●	○	○
	Capacity	●	●	●	○
Delivery	Non-wires alternatives	●	●	●	○
	Voltage support	○	○	○	●
Ancillary	Contingency reserves	○	●	●	○
	Frequency regulation	○	○	○	●
	Ramping	○	○	○	●

From the demand-side perspective, the value that demand reduction due to energy efficiency provides is a function of the amount, timing, and location of the savings. From the supply-side perspective, the value of energy efficiency is a function of the system's characteristics such as the peak demand timing (such as seasonal and diurnal on-peak periods), load factor, and reserve margin. For example, load reductions that occur during times of on-peak demand are more valuable than reductions that occur during off-peak times. From the T&D perspective, an important energy efficiency benefit is reduced marginal line losses, which are higher during times of peak demand due to resistive losses that scale non-linearly (i.e., vary with the square of current according to Joule's first law). These marginal losses must be covered by a utility's generating reserve, which makes on-peak energy efficiency more valuable.

2. *Demand response*: In contrast to energy efficiency, which focusses on reducing energy consumption, DR aims at lowering peak power requirements through the timing of energy use. DR is also the most common form of demand flexibility currently in use in commercial and residential buildings. From the perspective of building owners, DR can provide reduced utility costs by avoiding high peak demand charges and providing incentives for reducing demand during peak periods. From the perspective of the grid, DR can provide increased reliability by reducing operating costs and deferring or avoiding capital upgrade costs. DR can be classified into two types, dispatchable and nondispatchable, depending on who initiates the response to a peak demand event. Dispatchable DR (incentive-based) is initiated by the utility, grid operator, or third-party aggregator in order to directly control building systems to reduce demand during a peak event. Nondispatchable DR (price-based) is initiated at the building in response to price signals (Paterakis et al., 2017).

14.2.2 GEB Technologies

GEB technologies are organized from a building design perspective according to design disciplines as summarized in Table 14.2. Each technology is described briefly in the context of its ability to contribute to DSM strategies and thus provide services to the grid and ultimately be suitable for GEBs.

TABLE 14.2
Summary of Technologies by Category

Building Envelopes	Electrical Systems	HVAC Systems	Control Strategies
Thermally anisotropic systems	SSL displays	Liquid desiccant thermal energy storage	Smart thermostats
Thermal storage	Continuous-operation electronics	Hybrid evaporative precooling for AC	Advanced controls for HVAC equipment with embedded thermostats
Moisture storage and extraction	Battery-powered electronics	Dual-fuel HVAC systems	Water heaters with smart connected controls (electric and gas)
Variable radiative technologies	Electronic displays	Thermal energy storage	Advanced dishwasher/ clothes washer controls
Building-integrated photovoltaics	Modulating, advanced clothes dryers (electric)	Modulating capacity vapor compression	Advanced residential refrigerator/freezer controls
	Building-scale CHP	Dual fuel water heaters	
Dynamic glazing		Non-vapor-compression materials and systems	Advanced controls for commercial refrigeration

An overview of the various GEB technologies outlined in Table 14.1 is provided in the following sections.

14.2.21 Building Envelope Systems

Building envelope systems that are suitable for GEBs include systems that separate the outdoor environment from the conditioned indoor environment (collectively, they are specific to the building envelope). These systems can be divided into the opaque envelope (floors, roofs, walls) and windows, and further categorized as having static, dynamic, or both properties that provide grid services through DSM strategies. Architectural technologies affect cooling, heating, and lighting energy end uses by decreasing assembly thermal transmittance (U-factor), decreasing, or increasing solar heat gain through windows, or by allowing or blocking visible light to interior spaces.

1. *Static systems*: Static building envelope systems do not have dynamic or time-varying properties relying instead on inherent high-performance properties that manage heat transfer, thermal bridging, and air flow (infiltration) through the system assembly. These properties can be a result of a single system component that contributes to the assembly's overall performance, such as a layer of insulation with high thermal resistance (R-value), or the result of multiple system components, such as reduced thermal bridging and air flow. Because static envelope systems do not have dynamic properties, they are primarily an energy efficiency strategy, although they can passively shift load by delaying peak cooling and heating loads compared to lower performing systems. In some cases, a high-performance envelope (i.e., with

high thermal resistance, for instance) can be detrimental to energy use by trapping heat and increasing cooling load. Static architectural systems are generally not suitable for GEBs because they don't have dynamic properties enabling them to interact with the grid.

2. *Dynamic systems*: Dynamic building envelope systems have properties that can be actively modified to decrease or increase their performance to achieve desired DSM strategies and grid services. These technologies can be categorized as either opaque or window systems and include the following technologies (DOE, 2019b).

Opaque Systems

- Tunable Thermal Conductivity Materials
- Thermally Anisotropic Systems
- Thermal Storage
- Moisture Storage and Extraction
- Variable Radiative Technologies
- Building-Integrated Photovoltaics

Window Systems

- Dynamic Glazing
- Automated Attachments
- Photovoltaic Glazing

Tunable thermal conductivity materials have thermophysical properties that can be dynamically adjusted to produce a desired thermal performance. For example, during cooling periods these materials would have high thermal transmittance (low R-value) when the outdoor temperature is lower than the indoor temperature to precool the building effectively acting like night flushing without the need for mechanical ventilation. Conversely, in the heating season these materials would have low thermal transmittance (high R-value) to minimize heat loss to the outdoors. Several tunable insulation technologies have been reported in the literature and include dynamic insulation materials and adaptive insulation systems. In particular, dynamic insulation materials have been found to reduce annual cooling and heating energy use and cost in commercial buildings by an average of 17% and by up to 42% in residential buildings (Park et al., 2015; Shekar and Krarti, 2017; Menyhart and Krarti, 2017; Rupp and Krarti, 2019; Dehwah and Krarti, 2020).

Thermally anisotropic systems include components and assemblies with areas of high and low thermal conductivity that allows heat to be routed through the envelope to a heat sink such as a plumbing loop. These systems can achieve cooling and heating load reductions of 86% and 63% compared to a baseline wall assembly with traditional cavity insulation, and numerical simulation results showed cooling energy savings of 11% in cooling-dominated climates and heating energy savings of 21% in heating-dominated climates (Steen and Krarti, 2020).

Thermal storage materials have variable heat capacity properties that allow them to release or store heat. They behave similarly to passive thermal mass, such as strategically adding materials with high heat capacity, but with less volume and weight. Phase change materials (PCMs) are a type of thermal storage material that passively charge and discharge. PCMs have been widely reviewed and are currently available but are not widely used in buildings. However, PCMs have high potential to offer grid services and have been shown to reduce heating energy use by up to 63% in experimental studies (Nghana and Tariku, 2016).

Moisture storage and extraction systems use materials that can extract moisture from the indoors, store it, and then reject it to the outdoors. In humid climates and buildings with high internal latent loads, moisture control is a significant portion of the cooling load which drives peak electrical demand. Energy savings between 7% and 19.6% have been reported for phase change humidity control and hygroscopic materials (Wu et al., 2019).

Variable radiative technologies include materials that can reject heat during the daytime or even in direct sun. Cool roofs are a well established strategy to reduce cooling loads in warm/hot climates that are not appropriate for colder climates because they do not decrease energy use. These strategies are currently passive, but if they could be dynamically controlled to change their radiative heat transfer properties they could provide additional grid services through expanded DSM functions. A recently developed metamaterial showed noontime radiative cooling of 93 Watts per square meter under direct sun (Zhai et al., 2017).

Building-integrated photovoltaics (BIPV), although part of a building's electrical system, refer to PV integrated into the opaque portions of the envelope. BIPV applications include semi-transparent systems (skylights, exterior shading attachments), cladding systems (walls and curtain walls), solar tiles and shingles, and flexible laminates. Sehar et al. (2016) found that PV alone could reduce a building's demand during a DR event by 16.8%.

Dynamic glazing technologies include electrochromic glazing and thermochromic glazing that change tint (solar heat gain coefficient, SHGC) allowing more or less solar radiation into the building. Electrochromic glazing has more than one tint state (SHGC) compared to non-dynamic glazing, that can be changed by applying electric voltage to an electrochromic layer. Thermochromic glazing passively responds to temperature to change its SHGC. For commercial buildings located in the northern hemisphere, electrochromic glazing is estimated to reduce peak demand by 20%–30% for east, south, and west thermal zones with an estimated primary energy savings of 10%–20% (Steen and Krarti, 2020).

Automated attachments include exterior and interior devices that open or close to allow or block solar radiation into the building. Examples include interior devices such as blinds and shades, and exterior devices such as awnings or shutters. Tzempelikos and Shen (2013) found that interior shades controlled with a fixed incident solar radiation set-point reduced cooling loads by 15%–33% depending on the building's window-to-wall ratio. Dynamic exterior shading systems have been reported to result in energy savings up to 36% for commercial buildings.

Photovoltaic glazing generates electricity using the same mechanisms that traditional photovoltaic modules, but are semi-transparent or transparent allowing some portion of visible light to pass through. Thus, PV glazing integrates an electricity generating system and a window system. An optimized PV insulating glass unit has shown to provide 10.7% energy savings when compared to a traditional low-e IGU in whole-building energy simulations (Wang et al., 2016).

14.2.2.2 Electrical Systems

Electrical technologies that are suitable for GEBs include lighting systems, consumer electronics, and information technology (IT) equipment.

1. *Lighting*: Currently, the use of lighting for grid services is low, and estimated to represent 4% of commercial buildings reported demand responsive lighting (EIA, 2012). Historically, lighting systems have not been used for DR because they are not inherently capable of shifting their load. Moreover, lighting is generally considered a critical load for occupied spaces to enhance safety, productivity, and comfort so it cannot be completely turned off. The primary DR strategy of dimming lighting is restricted to a small range that will not compromise safety, productivity, and comfort of building occupants. Lighting is only valuable to the grid at the aggregate whole-building level because the average power of a single lamp is small relative to other individual loads, therefore the commercial building sector is the most viable market for grid-responsive lighting systems. In addition to DR, lighting systems can be used for DSM by reducing building peaks, which is typically accomplished by dimming to shed load. The following lighting technologies are suitable for GEBs (Steen and Krarti, 2020).
   - Hybrid Daylight SSL Systems
   - SSL Displays

   Hybrid daylight solid-state lighting (SSL) systems collect and redistribute daylighting to enhance their ability to provide light to spaces. These technologies include devices that concentrate daylight through the use of mirrored lenses, beam splitters that filter nonvisible light, and light guiding and diffusing systems that redistribute light to the building (i.e., fiber optic cables). These systems are connected by photosensors and controllers that automatically modulate electric lighting in real-time. Hybrid daylight SSL systems primarily provide the grid with efficiency but can also provide load shedding and modulation. These systems have been available for many years but have been adopted slowly by the market due to the high cost of fiber optic cable, low light transfer efficiency, and installation difficulty. Mayhoub and Carter (2012) found that hybrid solar lighting saved 33% of electricity compared to a base case.

   SSL displays are connected lighting displays that use electric lighting to replace natural light harvested from windows and skylights. These systems mimic the direct and diffuse light from the sun and can increase a building's envelope performance if their thermal performance is better compared to the windows and skylights that they replace. SSL displays can provide some DR capabilities, mainly by shedding load during peak periods through modulating light output and other energy consuming components. The market for this technology is currently small because of limited applications (i.e., only for spaces without windows or skylights), high costs, and customer preference for windows.
2. *Appliances and equipment*: Electronic equipment including consumer electronics and IT equipment technologies have the potential to provide grid

services. These devices are a subset of miscellaneous electric loads (MELs), which represent electrical loads other than those related to core building functions such as lighting, heating, ventilating, and air conditioning (HVAC), or service water heating (SWH). Historically, energy efficiency has focused on non-MELs loads because they have made up the largest proportion of energy use. However, as the efficiency of these end uses has improved, the proportion of energy use by MELs has increased in total energy use and percent of total. Current DR programs tend to focus on HVAC systems service commercial buildings or large industrial equipment not electronics, which are more typically the target of DSM programs that prescriptively improve energy efficiency. Additionally, electronic equipment does not typically include technologies that allow for automated DR. Moreover, electronics can have significant variation in their energy use profiles unlike other building loads with recommended diversity factors of 51%–75%, which limits their ability to provide DSM strategies. The primary market for grid-interactive electronics is large commercial office buildings and industrial data centers where a significant portion of the building's energy use is attributed to this end use. The following electrical equipment technologies are suitable for GEBs (DOE, 2019c).

- Continuous-Operation Electronics
- Battery-Powered Electronics
- Electronic Displays
- Modulating, Advanced Clothes Dryers (electric)

Continuous-operation electronics include stationary equipment that generally operate continuously and require a constant power supply. Examples of continuous-operation electronics include network equipment, stationary (desktop) computers, servers, and AV equipment. This type of electronic equipment generally uses a constant amount of energy, but some devices can operate in low power modes. This category of electronics has the highest relative potential to provide grid services because the equipment is constantly connected to a power supply and often operating continuously. Energy efficiency improvements offer the greatest DSM opportunities by integrating low power, standby, deep sleep, or power scaling modes. Staging the operation of this equipment also has the potential to reduce building peak demands. Jenkins et al. (2019) found a 58% reduction in average hourly electricity consumption for concession equipment from centralized controls that turned off equipment during non-event days. Jenkins et al. (2019) also found a 56% reduction for IT equipment using centralized controls. Additional grid services could include load shedding and modulation of grid-responsive servers in data centers and offices.

Battery-powered electronics include equipment with an internal rechargeable storage unit that requires a power supply to periodically recharge. Examples of battery-powered electronics include portable (laptop) computers, uninterruptible power supplies (UPS), and miscellaneous electronics such as mobile phones. The energy use of this type of equipment ranges from cycling in the case of a laptop computer to always on in the case of a UPS. This type of equipment has moderate relative potential to provide grid services because these devices are generally portable (disconnected from the grid), do not operate continuously, and generally are a smaller portion of total building loads. Additionally, these devices are usually energy efficient because they are designed to maximize battery life further limiting their potential. However, grid-responsive power supplies for these devices could provide load shifting to benefit the grid and building peak demand, especially UPS battery backups because they are usually always on and require more power than other devices in this category.

Electronic displays include monitors that are used for signage (i.e., flight information display systems), computers, and televisions, which can operate at different modes including off (sleep), low power, and always on. This category of electronics has low relative potential for grid services because they generally represent a very small portion of overall energy use in a building. However, there are opportunities for additional energy efficiency through dimming, automatic brightness control, and occupancy controls.

Appliances are an additional category of electric equipment that can provide grid services. For this review, natural gas appliances are discussed in the Plumbing section. Similar to consumer electronics and IT equipment, appliances are heterogeneous in load profiles and operation schedules. For example, some appliances operate relatively continuously (refrigerators) while others have finite cycles (dishwashers and clothes dryers). Appliances that operate continuously are more likely to benefit the grid by modulating their load. Appliances with finite cycles are most suitable for load shifting because their load can be easily rescheduled to avoid peak periods. These types of appliances have traditionally been targets for DR programs. For clothes dryers, technologies that allow for modulating or staging the heating cycle and delayed start options could be used to provide grid services for energy efficiency and load shifting respectively. Moreover, heat pump clothes dryers offer additional efficiency improvements over traditional electric resistance heating and can also be modulated when they incorporate variable speed compressors. Modulating the heating element allows the moisture removal rate to match the heat input rate more closely, which reduces over drying and improves energy efficiency. Longer cycle times with lower temperature drying allows for better energy efficiency along with load shifting and some load shedding. The highest grid service potential is modulation because electric resistance appliances can respond very quickly (i.e., within seconds) without damage (Steen and Krarti, 2020).

Combined heat and power (CHP), which combines electricity production with waste heat capture and use, can provide grid flexibility by serving on-site electric loads or exporting electricity to the grid while also serving on-site thermal loads (heating directly or cooling through absorption technology). These systems are often used by large buildings and campuses to reduce operating costs by avoiding high consumption and demand charges of grid-tied electricity. CHP systems can also be used with TES to shift loads to off-peak periods. Verda and Colella (2011) reported heating load reductions up to 33%. Operators of these systems can also adjust their dispatch schedule to take advantage of day-ahead and real-time electricity pricing in some markets as well as participate in capacity, energy, and DR markets. This technology has high GEB potential through energy efficiency by producing electricity on-site and avoiding losses, load shedding if able to increase production during a grid event, and load shifting when combined with TES (Nuytten et al., 2013)

14.2.2.3 HVAC Systems

Mechanical systems comprise the HVAC and refrigeration (HVACR) systems used to maintain thermal comfort and indoor air quality in buildings. These systems often make up a large portion of energy use in commercial buildings. The following HVACR technologies are suitable for GEBs (DOE, 2019d).

- Separate Sensible and Latent Space Conditioning
- Liquid Desiccant Thermal Energy Storage
- Hybrid Evaporative Precooling for AC
- Dual-Fuel HVAC Systems
- Thermal Energy Storage
- Modulating Capacity Vapor Compression
- Non-Vapor-Compression Materials and Systems
- Dual Fuel Water Heaters

Traditional heating and air conditioning systems couple sensible (temperature) and latent (moisture) control into the same component, that is a vapor-compression (direct expansion or DX) cooling system. These systems often have enlarged evaporators, operate at a lower temperature, or extend the operating cycle to remove moisture from the air stream, which can overcool the supply air below temperatures that are typical or comfortable resulting in the need to reheat the air before it is delivered to the space. Overcooling and reheating consequently increase energy use and demand during the cooling season. Decoupled sensible and latent air conditioning systems using liquid or solid desiccants, membrane dehumidifiers, and other technologies can remove moisture from the air without changing its temperature. Independently controlling sensible and latent cooling stages could provide grid flexibility by shifting from less efficient sensible cooling to more efficient latent cooling during peak periods. Decoupled sensible and latent space conditioning has high potential to provide grid services by primarily benefiting energy efficiency, but it can also provide load shedding and shifting opportunities (DOE, 2019d). Research evaluating this technology in packaged terminal air conditioners suggests that efficiency savings of 30%. Moreover, Kim et al. (2014) found that liquid desiccant systems incorporated with evaporatively-cooled dedicated outdoor air system (DOAS) saved from 51% to 68% of annual energy use compared to traditional variable air volume (VAV) systems.

Liquid desiccant thermal energy storage (TES) uses a chemical that absorb moisture from indoor air and then rejects it to the outdoors through a heating cycle known as regeneration. This type of TES stores energy chemically and does not require insulated storage tanks because liquid desiccants can be stored at ambient temperatures. The efficiency of these systems is not driven by round-trip efficiency losses as with other energy storage technologies (including thermal and electrical systems), but it is rather a function of the heating needed for the regeneration process. Solar thermal coupled with this technology can be an effective way to regenerate the liquid desiccant without the need for additional energy input because peak solar radiation and latent cooling often occur coincidentally during the day. This technology has high potential to provide grid services by providing load shifting during periods of high demand (DOE, 2019d). Load shed and efficiency have less potential because the desiccant will always need to be recharged later. Efficiency can be improved by using renewable energy to regenerate the desiccant.

Hybrid evaporative precooling for air conditioning (AC) combines evaporative cooling with vapor-compression cooling to increase AC energy efficiency in dry climates by shifting cooling from the high energy intensity vapor-compression cycle to a lower energy intensity evaporative process, thus reducing peak cooling demand. Evaporative modules can be packaged with vapor-compression systems or added to existing systems during a retrofit. This technology has low potential to provide grid services, primarily through efficiency by improving the cooling system's coefficient of performance. Delfani et al. (2010) found cooling load savings up to 75% and annual energy savings of 55% for indirect systems that pre-cool air for traditional mechanical cooling in Iran.

Dual-fuel HVAC systems temporarily switch fuels during heating or cooling to provide value to the grid through curtailment. Specifically, grid value is provided when the lower cost fuel is primarily used throughout the year and the more expensive fuel is only used during limited grid events. For example, in regions that experience electric peaks during winter due to the prevalence of electric heating, a heat pump system that switches from electric heat pump heating to natural gas heating during a peak event to shift demand to the natural gas grid. Similarly, absorption cooling, which uses a heat source to provide cooling, can also be used to shift cooling fuel from electricity to natural gas using gas-absorption cooling. Currently, the low cost of natural gas makes using it for electric curtailment infeasible. However, in areas where costly delivered fuel is used for heating (i.e., fuel oil or propane), using it to curtail electric heating can be more cost-effective, especially since the delivered fuel and electricity markets are not interactive unlike the natural gas and electricity markets which are often related. This technology has low potential to provide grid services by shedding load because the switch opportunities are infrequent, mostly during winter peak periods, and are most applicable when switching from heat pump heating to natural gas or delivered fuel heating. Low natural gas prices are a significant hindrance to the adoption of dual-fuel HVAC systems. Moreover, electrification of buildings, considered as a key strategy for decarbonization, may hinder the future adoption of dual-fuel HVAC systems (DOE, 2019d).

TES ranges from passive systems that use additional thermal mass or insulation within a building's envelope to active systems that rely on phase change media to store and release energy in an optimized sequence. These systems can be used for cooling-only, heating-only, or both cooling and heating. Active TES systems cool or heat materials to store energy during low-cost periods so that it can be used during high-cost periods with less energy input. The round-trip efficiency of these systems is typically 80% because of thermal losses to the surroundings. This technology has high potential to provide grid services primarily through daily load shifting to reduce demand charges. Load shedding is possible when the shift in energy use allows equipment to operate at a higher efficiency. Kamal et al. (2019) found annual average peak load shifting of 25%–78% for a large office model in a hot humid climate.

Modulating capacity vapor-compression systems can be used in ACs, heat pumps, heat pump water heaters (HPWHs), and heat pump dryers to more precisely control temperatures compared to systems that cycle all the way on and off. This technology has medium potential to provide grid services. It primarily offers energy efficiency improvements as well as thermal comfort, but it can provide load shifting by pre-cooling or preheating the building prior to a peak period thus shifting the load forward allowing the building's thermal inertia to carry it through the peak period. Additionally, a combination of load shifting, and shedding could be provided by changing set-point temperatures without precooling or preheating during a peak period, which sheds load and shifts it to a later time when the set-points return to their original states. This technology could also provide modulation services by providing frequency regulation and voltage support through power factor control using power electronics for variable speed drives (Steen and Krarti, 2020).

Non-vapor-compression (NVC) materials and systems include several space cooling and refrigeration technologies that use specialized materials or alternative system designs rather than the traditional vapor-compression cycle. These include solid-state NVC technologies such as thermoelectric, magnetocaloric, and electro-caloric systems that produce useful temperature differences when the solid-state material is activated by electrical input (Steen and Krarti, 2020). Other technologies include membrane, thermoelastic, Stirling, liquid desiccant, and thermoacoustic systems that use electrical or thermal input to change the property of a material (i.e., working fluid) to pump heat. These technologies could offer grid benefits by modulating capacity, separating sensible and latent cooling, and thermal storage. This technology has high potential to offer grid services primarily through energy efficiency and load shifting. However, load shedding and modulation are possible with variable capacity control. Cheon et al. (2019) indicated that a thermoelectric heat pump in a DOAS increased energy use by 23% compared to a reference system while Lim and Jeong (2018) found that a thermoelectric radiant system with a DOAS reduced annual energy consumption by 41% compared to a traditional VAV system.

Water heaters generally fall into two categories, storage and tankless. Tankless water heaters have limited potential to provide grid services because they are designed to operate on-demand and thus are not able to shift or shed load. In contrast, storage water heaters are well suited for GEBs because TES is inherently part of their fundamental operation, which allows them to decouple power demand from energy consumption and provide load shifting away from peak periods (Steen and Krari, 2020). Specifically, dual-fuel water heaters can temporarily switch fuels to provide value to the grid through curtailment. Like dual-fuel HVAC systems, these systems provide grid value when the lower cost fuel is mostly used throughout the year and the more expensive fuel is only used during limited periods. Markets or regions that use electricity and delivered fuels have the highest potential for these systems because delivered fuel costs are generally higher than electricity. Generally, this technology has low potential to provide grid services because it can only provide load shedding opportunities in limited viable markets due to the low cost of natural gas compared to electricity and regional differences in delivered fuel use. Given financial incentives this technology can provide load shedding for infrequent emergencies, but the electric grid will not benefit if natural gas costs less. Park et al. (2014) reported energy cost savings of 4% in one market for a hybrid gas fired HPWH.

14.2.2.4 Controls

A key part of GEBs are the control systems that connect the discrete systems within a building and allow for "smart" operation through sensors, actuators, and controllers. Advanced control systems have the potential to reduce site energy consumption by 29% in commercial buildings through the use of high-performance sequences of operation, occupancy-based optimization, and automated fault detection and diagnostics. Additionally, advanced control systems can reduce peak demands of commercial buildings by 10%–20%. Control systems cross multiple building systems and applications. The following controls technologies are suitable for GEBs (DOE, 2019d).

- Advanced Sensors and Controls (lighting)
- Smart Thermostats
- Advanced Controls for HVAC Equipment with Embedded Thermostats
- Water Heaters with Smart Connected Controls (electric and gas)
- Advanced Dishwasher/Clothes Washer Controls
- Advanced Residential Refrigerator/Freezer Controls
- Advanced Controls for Commercial Refrigeration
- EV Charging Management
- Energy Management and Information Systems (EMIS)

Advanced sensors and controls improve the ability of connected lighting systems to adjust their power-consuming features such as light levels and spectrum, sensors, or network interfaces through embedded control algorithms. These technologies would enable lighting systems to interact with the grid and other building-level sensors and controllers to reduce loads and provide the grid with contingency reserves and frequency regulation. The market for these technologies is expected to grow significantly in the commercial building sector, with one estimate predicting 35% penetration by 2035 (Penning et al., 2016). Snyder (2020) found simulated energy use savings of 43% for luminaire-level lighting controls compared to manual controls in open offices.

In addition to sensing the zone conditions and controlling the attached HVAC system, smart thermostats are connected to the internet, contain advanced control algorithms, and can connect to home automation systems. These features allow smart thermostats to perform more advanced controls for relatively simple HVAC systems. Smart thermostats are best suited for residential buildings and small commercial buildings where they can serve as a less complicated alternative to smart home energy management systems and building automation systems. Wang et al. (2020) estimated that smart occupancy-driven thermostats could save between 11%–34% of energy use.

Advanced controls for HVAC equipment with embedded thermostats include HVAC equipment with internal built-in sensors and control algorithms rather than external wall-mounted thermostats. Embedded thermostats are included in window ACs, portable ACs, packaged terminal ACs and heat pumps, and mini-split heat pumps where Wi-Fi is used to communicate directly with grid operators or utilities for DR programs. These controls allow the compressor to be temporarily turned off or the set-point to be changed with user overrides.

Smart water heaters include internal or external controls that take advantage of the equipment's inherent TES capability. Preheating the tank during off-peak periods allows for water draws with reduced or no power use during on-peak periods, shifting the water heater's load to benefit the grid without the loss of functionality to the consumer. Load shedding can be accomplished by turning the water heater off for emergency curtailment. Frequency regulation is also possible with water heaters that use electric resistance heating (solely or HPWH) and that allow direct utility control for the fast response time required for frequency regulation. However, frequency regulation may reduce the life of vapor-compression equipment. Pourmousavi et al. (2014) found that controlling electric water heaters with time-of-use (customer) and balancing reserve (utility) signals reduced on-peak energy use by 95%.

Clothes washers and dishwashers with advanced controls to delay or schedule their discrete cycles could shift loads to off-peak periods but require consideration of user preferences. For example, leaving wet clothes in a washing machine too long could cause concerns, but combined clothes washer-dryer is one technology that could alleviate concerns. This technology has medium potential to provide load shifting. Finn et al. (2013) found peak time load reduction between 28–70% for residential dishwashers under a control algorithm to optimize cost, wind generation, and carbon emissions. Perez et al. (2016) indicated that smart scheduling of washing machines and clothes dryers through model predictive control reduced peak electric demand by 5%.

Residential refrigerator and freezer controls can provide load shifting for the electric grid through low operation mode, defrost cycle delay, and freezer precooling. Low operation mode works by shutting off the compressor and anti-sweat heaters in response to a grid signal or by the consumer to reduce peak demand charges. During a grid event, delaying the defrost cycle can reduce the refrigerator's load. Freezer precooling works by overcooling the freezer prior to a curtailment period or during a scheduled peak load reduction period and then modulating the airflow from the freezer to the refrigerator to maintain its set-point. These technologies have low potential to provide load shifting. Farzamkia et al. (2016) reported that refrigerator peak load can be shifted by about 11% by precooling strategies and by over 40% with optimized algorithms.

Similar to residential refrigerators, commercial refrigeration equipment could benefit the grid using low operation mode, defrost cycle delay, and freezer precooling. However, in a commercial building such as a supermarket or refrigerated warehouse, these technologies can be staggered across multiple pieces of equipment to provide additional benefits. These technologies have high potential to offer grid services primarily through load shifting using scheduled precooling. Some load shedding is possible with corresponding load shifting to bring the equipment back to set point following the grid event. Energy efficiency gains are possible with smart controls, which can benefit owners through lower operating costs but with little benefit to the grid because the savings will most likely occur during off-peak times. Glavan et al. (2019) indicated that a peak load reduction of 18% can be achieved for commercial refrigeration in supermarkets due to precooling.

With higher adoption of electric vehicles (EV), smart chargers within buildings are needed to enhance demand flexibility and achieve efficient and reliable grid operation (Heredia et al., 2020). Various control techniques and scheduling algorithms have been proposed for peak shaving and valley filling of EV charging.

A field-testing of EV load management algorithms has indicated that up to 29% reduction in peak charging load demand can be achieved for a building if it has an on-site PV generation. Development of smart EV charging controllers are needed for future EV-to-building integration deployment projects.

Using monitored data obtained from energy management and information systems (EMIS), analytics and machine learning based tools can provide significant insights to enhance the energy efficiency of buildings. While there is no consensus on their definition and their categorization, EMIS include a set of tools that monitor, analyze, and control building energy consumption and end-uses (Granderson et al., 2015). They are typically categorized as energy information systems (EIS), fault detection and diagnostic (FDD) systems, and automated optimization systems (AOS). A study, based on a comprehensive set of US buildings retrofitted, has indicated that EMIS can be highly cost-effective resulting on average 13% energy savings including 9% associated with FDD tools after two years of operation (Kramer et al., 2019).

14.2.3 Energy Performance of GEB Technologies

Based on a literature review, Table 14.3 summarizes the reported energy efficiency, demand shifting, and load shedding of various GEB technologies (Steen and Krarti, 2020). While several potential GEB technologies have been discussed, most of the smart systems are at the early stage of research development. Additional evaluation and demonstration studies are needed to assess the true potential of GEB

TABLE 14.3
Reported Energy Performance for GEB Technologies

Category	Technology	Energy Efficiency	Demand Shedding	Load Shifting
Architectural	Thermal storage	63% (heating)	-	-
	Dynamic glazing	10%–20%	20%–30%	-
	Automated attachments	-	15%–33% (cooling)	-
Electrical	Continuous-operation electronics	56%–58%	-	-
Mechanical	Separate sensible and latent space conditioning	30%–68%	-	-
	Thermal energy storage	12%	-	25%–78%
Plumbing	Building-scale CHP	33% (heating)	-	-
Controls	Advanced sensors and controls (lighting)	43%	-	-
	Smart thermostats	11%–34%	-	-
	Water heaters with smart connected controls (electric and gas)	95% (on-peak)	-	-
	EV charging management	-	10%–30%	-

Note: Energy and demand savings are expressed in percent relative to the total building energy use/peak demand or relative to energy/demand end-use noted between parentheses.

technologies as well as to aid at their design and prototyping for both residential and commercial buildings.

14.3 NET ZERO ENERGY BUILDINGS

Over the last decade, various definitions and concepts of low energy or zero-energy buildings have been applied to design new buildings. The design procedures often utilize detailed simulation analysis tools coupled with optimization techniques. These same concepts are only now being extended to existing buildings to select the most cost-effective energy efficiency measures that minimize initial costs and maximize energy savings. In particular, the optimization approach entails typically the evaluation of various options until a set of energy efficiency measures and renewable energy technologies is identified that achieves minimum cost and/or maximum energy savings, zero-energy, or carbon-neutral conditions. While exhaustive enumeration can be performed, in which every possible combination of possible energy efficiency measures is evaluated, optimization techniques are often utilized to minimize the number of combinations that need to be evaluated and analysed.

Typically, zero net energy (ZNE) is defined either in terms of site energy or source energy. Site energy consists of energy produced and consumed at the building site. Source or primary energy includes site energy as well as energy used to generate, transmit, and distribute this site energy. Therefore, source energy provides a better indicator of the energy use of buildings and their impact on the environment and the society and thus is better suited for ZNE building analysis. Indeed, an analysis based on source energy effectively allows different fuel types, such as electricity and natural gas commonly used in buildings, to be considered and combined.

It is generally understood that ZNE buildings produce as much energy as they consume on-site annually. These buildings typically include aggressive energy efficiency measures and active solar water heating systems. Moreover, ZNE buildings employ grid-tied, net-metered renewable energy generation technologies, generally photovoltaic (PV) systems, to produce electricity. Effectively, ZNE buildings use the grid as "battery storage" to reduce the required generation system capacity.

Over the last two decades, few tools have been developed for selecting optimal packages of energy efficiency measures for residential and commercial buildings. Most of these tools are for designing new buildings and are based on the sequential search technique. The sequential search methodology was first applied for optimizing the energy efficiency level of the residential building sector (Meier, 1982). Then, the methodology was implemented to a spreadsheet-based analysis tool (DEG, 1993) and to the EnergyGauge-Pro software (FSEC, 2007). Recently, the sequential search methodology was used to find a path to design zero-energy homes using a building energy simulation-based computer program referred to as the building energy optimization or *BEopt* tool (Christensen et al., 2004). Starting from a reference building model, *BE*opt identifies a cost-optimal set of building energy options needed to achieve a variety of energy saving levels including the ZNE option (i.e., 100% energy savings). The building energy options are chosen from predefined or customized efficiency and renewable energy measures. Energy savings for each measure are calculated relative to a reference building on a source energy basis.

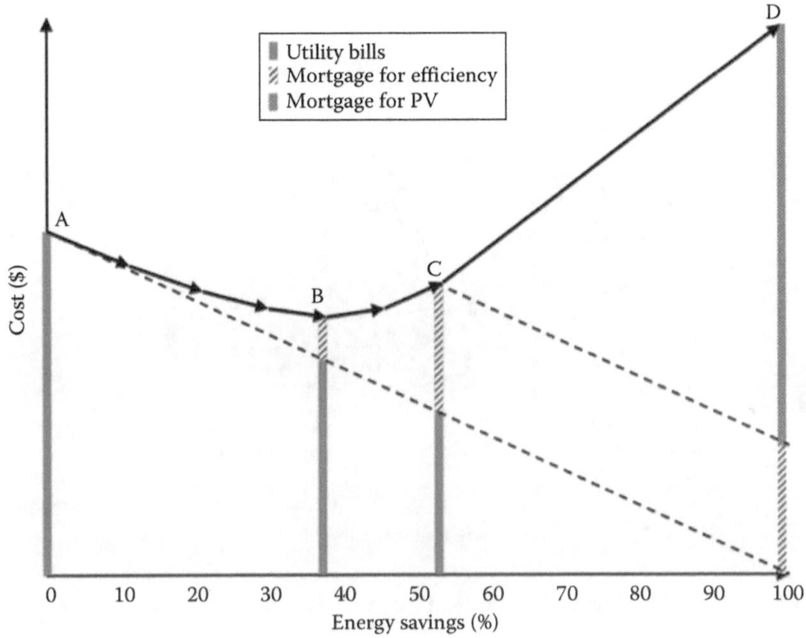

FIGURE 14.2 Pareto graph of the optimal path to ZNE for a building.

Originally, *BEopt* was used primarily to optimize design options for new homes. It can now be used for optimizing the energy retrofit options for existing residential buildings even though automatic calibration capabilities are not available in the current version of the tool. A conceptual plot of the cost/energy-saving Pareto graph from *BE*opt is illustrated in Figure 14.2. The *x*-axis in Figure 14.2 is the source energy savings accounting for any energy used by the building while the *y*-axis represents the energy-related costs, including the energy use cost and the capital cost for implementing any energy efficiency and/or renewable systems. At the starting point of the optimization, point A represents the baseline building conditions. As energy efficiency measures are introduced into the building, incremental mortgage costs increase and decrease until the marginal cost of saved energy equals the cost of utility power. At this point, the curve reaches a minimum and the global cost-optimal point is reached (point B). Additional efficiency measures with marginal costs more expensive than the cost of fuel are introduced until the marginal cost of saved energy equals the marginal cost of producing PV energy (point C). At this point, PV capacity is added until all source energy use is offset (point D). For building roofs where there is a limit to the number of PV panels that can be installed, additional energy efficiency measures are employed after the PV system selection until ZNE is reached. These additional energy efficiency measures may have a marginal cost of saved energy larger than that of the produced energy from PV. If all energy efficiency measures in the optimization search space are exhausted before ZNE is attained, the optimization stops at its maximum possible energy savings level.

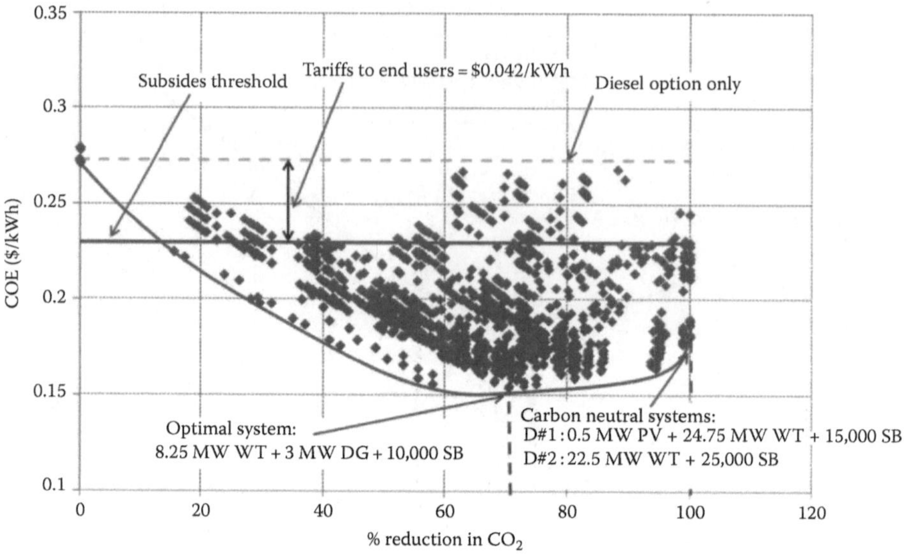

FIGURE 14.3 Pareto graph of the optimal path to carbon-neutral design for a community.

Similar analysis can be carried out to optimize hybrid distributed electricity generation technologies for large communities and urban centers as shown in Figure 14.3. Instead of the energy-related costs in the *y*-axis, the cost of energy (COE) is considered, while the *x*-axis indicates the percent savings in carbon emissions relative to the baseline design.

Other optimization techniques and algorithms have been considered to optimize building energy use (Wright et al., 2002; Wetter, 2004; Tuhus-Dubrow and Krarti, 2009). Some of these optimization techniques are briefly presented in the following section.

14.4 OPTIMIZATION APPROACHES

Inherently, high-performance building designs and retrofits consider several energy efficiency measures and renewable energy technologies and thus are multivariate. In particular, the energy efficiency measures encompass a wide range of options including building envelope insulation, HVAC (heating, ventilation, and air conditioning), equipment, appliances, lighting, and renewable energy systems, such as solar domestic hot water (DHW) heating. To accurately select the best package for the energy efficiency measures, all the possible combinations of measures would need to be evaluated. This approach, often referred to as the brute force optimization approach can be time consuming and, in several cases, not needed to select the optimal package. Instead, some optimization techniques and methodologies can be considered. Several optimization tools are now available and can be applied to design or retrofit buildings. For instance, GenOpt is a generic optimization program that can be connected with an external building energy simulation program in order to minimize a cost function. It includes several optimization methodologies suitable for continuous, discrete, univariate, or multivariate optimization problems (Wetter, 2004). In this section, a set of optimization techniques are briefly presented. These techniques can handle multivariate and discrete optimizations that have been successfully applied to select optimal packages of energy efficiency measures for buildings (Tuhus-Dubrow and Krarti, 2009).

14.4.1 Sequential Search

The sequential search technique, used in BEopt, is a direct search method that identifies the building energy option that will best decrease the cost function at each iterative step. It begins by simulating a user-defined reference building model. The reference model can be a calibrated energy simulation model for an audited building. A simulation is performed for each potential energy efficiency measure individually, and the most cost-effective measure is chosen and used in the building model description for the next point along the path. There are a number of discrete options for a given energy efficiency measure category such as glazing type, heating system, and ceiling insulation. The most cost-effective option is defined as the one that gives the largest reduction in annual costs for the smallest reduction in source energy use. Annual costs are a combination of mortgage costs (which increase as more expensive energy-efficient options are included) and utility costs. The process is repeated, ultimately defining a path from the reference building to the minimum cost point, and then to a ZNE building.

Without modifications, this simple algorithm would not reliably find the correct least-cost path due to the problem of interactive effects between different options. Three special cases have been identified for the sequential search technique as implemented in BEopt—invest/divest, large steps, and positive interactions (Christensen et al., 2004). The invest/divest case is a result of negative interactive effects. BEopt attempts to account for this by evaluating the removal of options in the current building design, which may result in a more cost-optimal point. For example, a highly efficient HVAC system may have been selected as the most cost-effective option at an early point in the process. Later on, however, the improvement of the building envelope may cause the efficient HVAC option to not be cost optimal, so it is removed from the building design.

The large steps case is another example of negative interaction among options. There may be a large energy-saving option that is available at a current point but is less cost-effective than another option that does not save as much energy. The latter option is chosen, and then the most cost-effective option is again chosen at that second point, which results in a third point. However, it is possible that the original large energy-saving option available at the first point is more cost optimal than the third point. In order to solve this problem, the sequential search keeps track of points from previous iterations and compares them to the current point. If a previous point is more cost optimal, it replaces the current point.

A positive interaction occurs if two options are more cost-effective when present together than they would be separately. An example could be the presence of both large south-facing windows and thermal mass for passive solar heating. The sequential search will only find these positive interactions if one of the options is first selected individually. This inability to always identify synergistic options is a potential deficiency with the sequential search method.

14.4.2 Genetic Algorithm

Genetic algorithm (GA) uses the evolutionary concept of natural selection to converge on an optimal solution over many generations. This technique differs from traditional optimization methods in a number of areas. First, rather than working with one potential solution at a time, the technique works with a set of solutions called a population. This ensures a global approach to optimization and helps the GA avoid getting stuck in local minima, which can be a problem with other methods. Second, the GA works with encodings of the parameters, not the parameters themselves. Parameters are traditionally encoded as binary strings, although other options are possible. Finally, GAs use probabilistic methods for determining the parameter values in each successive iteration, rather than deterministic rules. This means that each time a GA is run, the path toward convergence is different, and the end result may be different as well.

Each individual in the population represents a different solution to the problem. Every option for each parameter has a corresponding binary representation, and the parameters are concatenated to form the complete binary string. A new generation is formed at the end of each iteration of the algorithm, consisting of a new population, and this process is repeated until satisfactory convergence criteria are reached, or the maximum number of generations is reached. The algorithm uses only three operators to produce a new population for the next generation—selection, crossover, and mutation.

There are a number of different ways to handle selection. One method is to rank the population in ascending order by fitness value (after the cost function is evaluated for each individual) and assign probabilities for selection based on each individual's rank. This is called rank weighting. A virtual roulette wheel is spun (by generating a random number between 0 and 1) to determine the members in the new population selected for reproduction.

Once the population for reproduction is selected, the individuals are paired off and mated using the crossover process. A cross-point is selected at random for each pairing, and two new individuals are created by joining the first part of the first string with the second part of the second string, and vice versa.

Mutation is the last step in the formation of the population for the next generation, and involves flipping a bit at random in the population from 0 to 1 or vice versa. Mutation is intended to prevent the GA from converging prematurely and helps to maintain a global search. The mutation rate is set at the beginning of the algorithm. Finally, this mutated population becomes the population of the next generation, and the process is repeated until convergence is reached.

14.4.3 Particle Swarm Optimization

Particle swarm optimization (PSO) is one of the simplest techniques to implement that can deal with discrete options. PSO shares many similarities with GAs.

Like GAs, the technique works with a set of solutions called a population. Each potential solution is called a *particle.* Instead of using evolutionary methods, however, the PSO is based on the social behavior of bird flocks or fish schools. Each particle is characterized by a velocity with which it explores the cost function. The velocity and position of each particle are updated after successive iterations of the algorithm. The particle velocity and position are governed by Equations 14.1 and 14.2:

$$v_{new} = v_{old} + c_1 r_1 \left(p_{local\ best} - p_{old}\right) + c_2 r_2 \left(p_{global\ best} - p_{old}\right) \tag{14.1}$$

$$p_{new} = p_{old} + v_{new} \tag{14.2}$$

where

v is the particle velocity
p is the particle position
r_1, r_2 are independent uniform random numbers between 0 and 1
c_1 is the cognitive acceleration constant
c_2 is the social acceleration constant
$p_{local\ best}$ is the best local solution (best particle in current population)
$p_{global\ best}$ is the best global solution (best particle so far in all generations)

The two acceleration constants are usually numbers between 0 and 4. The PSO technique has become popular for the same reasons as the GA, in that it is easy to implement with relatively few parameters to adjust.

14.4.4 Applications

Optimization techniques have been used in a wide range of applications to improve the energy performance of buildings. Indeed, optimization-based analysis approaches have been widely used to improve the control and operation of buildings (Huang and Lam, 1997; Wright et al., 2002; Fong et al., 2006). More recently, optimization techniques have been proposed to select building envelope and heating and cooling system design features (Caldas and Norford, 2003; Christensen et al., 2004; Wang et al., 2005; Tuhus-Dubrow and Krarti, 2010; Bichiou and Krarti, 2011). In particular, Tuhus-Dubrow and Krarti (2010) developed a simulation environment using GA optimization technique to select the best combinations of several building envelope features in order to optimize energy consumption and life cycle costs (LCCs). Figure 14.4 illustrates the flowchart of the optimization simulation environment used to select building envelope features for a typical U.S. home. Table 14.4 shows the optimal building orientation, shape, and building envelope characteristics selected to minimize the LCC value obtained using the GA optimization technique. As indicated by the results of Table 14.4, insulating the walls and the attic, as well as reducing the air infiltration are selected for all the climates. Foundation insulation is not selected for hot climates such as Phoenix, AZ and Miami, FL. However, increasing the mass level within the home is not selected for any climate due to its high implementation cost.

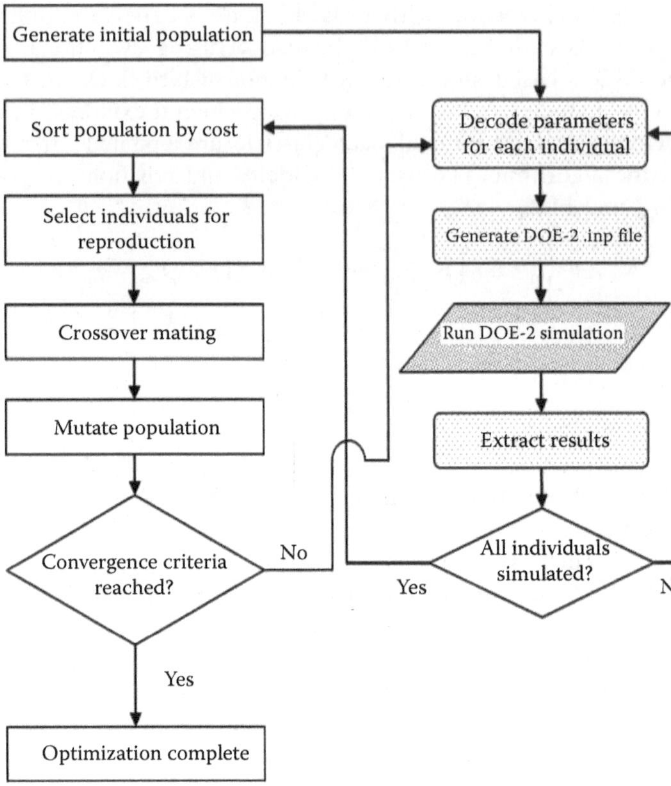

FIGURE 14.4 Flowchart diagram for the developed simulation/optimization tool.

TABLE 14.4
Specifications for a Residential Building Envelope Optimization for Five U.S. Climates

Climate	Boulder	Phoenix	Chicago	Miami	San Francisco
Minimum found (LCC in $)	38,104	36,569	41,450	38,267	46,967
Azimuth (°)	270	315	270	337.5	0
Shape	Rectangle	Rectangle	Rectangle	Rectangle	Rectangle
Aspect ratio	1.0	1.0	1.0	1.0	1.0
Wall construction	R21, 2 × 6	R21, 2 × 6	R21, 2 × 6	R21, 2 × 6	R21, 2 × 6
Ceiling insulation	30	30	30	30	30
Foundation level	4 ft R5 perimeter, R5 gap	Uninsulated	4 ft R15 perimeter, R5 gap	Uninsulated	4 ft R5 perimeter, R5 gap
Glazing type	Low-e very high SHGC argon	Low-e low SHGC argon	Low-e standard SHGC argon	Low-e low SHGC argon	Low-e very high SHGC argon
Infiltration	Typical	Tight	Typical	Tight	Typical
Mass level	Light	Light	Light	Light	Light

Table 14.5 summarizes the results of selecting the heating and cooling system and its efficiency for a typical home located in five U.S. climates (Bichiou and Krarti, 2011). The selection of the heating and cooling system is based on minimizing the LCC without a budget constraint. Several heating and cooling systems were considered in the optimization analysis including ground source heat pump (GSHP) with vertical borehole wells or horizontal loops, air conditioner, gas furnace, electric heating, and evaporative cooling. Figure 14.5 presents an example of the Pareto optimal plot that shows energy cost savings and LCC values for selecting both building envelope features as well as heating and cooling system type and settings for a two-story home located in Chicago, IL. The Pareto plot is obtained using the sequential search method. The optimal solution found by the sequential

TABLE 14.5
Optimized Selection of HVAC System Features for Five U.S. Sites

Site	Boulder, CO	Chicago, IL	Miami, FL	Phoenix, AZ	San Francisco, CA
Algorithm	Sequential search	Sequential search	Sequential search	Sequential search	Particle swarm optimization
System efficiency	98	–	80	98	98
HVAC system type	GSHP vertical rectangle 6 × 3	GSHP vertical 10 pipes	AC with electric resistance	Evaporative cooler with furnace	AC with furnace
Annual utility cost ($)	940	1,154	1,313	1,118	1,441
Life cycle cost ($)	45,258	43,886	38,924	38,036	41,181

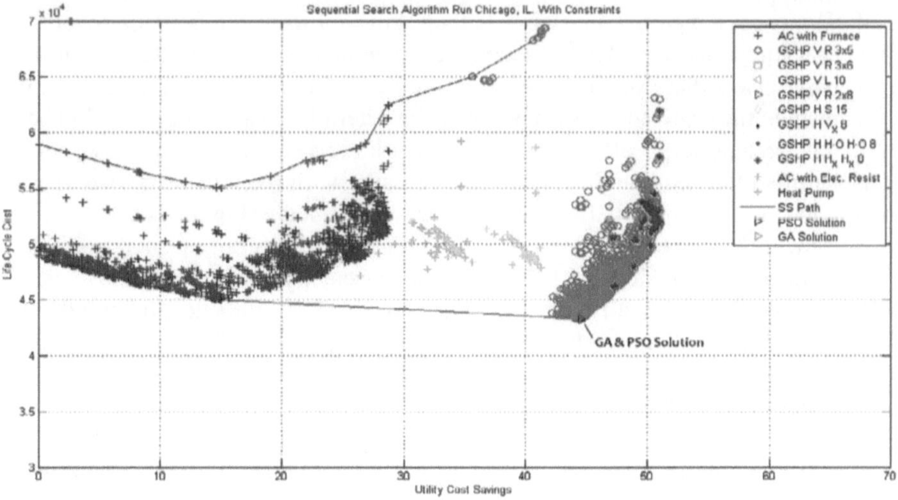

FIGURE 14.5 Pareto graph of LCC as a function of energy cost reduction for a house in Chicago, IL.

search method coincides with those obtained using the GA and PSO optimization techniques, as illustrated by Figure 14.5.

Alspector and Krarti (2009) have used GA-based optimization to both automatically calibrate a detailed building energy simulation model for an audited house using monthly utility data as well as select a package of energy efficiency measures subject to a budget constraint. Table 14.6 illustrates LCC-based optimization results to select energy retrofit measures for a residential building located in Denver, CO, when three budget levels are imposed: $1,000, $5,000, and $10,000. As shown in Table 14.6, more aggressive energy efficiency measures are selected when the retrofit budget increases. It is interesting to note as the budget increases and more retrofit measures are implemented, the payback period of the retrofit package is reduced.

14.5 NEAR-OPTIMAL ANALYSIS METHODOLOGY

14.5.1 General Methodology Description

Instead of using an optimization technique to find the best package energy and renewable energy measures to design or retrofit buildings or communities to ZNE or carbon-neutral conditions, a sequential search analysis procedure can be utilized. The basic sequential search process involves the following steps:

1. Start by the evaluation of different types of energy efficiency measures (e.g., wall insulation addition, window replacement, and HVAC equipment efficiency improvement) to determine the most cost-effective measure in terms of marginal cost. Using the Pareto graph similar to Figure 14.6, the LCC is shown in the y-axis and the percent saving of source energy is provided in the x-axis for each measure. The measure with the steepest negative slope (i.e., ratio cost over percent energy savings) is selected, as shown in Figure 14.6.
2. The chosen measure in Step 1 is then removed from the potential energy efficiency measures for the second iteration analysis. In this step, the new baseline building model consists of the initial building model modified to include the selected measure after the first iteration outlined in Step 1. The remaining energy efficiency measures are then simulated using the new baseline building model. Again, the measure with the most-effective marginal cost is selected after this iteration, as outlined in Figure 14.6.
3. Repeat the same procedure described for Step 2 until an optimal point is reached and the slope of LCC over percent energy savings becomes positive, as illustrated in Figure 14.7. The positive slope indicates that a reduction in building energy use results in increasing the LCC due to significant implementation costs. It should be noted that now the measure with the lowest positive slope is selected.
4. After reaching an optimal point, the most cost-effective efficiency measure's marginal cost is compared to the cost of PV energy. At the point where further improving the building has a higher marginal cost, PV is employed until the zero net source energy is achieved, as illustrated in Figure 14.6.

TABLE 14.6
Optimized Selection of Energy Retrofit Measures for a House in Denver, CO, Based on Three Budget Constraint Levels

Retrofit Budget ($)	List of Measures	Incremental Implementation Costs ($)	Electricity Savings (kWh)	Gas Savings (MMBtu)	Energy Cost Savings ($)	Simple Payback Period (Years)
0	Baseline case:	0	0	0	0	–
	Brick walls, no insulation					
	Single pane clear windows					
	Uninsulated slab on grade floor					
	Furnace efficiency = 82%					
1,000	Optimal case 1:	697	–43	12	116.10	6.0
	Brick walls, no insulation	0				
	Single pane clear windows	0				
	Slab, R-10 perimeter insulation	697				
	Furnace efficiency = 82%	0				
5,000	Optimal case 1:	4,941	608	100	1,055.1	4.7
	Brick walls, R-15 insulation	978				
	Low-E double pane windows	2,812				
	Slab, R-5 perimeter insulation	499				
	Furnace efficiency = 86%	652				
10,000	Optimal case 1:	5,889	994	123	1,320.10	4.5
	Brick walls, R-19 insulation	2,588				
	Low-E double Pane windows	2,812				
	Slab, R-5 perimeter insulation	499				
	Furnace efficiency = 82%	0				

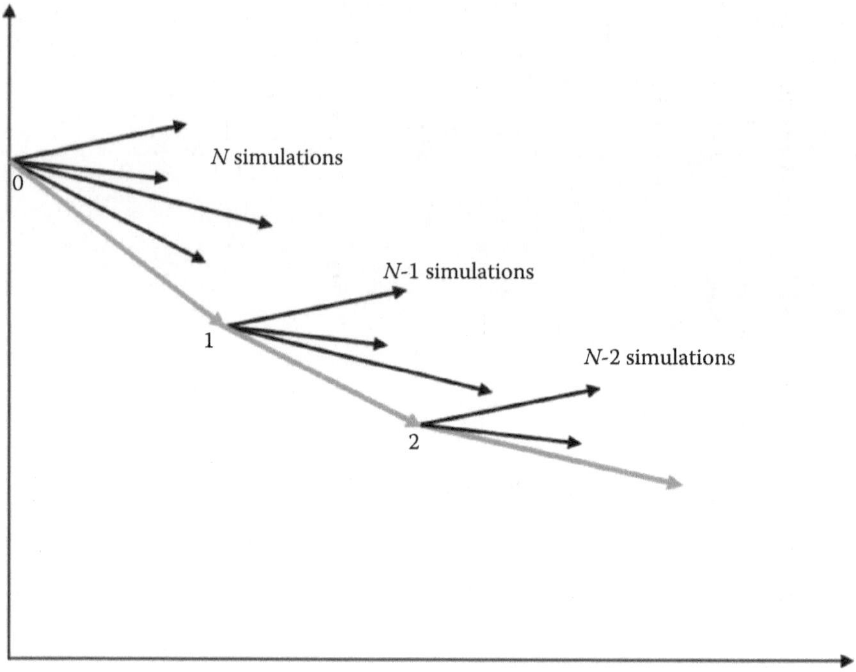

FIGURE 14.6 Sequential analysis procedure for selecting energy efficiency measures.

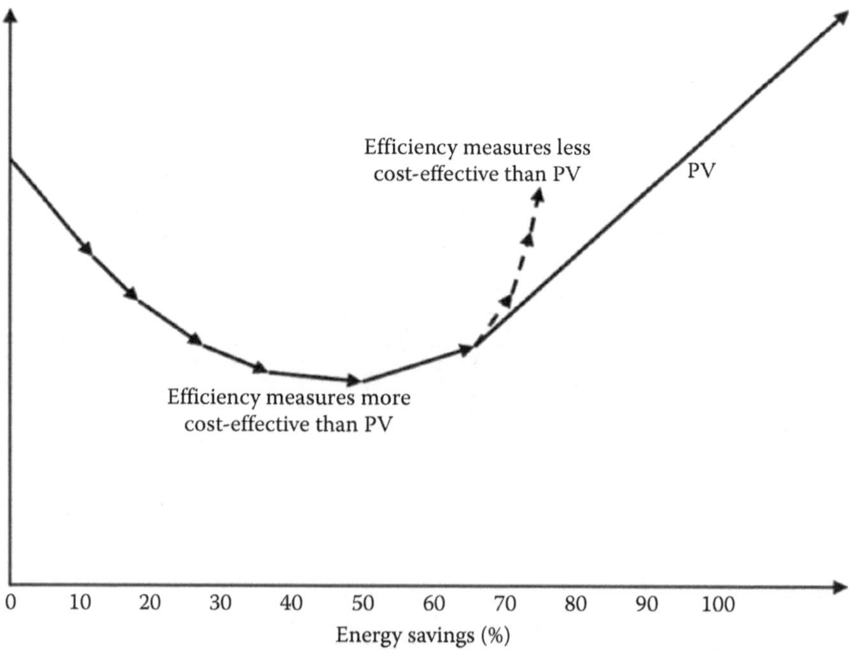

FIGURE 14.7 Sequential selection procedure for the path to ZNE.

5. A final check is needed to ensure that the size of the required PV system can be installed based on the constraints of the available spaces (for instance the house roof area). If no sufficient space is available to install the required PV system, then additional energy efficiency measures need to be selected until 100% source energy savings are achieved. In this case, the selection process is similar to that described for Step 4 without the need to compare with the marginal cost of the PV system.

14.5.2 Case Study for ZNE Design Approach

To illustrate the sequential selection procedure, high performance for a house is carried out. Several energy efficiency measures are evaluated individually and then using the optimization technique described in Section 14.5.1. Table 14.7 summarizes the list of building envelope, lighting, and HVAC system retrofit measures considered. Only the results for the energy and economic impacts of individual measures are provided in Table 14.7. Using the LCC analysis, all the measures can be cost-effective except adding wall insulation and foundation insulation as indicated in Table 14.7 by comparing the LCC values obtained for all the energy efficiency measures and the baseline building energy model.

Using the sequential selection procedure described in Section 14.5.1, an optimal set of energy conservation measures is determined. The results of the sequential selection procedure are summarized in the Pareto graph of Figure 14.8. The optimal package with the lowest LCC value consists of all of the energy efficiency measures listed in Table 14.7, except for the basement insulation measure. The optimal package achieves an annual energy use reduction of 23.63% and costs $26,879 to implement while reducing the annual electric bills by $3,638 per year. In addition, the optimal package can reduce LCCs from $199,133 to $180,675 over a 20-year period with a 5% discount rate. By adding a PV system in the roof, further reduction in source energy use can be achieved. In particular, for a neutral LCC point (as indicated by a filled circle in Figure 14.8), a 13.5 kW PV system needs to be installed. For this PV size, a reduction of 28% in the building energy use can be achieved, but an additional $40,500 to the total installed cost is needed. A ZNE design could be achieved with the addition of 50 kW of PV panels. However, this PV system would not be cost-effective and would be too large to install on the available roof area of the house.

Additional case studies for optimal design of residential homes and rural residential communities are described in the following sections.

14.6 CASE STUDY 1: OPTIMAL RETROFIT AND DESIGN OF HOMES

In this case study, various combinations of energy efficiency measures are evaluated in order to achieve optimum LCCs for retrofitting existing residential and designing new construction residential buildings in Salamanca, Mexico (Griego et al., 2012). Moreover, the required sizes of PV systems are determined through optimization to obtain ZNE buildings for various home configurations. Specifically, four separate optimizations are performed, the existing-unconditioned home, the

TABLE 14.7
Energy and Economic Analysis of Energy Efficiency Measures

	Only Individual Energy Efficiency Measures Are Evaluated						
EEM	Electricity Savings (kWh/Year)	Natural Gas Savings (Therm/Year)	Annual Costs ($/Year)	Total Cost Savings ($/Year)	Installed Cost ($)	Simple Payback (Years)	Life Cycle Costs[a] ($)
Thermostat management	407	498	15,629	350	0	0.00	$194,772
Reduce infiltration rates	50	366	15,738	241	1,200	4.98	$197,330
Install additional roof insulation	251	686	15,511	468	4,360	9.32	$197,661
Install wall insulation	398	1,772	14,798	1,181	19,155	16.22	$203,571
Install basement insulation	196	206	15,822	157	7,642	48.68	$204,819
Replace all of the incandescent bulbs with compact fluorescent (CFL) bulbs	13,636	−356	15,108	871	272	0.31	$189,150
Replace all of the incandescent flood lights with compact fluorescent (CFL) bulbs	3,585	−23	15,702	277	157	0.57	$195,959
Retrofit T-12 lighting with T-8 lighting and program start electronic ballasts	3,260	−51	15,749	230	450	1.96	$196,497
Install occupancy sensors in the bathrooms, laundry room, and living rooms	3,239	−62	15,757	222	1,285	5.79	$196,589
Baseline	0	0	15,979	0	0	—	$199,134

[a] LCC analysis assumes a life cycle of 20 years and a discount rate of 5%.

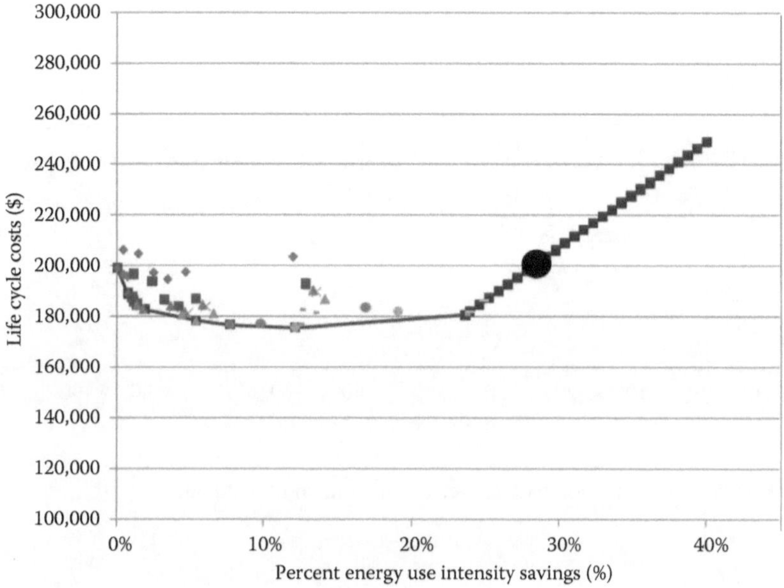

FIGURE 14.8 LCCs versus percent primary energy savings.

existing-conditioned home, the new construction-unconditioned home and the new construction-conditioned home. Only the analysis results are summarized in this section. For the analysis approach and simulation tools considered, the reader is referred to the study by Griego et al. (2012).

14.6.1 Retrofit of Existing Homes

For typical existing homes in Salamanca, Mexico, the baseline annualized energy-related costs for the conditioned and unconditioned homes are estimated to be $797 and $557, respectively (Griego et al., 2012). These costs are obtained before any energy efficiency and thermal comfort measures are applied. Using a detailed simulation analysis, it is determined that the cost of improved thermal comfort for a typical home in Salamanca is roughly $240 per year. The optimum point for the unconditioned case occurs at 17.3% annual energy savings and a corresponding minimum cost of $433 as shown in Figure 14.9. The conditioned case, on the other hand, has a greater opportunity for energy savings and achieves a minimum cost of $542 at 35.0% energy savings as indicated in Figure 14.10. The optimum point for the conditioned case includes implementing methods to reduce miscellaneous plug loads, R-1.4 m^2 K/W roof insulation, low-flow shower heads and sinks, an electric stove, 100% compact fluorescent lamps (CFLs), and R-0.35 m^2 K/W trunk-branch DHW pipe insulation. The optimum point for the unconditioned case includes all of the same measures as the conditioned model with the exception of the added roof insulation.

The comparison between the unconditioned and the conditioned optimization results, outlined in Figure 14.11, reveal that the optimum point for the conditioned

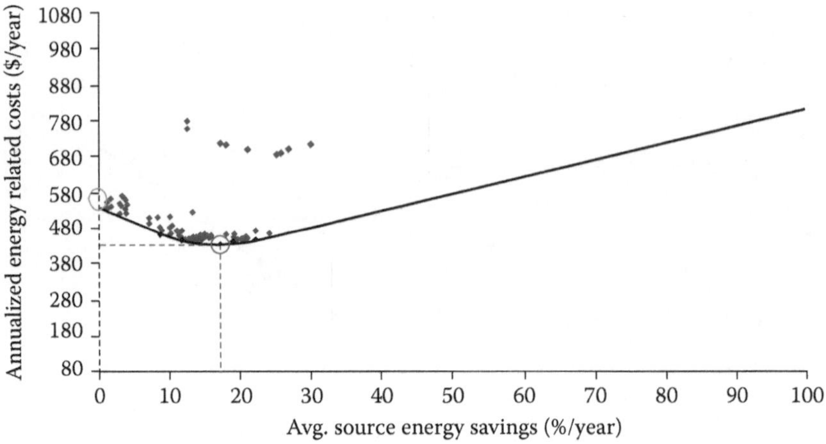

FIGURE 14.9 Optimization path for retrofitting unconditioned home.

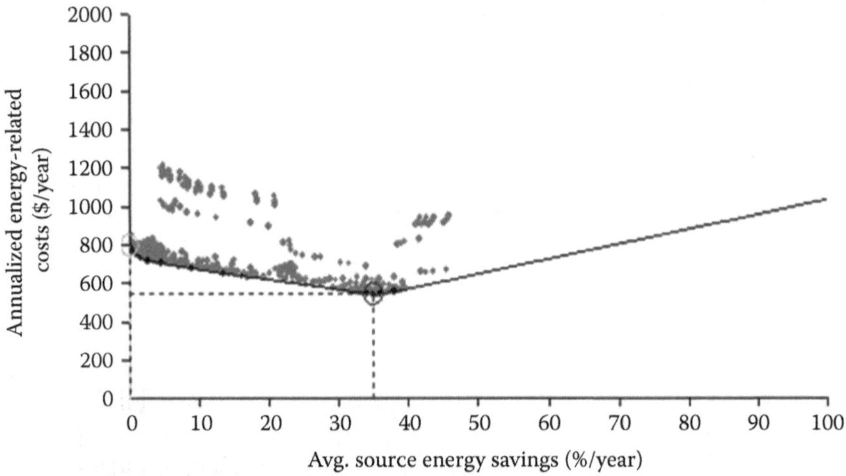

FIGURE 14.10 Optimization path for retrofitting conditioned home.

case ($542) is roughly the cost neutral point for the unconditioned case ($557). It is also useful to compare energy end uses to gauge measures with the highest potential of the energy savings. Figures 14.12 and 14.13 include a summary of the total annual source energy consumption by end use for the unconditioned and conditioned home models, respectively. The minimum cost option and the PV start options are compared with the baseline model. In the unconditioned case, the largest energy savings are obtained for the hot water, miscellaneous equipment, and lighting. The conditioned home has similar energy savings as the unconditioned home for miscellaneous equipment and DHW; however, the greatest end-use for energy savings is space cooling, which is primarily attributed to the use of roof insulation.

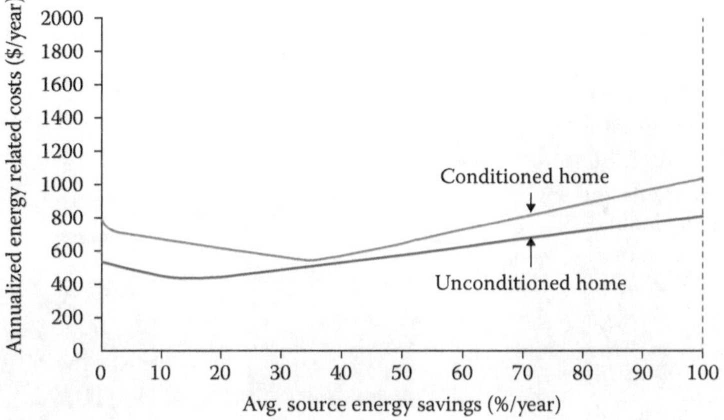

FIGURE 14.11 Retrofit optimization curve comparison between the unconditioned and conditioned cases.

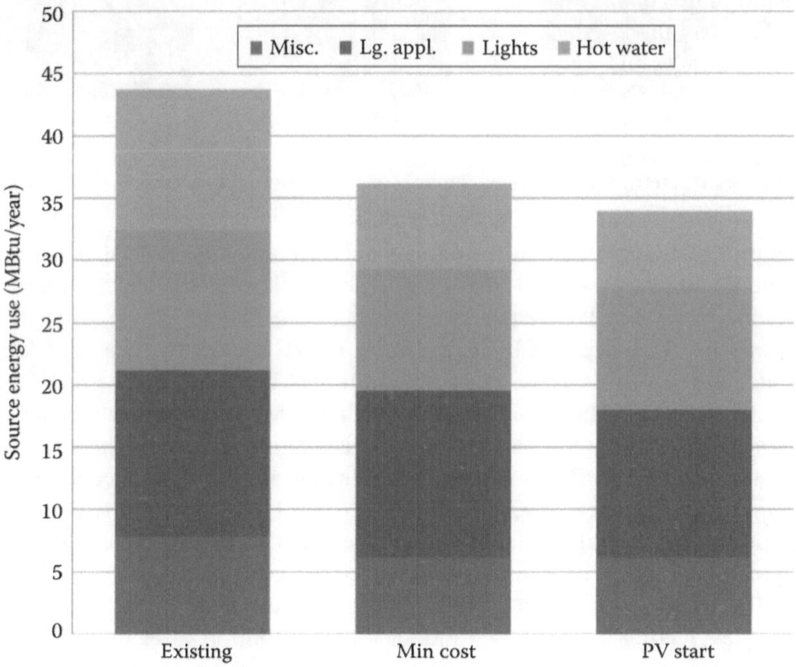

FIGURE 14.12 Retrofit-unconditioned annual source energy by end use.

14.6.2 PV Systems for Existing Homes

The two renewable energy technologies evaluated in this study are solar DHW systems and PV systems. Although solar DHW has a large potential for energy savings, the high implementation cost makes it unfeasible. Figures 14.9 and 14.10 show points hovering significantly above the optimization path; those points are associated with

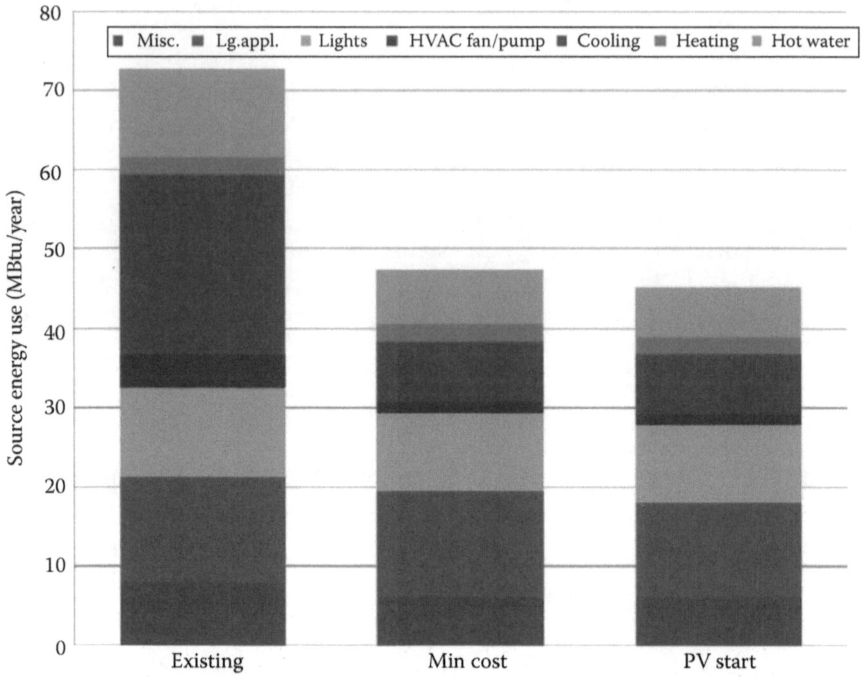

FIGURE 14.13 Retrofit-conditioned annual source energy by end use.

combinations of measures that include solar DHW systems. Note that the cost for labor and materials is assumed to be comparable to those in the United States.

The results for the PV system are shown by the sloped line leading to 100% energy savings. The size of a PV to arrive at a ZNE solution for the unconditioned home model is a 3 kW system and the conditioned home model is a 4 kW system. Both systems are south facing and installed in inclined panels to match the latitude in Salamanca. The slope of the line toward ZNE is relatively shallow where the annualized energy cost for 100% annual energy savings is $920 for the unconditioned and $1,185 for the conditioned cases. Similar to the solar DHW system, the U.S. costs are assumed for PV material and labor cost. PV technology may only be desirable with the appropriate subsidies for implementation costs.

The PMV thermal comfort analysis is used as verification for the optimization results by evaluating the improved indoor thermal comfort after implementing the recommended energy efficiency measures. First, the PMV ratings above and below the acceptable comfort range (i.e., PMV values between −1 and 1) are determined for the baseline and optimal cases for both conditioned and unconditioned building models. The unconditioned building model shows roughly 1,550 hours above 1 PMV and 150 hours below 1 PMV annually. This is in contrast to the conditioned building models where thermal comfort is maintained throughout the year, as expected.

The thermal comfort analysis is also applied at the optimum building models the impact of the final recommendations. The annual energy consumption in the

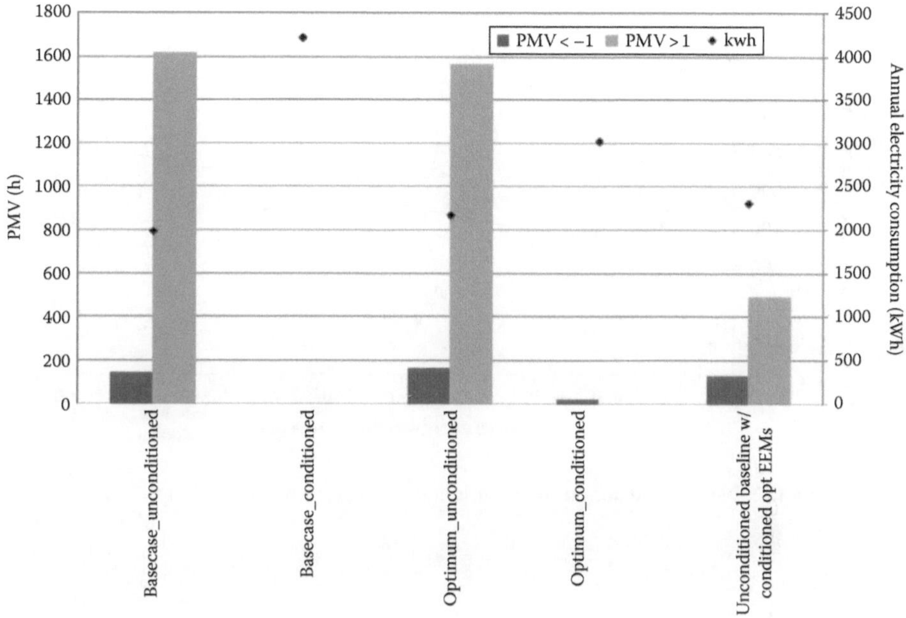

FIGURE 14.14 Retrofit construction case—PMV thermal comfort analysis.

conditioned existing building models decreases relative to the baseline when thermal insulation is added to the building and as predicted, the annual PMV ratings remain relatively constant. However, when the optimum set of EEMs from the conditioned case are applied to the unconditioned baseline home model, the number of hours outside of the thermal comfort zone decreases significantly as indicated in Figure 14.14.

In the conditioned baseline retrofit case, the cost of installing an electric heat pump is estimated at $4,394 for a 3.5-ton unit. However when roof insulation is added to the unconditioned retrofit building the number of hours outside of the thermal comfort zone decreases by almost 60% for a much lower initial cost of roughly $426.

14.6.3 Design of Energy-Efficient Homes

The results for the new construction case have a higher potential for cost and energy savings due to the reduced implementation costs and larger flexibility in selecting measures, especially those related to the building envelope. Similarly, the new construction baseline annualized energy costs are also lower when compared to the retrofit baseline case. The annualized energy-related costs for the conditioned and unconditioned new construction baseline homes are $764 and $495, respectively. Thus, without applying energy efficiency measures, the cost of improved thermal comfort for new homes in this part of Mexico is almost $270 per year.

In the new construction homes, the optimum energy efficiency combination for the unconditioned case occurs at 19.0% annual energy savings and a corresponding minimum annual cost of $381, as shown in Figure 14.15. The conditioned case on the

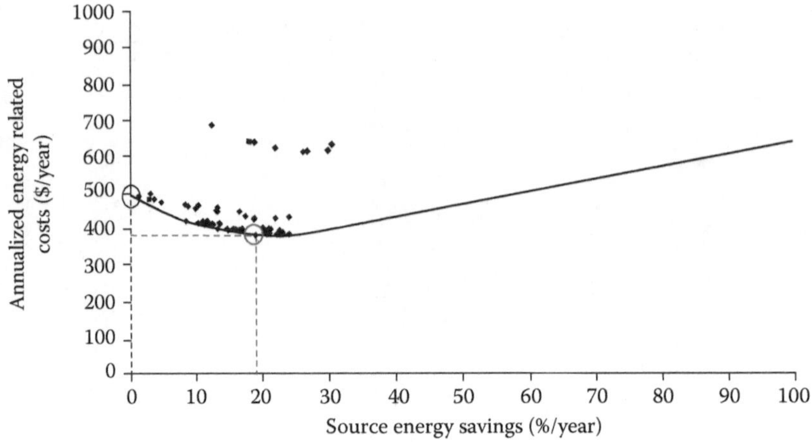

FIGURE 14.15 New construction-unconditioned optimization path.

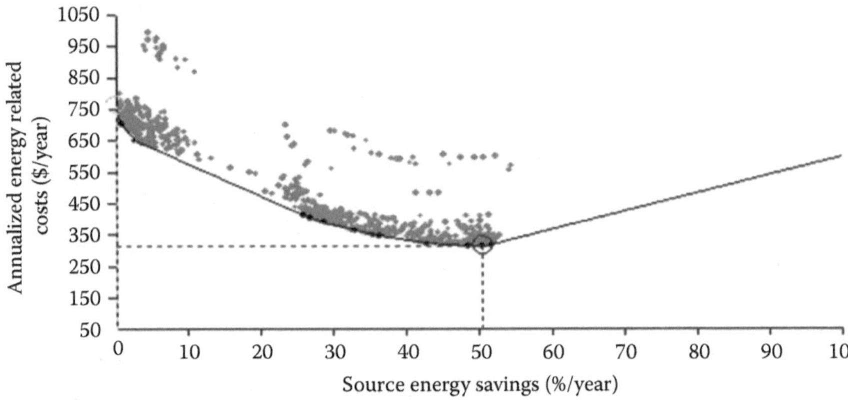

FIGURE 14.16 New construction-conditioned optimization path.

other hand has a higher opportunity for energy savings and achieves a minimum cost of \$315 at 50.6% annual energy savings, as outlined in Figure 14.16. The optimum point for the conditioned case includes reducing miscellaneous plug loads, R-1.4 m^2 K/W roof insulation, R-1 m^2 K/W wall insulation, white or cool white asphalt shingle roof finish, 15% window to wall ratio, high efficiency refrigerator and washing machine, low-flow shower heads and sinks, an electric stove, 100% CFLs, SEER 17 AC unit, and R-0.35 m^2 K/W trunk-branch DHW pipe insulation. The optimum combination of energy efficiency measures for the unconditioned analysis include an electric stove, 100% CFLs, reduced miscellaneous equipment loads, a high efficiency clothes washing machine, low-flow showers and sinks, and R-0.35 m^2 K/W trunk-branch DHW pipe insulation.

When comparing the unconditioned and the conditioned new construction buildings, as shown in Figure 14.17, the optimum point for the conditioned case (\$315)

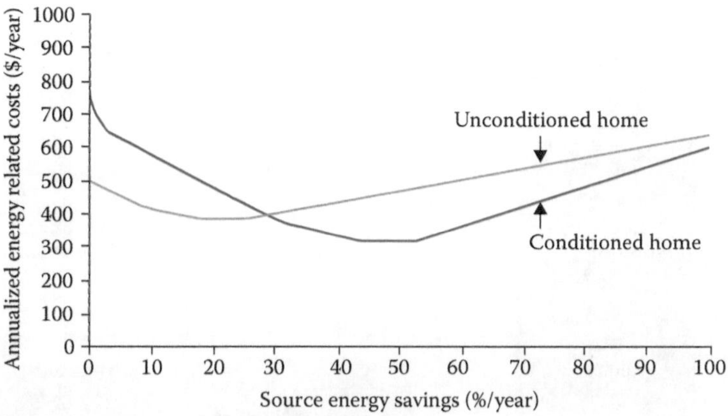

FIGURE 14.17 New construction optimization curve comparison: unconditioned and conditioned cases.

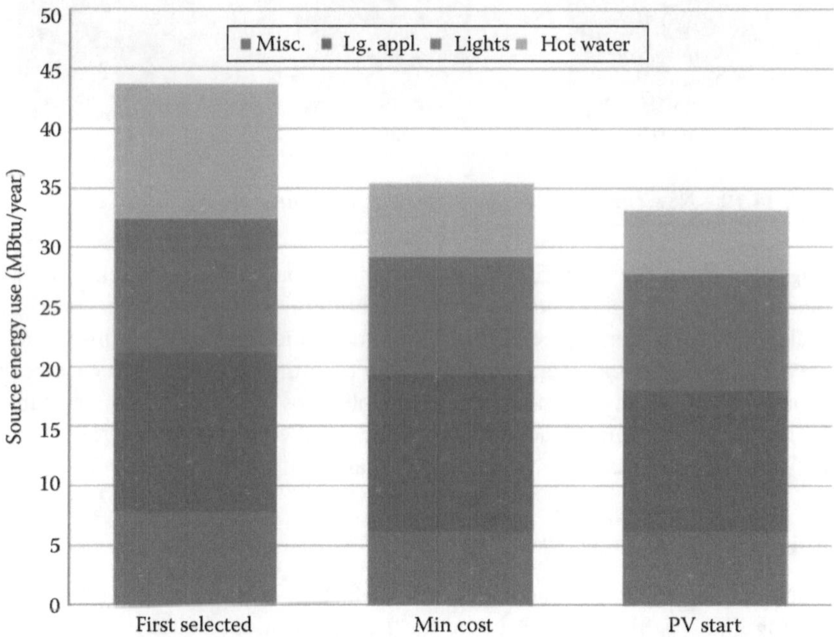

FIGURE 14.18 New construction-unconditioned annual source energy by end use.

has lower annualized energy cost than the optimum point for the unconditioned case ($381). Therefore, higher energy savings can be obtained for new construction at a low cost when the total implementation costs are included in the 30-year mortgage.

Figures 14.18 and 14.19 provide a summary of the total annual source energy consumption by end use for the unconditioned and conditioned home models, respectively. The results indicate that a significant reduction can be achieved in annual

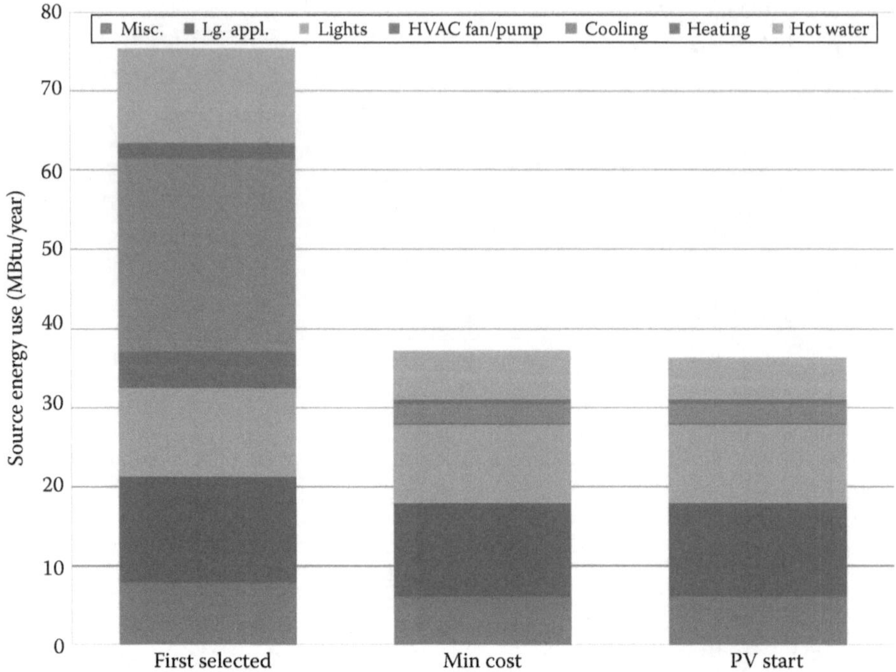

FIGURE 14.19 New construction-conditioned annual source energy by end use.

cooling energy for the conditioned case when compared to the baseline building. Similar to the retrofit study, the unconditioned building shows the largest energy savings related to hot water, miscellaneous equipment and lighting. The minimum cost option has a lower annual energy consumption than the unconditioned case because it is cost-effective to install more efficient appliances and use condensing tankless DHW heater. The cooling loads are primarily reduced by including roof and wall insulation to achieve assembly R-values of 1.4 and 1 m^2 K/W respectively.

14.6.4 PV Systems for New Homes

Similar to the retrofit case, installing a solar DHW system is not cost-effective. The size of the PV to achieve a ZNE home for the unconditioned case is a 3 kW system and a 4 kW system for the conditioned case. When implementing the PV system in the conditioned building, the annualized energy-related cost for the 30-year period is only $773, which is roughly the same annual cost when compared to the baseline. The annualized energy-related cost of implementing PV in the unconditioned home is $726.

The annual energy consumption in the new construction-conditioned model decreases relative to the baseline when thermal insulation is added to the exterior walls and roof as anticipated. However, the annual PMV ratings remain relatively constant. When thermal comfort measures associated with the conditioned case are

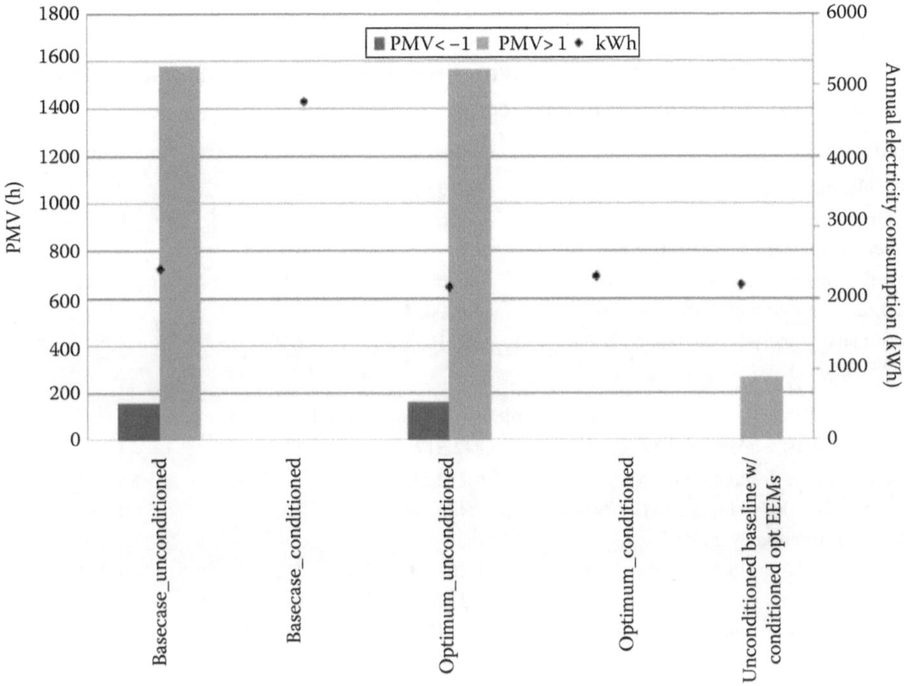

FIGURE 14.20 New construction case—PMV thermal comfort analysis.

applied to the unconditioned baseline home model, the number of hours outside of the thermal comfort zone decreases significantly, as indicated in Figure 14.20.

The cost of installing an electric heat pump in the new construction baseline conditioned model is approximately $4,622 for a four-ton unit and only costs $1,952 to install roof and wall insulation. Furthermore, the annual electricity consumption in the optimized new construction building is almost the same as the unconditioned baseline building with the optimum combination of EEMs from the conditioned case. The analysis results indicate that there is a strong correlation between the use of thermal insulation and improved thermal comfort in unconditioned homes. Implementing thermal insulation also proves to be a cost-effective way of improving thermal comfort when compared to the cost of installing a unit air conditioner.

14.6.5 Economic Analysis of Energy-Efficient Homes

If the optimum annual energy savings for the unconditioned home is applied to all existing residential buildings in the city of Salamanca, the city has the potential of saving 13,817 MWh of electricity annually, or nearly 17% of the total domestic energy consumption. Similarly, if the same energy savings is applied to the existing housing stock nationally, nearly 8,425,170 MWh is saved per year. This equates to avoiding 1.5 MW of power production in Salamanca annually and almost 914 MW per year in Mexico. Because it is unlikely for the entire country to adopt all of the

TABLE 14.8
Potential Electrical Energy Use Savings for the City of Salamanca, State of Guanajuato, and Country of Mexico

Data	Mexico	Guanajuato	Salamanca
Population	112,336,538	5,486,372	260,732
Housing units	28,607,568	1,276,584	64,073
Average number of occupants/home	3.9	4.3	4.1
Total domestic consumption (MWh/year)	48,700,400	1,481,564	79,865
Estimated number of homes by 2030	50,000,000	2,231,200	111,986
Estimated domestic consumption 2030 (MWh/year)	85,118,036	2,589,462	139,587
Existing buildings unconditioned			
Savings for 10% adoption (MWh/year)	842,517	25,631	1,382
Savings for 50% adoption (MWh/year)	4,212,585	128,155	6,908
Savings for 100% adoption (MWh/year)	8,425,169	256,311	13,817
New construction by 2030[a]			
Savings for 10% adoption (MWh/year)	1,267,334	38,555	2,078
Savings for 50% adoption (MWh/year)	6,336,669	192,774	10,392
Savings for 100% adoption (MWh/year)	12,673,337	385,548	20,783

[a] Only new construction buildings are included in this estimate—50% are conditioned and 50% are unconditioned.

recommended EEMs, estimates for 10% and 50% of the market penetration are also included in Table 14.8. Estimated annual energy savings are also included for the new construction housing units predicted by 2030. It is assumed that 50% of the new construction homes have air-conditioning units, while the other half are unconditioned. Table 14.8 also outlines pertinent data used for the market analysis along with other general characteristics for the city of Salamanca, the state of Guanajuato, and the country of Mexico (Griego et al., 2012).

14.7 CASE STUDY 2: DESIGN OF ELECTRIFIED AND RESILIENT RESIDENTIAL BUILDINGS AND COMMUNITIES

14.7.1 General Description

This case study assesses the impacts of such building design adjustments including energy efficiency measures (EEMs) and natural gas and electricity energy sources, on the generating capacity of on-site PV generation that is necessary for US residential buildings and communities to achieve net-zero energy and resiliency goals (Dafoe et al., 2022). In this study, six cases are considered for the energy resources for the residential buildings including natural gas operated space heating baselines, optimized and net zero energy residences, and electrified baselines, optimized, and net zero energy residences. These design cases for individual residential buildings are considered for residential communities to estimate the overall community PV generation needed to achieve net zero energy. In addition, the impact of building designs on the required PV system size given the grid power outage criteria used to enhance the energy resilience for the residential communities.

Brief descriptions are provided first for the design categories for US residential buildings and communities considered in the study (Dafoe et al., 2022):

1. *Baseline designs*: These baseline homes are considered to represent code-compliant US new-construction practices for residential buildings. Two cases for the baseline home designs are considered in the study including those relying on natural gas systems to provide space heating, water heating, and cooking and those that are totally electrified. Table 14.9 lists the main characteristics for the baseline designs when the homes are in Longmont, CO.
2. *Optimal energy efficiency home designs*: These home designs, also referred to as minimum cost options, integrate cost optimal sets of several energy efficiency measures based on life cycle cost analysis. These measures, applied to the baseline home designs, include system efficiencies and energy source types (i.e., gas and electricity), building envelope constructions, temperature setbacks, and appliance performances. Based on the optimization analysis, home designs are achieved with the lowest annualized energy related cost, henceforth referred to as the "gas minimum cost design", and "electrified minimum cost design".
3. *Net-zero energy home designs*: These designs combine both energy efficiency measures and PV systems to achieve net zero energy buildings. Specifically, net zero energy (NZE) designs are achieved through simultaneous optimization of energy efficiency measures and on-site generation (i.e., PV system) capacities to minimize life cycle costs. Net zero energy designs are henceforth referred to as "gas NZE" and "electrified NZE".
4. *Net-zero energy community designs*: These communities feature community-scale and centralized PV system to improve ease of installation and maintenance, as well as shared electricity infrastructure such as larger inverters that often cost more for individual homes. For the study, residential communities made up one-hundred single-family homes are evaluated.
5. *Energy resilient community designs*: Energy resilience for these communities is achieved through properly sized on-site energy provision to sustain a specified level of loads for a select duration of time. Communities may elect to prioritize resilience for several reasons, such critical loads and unreliable grid power supply. For all the communities considered in this study, it is assumed that all electrical loads are considered critical to avoid load shedding. Various outage durations ranging from six hours to three days are considered for the design analysis. Longer-term outages often require more on-site generation than short-term outages.

14.7.2 Design Specifications for Homes

Table 14.10 summarizes energy usage data associated with end-uses for both the baseline natural gas operated and electrified home designs. The results of Table 14.10 imply that an electrified home of comparable construction and system design would use more

TABLE 14.9
Main Features for Natural Gas and Electrified Baseline Designs for Homes

	Gas Baseline	Electrified Baseline
Orientation	North	North
Neighbors	Left/right at 15ft, front/back at 80ft	Left/right at 15ft, front/back at 80ft
Wood Stud	R-20 open cell spray foam, 2 × 6, 24 in o.c.	R-20 open cell spray foam, 2 × 6, 24 in o.c.
Wall sheathing	OSB	OSB
Exterior finish	Stucco, medium/dark	Stucco, medium/dark
Unfinished basement	Ceiling R-49 fiberglass, vented	Ceiling R-49 fiberglass, vented
Roof material	Asphalt shingles, medium	Asphalt shingles, medium
Unfinished basement	Whole wall R-19 fiberglass batt, 2 × 6, 24 in o.c.	Whole wall R-19 fiberglass batt, 2 × 6, 24 in o.c.
Attic	R-49 fiberglass batt insulation	R-49 fiberglass batt insulation
Carpet	0%	0%
Floor mass	Wood surface	Wood surface
Window areas	11% of all sides window area	11% of all sides window area
Windows	Low-E, double, insulated, air, L-gain	Low-E, double, insulated, air, L-gain
Door area	40 sqft	40 sqft
Doors	Fiberglass	Fiberglass
Air leakage	2 ACH	2 ACH
Mechanical ventilation	2013, Exhaust (102 cfm, 81.6 W/unit)	2013, Exhaust (102 cfm, 81.6 W/unit)
Air conditioner	EER 9.8	EER 9.8
Furnace	Gas, 98% AFUE	Electric, 100% AFUE
Ducts	15% Leakage, R-8	15% Leakage, R-8
Cooling setpoint	75 F	75 F
Heating setpoint	72 F	72 F
Water heater	Gas benchmark (0.59 energy factor)	Electric benchmark
Distribution	Uninsulated trunk branch, copper	Uninsulated trunk branch, copper
Lighting	100% CFL hardwired, 34% CFL plugin	100% CFL hardwired, 34% CFL plugin
Refrigerator	Side freezer, EF = 15.7	Side freezer, EF = 15.7
Cooking range	Gas, 120% usage, cooktop energy factor 0.3	Electric, 120% usage
Dishwasher	318 Rated kWh	318 Rated kWh
Clothes washer	Standard, 120% usage	Standard, 120% usage
Clothes dryer	Gas, 120% usage, energy factor 2.25 lb/kWh	Electric, 120% usage
Plug loads	250%	250%
Freezer	Upright, EF = 20	Upright, EF = 20
Gas grill	Energy multiplier 0.4	Converted from gas load to electric load

TABLE 14.10
Source Energy End-Uses for Baseline Natural Gas and Electrified Designed Homes (*E* Refers to Electric and *G* Natural Gas)

	Natural Gas Baseline [kBtu/Year]	Electrified Baseline [kBtu/Year]
Total source energy	197,600	311,600
Heating (E)	0	131,760
Heating (G)	45,950	0
Cooling (E)	11,770	11,160
Lighting (E)	15,420	15,420
Large appliances (E)	11,780	29,160
Misc. (E)	68,300	72,310
Large appliances (G)	10,180	0
Misc. (G)	1,390	0
Fans (E)	10,730	10,800
Water systems (G)	22,060	0
Water systems (E)	0	41,000
	Cost data	
Present value (initial cost)	$66,543.00	$66,322.00
Annualized energy related cost	$2,321.17	$3,751.89
Annualized bill (E)	$1,055.00	$2,965.00
Annualized bill (G)	$477.00	$0.00
	Carbon emissions	
CO_2 emissions (metric tons/year)	12.3	20.1

source energy each year than one that relies on natural gas for some of its functions. The electrified baseline uses nearly 58% more source energy usage than the gas-reliant baseline case. This result is attributed to the site-to-source coefficient of 3.15 for electricity generated by the grid which is heavily reliant on fossil-fuel based power plants with low conversion energy efficiencies (Dafoe et al., 2022). Consequently, the carbon emissions associated with the electrified home design are higher than that of the natural gas case.

Figure 14.21 illustrates the optimal path to achieve optimal energy efficiency designs for both natural gas and electrified homes. By implementing optimal energy efficiency measures, the source energy usage for natural gas home can be reduced by 16.99% each year, with an annualized energy related cost of $2,074.46, $246.71 less than the gas baseline design. In particular, the cooling loads are reduced by 71.45%, and heating loads decrease by approximately 20%. Large appliance loads are reduced by 35.23% for electrical and 13.36% for gas. Fans and pumps for heating and cooling systems are decreased by 9.23%. Hot water energy is decreased by 15.41%. These factors imply that heating and cooling systems are most impactful in energy usage reduction, followed by large appliances. The optimal design differs from the baseline design through a southern orientation, 2 ACH infiltration, and an added 2010 heat recovery unit at 70% operation. The optimal energy efficiency

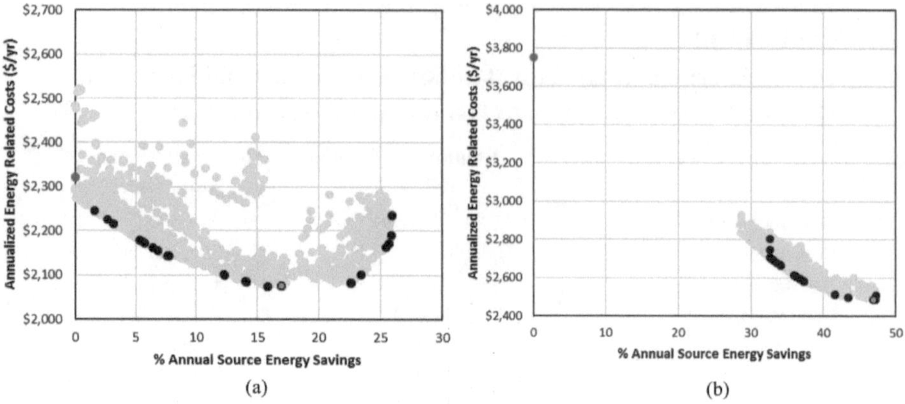

FIGURE 14.21 Optimal Pareto graphs for annualized energy related costs and source energy for (a) natural gas home design and (b) electrified home design

design makes use of natural ventilation during cooling months, a cooling setpoint of 78°F and heating setpoint of 71°F with a weekday setback of 65°F. Instead of a standard gas water heater, the optimal design uses a premium gas-fueled water heater. The lighting is 100% LED, and uses a top freezer refrigerator with 19.9% efficiency, chest freezer with 29% efficiency, EnergyStar 80% usage clothes washer, and 120% usage gas clothes dryer. The initial cost of this design is $66,054.

Similarly, the annualized energy related costs for the optimal design for electrified home are estimated to be $2,485.58 per year. The source energy savings amount to 46.91% relative to the baseline design. The energy efficiency optimization can achieve 81.71% reduction in heating energy usage, and 76.97% reduction in cooling. Hot water energy can be reduced by nearly 76.68%, and large appliance and lighting loads can be reduced by 23.18% and 23.67% respectively. The building characteristics used to achieve this optimal energy efficiency design vary from the baseline case through the deployment of the following measures: southern orientation, low-E double insulated argon high-gain windows, 1 ACH air leakage, 2010 HRV 70% mechanical ventilation, and natural ventilation. A SEER 18, 9.3 HSPF air source heat pump with 78°F cooling setpoint and 69°F heating setpoint make up the space conditioning systems. The water heater is a heat pump water heater with a 50 gallon capacity. The lighting is 100% LED, with a top freezer refrigerator of 19.9% efficiency, an EnergyStar clothes washer with 120% usage, and electric 120% usage clothes dryer. The extra freezer is a chest freezer of 29% efficiency.

Overall, the optimal design reduces both the initial construction costs by $489, and annualized energy related costs by $247 when compared to the baseline design for natural has operated homes. The initial construction costs for the electrified baseline case are less expensive than the gas baseline case by $221. However, the optimal design does not reduce the initial construction costs when compared to the baseline case for electrified homes. In fact, the optimal design increases the initial construction cost by $4,991, though the annualized energy related costs are reduced by $1,266 compared to the baseline design for electrified homes. Figure 14.22 summarizes the

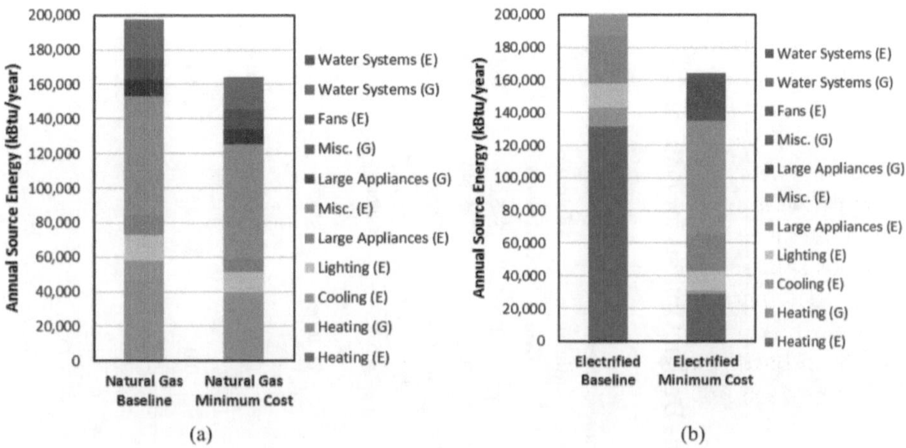

FIGURE 14.22 Comparison of source energy consumption for baseline and optimal designs for (a) natural gas homes and (b) electrified homes

end-use energy associated with baseline and optimized designs for both natural gas operated and electrified homes. For both natural gas and electrified cases, the energy-related costs and carbon emissions are reduced by optimizing their energy efficiency design, with a reduction of source energy of 17% for the natural gas homes and 47% for the electrified homes, as well as a reduction in annual carbon emissions of 2.1 metric tons for the natural gas operated designs and 9.4 metric tons for the electrified designs.

Table 14.11 summarizes the initial construction cost, source-energy use, annualized energy cost, carbon emissions, and PV capacity for various design configurations associated with both natural gas and electrified homes. The deployment of on-site power generation through PV rooftop systems for the optimal energy efficiency design (minimum cost with PV options) achieves carbon neutrality as well as net zero energy (NZE) and carbon neutrality for both natural gas and electrified homes. The initial construction costs for the baseline case are lower compared to all other design options for the electrified design. However, the initial construction costs for the optimal design are lower than those of the baseline case for natural gas homes. In general, the initial construction costs are higher for electrified homes than those for natural gas operated homes because electrified appliances tend to be more expensive, and the initial costs of PV systems tend to be more expensive for electrified cases due to the larger system size requirements. Although the electrified net zero energy designs and minimum cost designs rely on identical energy efficiency measures and utilize the same on-site generation sizes to achieve net zero energy, the annualized energy related costs are different between the two design options.

14.7.2 Resiliency Designs for Communities

The hourly load profiles for various community design configurations are illustrated in Figure 14.23 using duration curves of the total source energy demand. As indicated in Figure 14.23, gas communities generally require lower electricity demand

TABLE 14.11
Summary of Initial Construction Costs, Annual Energy Use, Carbon Emissions for Various Design Options for Natural Gas and Electrified Homes

	Gas Baseline	Gas Baseline (with PV)	Gas Minimum Cost	Gas Minimum Cost (with PV)	Gas NZE	Electric Baseline	Electric Baseline (with PV)	Electric Minimum Cost	Electric Minimum Cost (with PV)	Electric NZE
Total initial cost ($)	$66,543	$96,713	$66,054	$90,473	$91,135	$66,322	$112,181	$71,313	$96,656	$96,656
PV initial cost ($)	$0	$30,170	$0	$22,929	$25,343	$0	$45,858	$0	$25,343	$25,343
Annualized energy related costs ($/yr)	$2,321	$2,362	$2,074	$2,165	$2,171	$3,752	$3,663	$2,486	$2,428	$2,633
Source energy (kBtu/yr)	197,600	–7,300	164,000	–2,800	–5,800	311,600	–100	165,400	–6,600	–6,600
Net annualized utility bills ($/yr)	$1,532	$287	$1,307	$389	$335	$2,965	$921	$1,469	$537	$537
CO_2 emissions (metric tons/yr)	12.3	-0.9	10.2	–0.5	–0.8	20.1	0	10.7	–0.4	–0.4
PV size (kW)	0.0	12.5	0.0	9.5	10.5	0.0	19	0.0	10.5	10.5

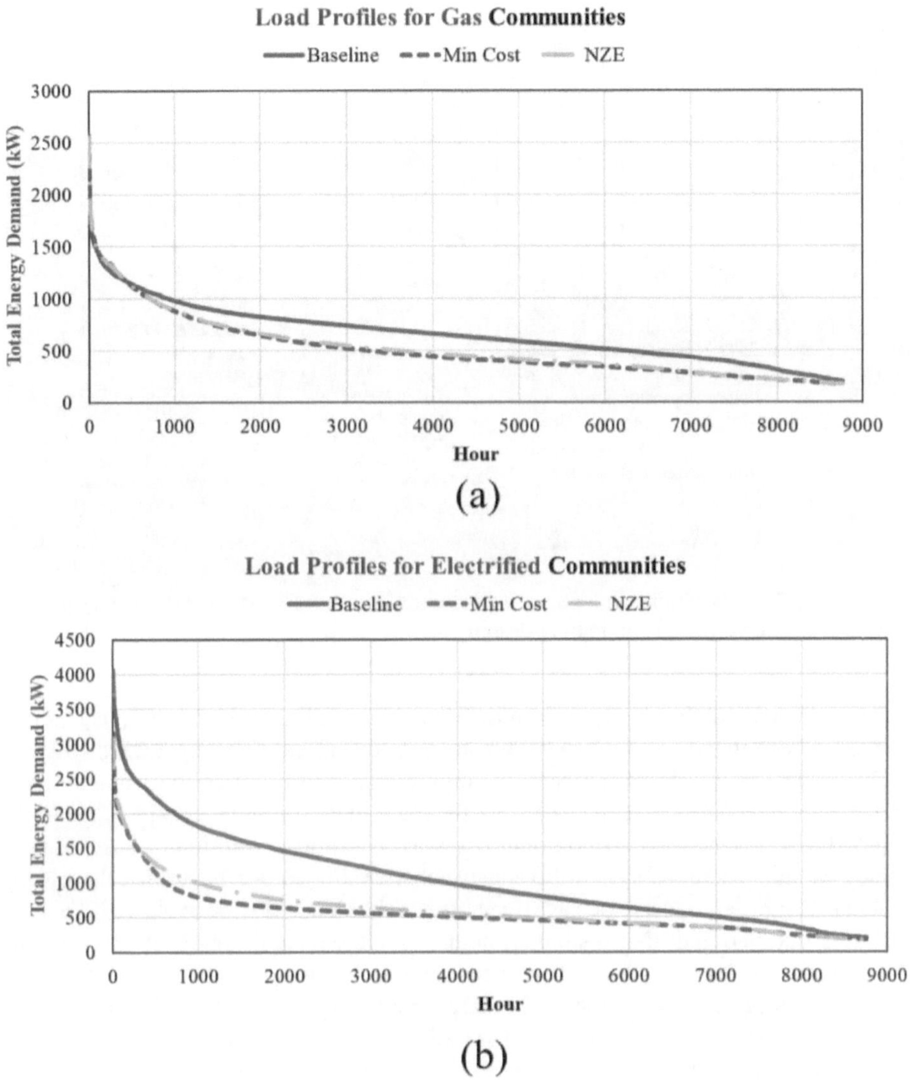

FIGURE 14.23 Duration curves for load profiles for various community designs made up of (a) natural gas operated homes and (b) electrified homes

than their electrified counterparts. For instance, the baseline gas community consumes 36.6% less than the electrified communities. For both community types, the energy demands for the baseline designs are the highest followed the NZE and minimum cost options. Note in Figure 14.23, the PV power generation is not accounted for the NZE communities (Dafoe et al., 2022).

Since the residential communities are made up of each of home designs (baseline, optimal, and net zero energy for gas and electrified energy types), their relative cost and energy performances are similar to those outlined for individual homes.

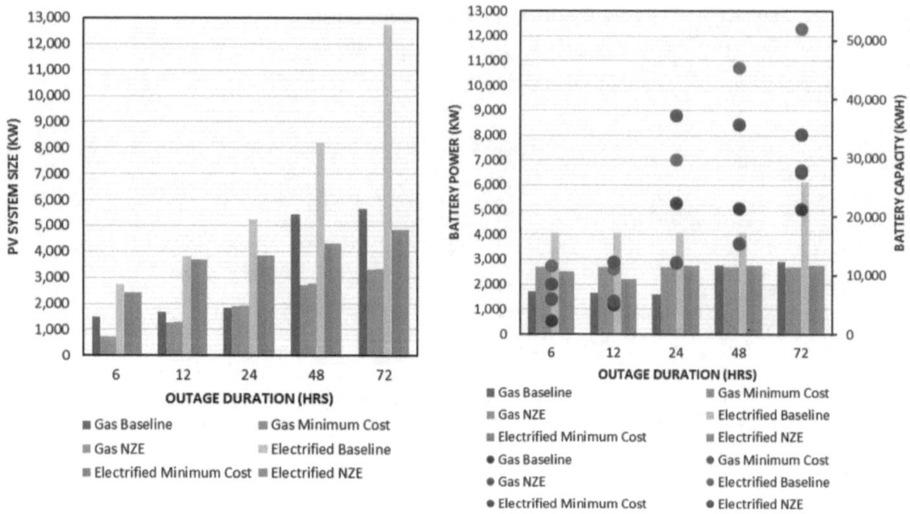

(a) (b)

FIGURE 14.24 Variations with power outage duration of (a) PV size and (b) battery power/capacity required for each community design

Specifically, natural gas NZE communities increase initial construction costs and annualized energy related costs relative the natural gas minimum cost by \$2,508,100 and \$9,700 respectively. Electrified NZE communities increase initial construction costs and annualized energy related costs from the electrified minimum cost by \$2,534,300 and \$14,700, respectively. If communities with baseline or minimum cost designs deploy on-site generation systems to achieve NZE and/or carbon neutrality, those with baseline designs are most expensive due to the higher loads (Dafoe et al., 2022).

Using detailed and hourly simulation analysis, the design specifications for both PV systems and batteries are determined using life cycle cost optimization for various community designs (Dafoe et al., 2022). Figure 14.24 shows the PV and battery sizes for various power outage durations and community designs. Because pursuing energy resilience requires reliance on on-site energy generation (i.e., PV system) and/or storage (i.e., batteries) with no grid support, it is expected that the size of PV will be larger for resiliency designs than for net zero energy designs as indicated by the results shown in Figure 14.24. Resiliency based designs require larger PV systems than those required to achieve annual net zero energy. For natural gas operated communities, these PV system sizes do not appear to increase significantly for power outage durations between 6 and 24 hours. However, after the 24-hour threshold, there is a large increase in the size required to meet baseline resiliency requirements. With longer outage durations, more electricity must be generated and stored to meet that the community energy needs during the grid power outage period. However, larger capacities for the PV and batteries require more capital and depending on the budgetary constraints. Table 14.12 lists the capacities of the PV systems and batteries as well as the capital costs, utility costs, and LCC values for various community designs

TABLE 14.12
System Capacities, Initial Costs, and Utility Costs Specific to Various Community Designs for Resiliency Requirements Associated with 24-Hour Power Outage

Community Design Case	System Initial Cost (Without Incentives) ($)	Utility Cost (Before Tax) ($/yr)	Total LCC	PV Size (kW)	Battery Power (kW)	Battery Capacity (kWh)
Gas baseline	$14,105,410	$188,474	$14,817,958	1,830	1,581	22,449
Gas minimum cost	$10,824,847	$126,897	$11,092,513	1,899	2,677	12,145
Gas NZE	$10,909,332	$129,254	$11,196,353	1,913	2,680	12,280
Electrified baseline	$25,511,206	$165,195	$24,572,058	5,251	4,059	29,745
Electrified minimum cost	$24,959,368	$37,468	$22,994,477	3,828	2,763	37,222
Electrified NZE	$24,959,368	$37,468	$22,994,477	3,828	2,763	37,222

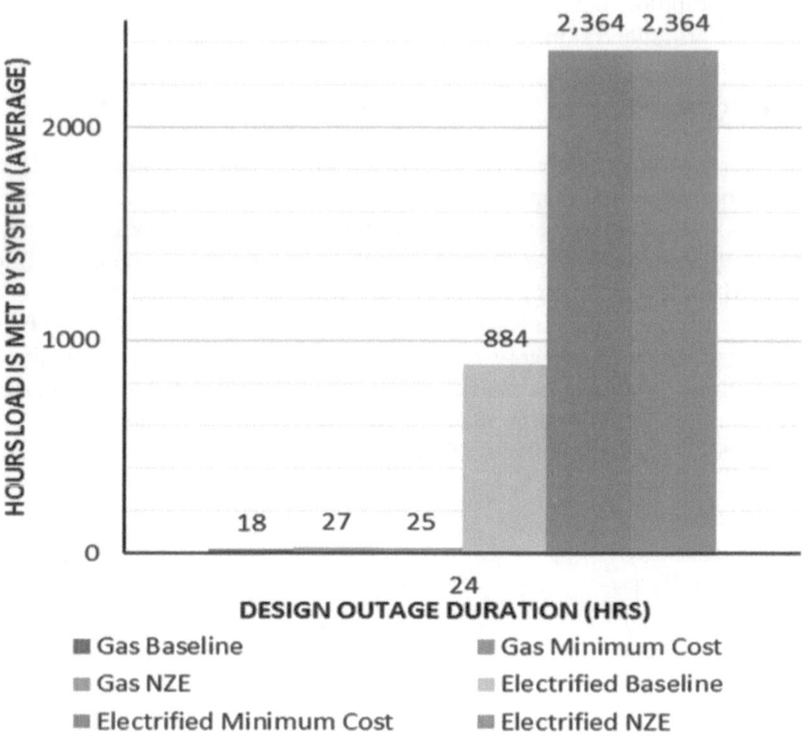

FIGURE 14.25 Average hours of power outage survival for various community types using a 24-hour resiliency design requirement

specific to 24-hour outage duration. The initial costs for on-site generation systems are generally larger for resiliency requirements than for net zero energy designs, with the most comparable costs being those associated with 6-hour outage durations.

For the evaluation analysis outlined by the results of Figure 14.24 and Table 14.12, the power outage duration is set to coincide with the period of highest peak demand (Dafoe et al., 2022). These specified PV systems and Batteries could support the communities during power outages that could occur during other periods of the year. In general, outages of 6 hours provide limited abilities of the community to meet its load during other power outages throughout the year. However, with design power outages of longer durations, the communities could survive with greater probability of success through other outages. Figure 14.25 summarizes the average survivability duration of various community designs given a 24-hour power resilience design. Longer design outage durations enable communities to survive longer periods since they have larger on-site generation and storage systems. The resilience generation sizes for minimum cost and net zero energy designs for electrified communities are identical, and for a 24-hour outage design case, the average number of hours that the system can survive an outage over the year performs better than the other cases. For a day-long outage, the resilience design can survive over 2,300 hours with the minimum cost and NZE designs for electrified communities. The baseline designs can survive only 884 hours for the same electrified communities. However, the natural gas operated communities cannot survive power outages longer than 27-hour periods, because the resilience-based on-site power and storage capacities are relatively small.

14.8 SUMMARY

In this chapter, optimization-based approach is presented to design and retrofit individual buildings and communities to achieve high energy efficiency and resiliency goals. Various energy efficiency and demand response technologies have been described suitable to design and operate not only grid-interactive efficient buildings but net-zero energy and energy resilient communities. The optimal designs for these buildings and communities involves economic analysis of various energy efficiency measures and renewable energy technologies. Through Pareto graphs, the multi objective optimization results can be visualized and utilized depending on the specific constraints. Using specific case studies, the optimization approach has been applied to design net-zero energy buildings as well as resilient and electrified residential communities.

Appendix

TABLE A.1
Conversion Factors (Metric to English)

Area	1 m^2	= 1550.0 $in.^2$
	1 m^2	= 10.764 ft^2
	1 mm^2	= 0.00155 $in.^2$
	1 mm^2	= 1973.5 cmil
Electrical resistivity	1 Ωm	= 3.281 Ω ft
Energy	1 J	= 9.4787×10^{-4} Btu
	1 kWh	= 0.03412 Therm
Force	1 N	= 0.2248 lbf
Length	1 m	= 3.2808 ft
	1 mm	= 0.03937 in.
	1 mm	= 39.370 mil
Mass	1 kg	= 2.2046 lb_m
Mass density	1 kg/m^3	= 0.062428 lb_m/ft^3
Power	1 kW	= 1.340 HP
	1 kW	= 3412 Btu/h
Rotation speed	1 rad/s	= 9.5493 RPM
Temperature	°C	= (5/9)(°F − 32)
Thermal resistance	1 K/W	= 0.52750°F/h Btu
Torque	1 Nm	= 0.7375 lbf ft

TABLE A.2
Conversion Factors (English to Metric)

	1 ft^2	= 0.0929 m^2
	1 ft^2	= 9.2903×10^4 mm^2
Area	1 cmil	= 5.06×10^{-4} mm^2
Electrical resistivity	1 Ωft	= 0.3048 Ωm
Energy	1 Btu	= 1055 J
	1 Therm	= 29.307 kWh
Force	1 lbf	= 4.4482 N
Length	1 ft	= 0.3048 m
	1 in.	= 25.4 mm
	1 mil	= 0.0254 mm
Mass	1 lb_m	= 0.4536 kg
Mass density	1 lb_m/ft^3	= 16.0185 kg/m^3
Power	1 HP	= 0.7457 kW
	1 Btu/h	= 0.2931 W
Rotation speed	1 RPM	= 0.1047 rad/s
Temperature	°F	= (9/5)°C + 32
Thermal resistance	1°F/h Btu	= 1.8960 K/W
Torque	1 lbf ft	= 1.356 Nm

References

ACEEE, 2022, *Building Electrification: Programs and Best Practices*. Report by Cohn, C., and N. W. Esram, American Council for an Energy-Efficient Economy, Washington, DC. http://aceee.org/research-report/b2201.

AEA, August 2007, *Alaska Energy Authority, Power Cost Equalization Program Guide*, Alaska Energy Authority, Anchorage, AK.

AEA, August 2011a, *Alaska Energy Authority, Renewable Energy Atlas of Alaska*, Alaska Energy Authority, Anchorage, AK.

AEA, March 2011b, *Alaska Energy Authority, Statistical Report of the Power Cost Equalization Program*, Alaska Energy Authority, Anchorage, AK.

Allen, E. and Jano, J., 2007, *The Architect's Studio Companion: Rules of Thumb for Preliminary Design*, Wiley, Hoboken, NJ.

Alspector, D. and Krarti, M., 2009, Simulation environment for automated calibration of building energy models, ASME Energy Sustainability Conference, San Francisco, CA.

American Gas, July 16, 1999, Distributing Power in the Future, *American Gas Journal*, *97*, 30–40.

Andreas, J., 1992, *Energy-Efficient Motors: Selection and Application*, Marcel Dekker, New York.

Arvizu, D., Balaya, P., Cabeza, L., Hollands, T., Jäger-Waldau, A., Kondo, M., Konseibo, C., August 2011, *IPCC Special Report on Renewable Energy Sources and Climate Change Mitigation—Direct Solar Energy*, Cambridge University Press, Cambridge.

ASHRAE, 2013, *Handbook of Fundamentals*, American Society of Heating, Refrigerating and Air-Conditioning Engineers Inc., Atlanta, GA.

AVEC, 2011, *Alaska Village Electric Cooperative AVEC Facilities*, AVEC Communities Served, Anchorage, AK, http://www.avec.org/communities/. Accessed April 10, 2013.

Cowern, E., 2017, *Baldor Motor Basics: Electric Motors and Power Systems and Electric Motors and Voltage, a Technical Report Written for Power Transmission Engineering*, pp. 52–55, September 2017. https://www.powertransmission.com/ext/resources/issues/0917/baldor-basics.pdf?1646352124. Accessed October 10, 2022.

Baldor, 2015, *Elevator Application Guide*, Baldor Motors and Drives, Fort Smith, AR. http://www.baldor.com. Accessed December 5, 2015.

Baring-Gould, I. and Dabo, M., 2009, Technology, performance, and market report of wind-diesel applications for remote and island communities, National Renewable Energy Laboratory, Preprint for the European Wind Energy Conference, Marseille, France, February 2009.

Bichiou, Y. and Krarti, M., 2011, Optimization of envelope and HVAC systems selection for residential buildings, *Buildings and Energy*, *43*(*12*), 3373–3382.

Biesemeyer, W.D. and Jowett, J., 1996, *Facts and Fiction of HVAC Motor Measuring for Energy Savings*, Arizona Department of Commerce Energy Office, Phoenix, AZ.

Bollen, M.H., 2000, *Understanding Power Quality Problems*, Institute of Electrical and Electronics Engineers, Piscataway, NJ.

BPA, 1990, *Energy Efficient Motor Selection Handbook*, Bonneville Power Administration, Portland, OR.

Brockway A.M. and P. Delforge, P., 2018, Emissions reduction potential from electric heat pumps in California homes, *The Electricity Journal*, *31*(9), 44–53.

Caldas, L.G. and Norford, L.K., 2003, Genetic algorithms for optimization of building envelopes and the design and control of HVAC systems, *Journal of Solar Energy Engineering*, *125*, 343–351.

Caramia, P., Carpinelli, G., and Verde, P., 2009, *Power Quality Indices in Liberalized Markets*, John Wiley & Sons Ltd, Chichester.

Cheon, S.Y., Lim, H., and Jeong, J.W., 2019, Applicability of thermoelectric heat pump in a dedicated outdoor air system, *Energy*, *173*, 244–262.

Christensen, C., Barker, G., and Horowitz, S., 2004, A sequential search technique for identifying optimal building designs on the path to zero net energy. Proceedings of the Solar 2004, Portland, OR, American Solar Energy Society, Boulder, CO.

Clark, W. and Isherwood, W., September 2004, Distributed generation: Remote power systems with advanced storage technologies, *Energy Policy*, *32*, 1573–1589.

Czarkowski, D. and Domijan, A., 1997, Performance of electric power meters and analyzers in adjustable speed drive applications, *ASHRAE Transactions*, *103*(1), 235–240.

Dafoe S., Krarti M., Baker K., 2022, Optimal Designs of Grid-Connected Energy Efficient and Resilient Residential Communities, *ASME Journal of Engineering for Sustainable Buildings and Cities*, *3*(1): 011004. https://doi.org/10.1115/1.4053908.

Dalziel, C.F., 1972, Electric shock hazard, *IEEE Spectrum*, *9*(2), 41–50.

DEG, 1993, *Coachella Valley Project: La Paloma Site*, Davis Energy Group, Davis, CA.

Dehwah, A.H.A. and M. Krarti, M., 2020, Impact of switchable roof insulation on energy performance of US residential buildings, *Building and Environment*, *177*, 106882.

Delfani, S., Esmaeelian, J., Pasdarshahri, H., and Karami, M., 2010, Energy saving potential of an indirect evaporative cooler as a pre-cooling unit for mechanical cooling systems in Iran, *Energy and Buildings*, *42*, 2169–2176.

DOD, 1999, *Foreign Voltages and Frequencies Guide*, U.S. Department of Army, Washington, DC.

DOE, 2022, *Heat Pump Systems*. Energy Saver, US Department of Energy, Washington DC. https://www.energy.gov/energysaver/heat-pump-systems.

DOE, 2019a, *Grid-Interactive Efficient Buildings, Overview of Research Challenges and Gaps*, US Department of Energy, Washington, DC.

DOE, 2019b, *Grid-Interactive Efficient Buildings, Windows and Opaque Envelope,* US Department of Energy, Washington, DC.

DOE, 2019c, *Grid-Interactive Efficient Buildings, Lighting and Electronics*, US Department of Energy, Washington, DC.

DOE, 2019d, *Grid-Interactive Efficient Buildings, HVAC, Water Heating, Appliances and Refrigeration,* US Department of Energy, Washington, DC.

DOE, 2009, *User Manual of MotorMaster*, Department of Energy, Washington, DC, http://www.energy.gov/sites/prod/files/2014/04/f15/motormaster_user_manual.pdf, Accessed July 10, 2014.

Domijan, A., Czarkowski, D., Abu-Aisheh, A., and Embriz-Sander, E., 1996, *Measurements of Electrical Power Inputs to Variable Speed Motors and Their Solid State Power Converters: Phase II*, ASHRAE, Atlanta, GA.

Dunlop, J., 1997, *Battery Cycle Life Testing for Photovoltaic Systems*, Solar Energy Forum, Albuquerque, NM.

Eaton, 2016, *Protection Devices Products: Fuses and Circuit Breakers*, Bussmann and Cooper Industries, Houston, TX, http://www.cooperindustries.com. Accessed July 20, 2016.

EIA, 2000, *The Changing Structure of the Electric Power Industry*, Energy Information Administration, Washington, DC, http://www.doe.eia.gov. Accessed January 26, 2013.

EIA, 2009, *Annual Energy Review, Department of Energy*, Energy Information Administration, Washington, DC, http://www.doe.eia.gov. Accessed on March 10, 2015.

EIA, 2012, *Commercial Buildings Energy Consumption Survey*, US Energy Information Administration, Washington, DC.

EIA, 2013, *Updated Capital Cost Estimates for Utility Scale Electricity Generating Plants*, Energy Information Agency, Department of Energy, Washington, DC, http://www.eia.gov/forecasts/capitalcost/pdf/updated_capcost.pdf. Accessed March 25, 2015.

EIA, 2017, *U.S. Households' Heating Equipment Choices are Diverse and Vary by Climate Region*. Energy Information Administration, Washington, DC, https://www.eia.gov/todayinenergy/detail.php?id=30672.

EIA, 2020, *Annual Energy Outlook 2020 with Projections to 2050*, US Energy Information Administration, Washington, DC.

EIA, 2021, *Annual Energy Outlook*, Energy Information Administration, Washington, DC, http://www.eia.gov/outlooks/aeo/.

Emerge Alliance, 2022, *Data and Telecom*. Available online: https://www.emergealliance.org/standards/data-telecom/ (accessed on September 15, 2022).

Energy Nexus Group, 2002a, *Technology Characterization: Steam Turbines,* Climate Protection Partnership Division, Washington, DC.

Energy Nexus Group, 2002b, *Technology Characterization: Reciprocating Engines*, Climate Protection Partnership Division, Washington, DC.

Energy Nexus Group, 2002c, *Technology Characterization: Gas Turbines*, Climate Protection Partnership Division, Washington, DC.

Energy Nexus Group, 2002d, *Technology Characterization: Micro-Turbines*, Climate Protection Partnership Division, Washington, DC.

Energy Nexus Group, 2002e, *Technology Characterization: Fuel Cells,* Climate Protection Partnership Division, Washington, DC.

Farzamkia, S., Ranjbar, H., Hatami, A., and Iman-Eini, H., 2016, A novel PSO (particle swarm optimization)-based approach for optimal schedule of refrigerators using experimental models, *Energy*, *107*, 707–715.

Fay, G., Keith, K., and Schwörer, T., June 2010, *Alaska Isolated Wind-Diesel Systems: Performance and Economic Analysis,* Institute of Social and Economic Research, University of Alaska Anchorage, Anchorage, AK.

Fay, G., Meléndez, A.V., and Converse, A., September 2011, *Alaska Energy Statistics 1960–2010,* Institute of Social and Economic Research, University of Alaska Anchorage, Anchorage, AK.

FERC, 1978, *Regulations Under Section 201 and 210 of the Public Utility Regulatory Policies Act of 1978 with Regard to Small Power Production of Cogeneration*, Washington, DC. https://www.ecfr.gov/current/title-18/chapter-I/subchapter-K/part-292

Ferris, J. and Johnson, A., 2014, *Selecting Energy Efficiency Transformers,* Consulting-Specifying Engineer, Oak Brook, IL, http://www.csemag.com. Accessed November 15, 2015.

Finn, P., O'Connell, M., and Fitzpatrick, C., 2013, Demand side management of a domestic dishwasher: Wind energy gains, financial savings and peak-time load reduction, *Applied Energy*, *101*, 678–685.

Fong, K., Hanby, V., and Chowa, T., 2006, HVAC system optimization for energy management by evolutionary programming, *Energy and Buildings*, *38*, 220–231.

FSEC, 2007, *Energy Gauge Pro Software*, Florida Solar Energy Center, Cocoa, FA, http://energygauge.com/FlaRes/features/pro.htm. Accessed April 15, 2014.

Fuchs, E.F. and Masoum, M.A., 2008, *Power Quality in Power Systems and Electrical Machines*, Elsevier, Boston, MA.

Gaur, A.S., Fitiwi, D.Z., and Curtis, J., 2021, Heat pumps and our low-carbon future: A comprehensive review, *Energy Research & Social Science*, *71*, 101764.

Glavan, M., Gradišar, D., Humar, I., and Vrančić, D., 2019, Refrigeration control algorithm for managing supermarket's overall peak power demand, *IEEE Transactions on Control Systems Technology*, *27*, 2279–2286.

Granderson, J., Lin, G., and Fernandes, S., 2015, *A Primer on Organizational Use of EMIS*, Lawrence Berkeley National Laboratory, Berkeley, CA.

Green Seal, 1993, *Proposed Environmental Standards for Major Household Appliances*, Green Seal, Washington, DC.

GRI, 1993, *Electric and Gas Rates for the Residential, Commercial, and Industrial Sectors*, Gas Technology Institute, Des Plaines, IL.

Griego, D., Krarti, M., and Hernandez-Guerrero, A., 2012, Optimization of energy efficiency and thermal comfort measures for residential buildings in Salamanca, *Mexico, Energy and Buildings*, *54*, 540–549.

Halbhavi, N., 2016, *The Benefits of Using Water-Source Heat Pumps*, Consulting-Specifying Engineer, Oak Brook, IL.

Heredia, W.B., Chaudhari, K., Meintz, A., Jun, M., and Pless, S., 2020, Evaluation of smart charging for electric vehicle-to-building integration: A case study, *Applied Energy*, *266*, 114803.

Homer, 2013, *Homer Energy Legacy: Sofyware*, downloaded March 2013, http://homerenergy.com/.

Hughes S.D., 1988, *Electrical Systems in Buildings*, Delmar Publishers Inc. Albany, New York.

Huang, W. and Lam, H.N., 1997, Using genetic algorithms to optimize controller parameters for HVAC systems, *Energy and Buildings*, *26*(3), 277–282.

IEA, 2011, *Energy Efficiency Series: Energy Efficiency Policy Opportunities for Electric Motor-Driven Systems*, International Energy Agency, Paris, France.

IEA, 2022, *IEA Statistics: World Energy Balances*, International Energy Agency, Paris, France. http://www.iea.org/statistics. Accessed December 15, 2022.

IPCC, 1996, *Intergovernmental Panel on Climate Change,* IPCC Technologies, Geneva, Switzerland.

Jenkins, C., Young, R., Tsau, J., Razavi, H., Kaplan, J., and Ibeziako, M.O., 2019, Effective management of plug loads in commercial buildings with occupant engagement and centralized controls, *Energy and Buildings*, *201*, 194–201.

Kamal, R., Moloney, F., Wickramaratne, C., Narasimhan, A., and Goswami, D.Y., 2019, Strategic control and cost optimization of thermal energy storage in buildings using EnergyPlus, *Applied Energy*, *246*, 77–90.

Kim, M.H., Park, J.Y., Sung, M.K., Choi, A.S., and Jeong, J.W., 2014, Annual operating energy savings of liquid desiccant and evaporative-cooling-assisted 100% outdoor air system, *Energy and Buildings*, *76*, 538–550.

Kramer, H., Lin, G., Granderson, J., Curtin, C., Crowe, E., and Tang, R., 2019, *Synthesis of Year Three Outcomes in the Smart Energy Analytics Campaign*, Lawrence Berkeley Laboratory, Berkeley, CA.

Krarti, M., 2010, *Energy Audit of Building Systems: An Engineering Approach*, Taylor & Francis Publishing/CRC-Press, Boca Raton, FL.

Krarti, M., 2012, *Weatherization and Energy Efficiency Improvement for Existing Homes: An Engineering Approach*, Taylor & Francis Publishing/CRC-Press, Boca Raton, FL.

Krarti, M., 2020, Energy Audit of Building Systems, An Engineering Approach. Third Edition, CRC Press, Taylor and Francis Group, Boca Raton, FL.

Krarti, M., Erickson, P., and Hillman, T., 2005, A simplified method to estimate energy savings of artificial lighting use from daylighting, *Building an Environment*, *40*, 747–754.

LBNL, 2004, Measurements of whole-house standby power consumption in California Homes. *International Energy Journal*, *27*(9), 861–868.

Lim, H. and Jeong, J.W., 2018, Energy saving potential of thermoelectric radiant cooling panels with a dedicated outdoor air system, *Energy and Buildings*, *169*, 353–365.

Lin Y. and Xiao W., 2019, Novel piecewise linear formation of droop strategy for DC microgrid, *IEEE Tranactions on Smart Grid, 10*, 6747–6755.

Lobodovsky, K.K., 1994, Motor efficiency management, *Energy and Engineering*, *91*(2), 32–43.

Lukanov B.R. and E. M. Krieger, 2019, Distributed solar and environmental justice: Exploring the demographic and socio-economic trends of residential PV adoption in California, *Energy Policy*, *134*, 110935.

Mayhoub, M. and Carter, D., 2012, A feasibility study for hybrid lighting systems, *Building and Environment*, *53*, 83–94.

McKenna, C., Shah, A., and Louis-Prescott, L., 2020, *All-Electric New Homes: A Win for the Climate and the Economy*, Rocky Mountain Institute, Boulder, CO, http://rmi.org/all-electric-new-homes-a-win-for-the-climate-andthe-economy/

McKinsey, 2022, *Building Decarbonization: How Electric Heat Pumps Could Help Reduce Emissions Today and Going Forward,* McKinsey Company, New York City, NY.

Meier, A.K., 1982, *Supply Curves of Conserved Energy*, Lawrence Berkeley National Laboratory, Berkeley, CA.

Menyhart, K. and Krarti, M., 2017, Potential energy savings from deployment of dynamic insulation materials for US residential buildings, *Building and Environment*, *114*, 203–218.

Mont, J.A. and Turner, W.C. 1999, A study on real-time pricing electric tariffs, *Energy Engineering, 96*(5), 7.

Nadel S., 2016, Comparative Energy Use of Residential Furnaces and Heat Pumps. *American Council for Energy Efficient Economy*, Washington D.C. http://www.aceee.org/research-report/a1602.

Nadel, S., Shepard, M., Greenberg, S., Katz, G., and de Almeida, A., 1991, *Energy-Efficient Motor Systems: A Handbook of Technology, Program, and Policy Opportunities*, American Council for Energy-Efficient Economy, Washington, DC.

NEEP, 2015, *Energy Efficiency as a T&D Resource: Lessons from Recent U.S. Efforts to Use Geographically Targeted Efficiency Programs*, Defer T&D Investments, Woburn, MA.

NEC, 2014, *National Electrical Code,* National Fire Protection Association, Quincy, MA.

NEMA, 2010, *NEMA Premium Efficiency Transformer Program Guidelines*, National Electrical Manufacturers Association, Rosslyn, VA, https://www.nema.org/Policy/Energy/Efficiency/Documents/NEMA_Premium_Efficiency_Transformer_Product_Specifications.pdf. Accessed September 5, 2015.

NFPA, 2022, *National Fire Alarm and Signaling Code 72*, National Fire Protection Association, Quincy, MA.

Nghana, B. and Tariku, F., 2016, Phase change material's (PCM) impacts on the energy performance and thermal comfort of buildings in a mild climate, *Building and Environment*, *99*, 221–238.

Nuytten, T., Claessens, B., Paredis, K., Bael, J., and Six, D., 2013, Flexibility of a combined heat and power system with thermal energy storage for district heating, *Applied Energy*, *104*, 583–591.

Park, B., Srubar, W.V., and Krarti, M., 2015, Energy performance analysis of variable thermal resistance envelopes in residential buildings, *Energy and Buildings*, *103*, 317–325.

Park, H., Nam, K.H., Jang, G.H., and Kim, M.S., 2014, Performance investigation of heat pump–gas fired water heater hybrid system and its economic feasibility study, *Energy and Buildings*, *80*, 480–489.

Paterakis N.G., Erdinç O., and Catalão J.P.S. An overview of Demand Response: Key-elements and international experience, *Renewable and Sustainable Energy Reviews*, *69*, 871–891.

Penning, J., Stober, K., Taylor, V., Yamada, M., 2016, *Energy Savings Forecast of Solid-State Lighting in General Illumination Applications*, US Department of Energy, Solid-State Lighting Program, Washington, DC.

Perez, K.X., Baldea, M., and Edgar, T.F., 2016, Integrated HVAC management and optimal scheduling of smart appliances for community peak load reduction, *Energy and Buildings*, *123*, 34–40.

Pourmousavi, S.A., Patrick, S.N., and Nehrir, M.H., 2014, Real-time demand response through aggregate electric water heaters for load shifting and balancing wind generation, *IEEE Transactions on Smart Grid*, *5*, 769–778.

Prabhala, V.A., Baddipadiga, B.P., Fajri, P., and Ferdowsi, M., 2018, An overview of direct current distribution system architectures and benefits, *Energies*, *11*, 2463.

RMI, 2021, Building Decarbonization Roadmap, Produce for the United States Climate Alliance, Rocky Mountain Institute, Boulder, CO.

Rupp, S. and Krarti, M., 2019, Analysis of multi-step control strategies for dynamic insulation systems, *Energy and Buildings, 204*, 109459.

Sanchez, M., Brown, E.R., Homan, G.K., and Webber, C.A., 2008, *Savings Estimates for the Energy Star Voluntary Labeling Program,* Lawrence Berkeley National Laboratory, Berkeley, CA.

Sankaran, C., 2002, *Power Quality*, The Electric Power Engineering Series, CRC Press, Boca Raton, FL.

Schwatz, E.K. and M. Krarti. 2022. Review of adoption status of sustainable energy technologies in the US residential building sector, *Energies, 15*, 6.

Sehar, F., Pipattanasomporn, M., and Rahman, S., 2016, An energy management model to study energy and peak power savings from PV and storage in demand responsive buildings, *Applied Energy*, *173*, 406–417.

Shekar V. and M. Krarti, M., 2017, Control strategies for dynamic insulation materials applied to commercial buildings, *Energy and Buildings*, *154*, 305–320.

Siraj, K. and Khan, H.A., 2020, DC distribution for residential power networks—A framework to analyze the impact of voltage levels on energy efficiency, *Energy Reports, 6*, 944–951.

Snyder, J., 2020, Energy-saving strategies for luminaire-level lighting controls, *Building and Environment*, *169*. https://doi.org/10.1016/j.buildenv.2018.10.026.

Steen, M. and Krarti, M., 2020, A review and categorization of grid-interactive efficient building technologies for building performance simulation, *ASME Journal of Engineering for Sustainable Buildings and Cities, 1*(4), 040801.

Surles, W.A., 2011, *Development and Application of a Control Analytic Tool for Evaluating Automated Residential Smart Grid Control Strategies*, University of Colorado, Boulder, CO.

Tuhus-Dubrow, D. and Krarti, M., 2009, Comparative analysis of optimization approaches to design building envelope for residential buildings, *ASHRAE Transactions*, *115*(2), 205–219.

Tuhus-Dubrow, D. and Krarti, M., 2010, Genetic algorithm based approach to optimize building envelope design for residential buildings, *Building and Environment*, *45*(7), 1574–1584.

Tzempelikos, A. and Shen, H., 2013, Comparative control strategies for roller shades with respect to daylighting and energy performance, *Building and Environment*, *67*, 179–192.

Usui, K., Babasaki, T., Hirose, K., Yoshida, Y., 2016, Dual-voltage output power supply system toward parallel use of 380 V DC and 48 V DC., *IEEE International Telecommunications, 8*, 1–5.

Vaught, D., 2007, *Deering Wind-Diesel Hybrid Feasibility*, V3 Energy, Eagle River, AK.

Vedam, R.S. and Sarma, M.S., 2009, *Power Quality: Var Compensation in Power Systems*, Taylor & Francis Group, Boca Raton, FL.

Verda, V. and Colella, F., 2011, Primary energy savings through thermal storage in district heating networks, *Energy, 36*, 4278–4286.

Verderber, R.R., Morse, O.C., and Alling, W.R., 1993, Harmonics from compact fluorescent lamps, *IEEE Transactions on Industry Applications*, *29*(3), 670–674.

Vossos V., Garbesi K., Shen H., 2014, Energy savings from direct-DC in US residential buildings, *Energy and Building, 68*, 223–231.

Waide, P., Lebot, B., and Hinnells, M., 1997, Appliance energy standards in Europe, *Energy and Buildings*, *26*(1), 45.

Wang, M., Peng, J., Li, N., Lu, L., Ma, T., and Yang, H., 2016, Assessment of energy performance of semi-transparent PV insulating glass units using a validated simulation model, *Energy*, *112*, 538–548.

Wang, C., Pattawi, K., and Lee, H., 2020, Energy saving impact of occupancy-driven thermostat for residential buildings, *Energy and Buildings*, *211*, 109791.

Wang, W., Zmeureanu, R., and Rivard, H., 2005, Applying multi-objective genetic algorithms in green building design optimization, *Building and Environment*, *40*, 1512–1525.

Washington State Energy Office (WSEO), 1992, *Motor Master Electric Motor Selection Software*, Washington State Energy Office, Olympia, WA.

Wetter, M., 2004, *Generic Optimization Program*, Lawrence Berkeley National Laboratory, Berkeley, CA, http://gundog.lbl.gov/GO/download/documentation.pdf. Accessed March 10, 2015.

Wiles, J., 2001, Voltage drop after NEC requirements, *Home Power*, *80*, 116–118.

Wiles, J., May-June 2005, *Permitting or Inspecting a PV System*, International Association of Electrical Inspectors, Richardson, TX, http://www.iaei.org. Accessed September 20, 2014.

Willam, L. and Krarti, M., 2013, Optimization of hybrid distributed generation systems for rural communities in Alaska, *Distributed Generation and Alternative Energy Journal*, *28*(4), 7–31.

Wright, J.A., Loosemore, H.A., and Farmani, R., 2002, Optimization of building thermal design and control by multi-criterion genetic algorithm, *Energy and Buildings*, *43*, 959–972.

Wu, W. and Skye, H.M., 2018, Net-zero nation: HVAC and PV systems for residential net-zero energy buildings across the United States, *Energy Conversion and Management*, *177*, 605–628.

Wu, Z., Qin, M., and Zhang, M., 2019, Phase change humidity control material and its impact on building energy consumption, *Energy and Buildings, 174*, 254–261.

Zhai, Y., Ma, Y., David, S.N., Zhao, D., Lou, R., Tan, G., Yang, R., and Yin, X., 2017, Scalable-manufactured randomized glass-polymer hybrid metamaterial for daytime radiative cooling, *Science, 355*, 1062–1066.

Index